Acoustics in Moving Inhomogeneous Media

Second Edition

Acoustics in Moving Inhomogeneous Media

Second Edition

Vladimir E. Ostashev

University of Colorado, Boulder, Colorado, USA

D. Keith Wilson

US Army ERDC, Hanover, New Hampshire, USA

CRC Press
Taylor & Francis Group
Boca Raton London New York

CRC Press is an imprint of the
Taylor & Francis Group, an **informa** business

A SPON PRESS BOOK

CRC Press
Taylor & Francis Group
6000 Broken Sound Parkway NW, Suite 300
Boca Raton, FL 33487-2742

First issued in paperback 2019

© 2016 by Taylor & Francis Group, LLC
CRC Press is an imprint of Taylor & Francis Group, an Informa business

No claim to original U.S. Government works

ISBN-13: 978-0-415-56416-8 (hbk)
ISBN-13: 978-0-367-87490-2 (pbk)

Visit the Taylor & Francis Web site at
http://www.taylorandfrancis.com

and the CRC Press Web site at
http://www.crcpress.com

Contents

III Numerical methods for sound propagation in moving media   311

9 Numerical representation of random fields   317

10 Ray acoustics and ground interactions   353

Preface

This book offers a complete and rigorous study of sound propagation and scattering in moving media with deterministic and random inhomogeneities in the sound speed, density, and medium velocity. This area of research is of great importance in many fields including atmospheric and oceanic acoustics, aeroacoustics, acoustics of turbulent flows, infrasound propagation, noise pollution in the atmosphere, theories of wave propagation, and even astrophysics, with regard to acoustic waves in extraterrestrial atmospheres. Over the past several decades, understanding of acoustics in moving media has grown rapidly, in response to its importance for practical applications such as prediction of sound propagation from highways, airports, and factories; acoustic remote sensing and tomography of the atmosphere and ocean; detection, ranging, and recognition of acoustic sources; and the study of noise emission by nozzles and exhaust pipes.

In the atmosphere, the wind velocity and its fluctuations usually lead to significant changes in sound and infrasound propagation, such as ducting in the downwind direction and scattering into shadow zones. Strong oceanic currents and tides can affect the phase and amplitude of acoustic signals. Sound propagation in gases or fluids are influenced by the mean flow. Propagation of sound waves emitted by moving sources is closely related to acoustics in moving media and is considered in this book. The bulk of the book presents systematic and rigorous formulations of sound propagation in inhomogeneous moving media, which may be applied in many areas of acoustics. Experimental data and numerical predictions considered in the book are pertinent mainly to atmospheric and oceanic acoustics. When studying outdoor sound propagation, the most advanced models for the vertical profiles of temperature and wind velocity and their fluctuations in the atmospheric surface layer and for the ground impedance are used.

Part I of the book considers sound propagation through moving media with deterministic inhomogeneities, such as vertical profiles of temperature and wind velocity in the atmosphere. Chapter 1 presents the history of acoustics in moving media, its applications, typical values of wind and current velocities in the atmosphere and ocean, and their effects on sound propagation. This chapter contains useful background for those new to the subject. In Chapter 2, classical and new equations for sound waves in inhomogeneous moving media are systematically derived from a set of linearized fluid-dynamic equations. This chapter provides appropriate starting equations for solving many particular problems. In Chapter 3, the main results of geometrical acoustics in an inhomogeneous moving medium are formulated systematically using the Debye series and Hamiltonian formalism. Among these are the law of acoustic energy conservation, the eikonal equation, refraction laws for the sound ray and the normal to the wavefront, and equations for the sound ray path. Geometrical acoustics is particularly useful for the visualization of sound propagation. Chapter 4 deals with the wave theory of sound propagation in stratified moving media (the atmosphere and ocean). The results presented elucidate the effects of the medium motion on propagation of plane and spherical sound waves. Chapter 5 covers the study of sound fields emitted by moving sources. The sound field due to a point source moving with an arbitrary velocity in a homogeneous, motionless medium is analyzed. The bulk of the chapter considers the effects of both source and medium motion on the sound field, such as the Doppler effect and sound aberration in moving media.

The classical theories of wave propagation in media with fluctuations in the sound speed (or light velocity) are well developed and presented in many books. However, in the turbulent atmosphere and ocean, in liquid marine sediments, and in the turbulent flows, the statistical moments of a sound field are affected not only by these fluctuations, but also by the density and medium velocity fluctuations. In Part II, we present rigorous and systematic formulations for the various statistical moments of a sound field propagating in a medium with random inhomogeneities in the sound speed, density, and medium velocity. In Chapter 6, the statistical description of random inhomogeneities in a medium is considered, including most widely used spectra of turbulence. The sound scattering cross section in a turbulent medium is calculated and applied to the analysis of sound scattering in the atmosphere. In Chapter 7, the variances and correlation functions of the phase and log-amplitude fluctuations, the mean sound field, and the mutual coherence function are considered for line-of-sight sound propagation. Multipath sound propagation in a random moving medium is analyzed in Chapter 8. This geometry can occur due to reflection of a sound wave from a surface (e.g., the ground), refraction, or sound scattering at large angles.

Part III describes numerical methods for performing calculations involving equations from the first two parts. Such numerical methods are often needed for practical problems involving sound propagation in the atmosphere, ocean, and other moving media, since the complex and dynamic nature of these environments often prevents the derivation of general, analytical results. Although the example calculations in Part III pertain to outdoor sound propagation near the ground, the techniques can be readily applied to other environments. Techniques for synthesizing realistic random media, as appropriate to wave propagation calculations, are described in Chapter 9. Chapter 10 describes implementation of ray-based methods for the atmosphere, and also provides some background material on boundary conditions for ground surfaces and interaction of sound waves with porous materials. Wave-based methods, in the frequency and time domains, are the subject of Chapters 11 and 12, respectively. The former includes solution of parabolic equations and wavenumber integration techniques. The emphasis of Chapter 12 is on finite-difference, time-domain (FDTD) calculations. Lastly, in Chapter 13, we explore incorporation of randomness and uncertainty in the outdoor environment (atmosphere and terrain) into propagation calculations.

When writing this book, the authors have endeavored to derive results systematically from first principles. Ranges of applicability are rigorously formulated before interpreting the physical meaning of the results. Such an approach is desirable since heuristic approaches for sound propagation in moving media have, in the past, led to some errors and misconceptions. The main quantities describing sound propagation have the same notation in all chapters of this book. Nevertheless, in each chapter all notations are introduced anew, so that it can be read independently of the other chapters.

This book has been significantly revised and extended from the first edition [290], which was published in 1997. Part I incorporates new results obtained since that time. Part II is significantly rewritten and extended with systematic formulations of sound propagation and scattering in random moving media. Part III, describing numerical methods, is entirely new with this edition.

This book should provide valuable background and a reference resource for engineers and scientists working in industry, government, and military laboratories on research problems involving outdoor noise control, acoustic detection and ranging in the atmosphere, and acoustic remote sensing of the atmosphere and ocean. The step-by-step approach and careful explanations should be useful to teachers and graduate students in universities, polytechnics and technical colleges, in departments of physics, mathematics, earth sciences, and engineering, who are interested in atmospheric and oceanic acoustics, aeroacoustics, acoustics of turbulent flows, outdoor noise, acoustic remote sensing of the atmosphere and ocean, and the theory of wave propagation in inhomogeneous media.

Vladimir Ostashev   *Boulder, Colorado*
Keith Wilson   *Hanover, New Hampshire*

Biographies

Dr. Vladimir E. Ostashev is a senior research scientist at the Cooperative Institute for Research in Environmental Sciences (CIRES) of the University of Colorado at Boulder (CU) and a government expert for the U.S. Army Engineer Research and Development Center. He earned a PhD in physics from the Moscow Physics and Technology Institute, Russia in 1979. His undergraduate and graduate advisor was Prof. Valerian I. Tatarskii. In 1992, Dr. Ostashev earned a Doctor of physical and mathematical sciences from the Acoustics Institute, Moscow, where his advisor was Prof. Yu. P. Lysanov. Since 1979, he has worked at the Institute of Atmospheric Physics (Moscow, Russia), Acoustics Institute (Moscow, Russia), and New Mexico State University (Las Cruces, New Mexico), before joining CIRES/CU in 2000. Dr. Ostashev has also held visiting positions at the University of Oldenburg (Oldenburg, Germany), Centre Acoustique, École Centrale de Lyon (Écully, France), Open University (Milton Keynes, United Kingdom), and NOAA/Environmental Technology Laboratory (Boulder, Colorado). He is a fellow of the Acoustical Society of America, and an associate editor of the *Journal of the Acoustical Society of America* and *JASA Express Letters*.

Dr. D. Keith Wilson is a research physical scientist with the U.S. Army Engineer Research and Development Center (ERDC), in Hanover, New Hampshire. He earned an M.S. in electrical engineering from the University of Minnesota in 1987, where he was advised by Prof. Robert F. Lambert, and a PhD in acoustics from the Pennsylvania State University in 1992, where he was advised by Prof. Dennis W. Thomson. Dr. Wilson was a research fellow at the Woods Hole Oceanographic Institution under the guidance of Prof. George V. Frisk from 1991–1993, and a research faculty member in the Pennsylvania State University Meteorology Department under Prof. John C. Wyngaard from 1993–1995. He joined the U.S. Army Research Laboratory in 1995 and ERDC in 2002. Dr. Wilson has been awarded U.S. Army Research and Development Achievement Awards on four occasions and received the U.S. Army Meritorious Civilian Service Award in 2012. He is a fellow and recipient of the Lindsay Award of the Acoustical Society of America, associate editor of the *Journal of the Acoustical Society of America*, and founding editor of *JASA Express Letters*. He is a member of the Institute for Noise Control Engineering and the American Meteorological Society.

Acknowledgments

The first edition of this book was prepared for publication with support from the German Acoustical Society, through arrangements by Prof. Volker Mellert (University of Oldenburg, Germany). Frank Gerdes (University of Oldenburg) undertook the printing of the manuscript in LaTeX. Prof. Keith Attenborough (Open University, United Kingdom) read both editions of the manuscript carefully and provided many useful comments. The authors sincerely thank all these people.

Both authors are grateful to their former research advisors for their mentorship, inspiration, and kind support lasting many years: V. Ostashev to Prof. V. I. Tatarskii (formerly with the National Oceanic and Atmospheric Administration) and Prof. Yu. P. Lysanov (deceased), and K. Wilson to Profs. Dennis W. Thomson, John C. Wyngaard, George V. Frisk, Robert F. Lambert, and Kenneth E. Gilbert.

The authors also acknowledge the long-term support of the United States Army, which was facilitated through the Army Research Office, the Engineer Research and Development Center, and the Army Research Laboratory. Indeed, a substantial portion of the research in outdoor sound propagation described in this book, whether conducted by the authors or by others, was sponsored by the U.S. Army.

We hope this book helps to demonstrate the tremendous progress that has been made by the larger research community during the past several decades.

Lastly, but most definitely foremost in mind, the authors are indebted to the unwavering support and patience of their spouses and families during this project.

Part I

Theoretical foundations of acoustics in moving media

In this part of the book, we analyze propagation of sound waves in moving media with deterministic inhomogeneities, such as the vertical profiles of temperature and wind velocity in the atmosphere and synoptic eddies in the ocean. Chapter 1 serves as introduction and acquaints readers with the history of acoustics in a moving medium and its modern applications. Parameters affecting outdoor sound propagation are discussed and a brief overview of atmospheric acoustics is presented. The effects of ocean currents on propagation of sound waves are outlined.

In Chapter 2, equations for acoustic and internal gravity waves in an inhomogeneous moving medium are systematically derived from first principles. An entire chapter is devoted to the derivation of these equations because (i) certain important equations have been obtained only recently, and (ii) the equations for sound waves are often presented in the literature without detailed analysis of their ranges of applicability. These equations are used in the subsequent chapters for analysis of sound propagation.

The main results of geometrical acoustics in an inhomogeneous moving medium are systematically presented in Chapter 3. Starting from a complete set of linearized fluid-dynamic equations, we formulate the law of acoustic energy conservation and derive the eikonal equation, refraction laws for the sound ray and the normal to the wavefront, and the equations for sound ray paths. The ray paths are particularly helpful in visualizing sound propagation. Examples of ray tracing in the atmosphere and ocean are presented.

Geometrical acoustics does not enable one to describe diffraction of sound waves and is not applicable to relatively low frequencies. These difficulties can be overcome with the wave theory of sound propagation in a stratified moving medium, which is considered in Chapter 4. This is a rigorous theory, which has a wider range of applicability than the effective sound speed approximation. The results in this chapter describe the effects of medium motion on the sound-pressure field.

Chapter 5 deals with the analysis of the sound fields emitted by moving sources and observed by moving receivers. The chapter begins with the study of the sound field of a point source moving with an arbitrary velocity in a homogeneous motionless medium. Then, the effects of source, receiver, and medium motion on the sound field are analyzed. In particular, sound aberration and the Doppler effect in an inhomogeneous moving medium are considered.

1

Introduction to acoustics in a moving medium

This introductory chapter begins our journey through the subject of acoustics in moving media: its history, theory, computational methods, and applications. By *moving media*, we mean a medium with an ambient flow; that is, a fluid in motion *prior* to the introduction of a sound wave. Such motion is called the *wind* in the atmosphere, or the *current* in the ocean. But, much of the underlying science applies to media other than the ocean or atmosphere, such as wave propagation through flows of fluids. In addition to winds and currents, the fluid motions may include random disturbances such as *turbulence* and *internal waves*.

Section 1.1 acquaints readers with the history of acoustics in a moving medium and with its modern applications. A brief overview of atmospheric acoustics follows in section 1.2. This section also describes parameters affecting sound propagation in the atmosphere, such as the vertical profiles of temperature and wind velocity. In section 1.3, the parameters of ocean currents are briefly overviewed and some experimental and theoretical results on sound propagation in the ocean with currents are presented. The subsequent chapters, where the theory of sound propagation in inhomogeneous moving media is systematically presented, also contain some historical perspectives, experimental data, and numerical results, but more specifically related to the subject of the chapter.

1.1 Historical review

Throughout this book, and particularly with regard to the following historical discussion, we endeavor to present the subject of acoustics in moving media based on the primary, original papers. Two review papers [26, 93], and the historical review sections found in references [52, 71, 76, 151], were primary references enabling us to connect the threads among the original papers on this subject.

1.1.1   First papers on acoustics in a moving medium

The study of acoustics in a moving medium initially emerged from interest in sound propagation in the atmosphere. Long before the advent of modern science, it had been observed that sound appeared to be louder downwind than upwind from a source. This phenomenon was also evident in the first experiments on sound propagation in the atmosphere, such as those performed by Delaroche [95] and Arago [8]. A correct qualitative explanation of this phenomenon was not provided, however, until 1857 by Stokes [366]. Since the wind velocity should increase with height in the near-ground atmosphere, sound propagating upwind bends upward and thus can pass over the head of an observer, who is then said to be in an acoustic shadow zone (in analogy to an optical shadow). But, in the downwind direction, the sound bends downward, so that the observer is in an insonified zone. (The ray paths for sound propagation downwind and upwind are illustrated in figures 3.4 and 3.5 of Chapter 3.)

Nearly the same explanation for the distinct audibility in the downwind and upwind directions was given by Reynolds [329], seventeen years after Stokes. But, unlike Stokes, Reynolds used the concept of sound rays for this explanation. Reynolds also verified experimentally Stokes's assumption that the bending of sound (a sound ray, according to Reynolds) upwind causes a decrease in audibility. In his experiments, the height below which no sound was heard was measured. Reynolds's experiments confirmed that the greater the distance from the source to the receiver, the greater this height. Furthermore, Reynolds assumed [329], and then verified experimentally [330], that a sound ray bends upward if the temperature decreases with height. This allowed Reynolds to explain why sound from a particular source can be heard better at night than in the daytime. In the daytime, the temperature and, hence, the speed of sound decreases with height, so that the paths of sound rays turn upward, similarly to sound traveling upwind. On the other hand, at night in the near-ground atmosphere, the temperature usually increases with height and the sound rays turn downward, as in propagation downwind. Independently of the study done by Reynolds, and practically at the same time, the effect of wind on sound propagation in the atmosphere was studied experimentally by Henry [164]. Based on his experimental results, Henry concluded that sound propagation is affected significantly by refraction due to wind velocity stratification.

Rayleigh [324] developed the first mathematical description of sound propagation in moving media and formulated the refraction law governing the normal to the wavefront in a stratified moving atmosphere. Using this law, he derived an equation for the ray path. However, Rayleigh's derivation was later recognized to be incomplete, because it did not distinguish between the unit vector $\mathbf{n}$ normal to the wavefront and the unit vector $\mathbf{s}$ tangential to the ray path. Barton [23] was the first to show that these vectors do not generally coincide in a moving medium. Barton also formulated a rule for calculating the group velocity of a sound wave propagating in a stratified atmosphere. Using this rule, Barton calculated the sound ray paths. While many papers on acoustics in moving media have pointed out Rayleigh's mistake, he was the first to formulate the refraction law for the normal **n** to the wavefront in a moving medium. Moreover, the distinction between the vectors **n** and **s** had not been made prior to Rayleigh; by revealing their difference, Barton made a key contribution to the development of acoustics in a moving medium.

Further development of this field of acoustics was motivated by two practical problems. First, for detection and ranging of artillery and airplanes, the corrections due to refraction of sound waves in the inhomogeneous atmosphere had to be obtained. This problem was considered in detail by Milne [253] in 1921, who revised formulas for calculating the refraction corrections. Furthermore, Milne presented the equations for the phase and group velocities of a sound wave in a three-dimensional moving medium, and also for the ray path. Detection and ranging of artillery and airplanes remained an important application in atmospheric acoustics until the end of World War II. Another important application was the study of sound propagation from large explosions, which resulted in the development of geometrical acoustics for moving media.

1.1.2 Sound propagation from large explosions

In 1904, Borne [40] was the first to detect so-called abnormal propagation of sound in the atmosphere produced by large explosions on the earth's surface. It follows from the observations by Borne and many others (see references cited in [76, 103, 277]) that the sound waves from large explosions, propagating at small elevation angles with respect to the horizon, have turning points in the upper atmosphere. The sound waves then return to the earth's surface, but at large horizontal distances from the explosive source. Therefore, a zone of silence occurs between the initial zone of audibility near the source, and the one resulting from the upper atmosphere return. The term *abnormal sound propagation* was widely used in the first half of the 20th century to describe this phenomenon, but is seldom used nowadays. Sound signals from supersonic aircrafts, rocket launches, and volcanic eruptions can also have turning points in the upper atmosphere and propagate over long ranges.

Analysis of the acoustic travel time from large explosions has shown that turning points for the sound waves can occur in the troposphere, near the ozone layer in the stratosphere at a height of 40–50 km, or in the thermosphere at a height above 100 km. Sound signals with turning points in the troposphere can propagate over long ranges. Such sound signals have been recorded reliably at a horizontal range of 500 km from a source [103]. The cause of sound propagation of this type is the increase in wind velocity $v(z)$ with the height z in the troposphere, which results in downward refraction. A waveguide is formed between the turning points of sound waves and the earth's surface, thus enabling long-range propagation of low frequency signals.

In the thermosphere, the temperature $T(z)$ rapidly increases with height. As a result, sound signals from large explosions propagating at certain elevation angles can have turning points in the thermosphere and return back to the earth's surface. Sound signals with turning points in the stratosphere are considered in the next subsection.

Interest in sound propagation from large explosions declined significantly after World War II because of the use of rockets in studies of the upper atmosphere. Nevertheless, some research in this area continued [7, 32, 85]. Remote sensing of the upper atmosphere by sound signals from large explosions on the earth's surface was a part of the "Mass" project carried out in the USSR in the 1980s [4, 59, 60]. Sound signals from the supersonic Concorde aircraft, which had the turning points in the thermosphere, were recorded at horizontal ranges from 165 km to 10^4 km [21]. Sound propagating through the upper atmosphere can also be used to detect and range nuclear explosions [311, 315]. In the mid-1990s, a network of 60 infrasound stations located worldwide was designed to comply with the Comprehensive Nuclear Test Ban Treaty. This has renewed interest in sound and infrasound propagation in the upper atmosphere [219].

1.1.3   Sound signals with turning points in the stratosphere

In this subsection, we consider only sound propagation from large explosions which have turning points in the stratosphere. Initially, many different explanations were suggested for the observed properties of sound propagation from large explosions; for example, by Obolenskii [277]. In 1912, Fujiwhara [124, 125] was the first to suggest that the observations could be explained by refraction from wind stratification in the upper atmosphere. For arbitrary profiles of the sound speed $c(z)$ in a motionless medium and the wind velocity $\mathbf{v}(z)$, Fujiwhara derived the equation of a sound ray and obtained the height z_t of the turning point, the travel time along the path, and the location of the audibility zones. Working independently and using a different approach, six years later, Emden [108] rederived the equations obtained by Fujiwhara, and calculated the sound ray paths for various profiles of $c(z)$ and $\mathbf{v}(z)$. The papers by Fujiwhara and Emden showed clearly that the effects of wind stratification $\mathbf{v}(z)$ on the refraction of sound in the atmosphere are significant. However, stratification of $\mathbf{v}(z)$ cannot by itself explain the turning points of sound signals in the stratosphere, because the zones of audibility often have an appearance similar to a ring [103, 277].

By the end of the 1930s, the turning points in the stratosphere were usually attributed to the increase in the sound speed $c(z)$ in the stratosphere, causing refraction of sound rays and their return to the ground. In the stratosphere, sound ray paths can be obtained from Snell's law: $c(z)/\cos\theta(z) = $ constant. Here, θ is the elevation angle (the angle between the direction of ray propagation and the horizontal plane), and the sound speed $c(z)$ is related to the temperature $T(z)$ by equation (1.1) below. It follows from this equation and

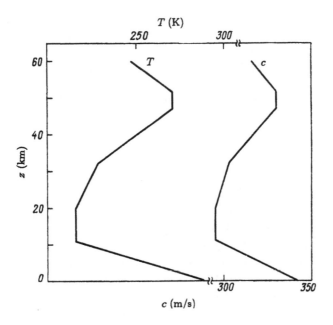

FIGURE 1.1
Vertical profiles of the temperature $T(z)$ and sound speed $c(z)$ of the standard atmosphere [365].

Snell's law, that a sound wave can have a turning point at a height z_t only if $c(z_t) > c(z = 0)$ and, hence, $T(z_t) > T(z = 0)$.

In the decades preceding World War II, sound propagation from large explosions was studied extensively. Hundreds of explosions on the earth's surface were made in many European countries [103]. Tens and sometimes hundreds of observers situated at different distances and azimuthal directions measured the travel time t_{tr} of sound propagation from the point of explosion to the point of observation. Such an interest in sound propagation in the upper atmosphere arose from the fact that direct measurements of temperature and wind velocity were not possible at heights above 20 km at that time. Therefore, scientists endeavored to reconstruct the vertical profiles of $c(z)$ and $T(z)$ from the measured travel times t_{tr} and the angles at which the sound wave arrived at the earth's surface, while usually assuming that $\mathbf{v} = 0$. The vertical profiles of $c(z)$, reconstructed by this method, indicated that at heights of 40–50 km the sound speed was much greater than its value $c(z = 0)$ near the ground (see, for example, references [103, 277]). But this height dependence of the sound speed differs qualitatively from that in the standard atmosphere [365] obtained using rocket data, as shown in figure 1.1. In the standard atmosphere $c(z) < c(z = 0)$ at heights $z \sim 40$–50 km so that, according to Snell's law, sound waves cannot have turning points (at least, in the upwind direction).

Thus, in the 1920s and 1930s, the vertical profiles of $c(z)$ and $T(z)$ in the stratosphere were not reconstructed correctly, and all causes of the return of sound signals from the stratosphere were not identified. Nevertheless, the study of sound propagation in the upper atmosphere did lead to many important new results for geometrical acoustics in a moving medium. Recent studies [80] explain the return of sound and infrasound signals from the stratosphere as scattering from the fine structure of the temperature and wind velocity fields when random inhomogeneities are significantly elongated in a horizontal direction.

1.1.4   Current applications of acoustics in a moving medium

Interest in atmospheric acoustics, and acoustics of moving media in general, was significantly reduced after World War II, because electromagnetic waves then became widely used for purposes such as detection and ranging, direction finding, and sounding. But, by the beginning of the 1970s, interest in acoustic methods reemerged in many fields of physics. This subsection describes many of the current applications where the theories of wave propagation in moving media are used.

Important applications of atmospheric acoustics include detection, recognition, and tracking of sound sources using microphone arrays; broadcasting over long ranges (loudspeakers can be installed near the ground and also on airplanes and helicopters); and prediction of sound propagation near the ground, given the temperature and wind velocity fields, terrain, and the properties of the ground. The latter is important in predicting noise levels near highways [328], railways [390], and airports [352], the peak and mean sound-pressure levels from small explosions and gunfire [148, 195], and in other practical concerns. The effect of temperature and wind velocity fluctuations on the rise time and shape of sonic booms has been examined [42, 312], as these relate to the annoyance of sonic booms from supersonic passenger aircraft, and aircraft designs to mitigate such annoyance.

Since the beginning of the 1970s, acoustic and radio-acoustic remote sensing techniques have rapidly evolved and entered into widespread usage for measuring the structure of the lower atmosphere [44]. These techniques are enabled by theories of wave propagation in moving media (section 6.4), which relate the sensed signals to atmospheric parameters. Acoustic and radio-acoustic sounding are also significantly affected by refraction of sound in the atmosphere. For example, the maximum height of radio-acoustic sounding is restricted mainly by sound beam advection due to the horizontal wind [188, 192]. Refraction of sound due to temperature and wind velocity stratification can affect acoustic sounding in the atmosphere [43, 134, 309, 310]. Some proposed techniques for remote sensing of the temperature of the atmosphere and the vertical profiles of the sound speed in the ocean are based on the refraction of sound [53, 290].

Acoustic tomography, including diffraction tomography [337], applied to the ocean [266, 267], the atmosphere [433], and other moving media, is a remote sensing technique for reconstruction of the sound speed and medium velocity fields. Acoustic travel-time tomography of the atmospheric surface layer is considered in Chapter 3.

Aeroacoustics, which deals with the radiation and propagation of sound waves due to aerodynamic forces and unsteady flows, is usually considered to be a field of acoustics distinct from the acoustics of moving media. However, the fields do have some overlapping research goals, such as investigation of the effects of the mean profile of a turbulent jet on the emission of sound waves [144].

Acoustics in moving media also pertains to studies of sound propagation in ducts, nozzles, and diffusers with gas flow [61, 166, 263, 265, 371]. This is relevant, for example, to the analysis of noise emitted by nozzles and exhaust pipes, and the stability of propellant combustion in rocket engines.

Until the beginning of the 1970s, the effects of currents on sound propagation in the ocean had usually been ignored. However, recent theoretical and experimental results have shown that in certain cases, currents can affect the sound field in the ocean quite significantly. (See section 1.3.)

Acoustics of inhomogeneous moving media has applications in astrophysics when studying acoustic and gravity waves in the solar atmosphere in the presence of laminar or random flows, e.g., references [268, 269].

1.2 Sound propagation in the atmosphere

In this section, parameters affecting sound propagation in the air are considered. The approximation of the effective sound speed, which is used widely in atmospheric acoustics, is introduced and discussed. A brief overview of near-ground sound propagation is presented.

1.2.1 Parameters affecting sound propagation in the air

Sound waves propagating in the atmosphere are attenuated by relaxation and dissipation processes in air. These phenomena have been well studied [52, 368]; the resulting absorption coefficient depends on the temperature, humidity, acoustic frequency, and, to a lesser extent, on atmospheric pressure.

Propagation of sound is also affected by the sound speed c and wind velocity $\mathbf{v}$. As will be derived later in this book (equation (6.84)), the sound speed in the atmosphere is given by

$$c = \sqrt{\gamma_a R_a T \left(1 + 0.511q\right)}. \tag{1.1}$$

Here, $\gamma_a = 1.40$ is the ratio of specific heats for dry air, $R_a = 287.058 \ \mathrm{m^2/(s^2 \, K)}$ is the gas constant for dry air, and q is specific humidity.

It follows from equation (1.1) that the sound speed c is primarily affected by temperature T, while the effect of humidity on c is smaller, but not necessarily negligible.

The temperature T, humidity q, and wind velocity $\mathbf{v}$ are among the most important parameters affecting sound propagation. Both their mean values and fluctuating components are important. The fluctuations in T, q, and $\mathbf{v}$ and their effect on sound propagation are studied in detail in Part II of this book. It is often reasonable, for modeling purposes, to view the atmosphere as a stratified, moving medium in which T, q, and $\mathbf{v}$ depend only on the vertical coordinate z. We refer to these quantities as the mean *vertical profiles*. The horizontal component of the vector $\mathbf{v}$ is usually much greater than its vertical component, which is thus often assumed to be zero.

The vertical profiles of $T(z)$, $q(z)$, and $\mathbf{v}(z)$ in the atmospheric surface layer (ASL) have been studied extensively. For neutral or unstable stratification, the ASL extends vertically to about 100–200 m from the ground; for stable stratification, the top of the ASL is usually lower. In the ASL, the vertical profiles of temperature, humidity, and wind velocity can be determined with the Monin–Obukhov similarity theory (MOST), which is considered in section 2.2.3.

Above the ASL (in the atmospheric boundary layer and free troposphere), these profiles can be obtained using direct measurements with weather balloons (tethersondes and radiosondes) and airplanes, and with the numerical weather prediction (NWP) computer models, such as the Weather Research and Forecasting (WRF) Model, which was developed through a collaboration of multiple organizations in the United States, or the European Centre for Medium-Range Weather Forecasts (ECMWF) Integrated Forecast System. NWP is improving in accuracy and resolution as a result of advances in numerical methods and computational hardware.

Jet flows often appear in the stratosphere and upper troposphere [245]. The height of the axis of the jet flows is about 10 km in the middle latitudes. The vertical and horizontal scales of the jet flow and the wind velocity on its axis vary significantly; the characteristic values of these quantities are 10 km, 1500 km, and 50 m/s, respectively. The wind velocity can reach a value of 200 m/s on the axis of the jet flow, corresponding to the Mach number $M = v/c \sim 0.6$.

The mean profiles of wind velocity and wind direction in January and July at latitude $30°$ are presented in figures 1.2 and 1.3 [146, 189]. These figures indicate that, in the stratosphere, the wind velocity reaches a few tens of meters per second and the wind direction can significantly vary with height. The vertical profile of the temperature in the standard atmosphere [365] is shown in figure 1.1. Significant spatial-temporal variability in the temperature and wind velocity in the atmosphere is caused by its general circulation, the seasonal modulations, and the day-to-day variations due to planetary waves and eddies, the tides produced by solar heating, and internal gravity waves [90].

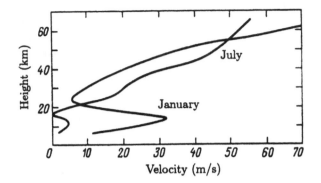

FIGURE 1.2
Vertical profiles of wind speed in January and July at latitude 30° [189, 146].

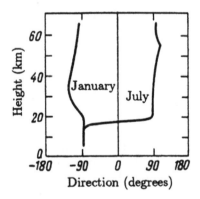

FIGURE 1.3
Vertical profiles of wind direction in January and July at latitude 30° [189, 146].

1.2.2 Effective sound speed approximation

Approximations are commonly employed to represent sound propagation through a moving medium using an effective, motionless medium. The most common of these is the effective sound speed approximation, which involves setting the sound speed to

$$c_{\text{eff}} = c + \mathbf{s} \cdot \mathbf{v}. \tag{1.2}$$

Here, $\mathbf{s}$ is the unit vector tangential to the sound ray path. This approximation was introduced by Rayleigh [324] and is still employed in atmospheric acoustics, ocean acoustics, and acoustics of tubes with flow. With this approximation, many analytical and numerical methods developed for acoustics in a motionless medium can be applied to a moving medium.

For sound propagation near the ground, the elevation angle of the vector **s** is usually small. In this case, the effective sound speed is often used in the following form

$$c_{\text{eff}} = c + v \cos \psi. \tag{1.3}$$

Here, ψ is the angle between the azimuthal direction of sound propagation and the horizontal wind velocity **v**.

It must be kept in mind that the effective sound speed approximation is a heuristic approach, which cannot describe many important effects that a moving medium has on sound propagation. The ranges of applicability of this approximation are studied in reference [139]. In particular, it is shown that this approximation is valid for calculations of the sound-pressure field only if $v/c \ll 1$, so that terms of order $(v/c)^2$ can be ignored. Other assumptions might also apply for this approximation to be valid [139].

As an example of the limitations of the effective sound speed approximation, note that, in a stratified moving medium, the actual sound ray path is generally a three-dimensional curve, even if it is calculated to order $O(v/c)$. On the other hand, with the effective sound speed, a ray path is always a two-dimensional curve. (This result does not contradict the ranges of applicability of this approximation obtained in reference [139], since a ray path is not directly related to the sound field.)

In this book, sound propagation in a moving medium is studied from first principles rather than with the effective sound speed approximation. In some approximate equations derived from the exact equations, the sound speed and medium velocity can be combined into c_{eff}. In addition to c_{eff}, the effective density ϱ_{eff} might be needed to approximately replace sound propagation in a moving medium with that in a motionless medium with c_{eff} and ϱ_{eff}. The effective density is considered in section 4.1. It has been used in the literature to a much lesser extent than the effective sound speed.

1.2.3   Sound propagation near the ground

Among the sources of acoustic waves propagating near the ground are cars, trains, aircraft, working factories, wind turbines, shots from artillery and small arms, and explosions. Sound from these sources can be heard or recorded at distances up to a few kilometers along the earth's surface. The frequency range of interest is typically from about 10 Hz up to a few kHz, as sound at higher frequencies is strongly attenuated by the air.

Propagation above a partially reflecting (impedance) ground has been studied extensively [16]. Analytical solutions are available for the case of a homogeneous atmosphere above flat ground. The solution is conveniently formulated as a complex image source, with corrections for spherical wave reflection [15, 106]. Extensions incorporate decorrelation between direct and

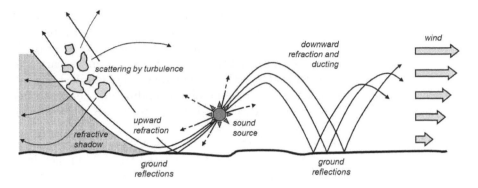

FIGURE 1.4
Outdoor sound propagation phenomena and their interactions.

ground-reflected ray paths resulting from turbulence [82, 86, 297, 347]. This phenomenon raises sound levels above what would normally be observed at locations where the direct and reflected rays interfere destructively.

The many complex effects of weather and terrain on outdoor sound propagation present great challenges for modeling. Refraction and turbulent scattering in the atmosphere vary dramatically in response to changing solar radiation and wind conditions. Sound waves interact with hills, man-made structures such as buildings, natural landcover cover (vegetation), soil, and near-surface geology. Landcover and elevation changes also locally modify the atmospheric flow, which in turn affects the propagation. Many of these interacting phenomena are illustrated in figure 1.4. Although this book, since it deals with the acoustics of moving media, addresses primarily the refraction and scattering effects, in Part III we also discuss numerical modeling of some other phenomena of interest.

1.2.3.1 Atmospheric stratification

Of particular importance for sound propagation is the stratification of the temperature and wind velocity fields, which leads to strong vertical gradients and the refraction of the sound. In the near-ground atmosphere, the temperature $T(z)$ and wind speed $v(z)$ profiles can be determined with MOST; see equations (2.45) and (2.44), respectively. The refraction effects can be most simply understood in terms of the effective sound speed c_{eff} defined with equation (1.3). A positive gradient in c_{eff} leads to downward refraction of sound, which normally enhances sound levels near the surface, whereas a negative gradient leads to upward refraction, which normally diminishes sound levels.

A negative vertical gradient, and hence upward refraction, may be caused by either a temperature lapse condition, which means that the temperature (and hence the sound speed) decreases with height, or by a negative wind shear. Negative wind shear (decreasing wind velocity with height) usually occurs in the upwind direction, meaning that the wind is blowing from the receiver to the source. A positive vertical gradient and downward refraction may be caused either by a temperature inversion condition, which means that the temperature increases with height, or by a positive wind shear, which usually occurs in the downwind direction. The temperature gradient and wind shear effects are both important, in general. Depending on the atmospheric state and propagation direction, they may combine to strengthen or diminish the overall refractive effect.

As mentioned, downward refraction is normally associated with enhanced sound levels, whereas upward refraction is associated with diminished levels. However, these expectations do not always hold. For example, sound levels may actually be elevated near the boundary of a refractive shadow zone, due to the presence of a caustic there. Downward refraction conditions are complicated by interference effects between ray paths or propagating modes, and by interactions between ducted sound and absorbing ground surfaces.

Given the importance of vertical refraction of sound in the near-ground atmosphere, it is important that propagation calculations incorporate this effect. Ray-based methods can calculate refraction very efficiently, and are also very helpful in visualizing propagation phenomena. Chapter 3 provides ray acoustics equations that correctly incorporate the effect of wind on refraction, whereas Chapter 9 describes the numerical solution of these equations.

The main drawback of ray acoustics is its unsuitability for low frequencies. Ordinary ray methods also do not describe diffraction and scattering into shadow regions, as occurs during strong upward refraction. The fast-field program (FFP) and parabolic equation (PE), which were introduced into atmospheric acoustics in the late 1980s, largely avoid the drawbacks of ray-tracing methods. The FFP [220, 411] solves a Helmholtz type equation (see sections 2.3 and 4.1) which has been Fourier-transformed with respect to the horizontal coordinates. The vertical coordinate is partitioned into a finite number of layers. The PE [137, 403] is based on a finite-angle (forward propagating) approximation to the full wave equation. Derivations of narrow-angle and wide-angle PEs are presented in section 2.5. A starting condition at the source is marched forward in the horizontal range coordinate. Chapter 11 further describes the numerical implementation of the FFP and PE and provides example calculations for sound propagation in the atmosphere.

Recently, there has been strong interest in finite-difference, time-domain (FDTD) methods, due to their ability to readily handle many complex signal generation and propagation phenomena. Refraction can be rigorously incorporated in FDTD calculations [38, 303]. Equations appropriate to FDTD calculations in a moving medium are derived in section 2.4, whereas their numerical solution is described in Chapter 11.

1.2.3.2 Turbulence

The lower atmosphere contains random motions on a variety of scales. Most of these motions are *turbulent*; that is, three-dimensional, rotational disturbances of temperature and wind velocity. Atmospheric turbulence is generated by wind shear, and by buoyancy instabilities resulting from unstable stratification. Large turbulent eddies span the depth of the atmospheric boundary layer, which is up to roughly 2 km thick, and have time scales of many minutes, whereas the smallest eddies have sizes less than 1 cm and produce variations shorter than 1 s. Therefore, the sound field also undergoes random variations on these time scales. While stable stratification suppresses turbulence, internal gravity waves are common in such conditions.

Atmospheric turbulence leads to scattering of sound energy into refractive shadow zones, amplitude and phase fluctuations in received sound signals, fluctuations in the angle of arrival, coherence loss, and changes in the interference pattern between the direct and ground-reflected waves. These effects are important for studies of noise propagation in the atmosphere, source localization with phased sensor arrays, and remote sensing techniques such as acoustic and radio-acoustic sounding.

Representation of the turbulence is often a major challenge. Ideally, we can derive closed-form equations for the sound-field statistics of interest (as considered in detail in Part II of this book), as this approach often enables a better understanding of the physics of the problem and can lead to faster numerical calculations. Such analytical results also aid the development of remote sensing techniques. However, suitable equations for the spectrum of the turbulence are still required. Alternatively, for numerical calculations, we may consider using a computational fluid dynamics (CFD) simulation, or synthesizing the turbulence kinematically. The relative benefits of these approaches are described in Chapter 9, which also discusses kinematic methods in detail, since they are widely used in outdoor sound propagation. Typically, numerical calculations are performed by a Monte Carlo approach, in which a sound field propagates through many realizations of the turbulence field.

1.2.3.3 Uncertainties in predictions

A useful analogy can be made between predicting sound propagation outdoors and predicting the weather. Weather forecasts are imperfect because the numerical forecast models are initialized with atmospheric observations that do not describe the atmospheric state with full accuracy and resolution. The forecast models themselves have a finite resolution and do not perfectly capture the atmospheric physics. Similarly, solutions for outdoor sound propagation are based on imperfect knowledge of the environment (natural and man-made ground cover, building materials, atmospheric wind and temperature profiles, terrain elevation variations, etc.) and cannot capture all of the pertinent physics, such as scattering from small-scale turbulence and vegetation, and coupling of vibrations into the ground.

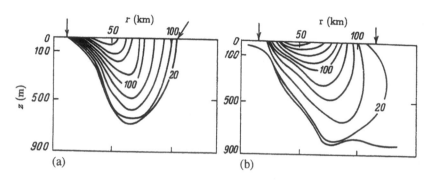

FIGURE 1.5
Isotachs of the Gulf Stream in vertical planes in two locations. (a)
$(60°09', 27°26')$ May–June 1967; (b) $(80°24', 30°20')$ July–August 1967 [333].

Uncertainties may be attributed to both the model itself, and to the model parameters. The literature on outdoor sound propagation modeling has typically focused on the former source of uncertainty, namely improving models to handle more complex physics and to improve numerical methods. However, numerical methods for sound propagation can be employed much more effectively when their predictive capabilities, relative to the inputs provided to them and the uncertainties inherent to sound propagation, are well understood. This topic is the primary concern of Chapter 13, which considers application of stochastic integration to problems involving sound propagation in the presence of refraction and turbulence.

1.3 Effects of currents on sound propagation in the ocean

1.3.1 Motion of oceanic water

Many types of motion occur in ocean waters. We consider here only the most typical types [91, 260].

The strongest currents in the world are the Antarctic circle current, the Gulf Stream, the Kurocio current, and Cromwell's current. The characteristic parameters of these currents are nearly constant in space and time. The vertical velocity of a current is much smaller than the horizontal velocity and is usually ignored in oceanic acoustics. The maximum velocity of currents reaches a value of 1.5–2 m/s.

Figure 1.5 shows isotachs of the velocity v of currents in the Gulf Stream measured in vertical planes perpendicular to the current axis at two locations [333]. The arrows indicate the horizontal range r at which the mean velocity of the current on the ocean surface is zero and z is the depth. The figure indicates that the current velocity has its maximum on the surface and the current does not reach the oceanic bottom. Although this dependence of v on z is typical for some other currents, this is not the case for all currents. For instance, the Antarctic current reaches the bottom, and the velocity of Cromwell's current has its maximum at a depth of 100–200 m in the eastern part of the Pacific Ocean.

Usually, the vertical profiles of currents have fine structure. The fine structure is caused by horizontal layers existing everywhere in the ocean, with nearly constant values of the sound speed c, temperature T, and velocity $\mathbf{v}$ in the layers and large vertical gradients of these functions near their boundaries. The vertical scales of the layers range from ten centimeters to few tens of meters. The horizontal scales are up to tens of kilometers, and they exist from a few hours to a few days. Near the boundaries of the layers, the vertical gradients of the current velocities are much greater than those in the layers and can reach a value of 5–10 cm/s per meter. The fine structure of the current velocity measured by a quick-response probe is seen clearly in figure 1.6 [260].

Synoptic eddies are unsteady objects in the ocean analogous to cyclones and anticyclones in the atmosphere. The synoptic eddies are usually subdivided into the eddies of the open ocean and the frontal eddies generated by frontal currents. For example, 5–8 cyclonic and anticyclonic frontal eddies typically break away each year from the Gulf Stream. In the Northern Hemisphere, the cyclonic eddies rotate anticlockwise and contain cold water relative to the surrounding water. The anticyclonic eddies contain warm water and rotate clockwise. The characteristic scales of the frontal eddies range from 100 km to 400 km, the mean velocity of the eddy center is of the order of a few cm/s, and the maximum velocity of water rotation in the eddy is about 1 m/s. (In some cases, this velocity can reach a value of 3 m/s.) The mean life span of a frontal eddy is a few years, its diameter is reduced in the course of time, and the eddy itself descends with a speed of nearly 0.6 m/day. The structure of the cyclonic eddies of the open ocean coincides qualitatively with that of the frontal eddies. However, the former have a smaller diameter and lower velocity of rotation.

Currents can also be caused by tides and internal gravity waves. The velocity of water motion in tides and internal gravity waves can reach a value of 1 m/s and a few tens of cm/s, respectively. The vertical velocity of water motion caused by an internal gravity wave may not be ignored.

1.3.2 Effects of currents on sound propagation

A typical variation of the sound speed in the ocean is $|c - c_0|/c_0 \sim 3 \times 10^{-2}$, where c_0 is a reference value of the sound speed c. On the other hand, the ratio v/c is of the order of 10^{-3} (i.e., 30 times smaller than the typical value of $|c - c_0|/c_0$), even for strong currents with $v \sim 1.5$ m/s. Nevertheless, currents can affect sound propagation as a result of at least three mechanisms.

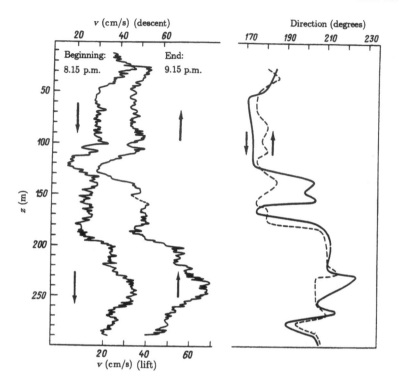

FIGURE 1.6
Magnitude and direction of current velocity measured by descent ($\downarrow$) and lift ($\uparrow$) of a quick-response probe [260].

First, currents can change the phase of a sound wave, and hence its travel time, if the propagation range is sufficiently large. It follows from equation (3.97) that the phase change caused by a current is given by $\Delta\Phi \approx -2\pi f R\bar{v}_R/c^2$. Here, R is the distance from the source to the receiver, f is the acoustic frequency, and $\bar{v}_R$ is the mean value (along the sound path) of the current velocity component $v_R(z)$ in the direction from the source to the receiver. This phase change is significant if $|\Delta\Phi| \gtrsim \pi/8$. Substituting the value of $\Delta\Phi$ into this inequality yields

$$R \gtrsim \frac{c^2}{16 f \bar{v}_R}. \tag{1.4}$$

Currents can contribute significantly to the phase change of a sound field if R satisfies the latter inequality. If $\bar{v}_R = 0.3$ m/s and $f = 100$ Hz, it follows from equation (1.4) that $R \gtrsim 4.7$ km.

Second, currents can cause a noticeable change in the amplitude of a sound field. Indeed, if two or more sound rays arrive at the observation point, and the phase change along at least one of the rays depends on $\bar{v}_R$, the amplitude of the resulting sound field also depends on $\bar{v}_R$. Note that the effect of $\bar{v}_R$ on the sound-pressure amplitude along a sound ray can usually be ignored.

Third, currents can lead to a qualitative change of sound propagation if $|\partial v/\partial z| \gtrsim |\partial c/\partial z|$.

In the remaining part of this section, we shall consider these effects for sound propagation through currents, synoptic eddies, and tides.

1.3.3 Currents

It has been shown that the currents can cause a significant change in the amplitude and phase of a sound field and its travel time [120, 364]. The profiles of $c(z)$ and $v(z)$ have been assumed to be linear or constant, and the sound field has been calculated using geometrical acoustics (see section 3.5).

Sound propagation in the direction of water motion in the Gulf Stream and in the opposite direction has been studied theoretically [149]. In these cases, the ocean can be considered a stratified moving medium. Figure 1.7 shows the vertical profiles of $c(z)$ and $v(z)$ that were adopted for the calculation of the sound field [149]. These profiles are typical of the Gulf Stream, where the ocean depth is large. It was assumed that the source and receiver are located at a depth of 250 m and the acoustic frequency is 50 Hz. Calculations were performed using the wave theory (section 4.1) and the effective sound speed c_{eff}, as defined with equation (1.3). The predicted sound intensity I versus the horizontal range r is shown in figure 1.8 for sound propagation in the direction of the current and in the opposite direction.

In this numerical example, $|\partial v/\partial z| > |\partial c/\partial z|$ for $z \lesssim 0.5$ km. As a result, sound propagation in the direction of the current differs qualitatively from that in the opposite direction. In the direction of the current and for $z \lesssim 0.5$ km, the value of c_{eff} decreases if the depth is increased, so that there is an antiwaveguide sound propagation and a shadow zone exists at a horizontal range 20 km $\lesssim r \lesssim 50$ km (see figure 1.8). But in the opposite direction, there is a waveguide near the ocean surface because c_{eff} increases with depth for $z \lesssim 0.5$ km. This waveguide contains one acoustic mode. Because of the mode, the sound intensity I is greater than that in the direction of the current by 15 dB. Since the profiles of $c(z)$ and $v(z)$ that have been adopted for calculations [149] are model profiles, it would be of interest to carry out experiments to confirm that the currents can qualitatively change the sound propagation in the ocean. (See also section 3.5.3.)

The effect of a geostrophic flow on sound propagation in the shallow ocean has been studied [119, 121]. To this end, a model of a geostrophic flow has been developed [121]. It is assumed in this model that the current velocity decreases linearly with depth and is directed to the north. Owing to a geostrophic balance, there is a horizontal gradient of the sound speed c in the eastern direction, which depends on the current velocity v_s at the ocean surface. The dependence of c on z is linear. This model of a geostrophic flow describes

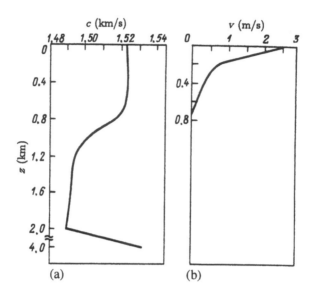

FIGURE 1.7
Vertical profiles of (a) sound speed and (b) current velocity used for calculation of the sound field [149] shown in figure 1.8.

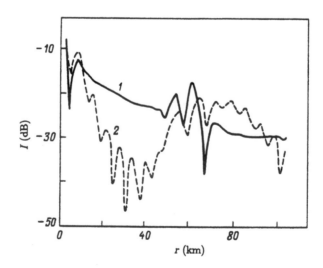

FIGURE 1.8
Sound intensity I versus horizontal range r for the profiles $c(z)$ and $v(z)$ shown in figure 1.7 [149]. Lines 2 and 1 correspond to sound propagation in the direction of the current and in the opposite direction, respectively.

approximately the profiles of c and $\mathbf{v}$ of the Gulf Stream in the Florida Strait. Sound propagation perpendicular to the axis of a geostrophic flow has been considered [119]. It has been shown that, to an accuracy of v/c, the current does not directly affect the sound field in the geometrical acoustics approximation. However, a variation of v_s causes a variation of $c(\mathbf{r}, z)$ (because the horizontal gradient of c depends on v_s) and, hence, a change of the sound field. Calculations [119] have shown that the phase change of a sound field depends linearly on the variation of v_s. This result is in good agreement with the experimental data [119] on sound propagation from Miami, Florida, to Bimini, Bahamas, perpendicular to the Florida Strait.

1.3.4 Synoptic eddies

Let us now consider the effects of water motion in synoptic eddies on sound propagation. Model profiles of the sound speed $c(r, z)$ and current velocity $\mathbf{v}(r, z)$ in the synoptic eddies have been constructed [163]. These profiles have been used for studying sound propagation through the synoptic eddies over horizontal ranges of few tens of kilometers [162, 179] and 1000 km [180]. This work has shown that the amplitude and phase of a sound field are affected mainly by variation of the sound speed in the eddy. Nevertheless, the water motion in the eddy can cause a sound intensity variation greater than 10–12 dB, and a phase change much greater than π. Thus, the effects of water motion on sound propagation should be taken into account if a sound wave passes through a synoptic eddy.

1.3.5 Tides

The effect of tidal currents on sound propagation in the shallow ocean has been considered [363]. It has been assumed that c and v do not depend on the spatial coordinates and that v varies slowly with time. It has been predicted that the sound-pressure amplitude depends on the current velocity significantly and that its phase is proportional to v_R. The linear dependence of the phase on v_R has been confirmed by measuring sound propagation between Block Island, Rhode Island, and Fishers Island, New York.

1.3.6 Reciprocal acoustic transmission

It has been argued above that the motion of oceanic water caused by currents, synoptic eddies, and tides can affect sound propagation significantly. Reciprocal transmission paths (parallel paths that are in close proximity, but in opposite directions) can be used to study this effect experimentally. Indeed, it is difficult to separate the effect of currents on a sound field from the effect of sound speed variations on the same sound field if sound signals propagate only in one direction. According to the reciprocity principle, sound signals propagating in opposite directions totally coincide if $\mathbf{v} = 0$, but they are different if $\mathbf{v} \neq 0$. Reciprocal acoustic transmission has been studied experimentally [266]. In reference [436], an experiment was described in which two sources were located at a depth of 1 km and the horizontal distance between them was 25 km. This experiment showed that the travel time of the sound impulse, its amplitude, and its shape depend significantly on the direction of sound propagation.

Currents can be reconstructed using reciprocal transmission and measuring the difference in the travel time of impulse propagation in opposite directions. Such a remote sensing technique would enable scientists to investigate the structures of unsteady currents, synoptic eddies, tides, and internal gravity waves. The significance of this application has resulted in several publications [271, 335, 336] where the effects of currents on reciprocal transmission have been studied numerically using the parabolic equation method. The results obtained in these papers show that currents can affect the amplitude, phase, and travel time of sound transmissions.

2

Equations for acoustic and internal gravity waves in an inhomogeneous moving medium

In this chapter, the equations for acoustic and internal gravity waves in inhomogeneous moving media are systematically derived and analyzed. The ranges of applicability of these equations are studied in detail and they are compared with results presented previously in the literature. The equations in this chapter are used in subsequent chapters for studies of sound propagation in moving media.

In section 2.1, a complete set of fluid dynamic equations is presented. These equations are then linearized resulting in a set of linearized equations of fluid dynamics which provides the most general description of both acoustic and internal gravity waves in a moving medium. The set of linearized equations is, however, rather involved for analytical or numerical studies of sound propagation. In a stratified moving medium (sections 2.2 and 2.3), this set reduces exactly to a single equation for the pressure of an acoustic or internal gravity wave, which is more convenient for analysis. In a three-dimensional moving medium (section 2.4), with some approximations, the linearized equations of fluid dynamics reduce to two coupled equations for the sound pressure and acoustic particle velocity. Using additional approximations, these coupled equations reduce to wave-type and Helmholtz-type equations for the sound pressure. In section 2.5, narrow-angle and wide-angle parabolic wave equations are derived, which are convenient for both analytical and numerical studies of sound propagation in a three-dimensional moving medium.

The flowchart of the most important equations considered in this chapter is shown in figure 2.1. It also provides with the ranges of applicability of these equations and indicates where they are used.

In this chapter, we also consider equations for the pressure, density, medium velocity, entropy, and concentrations of the components dissolved in the medium through which acoustic and internal gravity waves propagate. Section 2.1 presents such equations for a three-dimensional moving medium and section 2.2 for a stratified moving medium.

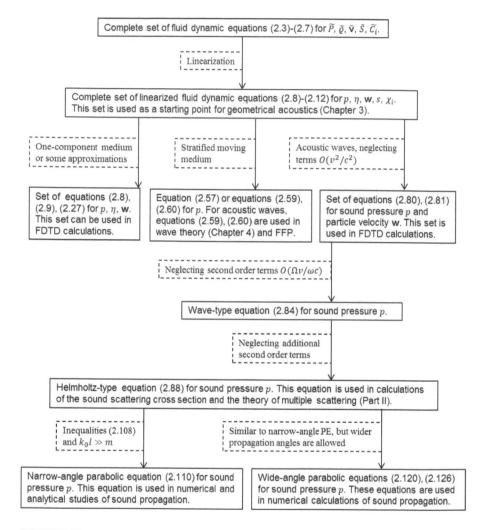

FIGURE 2.1

Flowchart of the most important equations considered in Chapter 2. Dashed boxes adjacent to the arrows indicate approximations/assumptions made in the derivations. $\widetilde{P}$, $\widetilde{\varrho}$, $\widetilde{\mathbf{v}}$, $\widetilde{S}$, and $\widetilde{C}_i$ are the total pressure, density, velocity, entropy, and the concentrations of the components dissolved in the medium; p, η, $\mathbf{w}$, s, and χ_i are their fluctuating components due to a propagating wave. FDTD stands for the finite-difference, time-domain method and FFP for the fast-field program.

2.1 Fluid dynamic equations and their linearization

2.1.1 Fluid dynamic equations

The most general possible approach to wave propagation in an inhomogeneous moving medium is based on the complete set of fluid dynamic equations:

$$\left(\frac{\partial}{\partial t} + \widetilde{\mathbf{v}} \cdot \nabla\right) \widetilde{\varrho}_i + \widetilde{\varrho}_i \nabla \cdot \widetilde{\mathbf{v}} = \widetilde{\varrho}_i Q, \quad i = 0, 1, \dots, n, \tag{2.1}$$

$$\widetilde{P} = \widetilde{P}(\widetilde{S}, \widetilde{\varrho}_0, \widetilde{\varrho}_1, \dots, \widetilde{\varrho}_n), \tag{2.2}$$

$$\left(\frac{\partial}{\partial t} + \widetilde{\mathbf{v}} \cdot \nabla\right) \widetilde{\mathbf{v}} + \frac{1}{\widetilde{\varrho}} \nabla \widetilde{P} - \mathbf{g} = \mathbf{F}, \tag{2.3}$$

$$\left(\frac{\partial}{\partial t} + \widetilde{\mathbf{v}} \cdot \nabla\right) \widetilde{S} = 0. \tag{2.4}$$

Here, $\widetilde{P}(\mathbf{R}, t)$ is the pressure in the medium, $\widetilde{\varrho}_i(\mathbf{R}, t)$ are the densities of the components of the medium, $\widetilde{\mathbf{v}}(\mathbf{R}, t)$ is the velocity vector, and $\widetilde{S}(\mathbf{R}, t)$ is the entropy, where $\mathbf{R} = (x, y, z)$ are the Cartesian coordinates and t is time. In equations (2.1)–(2.4), $\widetilde{\varrho} = \sum_{i=0}^{n} \widetilde{\varrho}_i$ is the total density of the medium, $\nabla = (\partial/\partial x, \partial/\partial y, \partial/\partial z)$, $\mathbf{g} = (0, 0, -g)$ the vector of the acceleration due to gravity (the direction of this vector is opposite to the direction of the vertical z-axis), and $\mathbf{F}(\mathbf{R}, t)$ and $Q(\mathbf{R}, t)$ characterize a force acting on the medium and a mass source, respectively. The tilde above the quantities $\widetilde{P}$, $\widetilde{\varrho}_i$, $\widetilde{\mathbf{v}}$, and $\widetilde{S}$ indicates that the medium is perturbed by a wave propagating through it. Equations (2.1)–(2.4) express, respectively, the law of mass conservation of the ith component of the medium, the equation of state, the law of momentum conservation, and the assumption of adiabatic motion of the medium. We consider that a medium consists of $n + 1$ components with densities $\widetilde{\varrho}_i$ since these components can affect sound propagation. For example, in the ocean, the sound speed depends on the concentration of salt dissolved in the water; in the atmosphere, fluctuations of water vapor cause fluctuations of the sound field.

Instead of the densities $\widetilde{\varrho}_0, \widetilde{\varrho}_1, \dots, \widetilde{\varrho}_n$, it is convenient to deal with the total density $\widetilde{\varrho}$ of the medium and the concentrations $\widetilde{C}_i = \widetilde{\varrho}_i/\widetilde{\varrho}_0$ of the components dissolved in the medium. Here, $\widetilde{\varrho}_0$ is the density of the basic component of the medium (the basic solvent) and the index i is redefined as $i = 1, 2, \dots, n$. In this case, equations (2.1) and (2.2) take the form

$$\left(\frac{\partial}{\partial t} + \widetilde{\mathbf{v}} \cdot \nabla\right) \widetilde{\varrho} + \widetilde{\varrho} \nabla \cdot \widetilde{\mathbf{v}} = \widetilde{\varrho} Q, \tag{2.5}$$

$$\left(\frac{\partial}{\partial t} + \widetilde{\mathbf{v}} \cdot \nabla\right) \widetilde{C}_i = 0, \quad i = 1, 2, \dots, n, \tag{2.6}$$

$$\widetilde{P} = \widetilde{P}(\widetilde{\varrho}, \widetilde{S}, \widetilde{C}_1, \widetilde{C}_2, \ldots, \widetilde{C}_n). \tag{2.7}$$

Equations (2.3)–(2.7) represent a complete set of equations for $\widetilde{P}$, $\widetilde{\varrho}$, $\widetilde{\mathbf{v}}$, $\widetilde{S}$, and $\widetilde{C}_i$. This set is referenced at the top of the flowchart of equations depicted in figure 2.1.

2.1.2   Linearized equations of fluid dynamics

For describing a wave propagating in the medium, in equations (2.3)–(2.7) we set $\widetilde{P} = P + p$, $\widetilde{\varrho} = \varrho + \eta$, $\widetilde{\mathbf{v}} = \mathbf{v} + \mathbf{w}$, $\widetilde{S} = S + s$, and $\widetilde{C}_i = C_i + \chi_i$. Here, P, ϱ, $\mathbf{v}$, S, and C_i are the ambient values (i.e., the values in the absence of a propagating wave) of the pressure, density, medium velocity, entropy, and the concentrations of the components dissolved in the medium, and p, η, $\mathbf{w}$, s, and χ_i are their fluctuations due to a propagating wave. All these quantities are functions of both $\mathbf{R}$ and t. In many cases, the wave propagating in the medium disturbs this medium only slightly. In such cases, equations (2.3)–(2.7) can be linearized with respect to p, η, $\mathbf{w}$, s, and χ_i. Rearranging the resulting linearized equations, we obtain

$$\frac{d\mathbf{w}}{dt} + (\mathbf{w} \cdot \nabla)\mathbf{v} + \frac{1}{\varrho}\nabla p - \frac{\eta \nabla P}{\varrho} = \mathbf{F}, \tag{2.8}$$

$$\frac{d\eta}{dt} + (\mathbf{w} \cdot \nabla)\varrho + \varrho\nabla \cdot \mathbf{w} + \eta\nabla \cdot \mathbf{v} = \varrho Q, \tag{2.9}$$

$$\frac{d\chi_i}{dt} + (\mathbf{w} \cdot \nabla)C_i = 0, \quad i = 1, 2, \ldots, n, \tag{2.10}$$

$$\frac{ds}{dt} + (\mathbf{w} \cdot \nabla)S = 0, \tag{2.11}$$

$$p = c^2\eta + hs + b_i\chi_i. \tag{2.12}$$

Here, the operator $d/dt = \partial/\partial t + \mathbf{v} \cdot \nabla$ is the full (material) derivative, repeated subscripts are summed from 1 to 3, $c^2 = \partial P/\partial\varrho$ is the square of the sound speed, $h = \partial P/\partial S$, and $b_i = \partial P/\partial C_i$. Hereafter, when calculating the partial derivatives of the ambient pressure P, we assume that P is a function of the thermodynamic variables ϱ, S, C_1, C_2,..., C_n (see equation (2.17) below). These thermodynamic variables are convenient for derivations of equations for acoustic and internal gravity waves. When deriving equations (2.8) and (2.9), we assumed that $\mathbf{F}$ and Q are of the same order of magnitude as p, η, $\mathbf{w}$, s, and χ_i. In other words, $\mathbf{F}$ and Q are the sources of waves propagating in the medium. The ambient quantities P, ϱ, $\mathbf{v}$, S, and C_i satisfy the set of equations (2.3)–(2.7) with $\mathbf{F} = Q = 0$. Rearranging these equations, we have

$$\frac{d\mathbf{v}}{dt} + \frac{1}{\varrho}\nabla P - \mathbf{g} = 0, \tag{2.13}$$

$$\frac{d\varrho}{dt} + \varrho\nabla \cdot \mathbf{v} = 0, \tag{2.14}$$

$$\frac{dC_i}{dt} = 0, \quad i = 1, 2, \ldots, n,$$ (2.15)

$$\frac{dS}{dt} = 0,$$ (2.16)

$$P = P(\varrho, S, C_1, C_2, \ldots, C_n).$$ (2.17)

The complete set of the linearized equations of fluid dynamics, equations (2.8)–(2.12), provides the most general description of wave propagation in an inhomogeneous moving medium. In order to calculate p, η, $\mathbf{w}$, s, and χ_i, one needs to know the ambient quantities P, ϱ, $\mathbf{v}$, S, and C_i. This set describes the propagation of both acoustic and internal gravity waves. We shall study these waves simultaneously where it is possible (in this and following sections). All equations for wave propagation in a moving medium are derived in this chapter from the set of equations (2.8)–(2.12). This set is also used in Chapter 3 as a starting point in the formulations of the geometrical acoustics. It is referenced close to the top in the flowchart of equations in figure 2.1. Equations (2.8)–(2.12) were first derived by Blokhintzev (sections 4 and 13 in reference [37]).

2.1.3   Set of three coupled equations

With some assumptions or approximations, equations (2.8)–(2.12) can be reduced to a set of three coupled equations for p, η, and $\mathbf{w}$. To this end, we apply the operator d/dt to both sides of equation (2.12). In the resulting equation, $d\chi_i/dt$ and ds/dt are expressed in terms of $\mathbf{w}$ using equations (2.10) and (2.11), respectively. As a result, we have

$$\frac{dp}{dt} = c^2 \frac{d\eta}{dt} + \eta \frac{dc^2}{dt} + s \frac{dh}{dt} + \chi_i \frac{db_i}{dt} - h\mathbf{w} \cdot \nabla S - b_i \mathbf{w} \cdot \nabla C_i.$$ (2.18)

The full derivative dc^2/dt appearing in this equation is recast in the form:

$$\frac{dc^2}{dt} = \frac{d}{dt}\frac{\partial P}{\partial \varrho} = \frac{\partial^2 P}{\partial \varrho^2}\frac{d\varrho}{dt} + \frac{\partial^2 P}{\partial S \partial \varrho}\frac{dS}{dt} + \frac{\partial^2 P}{\partial C_i \partial \varrho}\frac{dC_i}{dt} = \beta \frac{d\varrho}{dt},$$ (2.19)

where $\beta = \partial^2 P/\partial \varrho^2$. In equation (2.19), we took into account that $dS/dt = dC_i/dt = 0$ (see equations (2.15) and (2.16)). Analogously, we can derive equations for dh/dt and db_i/dt:

$$\frac{dh}{dt} = \alpha \frac{d\varrho}{dt}, \quad \frac{db_i}{dt} = \tau_i \frac{d\varrho}{dt},$$ (2.20)

where $\alpha = \partial^2 P/\partial \varrho \partial S$ and $\tau_i = \partial^2 P/\partial \varrho \partial C_i$.

We apply the operator $\mathbf{w} \cdot \nabla$ to both sides of the equation of state (2.17). Taking into account the definitions of c^2, h, and b_i, we obtain

$$\mathbf{w} \cdot \nabla P = c^2 \mathbf{w} \cdot \nabla \varrho + h\mathbf{w} \cdot \nabla S + b_i \mathbf{w} \cdot \nabla C_i.$$ (2.21)

Substituting the values of dc^2/dt, dh/dt, and db_i/dt into this equation and, then, adding equations (2.18) and (2.21), we have

$$\frac{dp}{dt} + \mathbf{w} \cdot \nabla P = c^2 \frac{d\eta}{dt} + c^2 \mathbf{w} \cdot \nabla \varrho + \tilde{c}^2 \frac{d\varrho}{dt}, \tag{2.22}$$

where $\tilde{c}^2 = \beta \eta + \alpha s + \tau_i \chi_i$. It can be shown that $\tilde{c}^2$ represents fluctuations in the sound speed squared, caused by a wave propagating in the medium. In the equation for $\tilde{c}^2$, we replace s with its value obtained from equation (2.12): $s = (p - c^2\eta - b_i\chi_i)/h$. As a result, we obtain the expression for $\tilde{c}^2$:

$$\tilde{c}^2 = \frac{1}{h} \left[\left(h\beta - \alpha c^2 \right) \eta + \alpha p + \Omega \chi_i \right], \tag{2.23}$$

where Ω is given by:

$$\Omega = h\tau_i - \alpha b_i = \frac{\partial P}{\partial S} \frac{\partial^2 P}{\partial \varrho \partial C_i} - \frac{\partial P}{\partial C_i} \frac{\partial^2 P}{\partial \varrho \partial S}. \tag{2.24}$$

We multiply equation (2.9) by c^2, add the resulting equation and equation (2.22), and replace $d\varrho/dt$ with $-\varrho \nabla \cdot \mathbf{v}$ using equation (2.14). As a result, we have

$$\frac{dp}{dt} + \varrho c^2 \nabla \cdot \mathbf{w} + \mathbf{w} \cdot \nabla P + \left(c^2 \eta + \varrho \tilde{c}^2 \right) \nabla \cdot \mathbf{v} = 0. \tag{2.25}$$

Substituting for $\tilde{c}^2$ using equation (2.23), we arrive at the following equation for dp/dt:

$$\frac{dp}{dt} + \varrho c^2 \nabla \cdot \mathbf{w} + \mathbf{w} \cdot \nabla P$$
$$+ \left\{ \left[\varrho\beta + c^2(1 - \alpha\varrho/h) \right] \eta + (\alpha\varrho/h) p + (\varrho\Omega/h)\chi_i \right\} \nabla \cdot \mathbf{v} = \varrho c^2 Q. \tag{2.26}$$

This equation is an exact consequence of the linearized equations of fluid dynamics (2.8)–(2.12). It contains the following unknown functions: p, η, $\mathbf{w}$, and χ_i.

Now let us omit the term $(\varrho\Omega/h)\chi_i$ appearing on the left-hand side of equation (2.26), the justification for which is discussed later. Then, this equation simplifies to [303]:

$$\frac{dp}{dt} + \varrho c^2 \nabla \cdot \mathbf{w} + \mathbf{w} \cdot \nabla P$$
$$+ \left\{ \left[\varrho\beta + c^2(1 - \alpha\varrho/h) \right] \eta + (\alpha\varrho/h) p \right\} \nabla \cdot \mathbf{v} = \varrho c^2 Q. \tag{2.27}$$

Equations (2.8), (2.9), and (2.27) comprise a desired set of three coupled equations for p, η, and $\mathbf{w}$. This set describes propagation of both acoustic and internal gravity waves. In the set, one needs to know the following ambient quantities: c, ϱ, $\mathbf{v}$, and P; the coefficients α, β, and h can be calculated with the equation of state (2.17). Equation (2.27) has been derived from the linearized equations of fluid dynamics assuming that $(\varrho\Omega/h)\chi_i = 0$. This assumption is valid if the medium consists of only one component ($C_i = \chi_i = 0$) or if $\Omega = 0$. The latter equality might hold for a particular medium; for example, it is valid for the medium consisting of two components with the equation of state $P(\varrho, S, C) = f_1(\varrho)f_2(S)f_3(C)$. Note that, generally, $\Omega \neq 0$. If neither of the equalities $C_i = 0$ and $\Omega = 0$ are valid, the set of equations (2.8), (2.9), and (2.27) describes approximately propagation of acoustic and internal gravity waves. This set can potentially be used for finite-difference, time-domain (FDTD) calculations of wave propagation in a moving medium, which are performed typically with partial differential equations that are first order in time. Figure 2.1 shows this set in the flowchart of equations.

Equation (2.27) can be simplified for an ideal gas. In this case, the equation of state is given by:

$$P = P_0(\varrho/\varrho_0)^\gamma \exp[(\gamma - 1)\mu(S - S_0)/R], \qquad (2.28)$$

where R is the universal gas constant, μ is the molecular weight of the gas, $\gamma = c_P/c_V$ is the ratio of the specific heat at constant pressure c_P to the specific heat at constant volume c_V, and P_0, ϱ_0, and S_0 are the reference values of the corresponding ambient quantities. Using equation (2.28), the sound speed c and the coefficients α, β, and h appearing in equation (2.27) can be calculated: $c^2 = \gamma P/\varrho$, $\alpha = \gamma(\gamma - 1)\mu P/(\varrho R)$, $\beta = \gamma(\gamma - 1)P/\varrho^2$, and $h = (\gamma - 1)\mu P/R$. Substituting these values into equation (2.27), we have

$$\frac{dp}{dt} + \varrho c^2 \nabla \cdot \mathbf{w} + \mathbf{w} \cdot \nabla P + \gamma p \nabla \cdot \mathbf{v} = \varrho c^2 Q. \qquad (2.29)$$

Equations (2.8), (2.9), and (2.29) together provide a complete set of three coupled equations for p, η, and $\mathbf{w}$ for the case of an ideal gas. To solve these equations, one needs to know the following ambient quantities: c, ϱ, $\mathbf{v}$, and P. With some approximations [303], this set can be reduced to the set of two coupled equations for p and $\mathbf{w}$ which was used in reference [346] for FDTD calculations of sound propagation in the atmosphere.

2.1.4 Energy considerations

Generally, a linear wave propagating in an inhomogeneous moving medium can exchange energy with the ambient medium so that the wave energy is not conserved. Indeed, in fluid dynamics, the law of energy conservation has the form

$$\frac{\partial}{\partial t}(\tilde{\varrho}\tilde{v}^2/2 + \tilde{\Xi}) + \nabla \cdot \left[\tilde{\varrho}\tilde{\mathbf{v}}(\tilde{v}^2/2 + \tilde{P}/\tilde{\varrho} + \tilde{\Xi})\right] = 0, \qquad (2.30)$$

where $\tilde{\Xi}$ is the internal energy. Let $\tilde{P} = P + p + p_2 + \ldots$, $\tilde{\varrho} = \varrho + \eta + \eta_2 + \ldots$, $\tilde{\mathbf{v}} = \mathbf{v} + \mathbf{w} + \mathbf{w}_2 + \ldots$, and $\tilde{\Xi} = \Xi + \varepsilon + \varepsilon_2 + \ldots$. The first two terms in these series have been introduced above (except for Ξ and ε) and describe the ambient medium and a linear wave propagating through the medium. The third terms p_2, η_2, w_2, and ε_2 describe a weakly nonlinear wave; note that $p_2 \sim p^2$, $\eta_2 \sim \eta^2$, $w_2 \sim w^2$, and $\varepsilon_2 \sim \varepsilon^2$. In the absence of a wave propagating in the medium, equation (2.30) takes the form

$$\frac{\partial}{\partial t}(\varrho v^2/2 + \Xi) + \nabla \cdot \left[\varrho \mathbf{v}(v^2/2 + \Xi + P/\varrho)\right] = 0. \tag{2.31}$$

The most consistent approach for deriving the law of energy conservation would be to subtract this equality from equation (2.30) and then to neglect terms of the order of p^3, η^3, w^3, and ε^3. The resulting equation would, however, contain not only the quantities p, η, and $\mathbf{w}$ characterizing a linear wave, but also the quantities p_2, η_2, $\mathbf{w}_2$, and ε_2, which are not considered in the linear theory.

Thus, generally, it is not possible to formulate a law of energy conservation and, hence, to define the energy density and its flux in a wave propagating in an inhomogeneous moving medium. Such a law can be derived only with certain assumptions about a moving medium or a propagating wave. In section 3.1, the law of acoustic energy conservation is formulated in the approximation of geometrical acoustics.

2.1.5   Reduction of the linearized equations of fluid dynamics to a single equation

The complete set of linearized equations of fluid dynamics, equations (2.8)–(2.12), describes, in principle, the propagation of acoustic and internal gravity waves in moving media. However, this set is rather involved and inconvenient for solving particular problems. In the general case of an inhomogeneous moving medium, equations (2.8)–(2.12) cannot be exactly reduced to a single equation, which would be more convenient for analysis.

Equations (2.8)–(2.12) can be reduced exactly to a single equation if we make certain assumptions about the ambient quantities P, ϱ, $\mathbf{v}$, S, and C_i. Such exact equations are the Andreev–Rusakov–Blokhintzev equation (section 2.4), derived assuming that $g = 0$, $S = $ constant, and rot $\mathbf{v} = 0$; the Goldstein equation for sound waves in a parallel shear flow (section 2.3); and the equation for acoustic and internal gravity waves in a stratified moving medium, derived by Ostashev [281, 283] and considered in detail in section 2.3. Moreover, the set of equations (2.8)–(2.12) can be simplified or reduced to a single equation using various approximations. The approximate equations for sound waves in a three-dimensional moving medium are considered in sections 2.4 and 2.5.

2.2 Stratified medium

A stratified time-independent moving medium, for which the ambient quantities P, ϱ, $\mathbf{v}$, S, and C_i depend only on the vertical coordinate z, is an important idealization of an inhomogeneous moving medium. The atmosphere and ocean can often be considered as stratified. This section provides with linearized equations of fluid dynamics and equations for the ambient quantities in a stratified medium. It also presents the vertical profiles of temperature, humidity, and wind velocity in the atmospheric surface layer obtained with the Monin–Obukhov similarity theory.

2.2.1 Linearized fluid dynamic equations

Since the horizontal component $\mathbf{v}_\perp(z)$ of the medium velocity in the atmosphere and ocean is usually greater than the vertical component $v_z(z)$ by factor of 10^1–10^2, it is often assumed that $v_z = 0$. When considering a stratified medium, we shall also make this assumption.

In a stratified moving medium, equations (2.8)–(2.12) can be simplified. Let $\mathbf{w} = (\mathbf{w}_\perp, w_z)$, where $\mathbf{w}_\perp$ and w_z are the horizontal and vertical components of the vector $\mathbf{w}$. We also denote the derivatives of the ambient quantities with respect to z by a prime:

$$\frac{\partial \mathbf{v}_\perp}{\partial z} = \mathbf{v}'_\perp, \quad \frac{\partial \varrho}{\partial z} = \varrho', \quad \frac{\partial P}{\partial z} = P', \quad \frac{\partial C_i}{\partial z} = C'_i, \quad \frac{\partial S}{\partial z} = S'.$$

With these notations, equations (2.8)–(2.12) become

$$\frac{dw_z}{dt} + \frac{1}{\varrho}\frac{\partial p}{\partial z} - \frac{\eta P'}{\varrho^2} = F_z, \tag{2.32}$$

$$\frac{d\mathbf{w}_\perp}{dt} + \mathbf{v}'_\perp w_z + \frac{1}{\varrho}\nabla_\perp p = \mathbf{F}_\perp, \tag{2.33}$$

$$\frac{d\eta}{dt} + \varrho' w_z + \varrho \frac{\partial w_z}{\partial z} + \varrho \nabla_\perp \cdot \mathbf{w}_\perp = \varrho Q, \tag{2.34}$$

$$\frac{d\chi_i}{dt} + C'_i w_z = 0, \tag{2.35}$$

$$\frac{ds}{dt} + S' w_z = 0, \tag{2.36}$$

$$p = c^2\eta + hs + b_i\chi_i. \tag{2.37}$$

Here, $\nabla_\perp = (\partial/\partial x, \partial/\partial y)$, and $\mathbf{F}_\perp$ and F_z are the horizontal and vertical components of the force $\mathbf{F} = (\mathbf{F}_\perp, F_z)$ acting on the medium. In section 2.3, equations (2.32)–(2.37) will be reduced exactly to a single equation for p.

2.2.2   Equations for ambient quantities

In a stratified moving medium, equations (2.13)–(2.17) for the ambient quantities P, ϱ, $\mathbf{v}_\perp$, S, and C_i also simplify. First, in a stratified medium, $\nabla \cdot \mathbf{v} = \partial v_z / \partial z + \nabla_\perp \cdot \mathbf{v}_\perp(z) = 0$. Second, the full derivatives of all ambient quantities with respect to t equal zero; for example, $dP/dt = (\partial / \partial t + \mathbf{v}_\perp \cdot \nabla_\perp)P(z) = 0$. Therefore, equations (2.14)–(2.16) become identities, equation (2.17) remains the same, and equation (2.13) reduces to the hydrostatic equation

$$P'(z) = -g\varrho(z). \tag{2.38}$$

Thus, in a stratified moving medium, the set of the fluid dynamic equations for the ambient quantities reduces to equations (2.17) and (2.38). If $C_i = 0$, equations (2.17) and (2.38) contain three unknown functions: P, ϱ, and S. Only one of these three functions is arbitrary; the other two functions, as well as the square of the sound speed $c^2 = \partial P / \partial \varrho$, can be expressed in terms of that function.

In a dry atmosphere, it is convenient to express all ambient quantities in terms of the vertical profile of the temperature $T(z)$. The atmosphere can be considered as an ideal gas with the equation of state (2.28), where the ambient pressure P is expressed in terms of the thermodynamic variables ϱ and S. For current purposes, it is also worthwhile to express P in terms of the thermodynamic variables ϱ and T:

$$P = R_a \varrho T. \tag{2.39}$$

Here, $R_a = R/\mu_a$ is the gas constant for dry air and μ_a is its molecular weight (the values of these parameters are given in section 6.3). Using equations (2.38) and (2.39) and equation (2.28), where we set $P_0 = P(z_0)$, $\varrho_0 = \varrho(z_0)$, $S_0 = S(z_0)$, the ambient quantities P, ϱ, c, and S can be expressed in terms of the temperature T:

$$P(z) = P_0 \exp\left(-\int_{z_0}^z \frac{g}{R_a T(z')} \, dz'\right), \tag{2.40}$$

$$\varrho(z) = \frac{\varrho_0 T_0}{T(z)} \exp\left(-\int_{z_0}^z \frac{g}{R_a T(z')} \, dz'\right), \tag{2.41}$$

$$c^2(z) = \gamma_a R_a T(z), \tag{2.42}$$

$$S(z) = S_0 + \frac{\gamma_a R_a}{\gamma_a - 1} \ln \frac{T(z)}{T_0} + \int_{z_0}^z \frac{g}{T(z')} \, dz'. \tag{2.43}$$

Here, $T_0 = T(z_0)$, z_0 is the reference height, and γ_a is the ratio of the specific heats for a dry air. The sound speed c given by equation (2.42) coincides with that determined by equation (1.1) for the considered case of a stratified atmosphere with a dry air.

2.2.3 Atmospheric surface layer

The vertical profiles of temperature $T(z)$, specific humidity $q(z)$, and horizontal wind velocity $\mathbf{v}_\perp(z)$ in the atmospheric surface layer (ASL) have been studied in many previous experimental and theoretical investigations. The characteristics of these profiles have important implications for near-ground sound propagation.

The defining characteristic of the ASL is that the vertical fluxes of quantities such as heat and momentum are approximately constant and equal to their values at the surface. The structure of the ASL depends on whether the air density exhibits unstable, neutral, or stable stratification. In the mid-latitudes, unstable stratification usually occurs on sunny days in spring, summer, and autumn. Near-neutral stratification is facilitated by cloudy and/or windy conditions. Stable stratification, which is caused by radiative cooling of the earth's surface, often occurs on clear nights. When stratification is neutral or unstable, the ASL may be regarded as the lowermost 10% of the atmospheric boundary layer (ABL), which in practice means the ASL is typically 100–200 m in depth. For stable stratification, the top of the ASL is usually lower.

The Monin–Obukhov similarity theory (MOST) is widely used to describe turbulence statistics, including the vertical profiles in the ASL. MOST enables determination of the entire profiles from a few near-ground observations, and can be applied to stratifications ranging from unstable to slightly stable; it does not apply to moderately or strongly stable conditions, since in that case the suppression of turbulent mixing disrupts the constant flux assumption.

MOST can be understood as a synthesis of surface-layer similarity theories for the limiting cases in which the turbulence is generated purely by wind shear or purely by free convection above a heated surface. For the turbulent shear surface layer, the appropriate length scale is the height z, the velocity scale is the friction velocity u_*, and the temperature scale is $T_* = -Q_H/(\varrho_0 c_P u_*)$, in which Q_H is the sensible heat flux from the surface to the overlying air, ϱ_0 is the air density, and c_P is the specific heat at constant pressure. (Note that $Q_H/\varrho_0 c_P$ has units of temperature times velocity. In this book, when we use the term "heat flux" by itself, we implicitly mean the sensible heat flux.) For the free-convection surface layer, z and Q_H continue to be dynamically relevant; however, the Boussinesq buoyancy parameter g/T_s (where g is gravitational acceleration and T_s is the temperature at the surface) influences the dynamics and u_* vanishes. Hence, based on dimensional arguments, the free-convection velocity scale is $w_{\text{fc}} = [z(g/T_s)(Q_H/\varrho_0 c_P)]^{1/3}$ and the temperature scale is $T_{\text{fc}} = -Q_H/(\varrho_0 c_P w_{\text{fc}})$. When a mixture of shear and convection forcings are present, we thus have four pertinent dynamical variables (e.g., z, u_*, $Q_H/\varrho_0 c_P$, and g/T_s), and three fundamental physical dimensions (length, time, and temperature). According to the well-known Buckingham-Π theorem [54], we may thus select three of these variables to normalize (non-dimensionalize) quantities of interest; these normalized quantities then depend upon a single nondimensional group involving the remaining fourth variable. In MOST, the normalizing variables are taken to be z, u_*, and T_*. The nondimensional group is taken to be $\xi = z/L_o$, where $L_o = -u_*^3 T_s \varrho_0 c_P / (g\kappa_v Q_H)$ is the Obukhov length, and $\kappa_v = 0.40$ is the von Kármán constant.

In unstable conditions, when $Q_H > 0$ and $L_o < 0$, $-\xi$ represents the ratio of production of turbulence by buoyancy (heating from the ground) to production by shear. In stable conditions, when $Q_H < 0$ and $L_o > 0$, ξ is proportional to the ratio of the suppression of turbulence by buoyancy to production by shear. For heights such that $\xi \gtrsim 1$, turbulence is largely suppressed; this condition can be interpreted as a limit on the applicability of MOST, as well as the height of the ASL in stable conditions. Conditions such that $|\xi| \ll 1$ are termed buoyantly *neutral*, because density stratification resulting from heat exchange between the ground and overlying air has a negligible effect on the atmospheric dynamics.

Based on the preceding discussion, the MOST form for the vertical gradient of the horizontal wind velocity profile is $\partial v_\perp(z)/\partial z = (u_*/z)\phi_m(\xi)$, where $\phi_m(\xi)$ is a universal function for the wind profile. The subscript m signifies momentum transfer. By integrating this equation, we arrive at the following result for the vertical profile (e.g., references [130, 306, 367]):

$$v_\perp(z) = \frac{u_*}{\kappa_v} \left[\ln\left(\frac{z}{z_0}\right) - \Psi_m\left(\frac{z}{L_o}\right) + \Psi_m\left(\frac{z_0}{L_o}\right) \right]. \tag{2.44}$$

Here, z_0 is the surface roughness length (the theoretical height at which the wind speed vanishes), and $\Psi_m(\xi)$ is found by integrating $\phi_m(\xi)$. Equation (2.44) is valid for $z \gg z_0$.

The temperature gradient is given by $\partial T(z)/\partial z = (T_*/z)\phi_h(\xi) - \Gamma_d$, where the subscript h signifies heat transfer and $\Gamma_d = g/c_P = 0.0098$ K/m is the dry adiabatic lapse rate. The lapse rate appears in this equation because MOST correctly applies to the *potential* temperature $\theta(z) = T(z) + \Gamma_d(z - z_s)$, rather than to the temperature itself. (The potential temperature and its relationship to atmospheric stability is described in textbooks on meteorology; see, for example, reference [367].) Integrating the gradient equation, we find (e.g., references [130, 306, 367])

$$T(z) = T_r - (z - z_r)\Gamma_d + \frac{P_t T_*}{\kappa_v} \left[\ln\left(\frac{z}{z_r}\right) - \Psi_h\left(\frac{z}{L_o}\right) + \Psi_h\left(\frac{z_r}{L_o}\right) \right]. \tag{2.45}$$

Here, z_r is a reference height, $T_r = T(z_r)$, $\Psi_h(\xi)$ is the universal profile function, and $P_t = 0.95$ is the turbulent Prandtl number in neutral stratification.

The specific humidity and other conserved scalar quantities are usually assumed to obey the same universal profile functions as potential temperature. Hence

$$q(z) = q_r + \frac{P_t q_*}{\kappa_v} \left[\ln\left(\frac{z}{z_r}\right) - \Psi_h\left(\frac{z}{L_o}\right) + \Psi_h\left(\frac{z_r}{L_o}\right) \right], \tag{2.46}$$

where $q_r = q(z_r)$, $q_* = -Q_E / (\varrho_0 L_v u_*)$ is the specific humidity scale for the surface layer, Q_E is the latent heat flux at the surface, and L_v is the latent heat of vaporization for water. Note that the concentration C of water vapor in the atmosphere, appearing in equations of fluid dynamics, is expressed in terms of specific humidity as $C = q/(1 - q)$. Since in the atmosphere $C \ll 1$, $C \approx q$ to a good degree of accuracy.

Various functional forms for the universal gradient functions $\phi_{h,m}(\xi)$ have been proposed in the literature; in practice, most of the forms provide similar agreement with experimental data. Here, we consider the relatively simple forms suggested in reference [416], which have the correct behavior in the limiting case of freely convective turbulence:

$$\phi_{h,m}(\xi) = \begin{cases} (1 + a_{h,m}|\xi|^{2/3})^{-1/2}, & \xi < 0, \\ 1 + b_{h,m}\xi, & \xi \geq 0, \end{cases} \qquad (2.47)$$

are universal profile functions. The a's and b's are constants with the values $a_h = 7.9$, $a_m = 3.6$, $b_h = 8.4$, and $b_m = 5.3$. For the corresponding integrated profile functions, we find [416]

$$\Psi_{h,m}(\xi) = \begin{cases} 2\ln[(1 + \sqrt{1 + a_{h,m}|\xi|^{2/3}})/2], & \xi < 0, \\ -b_{h,m}\xi, & \xi \geq 0. \end{cases} \qquad (2.48)$$

Equations (2.44)–(2.48) describe the vertical profiles $T(z)$, $q(z)$, and $v_\perp(z)$ as functions of the parameters T_r, q_r, u_*, z_0, L_o, T_*, and q_*. Typically, z_0 is estimated from tables of typical ground surfaces, such as provided in reference [306]. Observations from near-ground sensors can be used to determine T_r, q_r, u_*, Q_H, and Q_E, either by direct observation or by inferring the fluxes from profile measurements and numerical inversion of equations (2.44)–(2.48). Usually a wind velocity measurement is performed at one height (10 m being standard) and temperature at two heights (2 m and 10 m being standard) is used to infer u_*, Q_H, and T_r; humidity measurements at two heights can be used to infer q_r and Q_E if desired. Alternatively, one could measure wind velocity at one height, temperature at one height, and then use an estimate of the solar insolation based on solar angle and cloud cover. The reader is referred to references [306, 222] for more details. Finally, L_o, T_*, and q_* can be calculated from their definitions.

The values of u_*, Q_H, and Q_E span wide ranges in the ASL. For example, for summer weather at mid-latitudes, 0.05 m/s $\lesssim u_* \lesssim 0.7$ m/s and -50 W/m$^2 \lesssim Q_H \lesssim 600$ W/m^2. Representative values of the friction velocity u_* for light, moderate, and strong wind conditions can be selected as 0.1, 0.3, or 0.6 m/s, respectively. For mostly sunny conditions during the daytime (with moderate soil moisture), we may set $Q_H = 200$ W/m^2 and $Q_E = 50$ W/m^2. For mostly cloudy conditions, day or night, the heat fluxes become very small, and so we choose $Q_H = Q_E = 0$ W/m^2. For clear skies at night, we may set $Q_H = -4$ W/m^2, $Q_E = -1$ W/m^2 for light wind conditions, and $Q_H = -20$ W/m^2, $Q_E = -5$ W/m^2 for moderate or strong wind.

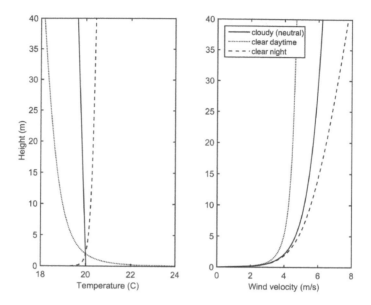

FIGURE 2.2
Vertical profiles of the mean temperature and wind velocity in the atmospheric surface layer as obtained from the Monin–Obukhov similarity theory. Shown are profiles for moderate wind conditions ($u_* = 0.3$ m/s), with values of the sensible and latent heat fluxes for clear daytime ($Q_H = 200$ W/m^2, $Q_E = 50$ W/m^2), mostly cloudy ($Q_H = Q_E = 0$ W/m^2), and clear nighttime ($Q_H = -20$ W/m^2, $Q_E = -5$ W/m^2) conditions.

Figure 2.2 depicts vertical profiles of the temperature $T(z)$ and wind velocity $v_\perp(z)$ obtained from equations (2.45) and (2.44). The cases shown are all for moderate wind ($u_* = 0.3$ m/s), with representative values for the heat fluxes as described in the preceding paragraph. The surface roughness was set to $z_0 = 0.01$ m, which is representative of flat ground with short grass.

2.3   Exact equation for acoustic and internal gravity waves in a stratified medium

In a stratified time-independent moving medium, the set of linearized equations of fluid dynamics, equations (2.32)–(2.37), can be reduced exactly to a single equation. In this section, we will derive and analyze this important equation.

2.3.1 Exact equation

The fluctuations of the entropy s and concentrations χ_i of dissolved components in the medium can be eliminated from equations (2.32)–(2.37). To this end, both sides of equation (2.35) are multiplied by the coefficient b_i and both sides of equation (2.36) by the coefficient h. Adding the resulting equations, we have

$$h\frac{ds}{dt} + b_i \frac{d\chi_i}{dt} = -w_z(hS' + b_i C_i').\tag{2.49}$$

On the right-hand side of this equation, the sum $hS' + b_i C_i'$ can be expressed in terms of the Brunt–Väisälä frequency N, which is one of the main characteristics of internal gravity waves. By definition, the square of the Brunt–Väisälä frequency is given by $N^2 = -(g^2/c^2 + g\varrho'/\varrho)$. Differentiating both sides of the equation of state (2.17) with respect to z, we have $P' = c^2\varrho' + hS' + b_i C_i'$. Substituting with equation (2.38), we express the sum $hS' + b_i C_i'$ in terms of the Brunt–Väisälä frequency: $hS' + b_i C_i' = \varrho c^2 N^2/g$. Using this formula, equation (2.49) becomes

$$\frac{d(hs + b_i\chi_i)}{dt} = -w_z \varrho c^2 N^2/g.\tag{2.50}$$

When deriving this equation, we took into account that in a stratified medium the coefficients h and b_i depend only on z; hence, $dh(z)/dt = (\partial/\partial t + \mathbf{v}_\perp \cdot \nabla_\perp)h(z) = 0$ and, analogously, $db_i(z)/dt = 0$.

Applying the operator d/dt to both sides of equation (2.37) and taking into account the fact that $dc^2(z)/dt = 0$, we obtain

$$\frac{dp}{dt} = c^2\frac{d\eta}{dt} + \frac{d(hs + b_i\chi_i)}{dt}.\tag{2.51}$$

Substituting the value of $(d/dt)(hs + b_i\chi_i)$ given by equation (2.50) into equation (2.51), we have:

$$\frac{d\eta}{dt} = c^{-2}\frac{dp}{dt} + w_z\varrho N^2/g.\tag{2.52}$$

The right-hand side of this formula does not contain s or χ_i.

Note that in products which contain the operators d/dt and $\nabla_\perp$ and functions which depend only on z, the order of multiplication can be changed, for example,

$$\frac{d}{dt}\nabla_\perp = \nabla_\perp\frac{d}{dt}, \quad \frac{d\varrho}{dt} = \varrho\frac{d}{dt}.$$

Using this rule, let us apply the operator d/dt to both sides of equation (2.32). In the equation obtained, P' and $d\eta/dt$ are replaced by the right-hand sides of equations (2.38) and (2.52), respectively. As a result, we have

$$\frac{d}{dt}\frac{\partial}{\partial z}\breve{p} + \left(\frac{d^2}{dt^2} + N^2\right)\breve{w}_z = \varrho\exp\left(\int_0^z \frac{g}{c^2}\,dz'\right)\frac{dF_z}{dt},\tag{2.53}$$

where, instead of p and w_z, it is convenient to use $\check{p}$ and $\check{w}_z$ given by

$$\check{p} = p\exp\left(\int_0^z \frac{g}{c^2}\,dz'\right), \quad \check{w}_z = \varrho w_z \exp\left(\int_0^z \frac{g}{c^2}\,dz'\right). \tag{2.54}$$

Equation (2.53) relates two functions, $\check{p}$ and $\check{w}_z$. Let us derive the second equation relating these functions. The operator $\varrho\nabla_\perp$ is applied to both sides of equation (2.33) and the operator d/dt to both sides of equation (2.34). Then, subtracting equation (2.34) from equation (2.33) and replacing $d\eta/dt$ with the right-hand side of equation (2.52), we obtain

$$\left(\Delta_\perp - \frac{1}{c^2}\frac{d^2}{dt^2}\right)\check{p} + \left(\mathbf{v}'_\perp\cdot\nabla_\perp + 2\Gamma\frac{d}{dt} - \frac{d}{dt}\frac{\partial}{\partial z}\right)\check{w}_z$$

$$= \varrho\exp\left(\int_0^z \frac{g}{c^2}\,dz'\right)\left(\nabla_\perp\cdot\mathbf{F}_\perp - \frac{dQ}{dt}\right). \tag{2.55}$$

Here, $\Delta_\perp = (\partial^2/\partial x^2, \partial^2/\partial y^2)$ and $\Gamma = \varrho'/2\varrho + g/c^2$ is the Eckart parameter.

Equations (2.53) and (2.55) comprise a set of the second-order partial differential equations for $\check{p}$ and $\check{w}_z$. In equations (2.52) and (2.55), the operator $\partial/\partial z$ does not commute with the operators d/dt and d^2/dt^2; for example,

$$\frac{\partial}{\partial z}\frac{d^2}{dt^2} = \frac{d^2}{dt^2}\frac{\partial}{\partial z} + 2(\mathbf{v}'_\perp\cdot\nabla_\perp)\frac{d}{dt}.$$

Therefore, $\check{w}_z$ can be eliminated between equations (2.53) and (2.55) only by increasing the order of these equations. We apply the operator $(d/dt)(\partial/\partial z)$ to both sides of equation (2.53) and the operator $(d^2/dt^2 + N^2)$ to both sides of equation (2.55). Adding the resulting equations yields

$$\left[\left(\frac{d^2}{dt^2} + N^2\right)\left(\Delta_\perp - \frac{1}{c^2}\frac{d^2}{dt^2}\right) + \left(\frac{d^2}{dt^2}\frac{\partial^2}{\partial z^2} + \mathbf{v}'_\perp\cdot\nabla_\perp\frac{d}{dt}\frac{\partial}{\partial z}\right)\right]\check{p}$$

$$+ \left[\left(\frac{d^2}{dt^2} + N^2\right)\left(\mathbf{v}'_\perp\cdot\nabla_\perp + 2\Gamma\frac{d}{dt}\right) + \left(2\mathbf{v}'_\perp\cdot\nabla_\perp\frac{d^2}{dt^2} + (N^2)'\frac{d}{dt}\right)\right]\check{w}_z$$

$$= \frac{d}{dt}\frac{\partial}{\partial z}\left[\varrho\exp\left(\int_0^z \frac{g}{c^2}\,dz'\right)\frac{dF_z}{dt}\right]$$

$$+ \left(\frac{d^2}{dt^2} + N^2\right)\left[\varrho\exp\left(\int_0^z \frac{g}{c^2}\,dz'\right)\left(\nabla_\perp\cdot\mathbf{F}_\perp - \frac{dQ}{dt}\right)\right], \tag{2.56}$$

where $(N^2)' = \partial(N^2)/\partial z$. In equation (2.56), the operator in front of $\check{w}_z$ is denoted by E. This operator does not contain $\partial/\partial z$ and, hence, commutes with the operator $(d^2/dt^2 + N^2)$, which appears in front of $\check{w}_z$ in equation (2.53). Therefore, applying the operator E to both sides of equation (2.53) and the operator $(d^2/dt^2 + N^2)$ to both sides of equation (2.56), and then subtracting the latter equation from the former, we eliminate $\check{w}_z$. Using equation (2.54) to replace $\check{p}$ with p, we then obtain the desired equation for the fluctuations in the pressure due to a propagating wave:

$$\left[\left(\frac{d^2}{dt^2}+N^2\right)^2+\left(\frac{1}{c^2}\frac{d^2}{dt^2}-\Delta_\perp\right)+\left(2\mathbf{v}'_\perp\cdot\nabla_\perp\frac{d}{dt}+(N^2)'\right)\frac{d^2}{dt^2}\frac{\partial}{\partial z}\right.$$

$$\left.-\frac{d^2}{dt^2}\left(\frac{d^2}{dt^2}+N^2\right)\left(\frac{\partial^2}{\partial z^2}-2\Gamma\frac{\partial}{\partial z}\right)\right]\exp\left(\int_0^z\frac{g}{c^2}\,dz'\right)p$$

$$=\varrho\exp\left(\int_0^z\frac{g}{c^2}\,dz'\right)\left\{\left[\left(\frac{d^2}{dt^2}+N^2\right)\left(\mathbf{v}'_\perp\cdot\nabla_\perp+\frac{g}{c^2}\frac{d}{dt}-\frac{d}{dt}\frac{\partial}{\partial z}\right)\right.\right.$$

$$\left.+2\mathbf{v}'_\perp\cdot\nabla_\perp\frac{d^2}{dt^2}+(N^2)'\frac{d}{dt}\right]\frac{dF_z}{dt}-\left(\frac{d^2}{dt^2}+N^2\right)^2\left(\nabla_\perp\cdot\mathbf{F}_\perp-\frac{dQ}{dt}\right)\right\},$$

$$(2.57)$$

which was derived previously by Ostashev [283, 286].

Equation (2.57) is a sixth-order partial differential equation with respect to the time t and horizontal coordinates $\mathbf{r}=(x,y)$, and a second-order differential equation with respect to the vertical coordinate z. This equation is an exact consequence of the linearized fluid dynamic equations (2.8)–(2.12) and is valid for a multicomponent medium. The concentrations C_i of the components dissolved in the medium appear in this equation through c^2, ϱ, N^2, and Γ. In order to solve equation (2.57), one needs to know the following ambient quantities: c, ϱ, and $\mathbf{v}$. (Note that N and Γ are expressed in terms of c and ϱ.) In the stratified dry atmosphere, c and ϱ are related and can be expressed in terms of the temperature T, according to equations (2.41) and (2.42). Equation (2.57) is referenced in the flowchart of the most important equations in this chapter (figure 2.1).

Equation (2.57) describes the propagation of both acoustic and internal gravity waves; in limiting cases, it becomes the equation for acoustic waves or the equation for internal gravity waves. Usually, the equation for internal gravity waves is derived from equations (2.32)–(2.37) using the Boussinesq approximation and assuming that the medium is incompressible, which correspond to $\Gamma=0$ and $c=\infty$ in equation (2.57), respectively.

The frequency of sound waves is always much greater than the Brunt–Väisälä frequency. Setting $N=0$ in equation (2.57) yields a third-order partial differential equation for acoustic waves in a stratified moving medium [283]

$$\left[\frac{1}{c^2}\frac{d^3}{dt^3}-\frac{d}{dt}\Delta+\left(2\mathbf{v}'_\perp\cdot\nabla_\perp+2\Gamma\frac{d}{dt}\right)\frac{\partial}{\partial z}\right]\exp\left(\int_0^z\frac{g}{c^2}\,dz'\right)p$$

$$=\varrho\exp\left(\int_0^z\frac{g}{c^2}\,dz'\right)\left[\frac{d^2Q}{dt^2}-\frac{d}{dt}\nabla_\perp\cdot\mathbf{F}_\perp\right.$$

$$\left.+\left(\frac{g}{c^2}\frac{d}{dt}-\frac{d}{dt}\frac{\partial}{\partial z}+2\mathbf{v}'_\perp\cdot\nabla_\perp\right)F_z\right]. \quad (2.58)$$

If $C_i=0$, $g=0$, $P=$ constant, and the medium velocity $\mathbf{v}_\perp$ is a parallel shear flow, equations (2.8)–(2.12) can also be reduced exactly to the single equation derived by Goldstein (equation (2.24) in reference [144]). Plane parallel shear flow, where the direction of the vector $\mathbf{v}_\perp(z)$ is constant, is a particular case of parallel shear flow. If in equation (2.58) $g = 0$ and the direction of the vector $\mathbf{v}_\perp$ is constant, this equation becomes Goldstein's equation for acoustic waves in plane-parallel shear flow. Thus, in a stratified moving medium, equation (2.58) has a wider range of applicability than Goldstein's equation.

2.3.2 Spectral representation of the exact equation

The coefficients of equation (2.57) depend only on the vertical coordinate z. Therefore, it is worthwhile to express the fluctuations in the pressure, $p(\mathbf{r}, z, t)$, as Fourier integrals with respect to the horizontal components $\boldsymbol{\kappa}$ of the wave vector and the angular frequency ω:

$$
p(\mathbf{r}, z, t) = \left(\frac{\varrho}{\varrho_1}\right)^{1/2} \int_{-\infty}^{\infty} \int_{-\infty}^{\infty} \left(1 - \frac{\boldsymbol{\kappa} \cdot \mathbf{v}_\perp}{\omega}\right) \left(1 - \frac{N^2}{\sigma^2}\right)^{1/2}
$$
$$
\times \, \tilde{p}(\boldsymbol{\kappa}, z, \omega) e^{i\boldsymbol{\kappa} \cdot \mathbf{r} - i\omega t} \, d\omega \, d^2\kappa. \quad (2.59)
$$

Here, $\varrho_1 = \varrho(z_1)$, where z_1 is a reference value of z; $\sigma = \omega - \boldsymbol{\kappa} \cdot \mathbf{v}_\perp$ is the acoustic frequency in the coordinate system moving with the medium; and $\tilde{p}$ is a new unknown function proportional to the spatial-temporal spectrum of p. We substitute p given by equation (2.59) into equation (2.57) and, for the sake of simplicity, assume that $\mathbf{F} = 0$. After a rather lengthy but straightforward manipulation, we obtain a second-order ordinary differential equation for the function $\tilde{p}$ [281]:

$$
\frac{d^2\tilde{p}}{dz^2} + \left[q^2 - \frac{N^2 q^2}{\sigma^2} + \frac{f''}{2f} - \frac{3}{4}\left(\frac{f'}{f}\right)^2\right] \tilde{p} = i\omega \left[\varrho \varrho_1 \left(1 - \frac{N^2}{\sigma^2}\right)\right]^{1/2} \hat{Q}, \quad (2.60)
$$

where $f' = \partial f/\partial z$, $f'' = \partial^2 f/\partial z^2$, and

$$
q^2 = \frac{\sigma^2}{c^2} - \kappa^2, \quad f = \varrho(\sigma^2 - N^2) \exp\left(2 \int_{z_1}^{z} \frac{g}{c^2} \, dz'\right),
$$

$$
\hat{Q}(\boldsymbol{\kappa}, z, \omega) = \frac{1}{8\pi^3} \int_{-\infty}^{\infty} \int_{-\infty}^{\infty} Q(\mathbf{r}, z, t) e^{i\omega t - i\boldsymbol{\kappa} \cdot \mathbf{r}} \, dt \, d^2r. \quad (2.61)
$$

Equation (2.60) does not contain the first derivative of $\tilde{p}$ with respect to z due to the factors $(\varrho/\varrho_1)^{1/2}$, $(1 - \boldsymbol{\kappa} \cdot \mathbf{v}_\perp/\omega)$, and $(1 - N^2/\sigma^2)$ in the integrand of equation (2.59). Equation (2.60) is an exact consequence of equations (2.32)–(2.37) and describes the propagation of both acoustic and internal gravity waves.

In a stratified moving medium, it is worthwhile reducing the calculation of

p to the solution of the second-order ordinary differential equation (2.60) and subsequent calculation of the Fourier integrals (2.59), because this enables use of well-known analytical and numerical methods. This approach for calculating p is the basis of the wave theory of sound propagation in a stratified moving medium, considered in detail in Chapter 4. Equations (2.59) and (2.60) are referenced in the flowchart in figure 2.1.

The terms in the square brackets on the left-hand side of equation (2.60) play different roles for acoustic and internal gravity waves. For high-frequency acoustic waves the main term in these square brackets is q^2; for short-wavelength internal gravity waves the main term is $N^2 q^2 / \sigma^2$. The terms containing the function f depend on the acceleration due to gravity and derivatives of the ambient quantities with respect to z:

$$
\frac{f''}{2f} - \frac{3}{4}\left(\frac{f'}{f}\right)^2 = \frac{2(\boldsymbol{\kappa} \cdot \mathbf{v}'_\perp)^2 - 2\sigma\boldsymbol{\kappa} \cdot \mathbf{v}''_\perp - (N^2)''}{2(\sigma^2 - N^2)}
$$

$$
- \frac{3}{4}\left(\frac{2\sigma\boldsymbol{\kappa} \cdot \mathbf{v}'_\perp + (N^2)'}{\sigma^2 - N^2}\right)^2 + \frac{\varrho''}{2\varrho} - \frac{3}{4}\left(\frac{\varrho'}{\varrho}\right)^2 - \frac{2gc'}{c^3} - \frac{g^2}{c^4}
$$

$$
- \frac{\varrho' g}{\varrho c^2} + \frac{\varrho'}{2\varrho}\frac{2\sigma\boldsymbol{\kappa} \cdot \mathbf{v}'_\perp + (N^2)'}{\sigma^2 - N^2} + \frac{g}{c^2}\frac{2\sigma\boldsymbol{\kappa} \cdot \mathbf{v}'_\perp + (N^2)'}{\sigma^2 - N^2}. \tag{2.62}
$$

For acoustic waves, equations (2.59)–(2.62) can be simplified. First, because the frequency ω of acoustic waves is much greater than the Brunt–Väisälä frequency N, in these equations we can set $N = 0$. Secondly, it follows from the inequality $\omega^2 \gg N^2 = -g^2/c^2 - g\varrho'/\varrho$ that $\omega^2/c^2 \gg -g^2/c^4 - g\varrho'/\varrho c^2$. The left-hand side of the latter inequality is of order q^2, while the right-hand side is a sum of the sixth and seventh terms on the right-hand side of equation (2.62). Therefore, this sum can be ignored in comparison with q^2 in the square brackets on the left-hand side of equation (2.60). Thirdly, let us define ω_g^2 as $\omega_g^2 = 2g\max(c'/c, v'_\perp/c)$ and assume that $\omega^2 \gg \omega_g^2$. Then, it can be shown that the fifth and ninth terms on the right-hand side of equation (2.62) can be ignored in comparison with q^2. In the atmosphere, $c' \lesssim 0.06$ s^{-1} and $v'_\perp \lesssim 0.3$ s^{-1} so that $\omega_g \approx 0.13$ s^{-1}; in the ocean, $\max(c', v'_\perp) \lesssim 0.08$ s^{-1} which yields $\omega_g \approx 0.03$ s^{-1}. Note that ω_g is usually greater than the Brunt–Väisälä frequency N, which is typically less than 2×10^{-2} s^{-2}.

As a result of these simplifications, for acoustic waves equations (2.59) and (2.60) become

$$
p(\mathbf{r}, z, t) = \left(\frac{\varrho}{\varrho_1}\right)^{1/2} \int_{-\infty}^{\infty}\int_{-\infty}^{\infty} \left(1 - \frac{\boldsymbol{\kappa} \cdot \mathbf{v}_\perp}{\omega}\right) \tilde{p}(\boldsymbol{\kappa}, z, \omega) e^{i\boldsymbol{\kappa}\cdot\mathbf{r} - i\omega t} \, d\omega \, d^2\kappa, \tag{2.63}
$$

$$
\frac{d^2\tilde{p}}{dz^2} + \left[q^2 + \frac{f''}{2f} - \frac{3}{4}\left(\frac{f'}{f}\right)^2\right]\tilde{p} = i\omega(\varrho\varrho_1)^{1/2}\hat{Q}. \tag{2.64}
$$

In the latter equation, q^2 and $\hat{Q}$ are given by equation (2.61), and

$$\frac{f''}{2f} - \frac{3}{4}\left(\frac{f'}{f}\right)^2 = -\frac{\boldsymbol{\kappa}\cdot\mathbf{v}''_\perp}{\sigma} - 2\left(\frac{\boldsymbol{\kappa}\cdot\mathbf{v}'_\perp}{\sigma}\right)^2 + \frac{\varrho'}{\varrho}\frac{\boldsymbol{\kappa}\cdot\mathbf{v}'_\perp}{\sigma} + \frac{\varrho''}{2\varrho} - \frac{3}{4}\left(\frac{\varrho'}{\varrho}\right)^2. \quad (2.65)$$

With some modifications, equations (2.63) and (2.64) are used in the fast-field program (FFP) for calculations of sound propagation in the atmosphere [220, 411].

2.3.3 Relationship between fluctuations due to a propagating wave

In certain cases, it is worthwhile knowing not only the fluctuations in the pressure, p, due to a wave propagating in the medium, but also the fluctuations in the density η, velocity $\mathbf{w}$, entropy s, and the concentrations of the components dissolved in the medium χ_i. In this subsection, the latter fluctuations are expressed in terms of p.

In equations (2.32)–(2.37), p is expressed in the spectral form

$$p(\mathbf{r}, z, t) = \int_{-\infty}^{\infty}\int_{-\infty}^{\infty} \hat{p}(\boldsymbol{\kappa}, z, \omega)e^{i\boldsymbol{\kappa}\cdot\mathbf{r}-i\omega t}\, d\omega\, d^2\kappa, \quad (2.66)$$

where $\hat{p}$ is the spatial-temporal spectrum of p. The functions η, w_z, $\mathbf{w}_\perp$, s, χ_i, F_z, $\mathbf{F}_\perp$, and Q are also expressed in the spectral forms analogous to equation (2.66); their spatial-temporal spectra are denoted as $\hat{\eta}$, $\hat{w}_z$, $\hat{\mathbf{w}}_\perp$, $\hat{s}$, $\hat{\chi}_i$, $\hat{F}_z$, $\hat{\mathbf{F}}_\perp$, and $\hat{Q}$. Starting with equations (2.32)–(2.37), a set of equations for these quantities can be derived [282]

$$-i\sigma\hat{w}_z + \frac{1}{\varrho}\frac{d\hat{p}}{dz} - \frac{P'}{\varrho^2}\hat{\eta} = \hat{F}_z, \quad (2.67)$$

$$-i\sigma\hat{\mathbf{w}}_\perp + \mathbf{v}'_\perp\hat{w}_z + \frac{i\boldsymbol{\kappa}}{\varrho}\hat{p} = \hat{\mathbf{F}}_\perp, \quad (2.68)$$

$$-i\sigma\hat{\eta} + \varrho'\hat{w}_z + \varrho\frac{d\hat{w}_z}{dz} + i\varrho\boldsymbol{\kappa}\cdot\hat{\mathbf{w}}_\perp = \varrho\hat{Q}, \quad (2.69)$$

$$-i\sigma\hat{\chi}_i + C'_i\hat{w}_z = 0, \quad i = 1, 2, \ldots, n, \quad (2.70)$$

$$-i\sigma\hat{s} + S'\hat{w}_z = 0, \quad (2.71)$$

$$\hat{p} = c^2\hat{\eta} + h\hat{s} + b_i\hat{\chi}_i. \quad (2.72)$$

Equations (2.68) and (2.70)–(2.72) are algebraic equations. Using these equations, $\hat{\mathbf{w}}_\perp$, $\hat{\chi}_i$, $\hat{s}$, and $\hat{\eta}$ are expressed in terms of $\hat{p}$ and $\hat{w}_z$:

$$\hat{\mathbf{w}}_\perp = \frac{\boldsymbol{\kappa}}{\varrho\sigma}\hat{p} - \frac{i\mathbf{v}'_\perp}{\sigma}\hat{w}_z + \frac{i\hat{\mathbf{F}}_\perp}{\sigma}, \quad (2.73)$$

$$\hat{\chi}_i = -\frac{iC_i'}{\sigma}\hat{w}_z, \quad i = 1, 2, \ldots, n, \tag{2.74}$$

$$\hat{s} = -\frac{iS'}{\sigma}\hat{w}_z, \tag{2.75}$$

$$\hat{\eta} = \frac{1}{c^2}\hat{p} + \frac{i\varrho N^2}{\sigma g}\hat{w}_z. \tag{2.76}$$

In deriving equation (2.76), we have taken into account the fact that $hS' + b_i C_i' = \varrho c^2 N^2/g$. Substituting $\hat{\mathbf{w}}_\perp$ and $\hat{\eta}$ given by equations (2.73) and (2.76) into equations (2.67) and (2.69) yields [282]

$$\frac{d\hat{p}}{dz} + \frac{g}{c^2}\hat{p} = i\varrho\hat{w}_z\frac{\sigma^2 - N^2}{\sigma} + \varrho\hat{F}_z, \tag{2.77}$$

$$\frac{d\hat{w}_z}{dz} - \left(\frac{g}{c^2} + \frac{\sigma'}{\sigma}\right)\hat{w}_z = i\hat{p}\frac{\sigma^2 - \kappa^2 c^2}{\sigma\varrho c^2} + \hat{Q} + \frac{\boldsymbol{\kappa}\cdot\hat{\mathbf{F}}_\perp}{\sigma}. \tag{2.78}$$

The set of two first-order differential equations (2.77) and (2.78) contains two unknown functions $\hat{p}$ and $\hat{w}_z$. For the case of a stratified moving atmosphere with the equation of state (2.39), this set was derived by Tatarskii [375].

Solving equation (2.77) for $\hat{w}_z$ and substituting the resulting value into equation (2.78) yields the second-order ordinary differential equation for the spatial-temporal spectrum $\hat{p}$

$$\sigma^2\nu^2\frac{d^2\hat{p}}{dz^2} - \frac{\sigma^2(\nu^2\varrho)'}{\varrho}\frac{d\hat{p}}{dz} + \left\{\nu^4 q^2 - \frac{\sigma^2 g}{c^2}\left[(\nu^2)' + \left(\frac{2c'}{c} + \frac{\varrho'}{\varrho} + \frac{g}{c^2}\right)\nu^2\right]\right\}\hat{p}$$

$$= i\varrho\nu^4(\boldsymbol{\kappa}\cdot\hat{\mathbf{F}}_\perp + \sigma\hat{Q}) - \varrho\sigma^2\left[\nu^2\left(\frac{g}{c^2} - \frac{d}{dz}\right) + (\nu^2)'\right]\hat{F}_z. \tag{2.79}$$

Here, $\nu^2 = \sigma^2 - N^2$ and $d\nu^2/dz = (\nu^2)'$. If $\hat{p}$ is known as a solution of equation (2.79), $\hat{w}_z$ can be obtained with equation (2.77). The other spatial-temporal spectra $\hat{\mathbf{w}}_\perp$, $\hat{\chi}_i$, $\hat{s}$, and $\hat{\eta}$ can then be calculated by using equations (2.73)–(2.76).

It follows from equations (2.59) and (2.66) that the quantities $\hat{p}$ and $\tilde{p}$ are related by:

$$\hat{p} = \frac{\sigma}{\omega}\left[\frac{\varrho}{\varrho_1}\left(1 - \frac{N^2}{\sigma^2}\right)\right]^{1/2}\tilde{p}.$$

Equation (2.79) for $\hat{p}$ is more involved than equation (2.60) for $\tilde{p}$. However, if $\sigma = 0$ or $\nu = 0$, the coefficients in equation (2.60) are undefined. On the other hand, the coefficients in equation (2.79) are always defined.

Pridmore-Brown [318] previously derived an equation for the spatial-temporal spectrum $\hat{p}$ from equations (2.32)–(2.37) using the following assumptions: (1) the medium (the atmosphere) is a ideal gas with the equation of state (2.39); (2) the medium contains only one component, i.e., $\chi_i = C_i = 0$; (3) the direction of the wind velocity vector does not depend on z; 4) $\mathbf{F} = 0$ and $Q = 0$. These assumptions are not used in the derivation of equation (2.79).

2.3.4   Assumption about the ambient entropy

The exact equations derived above allow us to analyze the assumption $S = $ constant, which is often used for deriving equations for sound waves (e.g., references [37, 210, 278]). If $C_i = 0$, it follows from equations (2.36) and (2.37) that $S = $ constant is equivalent to the equality $p = c^2\eta$, which is also used in the literature in the derivation of equations for sound waves. In section 2.2 it was shown that if in a stratified medium one of the vertical profiles $P(z)$, $\varrho(z)$, or $S(z)$ is known, the other two can be calculated with equations (2.17) and (2.38). In a stratified atmosphere with $S(z) = $ constant, it follows from equation (2.43) that the temperature decreases linearly with height with the adiabatic gradient $dT/dz = -(\gamma - 1)g/\gamma R_a$. Thus, strictly speaking, the assumptions $S = $ constant and $p = c^2\eta$ impose very strong restrictions on possible models of the medium.

In principle, one can use the assumption $S = $ constant only in the linearized equations (2.32)–(2.37), but not use it in equations (2.17) and (2.38). Although this approach is not consistent, it enables to remove restrictions on possible models of the medium. Let us assume that in equations (2.32)–(2.37), $S = $ constant and, for the sake of simplicity, that $g = 0$, $C_i = 0$, and $\mathbf{F} = 0$. Then, this set of equations can be reduced to the differential equation (2.60) and the Fourier integrals (2.59) where one should set $N = 0$, $\varrho = \varrho_1$, and $f = \sigma^2$. Comparing these equations with the exact ones reveals that the assumptions $S = $ constant and $p = c^2\eta$ are valid if the sound wavelength is much less than the scale of variation of $\varrho(z)$ and if in equation (2.59) the amplitude factor $(\varrho/\varrho_1)^{1/2}$ can be assumed to be equal to 1.

2.4   Equations for acoustic waves in a three-dimensional inhomogeneous medium

In certain cases, the model of a stratified moving medium is invalid. This is the case, for example, for sound propagation in a turbulent atmosphere, and for a sound wave passing through the boundaries of a current or a cyclonic eddy in the ocean. In such cases, the medium should be considered as a three-dimensional inhomogeneous moving medium; moreover, its characteristics can depend on the time t. It was mentioned in section 2.1 that, for this model of a moving medium, the set of the linearized fluid dynamic equations (2.8)–(2.12) cannot be reduced exactly to a single equation. In this section, we consider approximate equations for acoustic waves which have been derived from this set.

2.4.1   Set of two coupled equations

Equation (2.26) was derived from the linearized fluid dynamic equations (2.8)–(2.12) without any approximations. In this subsection, we will simplify this equation and also equation (2.8) so that the resulting equations will comprise a complete set of equations for p and $\mathbf{w}$.

To this end, we will consider only acoustic waves and omit terms of order v^2/c^2. These terms are almost always neglected in oceanic acoustics, because in the ocean $v/c \sim 10^{-4}$–10^{-3}. Often, they are also ignored in the study of the effects of wind velocity on sound propagation in the atmosphere near the ground, since in this case $v/c \lesssim 5 \times 10^{-2}$. However, the terms of order v^2/c^2 might be important in studying sound propagation in the upper atmosphere where the ratio v/c can be greater than 10^{-1}.

To simplify equations (2.8) and (2.26), we notice that [217] $\nabla\cdot\mathbf{v} \sim v^3/(c^2 l)$, where l is the length scale of variations in the density ϱ. Furthermore, in equation (2.26) the term proportional to $\nabla \cdot \mathbf{v}$ is of order v^2/c^2 and can be ignored. Second, in equations (2.8) and (2.26) the terms proportional to ∇P can also be ignored. Indeed, in an inhomogeneous moving atmosphere, ∇P is of order v^2/c^2 and can be omitted. In a stratified atmosphere, $\nabla P = -g\varrho$. In linearized equations of fluid dynamics, terms proportional to the acceleration due to gravity g are important for internal gravity waves and can be omitted for acoustic waves.

With these approximations, equations (2.26) and (2.8) become [288, 290]

$$\left(\frac{\partial}{\partial t} + \mathbf{v} \cdot \nabla\right) p + \varrho c^2 \nabla \cdot \mathbf{w} = \varrho c^2 Q, \tag{2.80}$$

$$\left(\frac{\partial}{\partial t} + \mathbf{v} \cdot \nabla\right) \mathbf{w} + (\mathbf{w} \cdot \nabla)\mathbf{v} + \frac{1}{\varrho}\nabla p = \frac{\mathbf{F}}{\varrho}. \tag{2.81}$$

Equations (2.80) and (2.81) comprise a complete set of two coupled equations for the sound pressure p and the acoustic particle velocity $\mathbf{w}$. In order to solve this set, one needs to know the following ambient quantities: c, ϱ, and $\mathbf{v}$. Though the concentrations C_i as well as the entropy S do not appear explicitly in these equations, c^2 and ϱ can depend on C_i and S. The set of equations (2.80) and (2.81) is referenced in the flowchart in figure 2.1.

Reference [303] compares equations (2.80) and (2.81) with those known in atmospheric acoustics [346, 387] and aeroacoustics [20, 110]. It is shown that the latter equations usually are particular cases of a set of coupled equations (2.80) and (2.81) and/or are derived using more assumptions. Since publication of reference [303], this set has often been used in finite-difference, time-domain (FDTD) calculations of sound propagation in the atmosphere (e.g., reference [67]). In atmospheric acoustics, FDTD simulations of sound propagation have drawn substantial interest due to their ability to handle complicated phenomena such as scattering from building and trees, dynamic turbulence fields, complex moving source distributions, and propagation of transient signals.

Equations (2.80) and (2.81) have been derived from linearized equations of fluid dynamics by omitting terms of order v^2/c^2 and considering only acoustic waves. Actually, the range of applicability of these equations can be much wider. Note that it is quite difficult to estimate with what accuracy one can ignore certain terms in differential equations. In reference [303], the range of applicability of equations (2.80) and (2.81) was studied by comparing them with various equations for acoustic waves, which have been most often used for analytical and numerical studies of sound propagation in a moving atmosphere and whose ranges of applicability are well known. It is shown that equations (2.80) and (2.81) have the same or a wider range of applicability than the latter equations and, in certain cases, describe sound propagation to any order in v/c. For example, equations (2.80) and (2.81) exactly describe sound propagation in a stratified moving medium and, hence, correctly account for terms of any order in v/c. In the geometrical acoustics approximation, these equations exactly describe the phase of a sound wave and, if the terms proportional to $\nabla \cdot \mathbf{v}$ are ignored, exactly describe the amplitude of the wave.

2.4.2   Equation for the sound pressure

Let us introduce a small parameter

$$\mu_1 = \left(\frac{\Omega v}{\omega c}\right)^{1/2}. \tag{2.82}$$

Here, Ω is the characteristic frequency of the medium motion ($\partial v/\partial t \sim \Omega v$) which is usually smaller than the acoustic frequency ω. Equations (2.80) and (2.81) can be reduced to a single equation for p if terms of order μ_1^2 are ignored.

The operator ∇ is applied to both sides of equation (2.80) and the operator d/dt to both sides of equation (2.81). Then, we subtract equation (2.81) from equation (2.80) and take into account the fact that

$$\left(\nabla \frac{d}{dt} - \frac{d}{dt}\nabla\right) \cdot \mathbf{w} = \frac{\partial v_i}{\partial x_j}\frac{\partial w_j}{\partial x_i} = \nabla \cdot [(\mathbf{w} \cdot \nabla)\,\mathbf{v}],$$

where $i, j = 1, 2, 3$ and v_i, x_i, and w_i are the components of the vectors $\mathbf{v}$, $\mathbf{R}$, and $\mathbf{w}$. (In deriving the latter equation, we use the equality $\partial v_i/\partial x_i = 0$.) As a result, we have

$$2\frac{\partial v_i}{\partial x_j}\frac{\partial w_j}{\partial x_i} + \nabla \cdot \left(\frac{\nabla p}{\varrho}\right) - \frac{d}{dt}\left(\frac{1}{c^2\varrho}\frac{dp}{dt}\right) = \nabla \cdot \mathbf{F} - \frac{dQ}{dt}. \tag{2.83}$$

We apply the operator $\partial/\partial t$ to both sides of this equation. In the resulting equation, $\partial \mathbf{w}/\partial t$ is replaced with its value obtained from equation (2.80) and terms of order v^2/c^2 and μ_1^2 are omitted. As a result, we obtain the desired, wave-type equation for the sound pressure p in a three-dimensional moving medium [288]:

$$\frac{d}{dt}\frac{\partial}{\partial t}\left(\frac{1}{c^2\varrho}\frac{dp}{dt}\right) - \nabla\cdot\frac{\partial}{\partial t}\left(\frac{\nabla p}{\varrho}\right) + 2\frac{\partial v_i}{\partial x_j}\frac{\partial}{\partial x_i}\left(\frac{1}{\varrho}\frac{\partial p}{\partial x_j}\right)$$
$$= \frac{d}{dt}\frac{\partial Q}{\partial t} - \nabla\cdot\frac{\partial\mathbf{F}}{\partial t} + 2\frac{\partial v_i}{\partial x_j}\frac{\partial F_j}{\partial x_i}. \quad (2.84)$$

This is a third-order differential equation with respect to the time t and a second-order differential equation with respect to the coordinates $\mathbf{R}$. Since $d/dt = \partial/\partial t + \mathbf{v}\cdot\nabla$, the first term on the left-hand side of equation (2.84) contains one component which is proportional to v^2 and, hence, should be ignored. In order to solve equation (2.84), one needs to know the ambient quantities c, ϱ, and $\mathbf{v}$. The entropy S and the concentrations C_i of the components dissolved in the medium do not appear explicitly in equation (2.84), but c^2 and ϱ can depend on S and C_i. The flowchart in figure 2.1 includes this equation.

Equation (2.84) is derived from linearized equations of fluid dynamics by considering only acoustic waves (i.e., setting $g = 0$) and omitting terms of order v^2/c^2 and μ_1^2. In a stratified moving medium, equation (2.84) coincides with equation (2.57), if the latter, exact equation is simplified using the same assumptions. Note that an equation for the sound field in a time-independent moving medium, similar to equation (2.84), was derived in reference [138].

Making certain additional assumptions about the ambient quantities, which are not restrictive, equation (2.84) can be simplified further. We begin by expressing c and ϱ in the forms: $c = c_0 + \tilde{c}$ and $\varrho = \varrho_0 + \tilde{\varrho}$. Here, c_0 and ϱ_0 are reference values of the sound speed and density, which do not depend on $\mathbf{R}$ and t, and $\tilde{c}$ and $\tilde{\varrho}$ are the deviations of c and ϱ from c_0 and ϱ_0, respectively. (Note that $\tilde{c}$ and $\tilde{\varrho}$ can be of the order of c_0 and ϱ_0.)

Let Ω, introduced above, also be the characteristic frequency of the functions $\tilde{c}$ and $\tilde{\varrho}$, i.e.,

$$\frac{\partial\tilde{c}}{\partial t}\approx\Omega\tilde{c} \quad\text{and}\quad \frac{\partial\tilde{\varrho}}{\partial t}\approx\Omega\tilde{\varrho}.$$

We shall assume that the following inequality,

$$\mu_2^2 = \max\left(\left|\frac{v\tilde{c}}{c_0^2}\right|, \left|\frac{v\tilde{\varrho}}{c_0\varrho_0}\right|, \left|\frac{\Omega\tilde{c}}{\omega c_0}\right|, \left|\frac{\Omega\tilde{\varrho}}{\omega\varrho_0}\right|\right) \ll 1, \quad (2.85)$$

is valid. Here, μ_2 is a new small parameter. Neglecting terms of order μ_2^2 in equation (2.84), we have

$$\left[\frac{1}{c^2}\frac{\partial^2}{\partial t^2} - \Delta + \frac{(\nabla\varrho)\cdot\nabla}{\varrho}\right]\frac{\partial p}{\partial t} + 2\left[\frac{(\mathbf{v}\cdot\nabla)}{c_0^2}\frac{\partial^2}{\partial t^2} + \frac{\partial v_i}{\partial x_j}\frac{\partial^2}{\partial x_i\partial x_j}\right]p$$
$$= \varrho\left[\frac{d}{dt}\frac{\partial Q}{\partial t} - \nabla\cdot\frac{\partial\mathbf{F}}{\partial t} + 2\frac{\partial v_i}{\partial x_j}\frac{\partial F_j}{\partial x_i}\right]. \quad (2.86)$$

We now express $p(\mathbf{R}, t)$ in the form

$$p(\mathbf{R}, t) = \int_{-\infty}^{\infty} \widehat{p}(\mathbf{R}, \omega) e^{-i\omega t} \, d\omega, \tag{2.87}$$

where $\widehat{p}$ is the temporal spectrum (frequency spectrum) of the sound pressure. Also, we express $Q(\mathbf{R}, t)$ and $\mathbf{F}(\mathbf{R}, t)$ in analogous forms and denote their temporal spectra as $\widehat{Q}(\mathbf{R}, \omega)$ and $\widehat{\mathbf{F}}(\mathbf{R}, \omega)$. Here we use a wider "hat" above the temporal spectra to distinguish them from the spatial-temporal spectra of the same quantities, introduced above: $\hat{p}(\boldsymbol{\kappa}, z, \omega)$, $\hat{Q}(\boldsymbol{\kappa}, z, \omega)$, and $\hat{\mathbf{F}}(\boldsymbol{\kappa}, z, \omega)$.

We now apply the integral operator $\int_{-\infty}^{\infty} \exp(i\omega t) \, dt$ to both sides of equation (2.86). It can be shown that if we neglect terms of order μ_1^2 and μ_2^2 when calculating the Fourier integrals in equation (2.86), the ambient quantities c^2, ϱ, and $\mathbf{v}$ can be considered as independent of t. As a result, we obtain a Helmholtz-type equation for $\widehat{p}$:

$$\left[\Delta - \frac{1}{\varrho}(\nabla \varrho) \cdot \nabla + \frac{\omega^2}{c^2} \right] \widehat{p} + 2i \left[\frac{k_0}{c_0}(\mathbf{v} \cdot \nabla) - \frac{1}{\omega} \frac{\partial v_i}{\partial x_j} \frac{\partial^2}{\partial x_i \partial x_j} \right] \widehat{p}$$

$$= \varrho \left[(i\omega - \mathbf{v} \cdot \nabla)\widehat{Q} + \nabla \cdot \widehat{\mathbf{F}} - \frac{2i}{\omega} \frac{\partial v_i}{\partial x_j} \frac{\partial \widehat{F}_j}{\partial x_i} \right], \tag{2.88}$$

where $k_0 = \omega/c_0$. In this equation, c, ϱ, $\mathbf{v}$ and, hence, $\widehat{p}$ can slowly (in comparison with the acoustic frequency ω) depend on t. The assumption of such slow dependence on the time is known as a quasi-static or frozen-medium approximation. The range of applicability of equation (2.88) is given by the inequality (2.85) and the range of applicability of equation (2.84). In equation (2.88), c and ϱ can differ significantly from c_0 and ϱ_0. If $\mathbf{v} = 0$, this equation becomes the well-known equation for acoustic waves in an inhomogeneous motionless medium (see, for example, section 8 in reference [72]), in which there are no restrictions on the deviations of c and ϱ from c_0 and ϱ_0. Equation (2.88) is a generalization of that equation for the case of a moving medium and allows us to account correctly for the terms of order v/c. Equation (2.88) is used in Part II of the book for calculations of the sound scattering cross section and in the theory of multiple scattering. It is shown close to the bottom in the flowchart of equations in figure 2.1.

Only a few assumptions were made when deriving equations (2.84) and (2.88) from the linearized equations of fluid dynamics. They are more general than most of other equations for acoustic waves in moving media known in the literature. In the remainder of this section, we discuss some of these previous equations and analyze their ranges of applicability.

2.4.3   Monin's equation

An equation derived by Monin has been used in atmospheric acoustics (see equation (26.46) in reference [262] or equation (15) in section 34 of reference [374]). Monin's equation can be derived from equation (2.88), if the following assumptions are valid. First, the medium should be an ideal gas so that $c^2 = \gamma R_a T$ and $\varrho = P/R_a T$. Second, in the latter equation for ϱ we should assume that $P = $ constant. Third, we should neglect terms of order $(\widetilde{T}/T_0)^2$, where $\widetilde{T} = T - T_0$ is the deviation of the temperature from its mean value T_0. For dry air, these assumptions are valid with good accuracy in the atmospheric boundary layer.

2.4.4 Equation for the velocity quasi-potential

In the acoustics of a moving medium, the equation for the velocity quasi-potential ψ is also used (e.g., [37, 77]):

$$d^2\psi/dt^2 = c^2\Delta\psi + (\nabla\Pi) \cdot (\nabla\psi) + (d\psi/dt)\mathbf{v} \cdot \nabla \ln c^2$$
$$+ c^2 \int (\nabla\psi) \cdot \Delta\mathbf{v}\,dt - (\nabla\Pi) \cdot \int (\mathrm{rot}\,\mathbf{v}) \times (\nabla\psi)dt. \qquad (2.89)$$

This equation was derived by Obukhov [278]. The acoustic particle velocity $\mathbf{w}$ is expressed in terms of ψ as $\mathbf{w} = \nabla\psi + \int \mathrm{rot}\,\mathbf{v} \times \nabla\psi\,dt$; therefore, ψ is termed the *velocity quasi-potential*. In equation (2.89), $\Pi = \int \varrho^{-1}dP$, the sound pressure p is related to ψ by $p = \varrho d\psi/dt$, and it is assumed that the medium consists of only one component.

In references [37, 278], the assumptions used to derive equation (2.89) are listed: (1) $\nabla \cdot \mathbf{v} = 0$; (2) $\Omega \ll \omega$; (3) $|\mathrm{rot}\,\mathbf{v}| \ll \omega$; (4) $S = $ constant; (5) the characteristic scale of variation in c is much greater than the sound wavelength λ; (6) terms of order v^2/c^2 are neglected. In accordance with the latter assumption, terms proportional to v^2 should be neglected in the operator d^2/dt^2 on the left-hand side of equation (2.89).

These assumptions do not determine the complete range of applicability of equation (2.89). To obtain additional assumptions, let us consider this equation in a stratified moving medium and compare it with the exact equation (2.60). Since in the stratified medium $dP = -\varrho g dz$, in equation (2.89) we have $\Pi = -gz$. We express p in the spectral form

$$p = \frac{\varrho}{\varrho_1} \exp\left(\int_0^z \frac{g}{2c^2}\,dz'\right) \int_{-\infty}^{\infty} \int_{-\infty}^{\infty} \left(1 - \frac{\boldsymbol{\kappa} \cdot \mathbf{V}_\perp}{\omega}\right) \tilde{p}(\boldsymbol{\kappa}, z, \omega)e^{i\boldsymbol{\kappa}\cdot\mathbf{r}-i\omega t}\,d\omega\,d^2\kappa,$$
$$(2.90)$$

where $\tilde{p}$ is proportional to the spatial-temporal spectrum of the sound pressure. Using the formula $p = \varrho d\psi/dt$ and equations (2.89) and (2.90), we obtain the equation for $\tilde{p}$:

$$\frac{d^2\tilde{p}}{dz^2} + \left[q^2 - \frac{\boldsymbol{\kappa} \cdot \mathbf{v}''_\perp}{\omega} + \frac{g\boldsymbol{\kappa} \cdot \mathbf{v}'_\perp}{\omega c^2} - \frac{g^2}{4c^4} - \frac{gc'}{c^3}\right]\tilde{p} = 0. \qquad (2.91)$$

Here, in q^2, the terms of order $v_\perp^2/c^2$ should be neglected. Let us now compare equation (2.91) with equation (2.60), where the term $f''/2f - \frac{3}{4}(f'/f)^2$ is given by equation (2.62). If the assumptions presented above are valid, the first and second terms in square brackets in equation (2.91) coincide with the first terms in square brackets in equations (2.60) and (2.62), respectively. The other three terms in square brackets in equation (2.91), which correspond to the second and fifth terms on the right-hand side of equation (2.89), do not coincide with the terms in square brackets in equation (2.60). Therefore, one should add two additional assumptions to the range of applicability of equation (2.89): $g = 0$ and the characteristic scale of variation in ϱ is greater than λ. Comparison of the Fourier integrals (2.59) and (2.90) reveals another discrepancy between the exact equation and the equation for the velocity quasi-potential: It follows from equation (2.59) that p is proportional to $\varrho^{1/2}$, while from equation (2.90) we conclude that p is proportional to ϱ.

Equation (2.84) is obtained from linearized equations of fluid dynamics using fewer assumptions than is equation (2.89) and, thus, has a wider range of applicability. In many respects, the latter equation is similar to equation (2.100) presented below, which describes sound propagation in the high-frequency approximation. But unlike equation (2.100), the equation for the quasi-potential (by means of the fourth term on the right-hand side of equation (2.89)) allows us to consider cases where the sound wavelength λ is smaller or greater than the characteristic scale of variation in v.

2.4.5   Andreev–Rusakov–Blokhintzev equation

If we retain only the first and second terms on the right-hand side of equation (2.89), we obtain the equation derived by Andreev and Rusakov [5]. If, additionally, we retain the third term, we get the equation obtained by Blokhintzev (section 4 in reference [37]). If the vorticity of the flow is zero (rot $\mathbf{v} = 0$) and the entropy of the medium is constant ($S =$ constant), this equation is an exact consequence of equations (2.8)–(2.12). The assumption $S =$ constant imposes strong limitations on possible models of the medium (see section 2.3.4). This is true also of the assumption that rot $\mathbf{v} = 0$. (In a stratified moving medium, rot $\mathbf{v}$ equals zero only if the medium velocity is constant.)

In a general case of an inhomogeneous moving medium, the Andreev–Rusakov–Blokhintzev equation is an approximation, which is valid for the high–frequency sound field when the wavelength λ is smaller than the characteristic scale l of variation in the ambient quantities ϱ, c, $\mathbf{v}$, S, and C_i. For example, in a stratified moving medium, this equation can be reduced to equations (2.90) and (2.91), if, in the latter equation, the terms proportional to $\mathbf{v}'_\perp$ and $\mathbf{v}''_\perp$ are omitted. Comparison of equations (2.90) and (2.91) with the exact equations (2.59)–(2.62) reveals that the Andreev–Rusakov–Blokhintzev equation is valid only in the high-frequency approximation.

2.4.6 Pierce's equations

Let us consider media with unsteady inhomogeneous flow where the characteristic length scale and characteristic time scale are larger than the corresponding scales for a sound field. For such media, Pierce [314] derived two sets of equations for the sound pressure p. These equations have been used as starting equations in several subsequent studies. To better understand the range of applicability of Pierce's equations, here we will consider their application to a stratified moving medium, and compare them with the exact equations (2.63)–(2.65) for acoustic waves.

The first set of Pierce's equations is given by

$$p = \varrho \frac{d\psi}{dt}, \tag{2.92}$$

$$\frac{1}{\varrho} \nabla \cdot (\varrho \nabla \psi) - \frac{d}{dt}\left(\frac{1}{c^2}\frac{d\psi}{dt}\right) = 0. \tag{2.93}$$

Here, ψ is the velocity quasi-potential, which has been introduced above. For a stratified moving medium, it can be shown that equations (2.92)–(2.93) are equivalent to the following:

$$p(\mathbf{r}, z, t) = \left(\frac{\varrho}{\varrho_1}\right)^{1/2} \int_{-\infty}^{\infty}\int_{-\infty}^{\infty}\left(1 - \frac{\boldsymbol{\kappa}\cdot\mathbf{v}_\perp}{\omega}\right)\tilde{p}(\boldsymbol{\kappa}, z, \omega)e^{i\boldsymbol{\kappa}\cdot\mathbf{r}-i\omega t}\,d\omega\,d^2\kappa, \tag{2.94}$$

$$\frac{d^2\tilde{p}}{dz^2} + \left[q^2 - \frac{\varrho''}{2\varrho} + \frac{1}{4}\left(\frac{\varrho'}{\varrho}\right)^2\right]\tilde{p} = 0. \tag{2.95}$$

The first of these equations coincides exactly with equation (2.63), while the second one differs from equation (2.64) by terms containing derivatives of $\mathbf{v}_\perp$ and ϱ with respect to z. Such terms can be ignored in the case $\lambda \ll l$, which is considered in reference [314].

The second set of Pierce's equations is given by

$$p = \frac{d\Phi}{dt}, \tag{2.96}$$

$$\nabla \cdot \left(\frac{1}{\varrho}\nabla\Phi\right) - \varrho\frac{d}{dt}\left(\frac{1}{\varrho^2 c^2}\frac{d\Phi}{dt}\right) = 0. \tag{2.97}$$

In a stratified moving medium, these equations can be written in the following form:

$$p(\mathbf{r}, z, t) = \left(\frac{\varrho}{\varrho_1}\right)^{1/2} \int_{-\infty}^{\infty}\int_{-\infty}^{\infty}\left(1 - \frac{\boldsymbol{\kappa}\cdot\mathbf{v}_\perp}{\omega}\right)\tilde{p}(\boldsymbol{\kappa}, z, \omega)e^{i\boldsymbol{\kappa}\cdot\mathbf{r}-i\omega t}\,d\omega\,d^2\kappa, \tag{2.98}$$

$$\frac{d^2\tilde{p}}{dz^2} + \left[q^2 + \frac{\varrho''}{2\varrho} - \frac{3}{4}\left(\frac{\varrho'}{\varrho}\right)^2\right]\tilde{p} = 0. \tag{2.99}$$

Again, the first of these equations coincides with equation (2.63). The second equation differs from equation (2.64) by terms containing derivatives of $\mathbf{v}_\perp$ with respect to z, which can be ignored if $\lambda \ll l$.

2.4.7   Equation for the high-frequency sound field

Consider another equation, which is often used (e.g., [271, 354, 355, 401]) in the acoustics of a moving medium:

$$\left[\frac{1}{c^2(R)}\left(\frac{\partial}{\partial t} + \mathbf{v}(\mathbf{R})\cdot\nabla\right)^2 - \Delta\right]p = \varrho(\mathbf{R})\left(\frac{dQ}{dt} - \nabla\cdot\mathbf{F}\right). \tag{2.100}$$

This is the exact equation for sound waves in a homogeneous, uniformly moving medium. Indeed, the exact equation (2.57) becomes equation (2.100) if ϱ, c, and $\mathbf{v}$ are constant and $g = 0$.

We now show that in an inhomogeneous moving medium, equation (2.100) describes approximately the high-frequency sound field. In equations (2.8)–(2.12), the terms containing the derivatives of p, η, $\mathbf{w}$, s, and χ_i are proportional to $1/\lambda$, while the terms containing the derivatives of P, ϱ, $\mathbf{v}$, S, and C_i are proportional to $1/l$, where l is a characteristic scale of the ambient quantities. The latter terms can be neglected in comparison with the former if $\lambda \ll l$. In this case, it follows from equations (2.10)–(2.12) that $s = \chi_i = 0$ and $p = c^2\eta$. If we also neglect the derivatives of the ambient quantities in equations (2.8) and (2.9), these equations can be reduced to equation (2.100). Note that, in accordance with the derivation of equation (2.100), one should omit the derivatives of $\mathbf{v}$ in the term $c^{-2}(\mathbf{v}\cdot\nabla)^2 p$ on the left-hand side of this equation.

It can be shown that equation (2.100) describes correctly the phase of the high-frequency sound-pressure field in an inhomogeneous moving medium, but it describes the amplitude of the field only approximately. For a stratified moving medium, it is possible to demonstrate the difference in the amplitudes of a quasi-plane wave calculated with equation (2.100) and with equation (2.60). In the high-frequency approximation, we can set $N = 0$ in the Fourier integral (2.59) and equation (2.60). In this case, the latter equation becomes $d^2\tilde{p}/dz^2 + q^2\tilde{p} = 0$ (for the sake of simplicity, we assume that $\hat{Q} = 0$). If in equation (2.59) we now set $(\varrho/\varrho_1)^{1/2} = 1$ and $1 - \boldsymbol{\kappa}\cdot\mathbf{v}_\perp/\omega = 1$, equations (2.59) and (2.60) can be reduced to the following equation:

$$\frac{1}{c^2}\frac{d^2p}{dt^2} - \Delta p = 0. \tag{2.101}$$

The left-hand side of this equation coincides with the left-hand side of equation (2.100). Thus, equation (2.100) does not allow us to take into account the factors $(\varrho/\varrho_1)^{1/2}$ and $(1 - \boldsymbol{\kappa}\cdot\mathbf{v}_\perp/\omega)$ in the amplitudes of the quasi-plane waves.

If terms of order $(v/c)^2$ are neglected, equation (2.100) can also be derived from equation (2.84). Indeed, if we assume that the derivatives of ϱ, c, and $\mathbf{v}$ (which are proportional to $1/l$) are equal to zero, equation (2.84) becomes equation (2.100).

2.5   Parabolic equations for acoustic waves

Parabolic equations enable significant simplification of analytical and numerical solutions for wave propagation in inhomogeneous media. In this section, starting with equation (2.88) we derive narrow- and wide-angle parabolic equations in a three-dimensional inhomogeneous moving medium.

Recall that equation (2.88) is valid in the Cartesian coordinate system $\mathbf{R} = (x, y, z)$, in which the direction of the z-axis is opposite to the direction of gravity. We rotate this coordinate system so that the x-axis is close to the direction of wave propagation. In equation (2.88), the operator ∇ is invariant under rotation of the Cartesian coordinate system and the term $(\partial v_i / \partial x_j) (\partial^2 \hat{p} / \partial x_i \partial x_j)$ can be expressed in the form $\nabla \cdot \{((\nabla p) \cdot \nabla) \mathbf{v}\}$. Therefore, equation (2.88) remains the same in the new coordinate system, in which all subsequent calculations are made.

Assuming that $\mathbf{F} = 0$ and $Q = 0$, equation (2.88) is expressed in the form

$$\Delta \hat{p} + k_0^2 (1 + \mathcal{E}) \hat{p} = 0, \tag{2.102}$$

where the operator $\mathcal{E}$ is given by

$$\mathcal{E} = \varepsilon - \left(\frac{\nabla \varrho}{\varrho} \right) \cdot \left(\frac{\nabla}{k_0^2} \right) + \frac{2i}{k_0^2} \left[\frac{k_0}{c_0} (\mathbf{v} \cdot \nabla) - \frac{1}{\omega} \sum_{i=1}^{3} \sum_{j=1}^{3} \frac{\partial v_i}{\partial x_j} \frac{\partial^2}{\partial x_i \partial x_j} \right]. \tag{2.103}$$

Here, $\varepsilon = c_0^2 / c^2 - 1$, $(x_1, x_2, x_3) = (x, y, z)$, and the vector $\mathbf{v} = (v_1, v_2, v_3) = (v_x, v_y, v_z)$.

2.5.1   High-frequency, narrow-angle approximation

We assume that a sound wave propagates in the positive direction of the x-axis and express the temporal spectrum of the sound pressure as

$$\hat{p}(\mathbf{R}, \omega) = A(\mathbf{R}, \omega) \exp \left(i k_0 x \right), \tag{2.104}$$

where $A(\mathbf{R}, \omega)$ is the *complex amplitude*. We substitute $\hat{p}(\mathbf{R})$ into equation (2.102) and neglect the term

$$\tilde{A} = \frac{1}{k_0^2} \left(1 - \frac{4i}{\omega} \frac{\partial v_x}{\partial x} \right) \frac{\partial^2 A}{\partial x^2}. \tag{2.105}$$

As a result, we obtain the following equation for the complex amplitude A, in which each term has been numbered:

$$\left[\underset{1}{\frac{2i}{k_0}\frac{\partial}{\partial x}} + \underset{2}{\frac{\Delta_\perp}{k_0^2}} + \underset{3}{\varepsilon} - \underset{4}{\frac{2v_x}{c_0}} + \underset{5}{\frac{2i}{\omega}\mathbf{v}_\perp \cdot \nabla_\perp} + \underset{6}{\frac{2i}{\omega}\left(v_x\frac{\partial}{\partial x} + \frac{\partial v_x}{\partial x}\right)}\right.$$

$$\underset{7}{-\frac{i}{k_0\varrho}\frac{\partial\varrho}{\partial x}} - \underset{8}{\left(\frac{\nabla_\perp\varrho}{k_0^2\varrho}\right)\cdot\nabla_\perp} + \underset{9}{\frac{2}{\omega k_0}\left(\nabla_\perp v_x + \frac{\partial\mathbf{v}_\perp}{\partial x}\right)\cdot\nabla_\perp}$$

$$\underset{10}{-\frac{2i}{\omega k_0^2}\sum_{i=2}^{3}\sum_{j=2}^{3}\frac{\partial v_i}{\partial x_j}\frac{\partial^2}{\partial x_i\partial x_j}} - \underset{11}{\frac{1}{k_0^2\varrho}\left(\frac{\partial\varrho}{\partial x}\right)\frac{\partial}{\partial x}}$$

$$\left.\underset{12}{+\frac{4}{\omega k_0}\left(\frac{\partial v_x}{\partial x}\right)\frac{\partial}{\partial x}} - \underset{13}{\frac{2i}{\omega k_0^2}\sum_{i=2}^{3}\left(\frac{\partial v_i}{\partial x} + \frac{\partial v_x}{\partial x_i}\right)\frac{\partial^2}{\partial x_i\partial x}}\right] A = 0. \tag{2.106}$$

Here, $\mathbf{v}_\perp = (v_y, v_z)$ and $\nabla_\perp = (\partial/\partial y, \partial/\partial z)$.

Equation (2.106) describes the propagation of sound waves in a moving medium in the high-frequency, narrow-angle approximation, when the term $\widetilde{A}$ given by equation (2.105) can be omitted. Let us study the conditions under which this can be done. First, since the variation of A along the x-axis is caused by inhomogeneities in the ambient quantities, $\partial A/\partial x \sim A/l$ and $\partial^2 A/\partial x^2 \sim A/l^2$, where l is a characteristic scale of these inhomogeneities. The term $\widetilde{A} \sim A/(k_0^2 l^2)$ can be neglected in comparison with the first term in equation (2.106), which is of order $A/(k_0 l)$, in the high-frequency approximation, when $k_0 l \gg 1$. Secondly, it is known in the theories of wave propagation [339] that the term $\partial^2 A/\partial x^2$ can be omitted in the narrow-angle approximation, when the angle θ between the direction of the wave propagation and the x-axis is small. The narrow-angle approximation also requires that the Green's function in a free space can be replaced with its Fresnel approximation; this is valid if $x \ll k_0^3 l^4$. Thirdly, since equation (2.106) is a one-way equation, the backscattering must be small. This condition is valid if $|\varepsilon| \ll 1$ and $x\sigma_b \ll 1$, where x is the propagation range and

$$\sigma_b = \int_0^{2\pi} d\varphi \int_{\pi/2}^{\pi} \sin\theta\, \sigma(\theta,\varphi)\, d\theta \tag{2.107}$$

is the total sound scattering cross section in the negative direction of the x-axis and $\sigma(\theta,\varphi)$ is the sound scattering cross section (see section 6.4). Combining these results, we obtain the conditions under which the term $\widetilde{A}$ can be omitted:

$$k_0 l \gg 1, \quad |\theta| \ll 1, \quad x \ll k_0^3 l^4, \quad |\varepsilon| \ll 1, \quad x\sigma_b \ll 1. \tag{2.108}$$

In reference [290], these inequalities are obtained by a different, more rigorous approach.

2.5.2 Narrow-angle parabolic equation

Equation (2.106) is rather involved. We will show that for a high enough acoustic frequency, most terms in this equation can be ignored. Let us consider the order of magnitude of these terms. We have

$$\frac{|\nabla \varrho|}{\varrho} \sim \frac{\nu}{l}, \quad c_0^{-1} \left| \frac{\partial v_i}{\partial x_j} \right| \sim \frac{M}{l}.$$

Here, $M = v/c_0$ is the Mach number and $\nu = |\varrho - \varrho_0|/\varrho_0$ characterizes the variation in the density. Note that, almost always, $\nu \ll 1$. As shown above, in equation (2.106) $\partial A/\partial x \sim A/l$.

Finally, let us estimate the order of magnitude of $|\nabla_\perp A|$. In the homogeneous medium, the phase factor of the function $\widehat{p}$ is given by $\exp[ik_0(x \cos \theta + \mathbf{e} \cdot \mathbf{r} \sin \theta)]$, where $\mathbf{e}$ is the unit vector in the direction of the projection of the wave vector $\mathbf{k}_0$ of the sound wave on the yz-plane. In the narrow-angle approximation, the complex amplitude A is proportional to the phase factor $\exp(ik_0\theta\mathbf{e}\cdot\mathbf{r})$, so that $|\nabla_\perp A| \sim Ak_0\theta_m$. Here, θ_m is the maximum angle θ between the direction of wave propagation and the x-axis. In an inhomogeneous medium, another estimate for $|\nabla_\perp A|$ is valid: $|\nabla_\perp A| \sim A/l$. Combining these two estimates, we have $|\nabla_\perp A|/k_0 \sim A \max(1/k_0l, \theta_m)$. In this equation, $1/k_0l$ is of order of the diffraction angle of the sound wave scattered (or refracted) by a medium inhomogeneity with the characteristic size l. If $x \gg l$, the wave is multiply scattered by inhomogeneities so that $\theta > 1/k_0l$. Therefore, $|\nabla_\perp A|/k_0 \sim \theta_m A$.

Using the estimations given above, we list the order of magnitudes of all thirteen terms in equation (2.106):

1. $2/(k_0l) \sim 3 \cdot 10^{-2} \ (3 \cdot 10^{-2})$.
2. $\theta_m^2 \sim 7 \cdot 10^{-2} \ (7 \cdot 10^{-2})$.
3. $\varepsilon \sim 2 \cdot 10^{-2} \ (3 \cdot 10^{-2})$.
4. $2M \sim 6 \cdot 10^{-2} \ (10^{-3})$.
5. $2M\theta_m \sim 2 \cdot 10^{-2} \ (3 \cdot 10^{-4})$.
6. $2M/(k_0l) \sim 10^{-3} \ (2 \cdot 10^{-5})$.
7. $\nu/(k_0l) \sim 3 \cdot 10^{-4} \ (4 \cdot 10^{-5})$.
8. $\nu\theta_m/(k_0l) \sim 7 \cdot 10^{-5} \ (10^{-5})$.
9. $2M\theta_m/(k_0l) \sim 2 \cdot 10^{-4} \ (5 \cdot 10^{-6})$.
10. $2M\theta_m^2/(k_0l) \sim 6 \cdot 10^{-5} \ (10^{-6})$.
11. $\nu/(k_0^2l^2) \sim 4 \cdot 10^{-6} \ (7 \cdot 10^{-7})$.
12. $4M/(k_0^2l^2) \sim 3 \cdot 10^{-5} \ (7 \cdot 10^{-7})$.
13. $2M\theta_m/(k_0^2l^2) \sim 4 \cdot 10^{-6} \ (8 \cdot 10^{-8})$. (2.109)

Here, we give first the number of the term in equation (2.106), then the order of magnitude of this term, followed by its numerical value in the atmosphere for typical values of v, ε, and ν. Values in parentheses refer to the ocean. In the numerical estimations, we have assumed that in the atmosphere, $v = 10$ m/s and $\varepsilon = \nu = 1.7 \times 10^{-2}$, which corresponds to a difference in temperature $|T - T_0| = 5$ K, and in the ocean $v = 1$ m/s, $\nu = 2.8 \times 10^{-3}$, and $\varepsilon = 2.7 \times 10^{-2}$, which corresponds to a difference in the sound speed $|c - c_0| = 20$ m/s. Furthermore, we have assumed that $\theta_m = \pi/12$ and $\lambda/l = 0.1$.

In equation (2.109), there are five small parameters, $1/k_0 l$, θ_m, M, ν, and ε, the values of which can differ significantly, but always $\theta_m > 1/k_0 l$. We assume that the angle θ_m between the direction of wave propagation and the x-axis is much greater than $1/k_0 l$ and does not depend on it. Then, it follows from the estimations (2.109) that the second to fifth terms in equation (2.106) do not contain the small parameter $1/k_0 l$. The first term is of the order of $2/k_0 l$, and the sixth to the thirteenth terms are of order $\mu/(k_0 l)^n$, where $n = 1$ or 2, and μ is one of the small parameters θ_m, M, ν, or their product. Therefore, for arbitrary values of the small parameters θ_m, M, ν, and ε, there is a large number m depending on these parameters, such that the first five terms in equation (2.106) are much greater than the sixth to the thirteenth terms if $k_0 l \gg m$.

In accordance with the above estimations for the values of v, ε, ν, $1/k_0 l$, and θ_m, typical of the atmosphere and ocean, the first to the fifth terms of equation (2.106) are greater by at least an order of magnitude than the sixth to the thirteenth terms. Using these estimations, it can be shown that $m > 1/\theta_m$. Therefore, if the inequality $k_0 l \gg m$ is valid, the inequality $\theta_m \gg 1/k_0 l$ is also valid.

Thus, for a high enough acoustic frequency ($k_0 l \gg m$), equation (2.106) can be simplified significantly [288]:

$$2ik_0 \frac{\partial A}{\partial x} + \Delta_\perp A + k_0^2 \varepsilon A - \frac{2k_0^2 v_x}{c_0} A + \frac{2ik_0}{c_0} \mathbf{v}_\perp \cdot \nabla_\perp A = 0. \tag{2.110}$$

This is the desired narrow-angle parabolic equation. In this equation, the ambient quantities are the sound speed c and medium velocity $\mathbf{v}$. If $\mathbf{v} = 0$, equation (2.110) becomes the well-known parabolic equation for sound waves in a motionless medium. Assuming that $k_0 l \gg m$, equation (2.110) can also be derived from equation (2.100). Equation (2.110) was derived from equation (2.88) assuming that inequalities (2.108) are valid and $k_0 l \gg m$. The parabolic equation (2.110) allows one to consider sound waves with $|\theta| \lesssim 20°$. Equation (2.110) is shown at the bottom of the flowchart in figure 2.1. Reference [290] compares equations (2.106) and (2.110) with the parabolic equations obtained elsewhere.

The fifth term in equation (2.110) describes advection of a sound wave due to the medium velocity component perpendicular to the x-axis. This term is always less than the fourth, but it can be of the order of or greater than the second and third terms, and one should keep it if $\theta_m \gg 1/k_0 l$. But if $\theta_m \sim 1/k_0 l$, the fifth term in equation (2.110) can be neglected. To the accuracy, with which this equation is valid, the third and fourth terms can be combined and expressed as $k_0^2 \left(c_0^2/c_{\text{eff}}^2 - 1 \right) A$, where $c_{\text{eff}} = c + v_x$ is the effective sound speed (see equation (1.3)). In this case, equation (2.110) takes the form

$$2ik_0 \frac{\partial A}{\partial x} + \Delta_\perp A + k_0^2 \left(\frac{c_0^2}{c_{\text{eff}}^2} - 1 \right) A + \frac{2ik_0}{c_0} \mathbf{v}_\perp \cdot \nabla_\perp A = 0. \tag{2.111}$$

The parabolic equation (2.110) is used in Chapter 7 for calculating the statistical characteristics of the sound field in a medium with random fluctuations in the sound speed c and velocity $\mathbf{v}$. In Part III, equation (2.111) is used for numerical simulations of sound propagation in the near-ground atmosphere. In this case, the x-axis is parallel to the ground and a two-dimensional version of the parabolic equation is usually solved [345] in the vertical plane (x, z). Since the mean value of the vertical wind velocity v_z can be set to zero, the last term in equation (2.111) vanishes. Thus, the two-dimensional parabolic equation does not account for the effect of the crosswind v_y on sound propagation. Equation (2.111) with the last term omitted is also used for numerical studies of sound propagation in the stratified ocean with currents [271, 335].

2.5.3 Wide-angle parabolic equation

For certain problems, there is a need to consider sound waves propagating at angles greater than $20°$ with respect to the x-axis and still use a one-way equation. This can be done with a wide-angle parabolic equation for acoustic waves in an inhomogeneous moving medium which is considered in this and next subsections [296].

In equation (2.102), $\mathcal{E}\widehat{p}$ contains the term $\xi = -(\partial v_x/\partial x)(\partial^2\widehat{p}/\partial x^2)$. The second-order derivative with respect to x appearing in this term should be transformed when deriving a wide-angle parabolic equation. To this end, equation (2.102) is written in the form

$$\frac{\partial^2 \widehat{p}}{\partial x^2} = - \left(\Delta_\perp + k_0^2 + \mathcal{E} \right) \widehat{p}. \tag{2.112}$$

In the formula for ξ, replacing $\partial^2\widehat{p}/\partial x^2$ with the right-hand side of equation (2.112), we have $\xi = (\partial v_x/\partial x)(\Delta_\perp + k_0^2 + \mathcal{E})\widehat{p}$. In this expression, taking into account that $\mathcal{E} = O\left(v/c, |\widetilde{c}/c_0|, |\varrho/\varrho_0|\right)$, the product $(\partial v_x/\partial x)\mathcal{E} \sim O(\mu_2^2, v^2/c^2)$ should be ignored since such terms have been ignored when deriving equation (2.102). Thus,

$$-\frac{\partial v_x}{\partial x} \frac{\partial^2 \widehat{p}}{\partial x^2} = \frac{\partial v_x}{\partial x}(\Delta_\perp + k_0^2)\widehat{p}. \tag{2.113}$$

The right-hand side of this formula does not contain $\partial^2\widehat{p}/\partial x^2$. We use this formula in equation (2.103) and substitute the resulting expression for $\mathcal{E}$ into equation (2.102). The latter equation can be written as

$$\left(\frac{\partial^2}{\partial x^2} + k_0^2 D^2 \right) \widehat{p} = 0. \tag{2.114}$$

Here, D is the pseudo-differential operator given by

$$D = \sqrt{1+L},\tag{2.115}$$

where

$$L = G + B\frac{\partial}{\partial x},\tag{2.116}$$

and

$$G = \varepsilon + \frac{2i}{\omega}\frac{\partial v_x}{\partial x} + \left(\frac{2i}{\omega}\mathbf{v}_\perp - k_0^{-2}\nabla_\perp \ln(\varrho/\varrho_0)\right)\cdot\nabla_\perp$$
$$+ k_0^{-2}\left(1 + \frac{2i}{\omega}\frac{\partial v_x}{\partial x}\right)\Delta_\perp - \frac{2i}{\omega k_0^2}\sum_{j=2}^{3}(\nabla_\perp v_j)\cdot\nabla_\perp\frac{\partial}{\partial x_j},\tag{2.117}$$

$$B = \frac{2i}{\omega}v_x - k_0^{-2}\frac{\partial \ln(\varrho/\varrho_0)}{\partial x} - \frac{2i}{\omega k_0^2}\left(\nabla_\perp v_x + \frac{\partial \mathbf{v}_\perp}{\partial x}\right)\cdot\nabla_\perp.\tag{2.118}$$

Note that the operator L does not contain the second-order derivative with respect to x, but contains the operator $B\partial/\partial x$, where $B\sim O\left(v/c_0,|\varrho/\varrho_0|\right)$.

If the operator D does not depend on x, equation (2.114) can be written as

$$\left(\frac{\partial}{\partial x} + ik_0 D\right)\left(\frac{\partial}{\partial x} - ik_0 D\right)\widehat{p} = 0.\tag{2.119}$$

From this equation, we obtain the wide-angle parabolic equation for acoustic waves

$$\left(\frac{\partial}{\partial x} - ik_0 D\right)\widehat{p} = 0.\tag{2.120}$$

This equation describes sound propagation in the positive direction of the x-axis and is also called a *one-way* parabolic equation. If the operator D depends on x, the wide-angle parabolic equation can be derived from equation (2.114) with some approximations.

The explicit form of the operator D is given by the Taylor series of the right-hand side of equation (2.115)

$$D = 1 + \frac{1}{2}L - \frac{1}{8}L^2 + \frac{1}{16}L^3 + \dots .\tag{2.121}$$

The third term in this series contains the operator $B^2\partial^2/\partial x^2$ which should not appear in a wide-angle parabolic equation. Since B is proportional to a small parameter, similarly to equation (2.113) the second-order derivative in the operator $B^2\partial^2/\partial x^2$ can be replaced with $-(\Delta_\perp + k_0^2)$. Similarly, the higher order derivatives $\partial^{2n}/\partial x^{2n}$ in equation (2.121) can be replaced with $\left(-(\Delta_\perp + k_0^2)\right)^n$, where $n = 2, 3, \dots$.

2.5.4 Padé approximation

In certain cases, the pseudo-differential equation (2.120) might be inconvenient for numerical calculations of the sound field. In such cases, the Padé (1,1) approximation of the operator D is often used [81]

$$D = \sqrt{1 + L} \cong \frac{a_1 + a_2 L}{a_3 + a_4 L}. \tag{2.122}$$

Here, $a_1 = 1$, $a_2 = 3/4$, $a_3 = 1$, and $a_4 = 1/4$ are numerical coefficients. In some papers (e.g., [201]) the values of a_1, a_2, a_3, and a_4 are chosen slightly differently in order to achieve a better approximation of the operator D. The Padé (1,1) approximation enables one to consider sound waves propagating at an angle up to about $40°$ with respect to the x-axis.

Substitution of equation (2.122) into equation (2.120) yields

$$\frac{\partial \widehat{p}}{\partial x} = i k_0 \frac{a_1 + a_2(G + B\partial/\partial x)}{a_3 + a_4(G + B\partial/\partial x)} \widehat{p}. \tag{2.123}$$

Applying the operator $a_3 + a_4(G + B\partial/\partial x)$ to both sides of this equation, we have

$$(a_3 + a_4 G - i a_2 k_0 B) \frac{\partial \widehat{p}}{\partial x} = i k_0 \left(a_1 + a_2 G + \frac{i a_4 B}{k_0} \frac{\partial^2}{\partial x^2} \right) \widehat{p}. \tag{2.124}$$

The operator $B\partial^2/\partial x^2$ on the right-hand side of this equation can be replaced with the operator $-B(\Delta_\perp + k_0^2)$ similarly to that in equation (2.113). Replacing $\widehat{p}$ with its value given by equation (2.104), we obtain the following equation for the complex amplitude $A(x, \mathbf{r})$

$$(a_3 + a_4 G - i a_2 k_0 B) \frac{\partial A}{\partial x}$$
$$= i k_0 \left[a_1 - a_3 + (a_2 - a_4)G + i(a_2 - a_4)k_0 B - \frac{i a_4 B}{k_0} \Delta_\perp \right] A. \tag{2.125}$$

Substituting with the values of the operators G and B, we obtain the desired wide-angle parabolic wave equation in the Padé (1,1) approximation

$$\Psi_1 \frac{\partial A}{\partial x} = i k_0 \Psi_2 A. \tag{2.126}$$

Here

$$\Psi_1 = a_3 + a_4 \varepsilon + \frac{2 i a_4}{\omega} \frac{\partial v_x}{\partial x} + \frac{i a_2}{k_0} \frac{\partial \ln(\varrho/\varrho_0)}{\partial x} + \frac{2 a_2 v_x}{c_0}$$
$$+ k_0^{-1} \left[\frac{2 i a_4 \mathbf{v}_\perp}{c_0} - \frac{2 a_2}{\omega} \left(\nabla_\perp v_x + \frac{\partial \mathbf{v}_\perp}{\partial x} \right) - \frac{a_4}{k_0} \nabla_\perp \ln(\rho/\varrho_0) \right] \cdot \nabla_\perp$$
$$+ a_4 k_0^{-2} \left[\left(1 + \frac{2 i}{\omega} \frac{\partial v_x}{\partial x} \right) \Delta_\perp - \frac{2 i}{\omega} \sum_{j=2}^{3} (\nabla_\perp v_j) \cdot \nabla_\perp \frac{\partial}{\partial x_j} \right], \tag{2.127}$$

and

$$\Psi_2 = a_1 - a_3 + (a_2 - a_4)\left[\varepsilon + \frac{2i}{\omega}\frac{\partial v_x}{\partial x} - \frac{i}{k_0}\frac{\partial \ln(\varrho/\varrho_0)}{\partial x} - \frac{2v_x}{c_0}\right]$$

$$+ (a_2 - a_4)k_0^{-1}\left[\frac{2i\mathbf{v}_\perp}{c_0} - k_0^{-1}\nabla_\perp \ln(\varrho/\varrho_0) + \frac{2}{\omega}\left(\nabla_\perp v_x + \frac{\partial \mathbf{v}_\perp}{\partial x}\right)\right]\cdot\nabla_\perp$$

$$+ k_0^{-2}\left[(a_2 - a_4)\left(1 + \frac{2i}{\omega}\frac{\partial v_x}{\partial x}\right) + a_4\left(\frac{i}{k_0}\frac{\partial \ln(\varrho/\varrho_0)}{\partial x} + \frac{2v_x}{c_0}\right)\right]\Delta_\perp$$

$$- \frac{2i(a_2 - a_4)}{\omega k_0^2}\sum_{j=2}^{3}(\nabla_\perp v_j)\cdot\nabla_\perp\frac{\partial}{\partial x_j} - \frac{2a_4}{\omega k_0^3}\left(\nabla_\perp v_x + \frac{\partial \mathbf{v}_\perp}{\partial x}\right)\cdot\nabla_\perp\Delta_\perp.$$

$$(2.128)$$

The wide-angle parabolic equations (2.120) and (2.126) contain many terms. These equations can, however, be solved relatively easily numerically by a marching technique because they are first-order differential equations with respect to x. These equations are referenced at the bottom of the flowchart in figure 2.1. The wide-angle parabolic equation (2.126) has been used in numerical studies of outdoor sound propagation (e.g., reference [35]). Other wide-angle parabolic equations were considered elsewhere [140, 228].

3

Geometrical acoustics in an inhomogeneous moving medium

The use of geometrical acoustics in an inhomogeneous moving medium enables
the identification of ray paths along which energy propagates and may assist
with the visualization of sound propagation. The basic ideas of geometrical
acoustics in a moving medium coincide with those for a motionless medium.
Nevertheless, the particular results in a moving medium cannot be obtained
from the analogous results for a motionless medium. The main results of geometrical acoustics in an inhomogeneous moving medium were obtained before
the mid-1940s. In this chapter, these results and those obtained after the 1940s
are systematically derived and presented.

In section 3.1, starting with a complete set of linearized equations of fluid
dynamics, the eikonal equation, the dispersion equation, and the law of acoustic energy conservation are derived for the case of a space- and time-varying
moving medium. The eikonal equation allows one to find the phase of a sound
wave and its ray path, while the law of acoustic energy conservation enables
calculation of the amplitude of the wave along this path. Particular forms of
these equations and conservation law in a time-independent moving medium
are presented in section 3.2. The phase and group velocities of a sound wave,
its ray path and eikonal, the wave amplitude, and the time of sound propagation along the path are considered in section 3.3. In the remainder of the
chapter (sections 3.4–3.7), sound propagation in a stratified moving medium is
studied. In section 3.4, the refraction laws for a sound ray and the normal to a
wavefront are derived, which generalize well-known Snell's law in a motionless
medium to the case of a moving medium. In section 3.5, analytical formulas for
the ray path, eikonal, and amplitude of a sound wave are presented; examples
of the sound ray paths in the atmosphere and ocean are also given and analyzed. Section 3.6 presents approximate formulas for the sound ray path and
eikonal in a medium with relatively small variations in the sound speed and
velocity. Acoustic travel-time tomography of the atmospheric surface layer is
overviewed in section 3.7.

3.1 Geometrical acoustics in a space- and time-varying medium

Propagation of sound waves in an inhomogeneous moving medium is described by the complete set of linearized equations of fluid dynamics (2.8)–(2.12). The most consistent derivation of the basic equations of geometrical acoustics in a moving medium is to use the Debye series for solving equations (2.8)–(2.12). This approach was used by Blokhintzev [37] to derive the eikonal equation and the law of acoustic energy conservation in a time-independent moving medium, the ambient quantities of which do not depend on the time t. In this section, this approach is generalized to the case of sound propagation in a time-dependent moving medium.

3.1.1 Debye series for the sound field

A sound wave propagating in a medium causes fluctuations in the pressure $p(\mathbf{R}, t)$, the velocity $\mathbf{w}(\mathbf{R}, t)$, the density $\eta(\mathbf{R}, t)$, the entropy $s(\mathbf{R}, t)$, and the concentrations $\chi_i(\mathbf{R}, t)$ of the components dissolved in the medium (e.g., water vapor in the atmosphere or salt dissolved in the water). Here $\mathbf{R} = (x, y, z)$ are the spatial coordinates and $i = 1, 2, ..., N$, where N is the number of the components. The functions p, $\mathbf{w}$, η, s, and χ_i satisfy equations (2.8)–(2.12). For a time-dependent inhomogeneous moving medium, these functions can be expressed in the form

$$p(\mathbf{R}, t) = \exp\left[ik_0\Theta(\mathbf{R}, t)\right] p_0(\mathbf{R}, t), \tag{3.1}$$

$$\mathbf{w}(\mathbf{R}, t) = \exp\left[ik_0\Theta(\mathbf{R}, t)\right] \mathbf{w}_0(\mathbf{R}, t), \tag{3.2}$$

$$\eta(\mathbf{R}, t) = \exp\left[ik_0\Theta(\mathbf{R}, t)\right] \eta_0(\mathbf{R}, t), \tag{3.3}$$

$$s(\mathbf{R}, t) = \exp\left[ik_0\Theta(\mathbf{R}, t)\right] s_0(\mathbf{R}, t), \tag{3.4}$$

$$\chi_i(\mathbf{R}, t) = \exp\left[ik_0\Theta(\mathbf{R}, t)\right] \chi_{i,0}(\mathbf{R}, t). \tag{3.5}$$

Here, k_0 is the reference value of the sound wavenumber, $\Theta(\mathbf{R}, t)$ is the phase function, and p_0, $\mathbf{w}_0$, η_0, s_0, and $\chi_{i,0}$ are the amplitudes of the corresponding acoustic quantities. Substituting equations (3.1)–(3.5) into equations (2.8)–(2.12), we obtain a set of equations for these amplitudes

$$ik_0\left[\mathbf{w}_0 \, d\Theta/dt + (p_0/\varrho)\nabla\Theta\right]$$
$$= (\eta_0/\varrho^2)\nabla P - d\mathbf{w}_0/dt - (\mathbf{w}_0 \cdot \nabla)\mathbf{v} - \varrho^{-1}\nabla p_0, \tag{3.6}$$

$$ik_0(\eta_0 \, d\Theta/dt + \varrho\mathbf{w}_0 \cdot \nabla\Theta) = -d\eta_0/dt - \mathbf{w}_0 \cdot \nabla\varrho - \varrho\nabla \cdot \mathbf{w}_0 - \eta_0\nabla \cdot \mathbf{v}, \tag{3.7}$$

$$ik_0\chi_{i,0} \, d\Theta/dt = -d\chi_{i,0}/dt - \mathbf{w}_0 \cdot \nabla C_i, \tag{3.8}$$

$$ik_0 s_0 \, d\Theta/dt = -ds_0/dt - \mathbf{w}_0 \cdot \nabla S, \tag{3.9}$$

$$p_0 - c^2\eta_0 - hs_0 - b_i\chi_{i,0} = 0. \tag{3.10}$$

Here, the pressure P, the density ϱ, the velocity of medium motion $\mathbf{v}$, the entropy S, the concentrations C_i of the components dissolved in the medium, the parameters $h = (\partial P/\partial S)_{\varrho,C_i}$ and $b_i = (\partial P/\partial C_i)_{\varrho,S,C_{j\neq i}}$, and the sound speed c are the quantities characterizing the ambient state of the medium. All these quantities depend on $\mathbf{R}$ and t. Furthermore, in equations (3.6)–(3.9), $d/dt = \partial/\partial t + \mathbf{v} \cdot \nabla$.

Since $\Theta \sim R - c_0 t$, on the left-hand sides of equations (3.6)–(3.9), $|\nabla\Theta| \sim 1$ and $|d\Theta/dt| \sim c_0$, where c_0 is the reference value of the sound speed. On the right-hand sides of these equations, the derivatives with respect to t are proportional to the characteristic frequency Ω of the variation of the ambient quantities in time ($\partial P/\partial t \sim \Omega P$, $\partial\varrho/\partial t \sim \Omega\varrho$, etc.), while the derivatives with respect to $\mathbf{R}$ are proportional to l^{-1}, where l is the characteristic scale of spatial variation of the ambient quantities. In a time-dependent inhomogeneous moving medium, geometrical acoustics is applicable if $k_0 l \gg 1$ and $k_0 c_0 \gg \Omega$. If these inequalities are valid, the terms on the left-hand sides of equations (3.6)–(3.9) are greater than those on the right-hand sides by factors of $k_0 l$ or $k_0 c_0/\Omega$. In this case, it is worthwhile to express the solutions of equations (3.6)–(3.10) as Debye series in the small parameters $1/(k_0 l)$ and $\Omega/(k_0 c_0)$:

$$p_0 = \sum_{n=1}^{\infty} \frac{p_n}{(ik_0)^{n-1}}, \quad \mathbf{w}_0 = \sum_{n=1}^{\infty} \frac{\mathbf{w}_n}{(ik_0)^{n-1}}, \quad \eta_0 = \sum_{n=1}^{\infty} \frac{\eta_n}{(ik_0)^{n-1}},$$

$$s_0 = \sum_{n=1}^{\infty} \frac{s_n}{(ik_0)^{n-1}}, \quad \chi_{i,0} = \sum_{n=1}^{\infty} \frac{\chi_{i,n}}{(ik_0)^{n-1}}, \tag{3.11}$$

where p_n, $\mathbf{w}_n$, η_n, s_n, and $\chi_{i,n}$ are the amplitudes of the terms of these series. We substitute series (3.11) into equations (3.6)–(3.10) and equate the coefficients of k_0^n. The coefficients of k_0^1 form a set of equations for $p_{n=1}$, $\mathbf{w}_{n=1}$, and $\eta_{n=1}$:

$$\mathbf{w}_1 \, d\Theta/dt + (p_1/\varrho)\nabla\Theta = 0, \tag{3.12}$$

$$\eta_1 \, d\Theta/dt + \varrho\mathbf{w}_1 \cdot \nabla\Theta = 0, \tag{3.13}$$

$$p_1 = c^2\eta_1, \tag{3.14}$$

while, according to equations (3.8) and (3.9), $\chi_{i,1} = s_1 = 0$. The coefficients of $(k_0)^0$ form a set of equations for p_2, $\mathbf{w}_2$, η_2, s_2, and $\chi_{i,2}$:

$$\mathbf{w}_2 \, d\Theta/dt + (p_2/\varrho)\nabla\Theta = (\eta_1/\varrho_2)\nabla P - d\mathbf{w}_1/dt - (\mathbf{w}_1 \cdot \nabla)\mathbf{v} - \varrho^{-1}\nabla p_1, \tag{3.15}$$

$$\eta_2 \, d\Theta/dt + \varrho\mathbf{w}_2 \cdot \nabla\Theta = -d\eta_1/dt - \mathbf{w}_1 \cdot \nabla\varrho - \varrho\nabla \cdot \mathbf{w}_1 - \eta_1\nabla \cdot \mathbf{v}, \tag{3.16}$$

$$\chi_{i,2} \, d\Theta/dt = -\mathbf{w}_1 \cdot \nabla C_i, \tag{3.17}$$

$$s_2 \, d\Theta/dt = -\mathbf{w}_1 \cdot \nabla S, \tag{3.18}$$

$$p_2 - c^2\eta_2 - hs_2 - b_i\chi_{i,2} = 0. \tag{3.19}$$

The set of equations for p_n, $\mathbf{w}_n$, η_n, s_n, and $\chi_{i,n}$ coincides with equations (3.15)–(3.19) if, in the latter equations, the subscripts 2 and 1 are replaced with n and $n - 1$, respectively, and the terms $-d\chi_{i,n-1}/dt$ and $-ds_{n-1}/dt$ are added to the right-hand sides of equations (3.17) and (3.18).

The set of homogeneous linear equations (3.12)–(3.14) is consistent only if its determinant $D = (d\Theta/dt)^2 - c^2|\nabla\Theta|^2$ equals zero. This condition yields the equation for the phase function $\Theta(\mathbf{R},t)$ of a sound wave propagating in a time-dependent moving medium:

$$\partial\Theta/\partial t + \mathbf{v} \cdot \nabla\Theta = -c|\nabla\Theta|. \tag{3.20}$$

Here, the sign in front of $|\nabla\Theta|$ is chosen in accordance with the time convention $\exp(-i\omega t)$, which is used for a monochromatic sound wave in this book. Equation (3.20) is called the eikonal equation. By definition, $\omega = -k_0\partial\Theta/\partial t$ is the angular frequency of a sound wave and $\mathbf{k} = k_0\nabla\Theta$ is its wave vector. Multiplying both sides of the eikonal equation (3.20) by k_0, we express it as the dispersion equation

$$\omega(\mathbf{R},t) = k(\mathbf{R},t)c(\mathbf{R},t) + \mathbf{k}(\mathbf{R},t) \cdot \mathbf{v}(\mathbf{R},t). \tag{3.21}$$

It follows from this equation that in a time-dependent inhomogeneous moving medium, the frequency of a sound wave and its wave vector depend on both the spatial coordinates and time.

3.1.2   Transport equation

In geometrical acoustics, the amplitude p_0 of the sound-pressure field is usually approximated with the first term p_1 of the asymptotic Debye series. Using this approximation in equation (3.1), we have

$$p(\mathbf{R},t) = p_1(\mathbf{R},t)\exp\left[ik_0\Theta(\mathbf{R},t)\right]. \tag{3.22}$$

In this subsection, we shall derive the transport equation for p_1.

Equations (3.12)–(3.14) enable us to express $\mathbf{w}_1$ and η_1 in terms of p_1:

$$\mathbf{w}_1 = -\frac{\nabla\Theta}{d\Theta/dt}\frac{p_1}{\varrho} = \frac{p_1\mathbf{n}}{\varrho c}, \quad \eta_1 = \frac{p_1}{c^2}. \tag{3.23}$$

Here, we have introduced the unit vector $\mathbf{n} = \nabla\Theta/|\nabla\Theta| = \mathbf{k}/k$ normal to the wavefront of a sound wave and taken into account the fact that $d\Theta/dt = -c|\nabla\Theta|$ (see equation (3.20)).

We multiply equations (3.15)–(3.19) by $-\varrho c^2(d\Theta/dt)^{-1}\nabla\Theta$, c^2, b_i, h, and $d\Theta/dt$, respectively, and then add these equations. Using the eikonal equation (3.20), it can be shown that in the resulting equation the sum of terms containing p_2, $\mathbf{w}_2$, η_2, $\chi_{i,2}$, and s_2 is zero. In this equation, we take into account the equality $\mathbf{w}_1 \cdot (h\nabla S + b_i \nabla C_i) = \mathbf{w}_1 \cdot (\nabla P - c^2 \nabla \varrho)$ (this follows from the equation of state $P = P(\varrho, S, C_i)$ if we apply the operator $\mathbf{w}_1 \cdot \nabla$ to both sides of this equation) and express $\mathbf{w}_1$ and η_1 in terms of p_1 making use of equation (3.23). As a result, we obtain the transport equation for the amplitude p_1 of the sound-pressure field:

$$\frac{\varrho \mathbf{n}}{c} \frac{d}{dt}\left(\frac{p_1 \mathbf{n}}{\varrho c}\right) + \frac{d}{dt}\left(\frac{p_1}{c^2}\right) + \varrho \nabla \cdot \left(\frac{p_1 \mathbf{n}}{\varrho c}\right)$$

$$+ \frac{\mathbf{n} \cdot \nabla p_1}{c} + \frac{p_1}{c^2}\mathbf{n}(\mathbf{n} \cdot \nabla) \cdot \mathbf{v} + \frac{p_1 \nabla \cdot \mathbf{v}}{c^2} = 0.$$

Removing the brackets in this equation and combining the terms proportional to $\partial p_1/\partial t$, ∇p_1, and p_1, the transport equation is expressed in the form:

$$2\frac{\partial p_1}{\partial t} + 2\mathbf{u} \cdot \nabla p_1 + p_1 \left[\nabla \cdot \mathbf{u} - \mathbf{u} \cdot \nabla \ln(\varrho c^2)\right.$$

$$\left. - \frac{\partial}{\partial t} \ln(\varrho c^3) + \mathbf{n}(\mathbf{n} \cdot \nabla) \cdot \mathbf{v} - \frac{\mathbf{v} \cdot \nabla c}{c}\right] = 0, \quad (3.24)$$

where $\mathbf{u} = c\mathbf{n} + \mathbf{v}$. It will be shown below that $\mathbf{u}$ is the group velocity of the sound wave, i.e., the velocity of acoustic energy propagation. Using the dispersion equation (3.21), we express the sum of the two last terms in square brackets in equation (3.24) in the form $(\mathbf{n}/k)\cdot\nabla\omega - \mathbf{u}\cdot\nabla\ln\sigma$, where $\sigma = \omega - \mathbf{k}\cdot\mathbf{v}$ is the acoustic frequency in the coordinate system moving with the medium. It follows from the equalities $\partial(\nabla\Theta)/\partial t = \nabla\partial\Theta/\partial t$ and $k = \sigma/c$ that

$$\frac{\mathbf{n} \cdot \nabla\omega}{k} = -\frac{\partial k/\partial t}{k} = -\frac{\partial}{\partial t}\ln(\sigma/c).$$

Using these equalities, the transport equation is expressed in the desired form

$$2\frac{\partial p_1}{\partial t} + 2\mathbf{u} \cdot \nabla p_1 + p_1 \left[-\frac{\partial}{\partial t}\ln(\varrho c^2 \sigma) + \nabla \cdot \mathbf{u} - \mathbf{u} \cdot \nabla\ln(\varrho c^2 \sigma)\right] = 0. \quad (3.25)$$

The transport equation allows us to calculate the amplitude $p_1(\mathbf{R}, t)$ of the sound-pressure field if the phase function $\Theta(\mathbf{R}, t)$ is known.

3.1.3 Acoustic energy conservation

Generally, it is not possible to formulate a law of acoustic energy conservation in an inhomogeneous moving medium (section 2.1.4). We show now that the law of acoustic energy conservation can be formulated in the approximation of geometrical acoustics. Multiplying the transport equation (3.25) by $p_1 \omega_0/(2\varrho c^2 \sigma)$, where $\omega_0 = k_0 c_0$ is the reference value of the sound frequency, and taking into account the fact that $\sigma = \omega(1 + \mathbf{n} \cdot \mathbf{v}/c)^{-1}$, this equation is expressed in the form

$$\partial E_1/\partial t + \nabla \cdot (E_1 \mathbf{u}) = 0, \quad (3.26)$$

where

$$E_1 = \frac{p_1^2}{2\varrho c^2} \frac{\omega_0}{\omega} \left(1 + \frac{\mathbf{n} \cdot \mathbf{v}}{c}\right). \tag{3.27}$$

Equation (3.26) is the law of energy conservation in the geometrical acoustics approximation for a time-dependent inhomogeneous moving medium. In this equation, E_1 is the mean acoustic energy density (averaged over the period of sound oscillations $T = 2\pi/\omega$) and $\mathbf{u}$ is the velocity of energy propagation. The mean acoustic energy flux (the *intensity*) is given by $\mathbf{I}_1 = E_1 \mathbf{u}$. It follows from the equations for E_1 and $\mathbf{I}_1$ that the acoustic energy and its flux depend on the velocity of the medium $\mathbf{v}$. In other words, this velocity contributes to the acoustic energy.

Using equation (3.22), we have $p_1 = \exp(-ik_0\Theta)p$. Substituting this value of p_1 into equation (3.27) and taking into account the equality $\partial\Theta/\partial t + \mathbf{u} \cdot \nabla\Theta = 0$, which follows from equation (3.20), the law of acoustic energy conservation can be expressed in the form

$$\partial E/\partial t + \nabla \cdot \mathbf{I} = 0. \tag{3.28}$$

Here,

$$E = \frac{p^2}{\varrho c^2} \frac{\omega_0}{\omega} \left(1 + \frac{\mathbf{n} \cdot \mathbf{v}}{c}\right) \tag{3.29}$$

is the instantaneous value of the acoustic energy density and $\mathbf{I} = E\mathbf{u}$ is its flux. The formula for E is consistent with equation (3.27) for E_1 if we take into account the fact that the mean value of p^2 over the period T is equal to $p_1^2/2$. It follows from equations (3.26) and (3.28) that both the mean and instantaneous values of the acoustic energy are conserved in the geometrical acoustics of a time-dependent inhomogeneous moving medium. Multiplying equation (3.23) by $\exp(ik_0\Theta)$, we obtain $\mathbf{w} = p\mathbf{n}/\varrho c$ and $\eta = p/c^2$. Using these equalities, the formulas for E and $\mathbf{I}$ can be expressed in the form

$$E = \frac{\omega_0}{\omega} \frac{p^2}{\varrho c^2} \left(1 + \frac{\mathbf{n} \cdot \mathbf{v}}{c}\right) = \frac{\omega_0}{\omega} \left(\frac{p\eta}{2\varrho} + \frac{\varrho w^2}{2} + \eta\mathbf{w} \cdot \mathbf{v}\right), \tag{3.30}$$

$$\mathbf{I} = \frac{\omega_0}{\omega} \frac{p^2}{\varrho c} \left(1 + \frac{\mathbf{n} \cdot \mathbf{v}}{c}\right) \left(\mathbf{n} + \frac{\mathbf{v}}{c}\right) = \frac{\omega_0}{\omega} [p + (\mathbf{w} \cdot \mathbf{v})\varrho] \left(\mathbf{w} + \frac{\eta\mathbf{v}}{\varrho}\right). \tag{3.31}$$

The formulas for E_1 and $\mathbf{I}_1$ can also be expressed in the form of the right-hand sides of equations (3.30) and (3.31) if we multiply them by the factor $1/2$ and replace p, η, $\mathbf{w}$ with p_1, η_1, $\mathbf{w}_1$, respectively.

The laws of acoustic energy conservation (3.26) and (3.28) have been derived previously [51, 157] using different approaches from the one described above. In these papers, equations (3.26) and (3.28) were expressed in the form

$$(\partial/\partial t)(E_0/\sigma) + \nabla \cdot (E_0\mathbf{u}/\sigma) = 0. \tag{3.32}$$

Here, E_0 is the mean value of the acoustic energy density $p_1^2/(2\varrho c^2)$ or the instantaneous value of the acoustic energy density $p^2/(\varrho c^2)$ in the coordinate system moving with the medium velocity $\mathbf{v}$. In equation (3.32), the quantity E_0/σ is analogous to the adiabatic invariant in mechanics and is sometimes called the wave action.

3.2 Eikonal equation and acoustic energy conservation in a time-independent medium

In the previous section, the eikonal equation (3.20), dispersion equation (3.21), and laws of acoustic energy conservation (3.26) and (3.28) have been derived for sound propagation in a time-dependent inhomogeneous moving medium. In the remainder of this chapter, we shall mainly consider the case in which the moving medium is time independent, i.e., c, $\mathbf{v}$, and ϱ do not depend on time t. For a time-independent moving medium, it is worthwhile considering equations (3.20), (3.21), (3.26), and (3.28) in detail.

3.2.1 Eikonal equation

In a time-independent moving medium, the solution of the eikonal equation (3.20) can be sought in the form $\Theta(\mathbf{R}, t) = -\omega t/k_0 + \Psi(\mathbf{R})$, where Ψ is the eikonal of a sound wave depending only on $\mathbf{R}$, and ω is its frequency which is independent of $\mathbf{R}$ and t. In this case, the sound pressure p is related to its amplitude p_1 by

$$p(\mathbf{R}, t) = p_1(\mathbf{R}) \exp\left[-i\omega t + ik_0 \Psi(\mathbf{R})\right]. \tag{3.33}$$

Substituting $\Theta = -\omega t/k_0 + \Psi$ into equation (3.20) and introducing the reference sound speed $c_0 = \omega/k_0$, we obtain the equation for the eikonal $\Psi(\mathbf{R})$

$$\frac{c_0}{c} - \frac{\mathbf{v} \cdot \nabla \Psi}{c} = |\nabla \Psi|. \tag{3.34}$$

This equation was derived by Blokhintzev in 1944 [37] and has since been used widely. This equation also describes the propagation of the sound wavefront in an inhomogeneous moving medium [71, 160, 209].

3.2.2 Dispersion equation

In a time-independent moving medium, the wave vector $\mathbf{k}$ depends only on $\mathbf{R}$, so that the dispersion equation (3.21) takes the form

$$\omega = k(\mathbf{R})c(\mathbf{R}) + \mathbf{k}(\mathbf{R}) \cdot \mathbf{v}(\mathbf{R}). \tag{3.35}$$

It is worthwhile to derive this important equation by using a different, qualitative approach. Let us consider a homogeneous uniformly moving medium. In this case, in the coordinate system moving with the medium velocity $\mathbf{v}$, the dispersion equation is given $\omega = kc$. In a fixed coordinate system, this dispersion equation takes the well-known form [217] $\omega = kc + \mathbf{k} \cdot \mathbf{v}$, where k, c, and $\mathbf{v}$ are constant. In an inhomogeneous moving medium in which the scales of the inhomogeneities are greater than λ, let us consider a small spatial domain inside which the medium can be considered as homogeneous and moving uniformly. In this spatial domain, the dispersion equation is given by $\omega = kc + \mathbf{k} \cdot \mathbf{v}$. If we consider another small spatial domain located at some distance from the first one, the form of the dispersion equation will remain the same but the values of k, c, and $\mathbf{v}$ will be different because they depend on $\mathbf{R}$. Therefore, in an inhomogeneous moving medium in which the scales of the inhomogeneities are greater than λ, the dispersion equation is given by equation (3.35).

In a time-independent moving medium, a solution of the linearized equations of fluid dynamics (2.8)–(2.12) can depend on time as $\exp(-i\omega t)$. Therefore, the frequency of a monochromatic sound wave remains constant in an inhomogeneous moving medium regardless whether geometrical acoustics is valid. In particular, the frequency of the sound wave is not changed due to sound scattering by inhomogeneities in the medium. One would think that the last result contradicts well-known facts, for instance, the frequency change resulting from sound scattering in a turbulent atmosphere. This contradiction is easily resolved if we take into account the fact that in a turbulent atmosphere, this frequency change is due to the dependence of c and $\mathbf{v}$ on time t.

3.2.3   Acoustic energy conservation

In a time-independent moving medium, the laws of acoustic energy conservation are still given by equations (3.26)–(3.28) if we set $\omega_0/\omega = 1$ in the equations for E_1, $\mathbf{I}_1$, E, and $\mathbf{I}$. For example, the formula for the mean acoustic energy flux, which will be needed below, is given by

$$\mathbf{I}_1 = \frac{p_1^2}{2\varrho c^3}(c + \mathbf{n} \cdot \mathbf{v})(c\mathbf{n} + \mathbf{v}). \tag{3.36}$$

The law of mean acoustic energy conservation in the geometrical acoustics approximation was first derived by Blokhintzev [37] for the case of a time-independent moving medium.

In an isentropic ($S = $ constant), irrotational ($\nabla \times \mathbf{v}$), time-independent moving medium, the law of acoustic energy conservation can be formulated without the assumption of the geometrical acoustics approximation. To derive this law, let us multiply both sides of equation (2.9) by $p/\varrho + \mathbf{w} \cdot \mathbf{v}$ and both sides equation (2.8) by the vector $\mathbf{w} + \eta \mathbf{v}/\varrho$. Adding the resulting equations and considering the spatial domain free of acoustic sources, we have

$$\partial \Xi / \partial t + \nabla \cdot \mathbf{J} = (1/\varrho)(\eta \nabla P - p \nabla \varrho) \cdot (\mathbf{w} + \eta \mathbf{v}/\varrho)$$
$$+ (1/2\varrho)(\eta \partial p / \partial t - p \partial \eta / \partial t) + \mathbf{w} \cdot [\mathbf{v} \times (\varrho \nabla \times \mathbf{w} - \eta \nabla \times \mathbf{v})], \quad (3.37)$$

where

$$\Xi = p\eta/2\varrho + \varrho w^2/2 + \eta \mathbf{w} \cdot \mathbf{v}, \tag{3.38}$$

$$\mathbf{J} = (p + \varrho \mathbf{w} \cdot \mathbf{v})(\mathbf{w} + \eta \mathbf{v}/\varrho). \tag{3.39}$$

If the medium consists of only one component ($C_i = 0$) and $S = $ constant, it follows from equations (2.11) and (2.12) and the equation of state $P = P(\varrho, S)$ that $p = c^2\eta$ and $\nabla P = c^2\nabla \varrho$. Using the latter two formulas, it can be shown that the first two terms on the right-hand side of equation (3.37) vanish. Then, using these formulas and the identity

$$\nabla(\mathbf{v} \cdot \mathbf{w}) = (\mathbf{v} \cdot \nabla)\mathbf{w} + (\mathbf{w} \cdot \nabla)\mathbf{v} + \mathbf{v} \times (\nabla \times \mathbf{w}) + \mathbf{w} \times (\nabla \times \mathbf{v}),$$

equation (2.8) can be expressed in the form

$$\partial \mathbf{w}/\partial t + \nabla(\mathbf{v} \cdot \mathbf{w}) - \mathbf{v} \times (\nabla \times \mathbf{w}) - \mathbf{w} \times (\nabla \times \mathbf{v}) + \nabla(p/\varrho) = 0. \tag{3.40}$$

If $\nabla \times \mathbf{v} = 0$, the solution of equation (3.40) can be sought in the form $\mathbf{w} = -\nabla\varphi$, where φ is the velocity potential. In this case, $\nabla \times \mathbf{w} = 0$ and the last term on the right-hand side of equation (3.37) also vanishes.

Thus, in an isentropic, irrotational, time-independent moving medium, equation (3.37) can be written as the law of acoustic energy conservation

$$\partial \Xi / \partial t + \nabla \cdot \mathbf{J} = 0. \tag{3.41}$$

Here, Ξ is the instantaneous value of the acoustic energy density and $\mathbf{J}$ is its flux. The quantities Ξ and $\mathbf{J}$ are given by equations (3.38) and (3.39), which are similar, but not identical to equations (3.30) and (3.31) for E and $\mathbf{I}$. These equations do not completely coincide even in the case of a time-independent medium where $\omega = \omega_0$. Indeed, since $\mathbf{w} \neq p\mathbf{n}/\varrho c$ in an isentropic, irrational moving medium, the acoustic energy density Ξ cannot be expressed in the form $\Xi = (p^2/\varrho c^2)(1 + \mathbf{n} \cdot \mathbf{v}/c)$.

Derivations of the law of acoustic energy conservation (3.41) and the analogous law for the mean values of Ξ and $\mathbf{J}$ have been given elsewhere [62, 263]. The assumptions $S = $ constant and $\nabla \times \mathbf{v} = 0$, which were used to derive equation (3.41), impose strong limitations on possible models of a moving medium (see sections 2.3 and 2.4). For an arbitrary moving medium, these assumptions are valid only in the geometrical acoustics approximation. In this approximation, the Andreev–Rusakov–Blokhintzev equation, which was derived using these assumptions (section 2.4), is in agreement with equation (2.100) for high-frequency sound waves.

The conservation laws in this and previous sections enable calculation of the acoustic energy and its flux, and hence of the sound pressure, in a moving medium. Usually, the sound pressure is the quantity that is measured experimentally. This approach for calculating E, $\mathbf{I}$, and p is used when considering sound propagation in ducts, nozzles, and diffusers (e.g., references [61, 256, 257]).

3.3   Sound propagation in a three-dimensional inhomogeneous medium

In this section, we consider sound propagation in a three-dimensional moving medium using the eikonal and dispersion equations and the law of acoustic energy conservation derived in sections 3.1 and 3.2.

3.3.1   Phase and group velocities

It is well known that the phase velocity of a wave (i.e., the velocity of the wavefront) is given by $\mathbf{U} = \mathbf{k}\omega/k^2$, while the group velocity (the velocity of energy propagation) is given by $\mathbf{u} = \partial\omega/\partial\mathbf{k}$. Substituting ω given by equation (3.35) into these formulas, we obtain the phase and group velocities of a sound wave in a moving medium:

$$\mathbf{U} = (c + \mathbf{n}\cdot\mathbf{v})\mathbf{n}, \tag{3.42}$$

and

$$\mathbf{u} = c\mathbf{n} + \mathbf{v}. \tag{3.43}$$

These equations are also valid in a time-dependent moving medium. The group velocity $\mathbf{u}$ obtained in this section using a kinematic approach coincides with the velocity of acoustic energy propagation appearing in the transport equation (3.24) and the law of energy conservation (3.26). It follows from equations (3.42) and (3.43) that the phase velocity equals to the projection of the group velocity in the direction of the normal to the wavefront (also see figure 3.1). Therefore, in a moving medium $U \leq u$ and, generally, the directions of the vectors $\mathbf{U}$ and $\mathbf{u}$ do not coincide.

It follows from equation (3.43) that the velocity of sound wave propagation in the direction of medium motion is greater than that in the opposite direction. This result was obtained by Derham in 1708 and published in reference [96], which probably was the first paper on acoustics in a moving medium. Derham measured the propagation time of sounds, generated by gunshots, from neighboring churches to the bell tower in Upminster. As a result of such measurements, it was revealed that the velocity of a sound wave downwind is

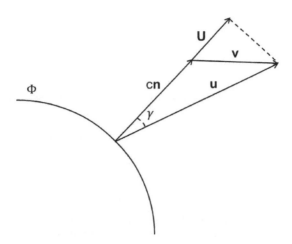

FIGURE 3.1
Phase velocity **U** and group velocity **u** of a sound wave in a moving medium. Φ is the wavefront.

greater than it is upwind. However, it was not feasible to measure the wind velocity in those experiments because an anemometer had not yet been invented. Therefore, Derham could not determine that the velocities of sound waves downwind and upwind were $c + v$ and $c - v$, respectively. These formulas were obtained experimentally by Cassini [63] in 1738. In a stratified moving medium, formulas for the phase velocity **U** and the group velocity **u** of a sound wave were proposed by Rayleigh [324] and Barton [23], respectively. Blokhintzev [37] was the first to show rigorously that in the approximation of geometrical acoustics, **U** and **u** are given by equation (3.42) and (3.43).

The unit vector tangential to the sound ray path is given by $\mathbf{s} = \mathbf{u}/u$. Substituting with equation (3.24), the vector **s** can be expressed in terms of the unit vector **n** normal to the wavefront:

$$\mathbf{s} = \frac{\mathbf{n} + \mathbf{v}/c}{|\mathbf{n} + \mathbf{v}/c|} = \frac{\mathbf{n} + \mathbf{v}/c}{(1 + 2\mathbf{n} \cdot \mathbf{v}/c + v^2/c^2)^{1/2}}. \tag{3.44}$$

It follows from this equation that the vectors **s** and **n** do not coincide in a moving medium. The vector **n** appears in many formulas pertinent to geometrical acoustics. In certain cases, it is worthwhile expressing these formulas in terms of **s**. This can be done if, using equation (3.44), **n** is expressed in terms of **s**. To this end, let us obtain the term $\mathbf{n} \cdot \mathbf{v}/c$ appearing in equation (3.44). Taking the scalar product of both sides of equation (3.44) with the vector $(\mathbf{v}/c)(1 + 2\mathbf{n} \cdot \mathbf{v}/c + v^2/c^2)^{1/2}$ and squaring both sides of the resulting equation yields a quadratic equation for $\mathbf{n} \cdot \mathbf{v}/c$. Solving this equation and substituting the value of $\mathbf{n} \cdot \mathbf{v}/c$ into equation (3.44), we obtain the desired equation for the vector $\mathbf{n}$:

$$\mathbf{n} = \mathbf{s}\left[(1 + (\mathbf{s}\cdot\mathbf{v}/c)^2 - v^2/c^2))^{1/2} + \mathbf{s}\cdot\mathbf{v}/c\right] - \mathbf{v}/c. \tag{3.45}$$

3.3.2   Ray path, eikonal, and travel time of sound propagation

One of the main goals of geometrical acoustics is to obtain the ray path along which the energy of a sound wave is propagated. Since $\mathbf{u}$ is the velocity of acoustic energy propagation, the unit vector $\mathbf{s} = \mathbf{u}/u$ is tangential to the ray path. The ray path can be obtained [73, 384] by making use of the Hamiltonian formalism.

Equation (3.34) is a nonlinear first-order partial differential equation (i.e., the Hamilton-Jacobi equation). This equation can be expressed in the form $H(\mathbf{b}, \mathbf{R}) = 0$, where $H(\mathbf{b}, \mathbf{R})$ is the Hamiltonian and the vector $\mathbf{b} = \nabla\Psi$. (Note that the vectors $\mathbf{b}$ and $\mathbf{k}$ are related by $\mathbf{b} = \mathbf{k}/k_0$.) The solution of the equation $H(\mathbf{b}, \mathbf{R}) = 0$ can be reduced to a set of the ordinary differential equations

$$\frac{d\mathbf{R}}{d\tau} = \frac{\partial H}{\partial \mathbf{b}}, \tag{3.46}$$

$$\frac{d\mathbf{b}}{d\tau} = -\frac{\partial H}{\partial \mathbf{R}}, \tag{3.47}$$

$$\frac{d\Psi}{d\tau} = \mathbf{b}\cdot\frac{\partial H}{\partial \mathbf{b}}, \tag{3.48}$$

where τ is the independent variable. There is some freedom in choosing the Hamiltonian H. The form of the sound ray equations depends on this choice. (These equations are, of course, equivalent because they determine the same ray path.) Following [290], we choose $H = b - c_0/c + \mathbf{b}\cdot\mathbf{v}/c$. In this case, a set of the differential equations (3.46)–(3.48) takes the form

$$\frac{d\mathbf{R}}{d\tau} = \frac{\mathbf{b}}{b} + \frac{\mathbf{v}}{c}, \tag{3.49}$$

$$\frac{d\mathbf{b}}{d\tau} = -\frac{1}{c}\left[b\nabla c + \mathbf{b}\times(\nabla\times\mathbf{v}) + (\mathbf{b}\cdot\nabla)\mathbf{v}\right], \tag{3.50}$$

$$\frac{d\Psi}{d\tau} = b + \frac{\mathbf{b}\cdot\mathbf{v}}{c}. \tag{3.51}$$

Derivation of equations for $\mathbf{R}$ and Ψ is straightforward. To obtain equation (3.50), we notice that when calculating the partial derivative $\partial H/\partial\mathbf{R}$, the vector $\mathbf{b}$ should not be differentiated:

$$\frac{d\mathbf{b}}{d\tau} = \nabla\left(\frac{c_0}{c}\right) - \nabla\left(\frac{\mathbf{b}\cdot\mathbf{v}}{c}\right) = -\frac{1}{c}\left[\left(\frac{c_0}{c} - \frac{\mathbf{b}\cdot\mathbf{v}}{c}\right)\nabla c + \nabla(\mathbf{b}\cdot\mathbf{v})\right]. \tag{3.52}$$

In this equation, $c_0/c - (\mathbf{b}\cdot\mathbf{v})/c = b$, which follows from equation (3.34) and the definition of **b**. Using a formula for the gradient of a dot product, equation (3.52) reduces to equation (3.50).

Since $\mathbf{b}/b = \mathbf{n}$, the right-hand side of equation (3.49) is equal to $\mathbf{u}/c$. Therefore, the directions of the vectors $d\mathbf{R}/d\tau$ and $\mathbf{u}$ coincide and, hence, the ray path in a moving medium is given by the vector $\mathbf{R}(\tau)$. To obtain the ray path, one should solve the set of equations (3.49) and (3.50). Solution is possible in analytical form for some particular cases, for example, in a stratified medium (see section 3.5). Equations (3.49) and (3.50) should be supplemented by the initial conditions for the vectors $\mathbf{R}$ and $\mathbf{b}$. As a result, the ray path can be expressed in the form $\mathbf{R} = \mathbf{R}(\zeta, \nu, \tau)$, where ζ and ν are the ray parameters which determine a certain ray. The ray parameters ζ and ν remain constant along the ray path while τ varies. For a point source, ζ and ν can be chosen, for instance, as the elevation and azimuthal angles of the ray near the source.

The independent variable τ can be related to the path length l. From equation (3.46), it follows that $(dl)^2 = (d\mathbf{R})^2 = (\partial H/\partial \mathbf{b})^2 (d\tau)^2$. The latter equation enables one to express $d\tau$ in terms of dl:

$$d\tau = \frac{dl}{|\partial H/\partial \mathbf{b}|} = \frac{c\,dl}{u}. \tag{3.53}$$

The travel time dt of sound propagation along an increment dl of the ray path is given by $dt = dl/u$. This allows us to express $d\tau$ in terms of dt:

$$d\tau = c\,dt. \tag{3.54}$$

If the time t is used instead of the independent variable τ, the ray equations (3.49) and (3.50) take the form:

$$\frac{d\mathbf{R}}{dt} = \frac{\mathbf{b}c}{b} + \mathbf{v}, \tag{3.55}$$

$$\frac{d\mathbf{b}}{dt} = -b\nabla c - \mathbf{b} \times (\nabla \times \mathbf{v}) - (\mathbf{b}{\cdot}\nabla)\,\mathbf{v}. \tag{3.56}$$

These equations can be reduced to equations (8-1.10a) and (8-1.10b) in reference [313]. For the case of sound propagation in the atmosphere and ocean, $|c - c_0| \ll c$ and $v/c \ll 1$. In this case, b is of order 1 so that equation (3.55) does not have singularities.

If the ray path has been obtained, the eikonal $\Psi(\mathbf{R})$ can be calculated using equation (3.51). Changing from τ to l, integrating the resulting equation, and taking into account the fact that $\mathbf{b} = \mathbf{k}/k_0$ and $k + \mathbf{k} \cdot \mathbf{v}/c = k_0 c_0/c$, we obtain the following equation for the eikonal of a sound wave propagating in an inhomogeneous moving medium:

$$\Psi(l) = \Psi_1 + \int_{l_1}^{l} \frac{c_0}{u}\,dl'. \tag{3.57}$$

Here, the integration is taken along the ray path, l_1 and l are the initial and final points of the path, and Ψ_1 is the value of the eikonal at $l = l_1$.

Integrating the equation $dt = dl/u$, we obtain the travel time of sound propagation along the ray path

$$t_{tr} = \int_{l_1}^{l} \frac{1}{u} \, dl' = \frac{\Psi - \Psi_1}{c_0}. \tag{3.58}$$

The latter equality in this formula is obtained with equation (3.57).

In the acoustics of a moving medium, the principle of least action can be expressed in the form [217]

$$\delta \int_{l_1}^{l} \mathbf{k} \cdot d\mathbf{l}' = 0,$$

where $d\mathbf{l}' = \mathbf{s} \, dl'$ and δ is the variation of the integral. Substituting the values of the vectors $\mathbf{k} = k\mathbf{n}$ and $\mathbf{s}$ (see equation (3.44)) into this equation yields

$$\delta \int_{l_1}^{l} \frac{\omega}{u} \, dl' = \omega \delta t_{tr} = 0.$$

This equation represents Fermat's principle.

3.3.3 Refractive index

The refractive index $N_c = c_0/c$ is widely used in the acoustics of a motionless medium. In a moving medium, the refractive index N could be defined as a function depending on $\mathbf{c}$ and $\mathbf{v}$, the substitution of which for N_c in all the equations of geometrical acoustics in a motionless medium yields the analogous equations for the case of a moving medium. However it is only possible to introduce such a function N for some equations of geometrical acoustics. Moreover, N can differ for different equations and, furthermore, it depends on the direction of wave propagation.

In the literature (e.g., references [37, 73]) N is sometimes defined as

$$N_{ph} = \frac{c_0}{U} = \frac{c_0}{c + \mathbf{n} \cdot \mathbf{v}}. \tag{3.59}$$

Here, the subscript ph indicates that N_{ph} is pertinent to the phase of a sound wave. This definition of N_{ph} enables to express some of the equations for the wavefront in a moving medium in the form that they have for a motionless medium. For example, it follows from equations (3.42) and (3.59) that the phase velocity of a sound wave in a moving medium is given by $U = c_0/N_{ph}$, i.e., it is given by the well-known equation for a motionless medium. Moreover, taking into account the fact that $(c_0 - \mathbf{v} \cdot \nabla \Psi)/c = (\omega - \mathbf{k} \cdot \mathbf{v})/ck_0$ and making use of the identity

$$1 - \frac{\mathbf{k} \cdot \mathbf{v}}{\omega} = \frac{1}{(1 + \mathbf{n} \cdot \mathbf{v}/c)}, \tag{3.60}$$

which can be proved using equation (3.35), the eikonal equation (3.34) can be expressed in the form $|\nabla \Psi| = N_{ph}$. It follows from this formula that $N_{ph} = k/k_0$. In a motionless medium, N_{ph} is given by the same equation.

However, when considering a sound ray, the refractive index in a moving medium should be defined by the formula

$$N_r = \frac{c_0}{u} = \frac{c_0}{(c^2 + 2c\mathbf{n} \cdot \mathbf{v} + v^2)^{1/2}}, \tag{3.61}$$

which differs from equation (3.59). From equations (3.43), (3.57), and (3.61), it follows that the expressions for the group velocity $u = c_0/N_r$, the phase increment

$$\Phi = k_0(\Psi - \Psi_1) = k_0 \int_{l_1}^{l} N_r \, dl',$$

and the travel time of sound propagation along the ray path

$$t_{tr} = \int_{l_1}^{l} \frac{N_r}{c_0} \, dl'$$

have the same form as those for a motionless medium.

The functions N_{ph} and N_r coincide to order v/c. Since the refractive index cannot be rigorously introduced in a moving medium, N_{ph} and N_r are used in the literature relatively seldom.

3.3.4 Sound-pressure amplitude

The sound-pressure amplitude p_1 can be calculated using the law of acoustic energy conservation (3.26). If p_1 does not depend on t, it follows from equation (3.26) that $\nabla \cdot (E_1 \mathbf{u}) = \nabla \cdot (I_1 \mathbf{s}) = 0$, where I_1 is given by equation (3.36). The solution of this equation can be expressed in the form $I_1 d\sigma = $ constant, where $d\sigma$ is the cross-sectional area of an infinitesimally narrow ray tube. In this formula, substituting I_1 with equation (3.36), we have

$$\frac{p_1^2(1 + \mathbf{n} \cdot \mathbf{v}/c)(1 + 2\mathbf{n} \cdot \mathbf{v}/c + v^2/c^2)^{1/2} \, d\sigma}{2\varrho c} = \text{constant.} \tag{3.62}$$

This equation has been derived by Blokhintzev [37]. The vector $\mathbf{n}$ and the cross-sectional area $d\sigma$, which appear in equation (3.62), can be obtained if the ray path $\mathbf{R} = \mathbf{R}(\zeta, \nu, \tau)$ has been calculated, where ζ and ν are the ray parameters (section 3.3.2). Indeed, using equation (3.45), the vector $\mathbf{n}$ can be expressed in terms of the vector $\mathbf{s} = (\partial \mathbf{R}/\partial \tau)/(|\partial \mathbf{R}/\partial \tau|)$. To calculate $d\sigma$ note that the infinitesimally narrow ray tube cuts out a parallelogram in the wavefront with sides $(\partial \mathbf{R}/\partial \zeta) \, d\zeta$ and $(\partial \mathbf{R}/\partial \nu) \, d\nu$. The area of the parallelogram is given by

$$dA = \left| \frac{\partial \mathbf{R}}{\partial \zeta} \times \frac{\partial \mathbf{R}}{\partial \nu} \right| d\zeta d\nu. \tag{3.63}$$

The cross-sectional area of the ray tube is given by $d\sigma = \mathbf{n} \cdot \mathbf{s}\, dA$. Here, we take into account the fact that the normal to the wavefront $\mathbf{n}$ does not coincide with the unit vector $\mathbf{s}$ tangential to the ray path.

Thus, equation (3.62) allows us to calculate the sound-pressure amplitude p_1 along the ray if we know the ray path $\mathbf{R} = \mathbf{R}(\zeta, \nu, \tau)$ and the value of p_1 at some point (this is necessary for calculating the value of the constant appearing on the right-hand side of this equation). This approach for calculating the sound-pressure amplitude is used in the literature [215, 354, 355].

3.4   Refraction laws for a sound ray and the normal to a wavefront in a stratified medium

The ray theory of sound propagation in a stratified moving medium (e.g., the atmosphere and ocean) is an important particular case of the theory considered in the previous section. We shall consider this theory in the remainder of this chapter. In a stratified moving medium, the ambient quantities c, ϱ, and $\mathbf{v}$ are functions only of the vertical coordinate z. Therefore, the quantities considered in the ray theory also depend only on z and do not depend on the horizontal coordinates $\mathbf{r} = (x, y)$. In this case, it is worthwhile to express the medium velocity as $\mathbf{v} = (v_z, \mathbf{v}_\perp)$, where v_z is the vertical component and $\mathbf{v}_\perp = (v_x, v_y)$ are the horizontal components.

For a stratified motionless medium, the refraction law for a sound ray (Snell's law) is given by $N_c \cos\alpha = \text{constant}$, where α is the elevation angle of a ray. Snell's law plays an important role in the acoustics of a motionless medium because this law enables calculation of the ray path. In this section, we consider a generalization of this law to the case of a stratified moving medium.

3.4.1   Vertical wavenumber

We first obtain an explicit formula for the vertical wavenumber. In a stratified moving medium, the wave vector $\mathbf{k} = k_0 \mathbf{b}$ can be expressed in terms of the vertical profiles $c(z)$ and $\mathbf{v}(z)$. Let $\mathbf{k} = (\boldsymbol{\kappa}, q)$ and $\mathbf{b} = (\mathbf{b}_\perp, b_z)$, where $\boldsymbol{\kappa}$ and $\mathbf{b}_\perp$ are the horizontal components of the vectors $\mathbf{k}$ and $\mathbf{b}$, respectively, and q and b_z are the vertical components of these vectors. Then, $\boldsymbol{\kappa}/k_0 = \mathbf{b}_\perp$. Applying the operator $d/d\tau$ to both sides of this formula, using equation (3.47), and recalling that the Hamiltonian $H = b - c_0/c + \mathbf{b} \cdot \mathbf{v}/c$, we obtain

$$\frac{1}{k_0}\frac{d\boldsymbol{\kappa}}{d\tau} = \frac{d\mathbf{b}_\perp}{d\tau} = -\frac{\partial H}{\partial \mathbf{r}} = \frac{\partial}{\partial \mathbf{r}}\left[\frac{c_0 - \mathbf{b} \cdot \mathbf{v}(z)}{c(z)}\right] = 0. \tag{3.64}$$

It follows from this expression that the vector $\boldsymbol{\kappa}$ remains constant along a ray path.

Substituting $k = (\kappa^2 + q^2)^{1/2}$ and $\mathbf{k} \cdot \mathbf{v} = q v_z + \boldsymbol{\kappa} \cdot \mathbf{v}_\perp$ into equation (3.35) and solving this equation for q, we obtain the explicit formula for the vertical wavenumber

$$q(z) = \frac{c\chi - (\omega - \boldsymbol{\kappa} \cdot \mathbf{v}_\perp)v_z}{c^2 - v_z^2}, \tag{3.65}$$

where

$$\chi = \pm \left[(\omega - \boldsymbol{\kappa} \cdot \mathbf{v}_\perp)^2 - (c^2 - v_z^2) \, \kappa^2 \right]^{1/2}. \tag{3.66}$$

In this formula, the upper and lower signs correspond to the ray propagating upwards (in the positive direction of the z-axis) and downwards, respectively.

3.4.2 Refraction law for the normal to a wavefront

Now we formulate the refraction law for the normal $\mathbf{n}(z)$ to a wavefront. It is worthwhile expressing $\mathbf{n}$ in the form $\mathbf{n} = (\mathbf{e}\cos\theta, \sin\theta)$. Here, θ is the elevation angle between $\mathbf{n}$ and the horizontal plane, and $\mathbf{e}$ is the unit vector in the direction of the projection of $\mathbf{n}$ on the horizontal plane (figure 3.2). The vector $\mathbf{n}$ can also be expressed as $\mathbf{n} = \mathbf{k}/k = (\boldsymbol{\kappa}, q)/(\kappa^2 + q^2)^{1/2}$. Eliminating $\mathbf{n}$ between these equations, solving the resulting equation for $\mathbf{e}$ and θ, and replacing q by the right-hand side of equation (3.65), we obtain

$$\mathbf{e} = \frac{\boldsymbol{\kappa}}{\kappa}, \tag{3.67}$$

and

$$\cos\theta(z) = \frac{\kappa(c^2 - v_z^2)}{c(\omega - \boldsymbol{\kappa} \cdot \mathbf{v}_\perp) - v_z \chi}. \tag{3.68}$$

These equations define $\mathbf{e}$, θ, and, hence, $\mathbf{n}$, along the path of a sound wave. Therefore, equations (3.67) and (3.68) can be considered as the refraction law for the normal to the wavefront in a stratified moving medium. It follows from equation (3.67) that the unit vector $\mathbf{e}$ remains constant along the path, i.e., the direction of the projection of $\mathbf{n}$ on the horizontal plane does not change. Equation (3.68) can be expressed in the form

$$\frac{c}{\cos\theta} + \mathbf{e} \cdot \mathbf{v}_\perp + v_z \tan\theta = \frac{\omega}{\kappa}. \tag{3.69}$$

The refraction law for $\theta(z)$ is usually formulated in such a form. If $v_z = 0$, equation (3.69) can be expressed as $N_{ph}\cos\theta = $ constant, i.e., in the same form as Snell's law. For $v_z = 0$, equation (3.69) was first obtained by Rayleigh [324].

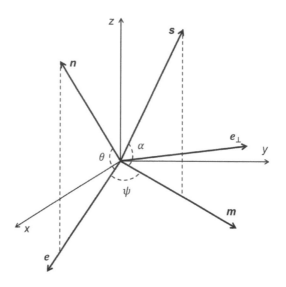

FIGURE 3.2
Unit vector **n** normal to the wavefront and unit vector **s** tangential to the ray path.

3.4.3   Refraction law for a sound ray

Now we proceed to the derivation of the refraction law for a sound ray. Similarly to the form used for **n**, the unit vector **s** is expressed as: $\mathbf{s} = (\mathbf{m}\cos\alpha, \sin\alpha)$. Here, α is the elevation angle between **s** and the horizontal plane and **m** is the unit vector in the direction of the projection of **s** on the horizontal plane (figure 3.2). It is worthwhile to introduce the azimuthal angle ψ between the vectors **e** and **m**. The angle ψ is positive if the shortest rotation required to make the vector **e** coincide with the vector **m** is anticlockwise, and $-\pi < \psi \leq \pi$. The vector **s** can also be expressed as:

$$\mathbf{s} = \frac{c\mathbf{n}+\mathbf{v}}{|c\mathbf{n}+\mathbf{v}|} = \frac{c\mathbf{k}+k\mathbf{v}}{|c\mathbf{k}+k\mathbf{v}|} = \frac{(c\boldsymbol{\kappa}+k\mathbf{v}_\perp,\, cq+kv_z)}{(c^2k^2+2ck\mathbf{k}\cdot\mathbf{v}+k^2v^2)^{1/2}}. \qquad (3.70)$$

From this formula and the equality $\mathbf{s} = (\mathbf{m}\cos\alpha, \sin\alpha)$, it follows that $\sin\alpha = (cq+kv_z)/|c\mathbf{k}+k\mathbf{v}|$ and $\mathbf{m} = (c\boldsymbol{\kappa}+k\mathbf{v}_\perp)/|c\boldsymbol{\kappa}+k\mathbf{v}_\perp|$.

Let $\mathbf{e}_\perp$ be the unit azimuthal vector, which is perpendicular to the vector **e** (figure 3.2). In this case, $\sin\psi = \mathbf{e}_\perp\cdot\mathbf{m}$ and $\cos\psi = \mathbf{e}\cdot\mathbf{m}$. Substituting the values of q and k into the equations for $\sin\alpha$,

$$\sin\alpha$$
$$= \frac{(c^2-v_z^2)\chi}{c\left\{\left[\omega - \boldsymbol{\kappa}\cdot\mathbf{v}_\perp - \frac{v_z\chi}{c}\right]\left[(c^2+v_\perp^2-v_z^2)\omega + (c^2-v^2)(\boldsymbol{\kappa}\cdot\mathbf{v}_\perp + \frac{v_z\chi}{c})\right]\right\}^{1/2}}, \qquad (3.71)$$

$$\cos \psi = \frac{1}{[1 + (\mathbf{e}_\perp \cdot \mathbf{v}_\perp)^2/(c \cos \theta + \mathbf{e} \cdot \mathbf{v}_\perp)^2]^{1/2}}, \tag{3.72}$$

$$\sin \psi = \frac{\mathbf{e}_\perp \cdot \mathbf{v}_\perp}{(c^2 \cos^2 \theta + 2c \cos \theta \mathbf{e} \cdot \mathbf{v}_\perp + v_\perp^2)^{1/2}}. \tag{3.73}$$

These equations determine the angles $\alpha(z)$ and $\psi(z)$ in terms of the vertical profiles $c(z)$ and $\mathbf{v}(z)$ and, hence, can be considered as the refraction law for a sound ray in a stratified moving medium. If $\mathbf{v} = 0$, these equations become Snell's law: $c/\cos \alpha = $ constant and $\psi = 0$. If $\mathbf{v} \neq 0$, equations (3.71)–(3.73) generalize Snell's law to the case of a moving medium. For $v_z = 0$, equation (3.71) can be expressed in the form

$$\frac{c}{\{\cos^2 \alpha + [(\mathbf{e} \cdot \mathbf{v}_\perp/c)^2 \sin^2 \alpha - v_\perp^2/c^2] \sin^2 \alpha\}^{1/2} - (\mathbf{e} \cdot \mathbf{v}_\perp/c) \sin^2 \alpha}$$
$$+ \mathbf{e} \cdot \mathbf{v}_\perp = \frac{\omega}{\kappa}. \tag{3.74}$$

Let us consider the case in which $v_z = 0$, the direction of the vector $\mathbf{v}_\perp$ does not depend on z, and the azimuthal direction of the sound ray is collinear with the vector $\mathbf{v}_\perp$. For this particular case, the ray path is located in a vertical plane and, hence, is a plane curve; the angle $\psi = 0$; and the angle $\alpha(z)$ satisfies equation (3.74) with $\mathbf{e} \cdot \mathbf{v}_\perp = \pm v_\perp$ (where the plus sign corresponds to sound propagation in the direction of the vector $\mathbf{v}_\perp$, and the minus sign corresponds to propagation in the opposite direction). For this particular case, the refraction law for the angle $\alpha(z)$ was first formulated by Barton [23] in 1901, who derived the following equation

$$\cot \alpha = \cot \theta + \frac{v}{c \sin \theta}. \tag{3.75}$$

If we substitute the value of $\cos \theta = \kappa c/(\omega - \boldsymbol{\kappa} \cdot \mathbf{v}_\perp)$ into this equation, it becomes equation (3.74) where $\mathbf{e} \cdot \mathbf{v}_\perp$ is equal to $\pm v_\perp$.

Equations (3.72) and (3.73) enable determination of the azimuthal angle ψ along the ray path. If $v/c \sim 1$, the angle ψ can be greater than π. If $v_\perp/c \ll 1$ and $\cos \theta \gg v_\perp/c$ (i.e., the angle θ is not too close to $\pi/2$), it can be shown from equations (3.72) and (3.73) that $\psi \approx \mathbf{e}_\perp \cdot \mathbf{v}_\perp/(c \cdot \cos \theta) \ll 1$. From this equation, it follows that ψ is affected mainly by the term $\mathbf{e}_\perp \cdot \mathbf{v}_\perp$. If $v_\perp/c \ll 1$ and $\cos \theta \leq v_\perp/c$ (i.e., the angle θ is close to $\pi/2$ or $-\pi/2$) the angle ψ is not small and can be of order π. In this case, ψ can depend significantly not only on $\mathbf{e}_\perp \cdot \mathbf{v}_\perp$, but also on v_z, $\mathbf{e} \cdot \mathbf{v}_\perp$, and c.

In the literature, there have been many incorrect formulations of the refraction law for a sound ray. For example, in reference [154], the refraction law for the normal to the wavefront (i.e., equation (3.69) where it is assumed that $v_z = 0$) is used incorrectly as the refraction law for a sound ray. In other publications [84, 327], equation (3.69) with $v_z = 0$ is replaced approximately by the equation $c + \mathbf{e} \cdot \mathbf{v}_\perp = (\omega/\kappa) \cos \theta$ which is also used incorrectly as the refraction law for a sound ray. An incorrect refraction law for a sound ray

$(c^2 - v_\perp^2 \sin^2 \alpha)^{1/2} / \cos \alpha + v_\perp = $ constant is formulated in reference [103]. (Compare this equation with equation (3.74) given $|\mathbf{e} \cdot \mathbf{v}_\perp| = v_\perp$.)

Reference [37] discriminates between the vectors $\mathbf{n}$ and $\mathbf{s}$. Nevertheless, in this reference, the ray path is determined approximately by the refraction law for the normal to the wavefront. This is motivated by the fact that the difference between the vectors $\mathbf{n}$ and $\mathbf{s}$ is small. Indeed, it can be shown from equation (3.45) that if $v/c \ll 1$, the angle γ between the vectors $\mathbf{n}$ and $\mathbf{s}$ is always small (of order v/c):

$$\gamma = \arccos(\mathbf{s} \cdot \mathbf{n}) \approx \frac{\left[v^2 - (\mathbf{s} \cdot \mathbf{v})^2\right]^{1/2}}{c}. \tag{3.76}$$

But in many cases, the refractive bending of the ray path can be of the order of or smaller than the angle γ. In such cases, the replacement of $\mathbf{s}$ by $\mathbf{n}$ is not valid.

Now we consider an example that reveals the clear difference between the vectors $\mathbf{n}$ and $\mathbf{s}$ in a moving medium. Let $c = $ constant, $v_z = 0$, and the direction of the vector $\mathbf{v}_\perp$ does not depend on z. Consider also the sound ray with a vector $\mathbf{e}$ perpendicular to the vector $\mathbf{v}_\perp$. (Recall that $\mathbf{e}$ characterizes the azimuthal direction of $\mathbf{n}$.) It follows from equation (3.69) that the angle θ is constant for the considered ray, and, hence, the vector $\mathbf{n}$ does not change along the ray path. On the other hand, from equations (3.71)–(3.73), it can be shown that the angles α and ψ depend on z, so that the sound ray path is a three-dimensional curve.

3.5   Sound propagation in a stratified medium

3.5.1   Ray path and eikonal

For a stratified medium, equations (3.49) and (3.50) for the ray path can be solved analytically. To do that, let $\mathbf{b} = \mathbf{k}/k_0 = (\kappa, q)/k_0$ and $\mathbf{R} = (\mathbf{r}, z)$. Using equation (3.47) (which is equivalent to equation (3.50)), we showed in section 3.4.1 that $\kappa = $ constant. Equation (3.49) can be written as

$$\frac{d\mathbf{r}}{d\tau} = \frac{\boldsymbol{\kappa}}{k} + \frac{\mathbf{v}_\perp}{c}, \tag{3.77}$$

$$\frac{dz}{d\tau} = \frac{q}{k} + \frac{v_z}{c}. \tag{3.78}$$

Eliminating $d\tau$ between these equations, integrating the resulting equation, and substituting the value of q yields

$$\mathbf{r} = \mathbf{r}_1 + \int_{z_1}^z \frac{c\boldsymbol{\kappa} + [(\omega - \boldsymbol{\kappa} \cdot \mathbf{v}_\perp)c - v_z \chi] \mathbf{v}_\perp / (c^2 - v_z^2)}{\chi} \, dz'. \tag{3.79}$$

This equation determines the path $\mathbf{r} = \mathbf{r}(z, \kappa/k_0)$ of a sound ray in a stratified moving medium. It follows from equation (3.79) that the ray path $\mathbf{r} = \mathbf{r}(z, \kappa/k_0)$ is generally a three-dimensional curve. It is located in a vertical plane only if the vectors κ and $\mathbf{v}_\perp(z)$ are collinear.

Dividing the numerator and denominator of the integrand in equation (3.79) by ω/c_0 reveals that $\mathbf{r}$ depends on the two-dimensional dimensionless vector $\kappa c_0/\omega = \kappa/k_0$. In this equation, the direction and magnitude of the vector κ/k_0 play the role of the ray parameters ζ and ν. Usually in the literature, either the elevation angle α and the azimuthal angle ψ of the vector $\mathbf{s}$ near a source, or the elevation angle θ and the azimuthal direction $\mathbf{e}$ of the normal $\mathbf{n}$ to the wavefront near a source, are used as the ray parameters. Let z_1 be the height of the source. Equations (3.71)–(3.73), (3.67), and (3.68) allow us to express the ray parameters $\alpha(z_1)$, $\psi(z_1)$, and $\theta(z_1)$, $\mathbf{e}$ in terms of the vector κ/k_0. The relationship between the different ray parameters becomes simplest if $\mathbf{v}(z_1) = 0$. In this case, $\mathbf{s} = \mathbf{n}$, $\alpha(z_1) = \theta(z_1)$, and the azimuthal direction of the vector $\mathbf{s}$ coincides with the direction of the vector κ/k_0. Furthermore, it follows from equation (3.69) that $\kappa/k_0 = c_0 \cos\alpha(z_1)/c(z_1)$. Setting $c_0 = c(z_1)$ in this equation, we obtain the relationship between the ray parameters

$$\kappa/k_0 = \cos\alpha(z_1). \tag{3.80}$$

In this book, the direction and magnitude of the vector κ/k_0 are usually used as the ray parameters. We choose these ray parameters because, first, the sound ray equation (3.79) takes the simplest form in this case. Second, these ray parameters are very convenient if the ray and wave theories are used simultaneously for studying sound propagation. In the wave theory of sound propagation in a stratified moving medium, the sound pressure p is expressed in the form of the Fourier integrals with respect to κ and $\omega = k_0 c_0$ (see equation (2.59) and Chapter 4).

In a stratified medium, the eikonal Ψ can be expressed analytically in terms of the profiles $c(z)$ and $\mathbf{v}(z)$. Indeed, since $\mathbf{k} = k_0 \mathbf{b}$, it follows from equation (3.51) that $d\Psi/d\tau = k/k_0 + \mathbf{k} \cdot \mathbf{v}/(k_0 c) = \omega/(k_0 c)$. Eliminating $d\tau$ between this equation and equation (3.78) and integrating the resulting equation yields the following formula for the eikonal:

$$\begin{aligned}
\Psi(\kappa, z) &= \Psi_1 + \frac{\omega}{k_0} \int_{z_1}^z \frac{(\omega - \kappa \cdot \mathbf{v}_\perp)c - v_z \chi}{(c^2 - v_z^2)\chi} \, dz' \\
&= \Psi_1 + \frac{(\mathbf{r} - \mathbf{r}_1) \cdot \kappa}{k_0} + \int_{z_1}^z \frac{q}{k_0} \, dz'.
\end{aligned} \tag{3.81}$$

Here, Ψ_1 is the value of Ψ at $\mathbf{R} = \mathbf{R}_1$, and $\mathbf{r}$ is given by the ray equation (3.79). Equation (3.81) determines $\Psi(\kappa, z)$ along a fixed ray, with the ray parameters given by the vector κ/k_0. To obtain $\Psi(\mathbf{r}, z)$ at a fixed point $\mathbf{R} = (\mathbf{r}, z)$, equation (3.79) should be solved (for example, numerically) for κ and, then, the value of this vector should be substituted into equation (3.81).

Making use of equations (3.65) and (3.43), equation (3.78) can be expressed in the form $dz/d\tau = \chi/(kc) = u_z/c$, where u_z is the vertical component of the group velocity $\mathbf{u}$. The quantities $dz/d\tau$, u_z, and χ can become zero at some level z_t, which is the level of the turning point of a sound ray (see figure 3.3 below). The value of z_t can be obtained from the equality $\chi(z_t) = 0$, where χ is given by equation (3.66). This results in the following equation for z_t:

$$\left[c^2(z_t) - v_z^2(z_t)\right]^{1/2} + \mathbf{v}_\perp(z_t) \cdot \boldsymbol{\kappa}/\kappa = \omega/\kappa. \tag{3.82}$$

Equations (3.79) and (3.81) are valid up to the turning point of a sound ray. The generalization of these equations to include the presence of a turning point can be carried out analogously to the method used for a motionless medium. For example, if there is only one turning point, the integral $\int_{z_1}^z dz'$ in equations (3.79) and (3.81) should be replaced by

$$\left(\int_{z_1}^{z_t} + \int_{z_t}^z\right) dz'.$$

(Here, we take into account the rule for choosing the sign in the equation for χ.) For $v_z = 0$, the ray path is symmetrical with respect to the turning point. If $v_z \neq 0$, the ray path is not symmetrical with respect to this point since in the numerator of the integrand in equation (3.79), the sign of the term $v_z\chi$ is different for the rays propagating upwards and downwards.

The sound ray equation (3.79) determines the ray path for arbitrary profiles $c(z)$ and $\mathbf{v}(z)$. We now present examples of the sound ray paths in a moving medium. For all these examples $v_z = 0$.

3.5.2 Ray path for the linear velocity profile

First we obtain the ray path for $c = c_0 = $ constant and a linear profile of the horizontal velocity $\mathbf{v}_\perp = \mathbf{B}c_0 z$, where the vector $\mathbf{B}$ is constant. For this profile, calculating the integral on the right-hand side of equation (3.79) yields the ray path:

$$\mathbf{r} = \mathbf{r}_1 - (1/2)(\boldsymbol{\kappa} \cdot \mathbf{B})^{-2} \left[\kappa^2(2\mathbf{e}(\mathbf{e} \cdot \mathbf{B}) - \mathbf{B}) \ln(k_0 - \boldsymbol{\kappa} \cdot \mathbf{B}z' + q(z'))\right.$$
$$\left. + \mathbf{B}q(z')(k_0 + \boldsymbol{\kappa} \cdot \mathbf{B}z')\right]\Big|_{z_1}^z. \tag{3.83}$$

Here, $q(z) = \left[(k_0 - \boldsymbol{\kappa} \cdot \mathbf{B}z)^2 - \kappa^2\right]^{1/2}$ is the vertical wavenumber, and the function f between the square brackets before the symbol $\big|_{z_1}^z$ is determined as $f(z) - f(z_1)$. In deriving equation (3.83), it is assumed that there is no turning point and that $z > z_1$. Equation (3.83) was obtained by Barton [23] for the case when the vectors $\mathbf{e}$ and $\mathbf{B}$ are collinear and, hence, the ray path is located in a vertical plane.

If the medium is motionless and the profile $c(z)$ is linear, it is well known [49] that the ray path is an arc of a circle. For the linear profile $\mathbf{v}_\perp(z)$, it follows from equation (3.83) that the ray path is a more complicated curve.

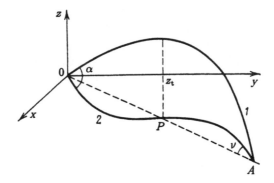

FIGURE 3.3
Qualitative sketch of the ray path (curve 1) and its projection on the plane $z = 0$ (curve 2) for a monotonically increasing medium velocity. The direction of the velocity coincides with the y-axis. Source is located at the center $\mathbf{R} = 0$ of the coordinate system. P indicates the projection of the turning point on the plane $z = 0$.

Equation (3.83) can be readily generalized to cases of ray reflection from the surface or the presence of a turning point. Let us consider a ray path with one turning point at the level z_t. In this case, in equation (3.83), the symbol $|_{z_1}^z$ should be replaced by $\left(|_{z_1}^{z_t} + |_z^{z_t}\right)$. Figure 3.3 is a sketch of a ray path with one turning point if the medium velocity $\mathbf{v}_\perp(z)$ increases monotonically (for instance, linearly). This sketch can be obtained by making use of equation (3.79) for a sound ray and the symmetry of the ray path with respect to the turning point.

For the linear profile $v_\perp(z)$, the level z_t of the turning point, which can be obtained from the equality $q(z_t) = 0$, is given by $z_t = (k_0 - \kappa)/\kappa \cdot \mathbf{B}$. Substituting the value of κ given by equation (3.80) into the latter equation, we obtain the relationship between the level z_t, the velocity $\mathbf{v}_t = \mathbf{v}_\perp(z_t)$ at this level, and the elevation angle α of a sound ray near the source

$$\cos \alpha = \frac{1}{1 + \mathbf{e} \cdot \mathbf{B} z_t} = \frac{1}{1 + \mathbf{e} \cdot \mathbf{v}_t/c_0}. \tag{3.84}$$

Let A denote the point of intersection of the ray path with the plane $z = 0$ (figure 3.3). The magnitude of the vector $\mathbf{d}$ from the point $\mathbf{R} = 0$ to the point A determines the ray skip distance. Setting $\mathbf{r}_1 = 0$ in equation (3.83), replacing the symbol $|_{z_1}^z$ by $2|_0^{z_t}$, and substituting the value of κ from equation (3.80), we obtain the vector $\mathbf{d}$:

$$\mathbf{d} = \frac{[2\mathbf{e}(\mathbf{e} \cdot \mathbf{B}) - \mathbf{B}] \ln (1/\cos \alpha + \tan \alpha) + \mathbf{B} \sin \alpha/(\cos^2 \alpha)}{(\mathbf{e} \cdot \mathbf{B})^2}. \tag{3.85}$$

Since the ray path is symmetrical with respect to the turning point in the case

$v_z = 0$, the vector $\mathbf{d}$ is twice the vector from the point $\mathbf{R} = 0$ to the point P, where P is the projection of the turning point on the plane $z = 0$ (figure 3.3). The phase change $\Phi_d = (\Psi - \Psi_1)/k_0$ along the ray path from the source to the point A can be calculated using equation (3.59), where the integral $\int_{z_1}^{z} dz'$ should be replaced by $2 \int_{0}^{z_t} dz'$:

$$\Phi_d = \frac{2k_0 \tan \alpha}{\mathbf{e} \cdot \mathbf{B}}. \tag{3.86}$$

The ray path is not located in the vertical plane containing the points $\mathbf{R} = 0$ and A if the vector $\mathbf{v}_\perp$ has a component perpendicular to the vector $\mathbf{d}$. That is seen clearly in figure 3.3 where the ray path projection on the plane $z = 0$ is shown. The figure shows that, at the point A, the horizontal direction of the sound ray differs from the horizontal direction to the source by the angle ν. This difference should be taken into account when taking bearings on sound sources such as mortars and artillery, explosions on the ground, or strong storms in the ocean which are sources of infrasound.

3.5.3   Numerical examples of the ray paths

Now we present numerical examples of the ray paths in the atmosphere and ocean. Let us first consider the effect of the wind velocity rotation with height on the ray path in the atmosphere [73]. The vertical profiles of the magnitude and direction of the wind velocity are exponential:

$$v_\perp(z) = v_0 \left[1 - \exp\left(-\frac{z}{h_1} \right) \right], \quad \psi(z) = \psi_0 \left[1 - \exp\left(-\frac{z}{h_2} \right) \right].$$

Here, $\psi(z)$ is the angle between the vector $\mathbf{v}_\perp(z)$ and the x-axis, the vector $\mathbf{v}_\perp(z = 0)$ coincides with the direction of the x-axis, $v_0 = 20$ m/s, $h_1 = 55.6$ m, $\psi_0 = 45°$, and $h_2 = 111.1$ m. A source is located at the point $\mathbf{R}_1 = 0$ on the ground. For this atmospheric stratification, the ray path projections on the xz and yz planes are shown in figure 3.4. The source emits sound rays only in the xz-plane, and the directions of these rays near the source make various angles φ with respect to the z-axis.

The ray paths would be located in the vertical xz-plane if the direction of the wind velocity were constant. The deviation of sound rays from this plane is caused by the rotation of the vector $\mathbf{v}_\perp$ with height. Furthermore, since $v_\perp$ increases as a function of height, the sound rays with $\varphi \geq 73.02°$ have turning points and come back to the ground at various horizontal distances from the source.

In figure 3.4, the ray paths are interrupted at the earth's surface. The ray paths with reflections from the ground are shown in figure 3.5 [326]. The point source is located at the height $z_1 = 50$ m, c is constant, the direction of the vector $\mathbf{v}_\perp(z)$ coincides with the direction of the x-axis, and the wind velocity increases logarithmically with height: $v_\perp(z) = v_2 \ln(z/z_0)$. Here,

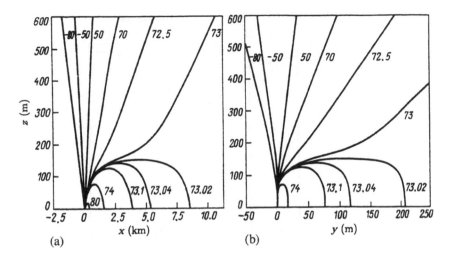

FIGURE 3.4

Ray path projections on the (a) xz and (b) yz planes in an atmosphere with exponential profiles of $v_\perp(z)$ and $\psi(z)$ [73]. The numbers beside the curves indicate the value (in degrees) of the angle φ between the sound rays near the source and the z-axis.

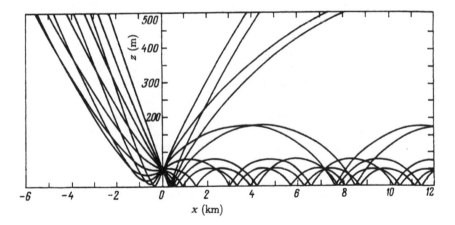

FIGURE 3.5

Ray path in the xz-plane for sound propagation in an atmosphere with a logarithmic velocity profile $v_\perp(z)$ [326]. The direction of the vector $\mathbf{v}_\perp$ coincides with the direction of the x-axis.

$v_2 = v_\perp(z = 10\text{m}) = 5$ m/s, and the surface roughness parameter $z_0 = 8$ cm. The angles φ between the sound rays near the source and the z-axis vary from $81°$ to $99°$ at intervals of $2°$. Figure 3.5 shows that there is waveguide propa-

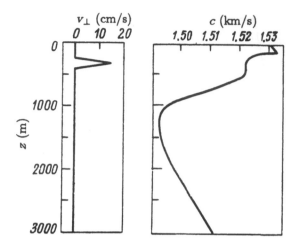

FIGURE 3.6
Vertical profiles $v_\perp(z)$ and $c(z)$ used for calculation of ray paths in the ocean
[348] shown in figure 3.7.

gation of sound downwind. On the other hand, upwind there is antiwaveguide
sound propagation causing an acoustic shadow zone at certain ranges from
the source.

Finally, let us consider the effects of oceanic currents on sound ray paths
[348]. It has been shown experimentally that in the ocean $|\partial v_\perp/\partial z|$ can be
greater than $|\partial c/\partial z|$ over some range of depth. This inequality is valid for the
profiles of $c(z)$ and $v_\perp(z)$ shown in figure 3.6 in the range 250 m $\lesssim z \lesssim$ 400 m.
The profile of $c(z)$ was measured in the Sargasso Sea. The profile of $v_\perp(z)$ is
an idealized one, but it agrees qualitatively with the profiles of the horizontal
currents measured in the Sargasso Sea in the indicated depth range.

For these profiles of $c(z)$ and $v_\perp(z)$, ray paths in the xz-plane are depicted
in figure 3.7. The elevation angles α of the rays near the source do not exceed 1°. The source is located at a depth $z = 350$ m, and the direction of
the vector $\mathbf{v}_\perp$ coincides with the direction of the x-axis. The figure shows that
there is antiwaveguide sound propagation downstream. On the other hand, upstream there is waveguide propagation of sound. Thus, the current leads to a
qualitative difference between sound propagation downstream and upstream.
Nevertheless the presence of the waveguide and antiwaveguide leads only to
a small spatial redistribution of the sound energy emitted by the source. Furthermore, these weak waveguides and antiwaveguides can be destroyed by
features such as horizontal inhomogeneity of the ocean or fine structure of the
profile $c(z)$. Therefore, it would be worthwhile to carry out experiments to
determine whether currents have such an effect on sound propagation.

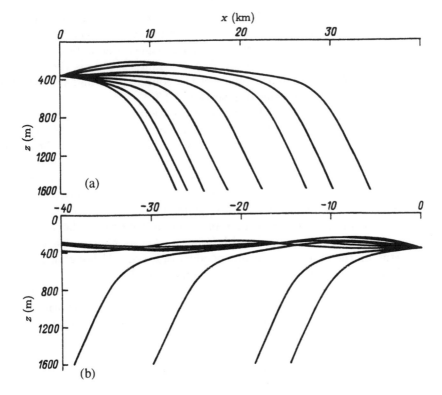

FIGURE 3.7
Ray paths for the profiles $v_\perp(z)$ and $c(z)$ shown in figure 3.6 [348]; (a) downstream and (b) upstream sound propagation. The direction of the vector $\mathbf{v}_\perp$ coincides with the direction of the x-axis.

3.5.4 Refraction in the atmospheric surface layer

As described in section 2.2.3, Monin–Obukhov similarity theory (MOST) is commonly used to describe the wind and temperature profiles in the atmospheric surface layer (ASL) [306, 367]. Here, we consider the systematic application of MOST to refraction of sound rays. For simplicity, we assume that the rays are oriented within about 20° of horizontal, and thus refraction may be modeled on the basis of vertical gradients of the effective sound speed $c_{\text{eff}}(z, \psi) = c(z) + v_\perp(z) \cos \psi$, as given by equation (1.3). Here, ψ is the angle between the wind and the nominal horizontal propagation directions. A negative vertical gradient in c_{eff} causes upward refraction (away from the ground), whereas a positive gradient causes downward refraction (toward the ground).

Differentiation of equations (2.44) and (2.45) leads to the following equation for the gradient [421]:

$$\frac{\partial c_{\text{eff}}}{\partial z} = \frac{1}{k_v z} \left[P_t c_* \phi_h(\xi) + u_* \cos(\psi) \phi_m(\xi) \right] - \frac{g(\gamma - 1)}{2c_0}, \qquad (3.87)$$

in which $k_v \simeq 0.40$ is von Kármán's constant, $P_t = 0.95$ is the turbulent Prandtl number in neutral stratification, γ is the ratio of specific heats, $c_* = (c_0/2T_0) T_*$ is a scale characteristic of the sound-speed fluctuations, $\xi = z/L_o$ is the normalized height, $L_o = u_*^2 T_0/k_v g T_* = c_0 u_*^2/2k_v g c_*$ is the Obukhov length, c_0 and T_0 are reference values of the sound speed and temperature, and other parameters are defined in section 2.2.3.

The first term on the right-hand side of equation (3.87) is the contribution from temperature gradients. It can be positive or negative, depending on the sign of c_* (which is determined by stability in the atmosphere). The second term is the contribution from wind gradients. Although $u_* \geq 0$, this term can also be positive or negative, depending on the sign of $\cos \psi$. The third term is the constant background gradient due to adiabatic compression (the adiabatic lapse rate for temperature) of the air column. The first and second terms on the right, when integrated, have an approximately logarithmic behavior; they change rapidly near the ground and become weaker as z increases. The adiabatic term, on the other hand, is independent of height. Although it produces a relatively small gradient, it becomes dominant for large z.

The validity of equation (3.87) depends on the range of values of ξ encountered in a particular scenario, which in turn depends on the atmospheric state (the value of L_o) and the heights at which the sound propagates. Values of $|L_o|$ less than several m occur only for very low wind speeds and are rather rare [367]. As discussed in section 2.2.3, MOST yields unreliable results $\xi \gtrsim 1$. Furthermore, MOST should be applied only within the surface layer, which is normally defined as the lowermost 1/10 of the atmospheric boundary layer (ABL) [367]. In practice, this means the propagation must be confined to altitudes below about 100 m to 200 m, depending on the vertical development of the ABL.

A useful recasting of equation (3.87) involves generalizing the definition of the turbulent Prandtl number to include dependence on the stability parameter ξ, specifically $P_{t,h}(\xi) \equiv P_t \phi_h(\xi)/\phi_m(\xi)$. We thus have

$$\frac{\partial c_{\text{eff}}}{\partial z} = \frac{1}{k_v z} \left[P_{t,h}(\xi) c_* + u_* \cos \psi \right] \phi_m(\xi) - \frac{g(\gamma - 1)}{2c_0}. \tag{3.88}$$

Based on equation (2.47), $P_{t,h}(\xi)$ varies from $P_t \sqrt{a_m/a_h} \approx 0.64$ in the unstable stratification limit ($\xi \to -\infty$) to $P_t b_h/b_m \approx 1.5$ in the stable stratification limit ($\xi \to \infty$). Thus, it is often reasonable to approximate $P_{t,h}(\xi)$ with P_t (or simply 1), particularly given the random scatter in experimental observations of the wind and temperature gradients, and the resulting uncertainties in the functions ϕ_m and ϕ_h. Then equation (3.88) simplifies to

$$\frac{\partial c_{\text{eff}}}{\partial z} \simeq \frac{P_t c_{\text{eff}}^*}{k_v z} \phi_m(\xi) - \frac{g(\gamma - 1)}{2c_0}, \tag{3.89}$$

where

$$c_{\text{eff}}^* \equiv c_* + (u_*/P_t) \cos \psi = (u_*/P_t)(A + \cos \psi) \tag{3.90}$$

and $A = P_t c_* / u_*$ is a dimensionless parameter indicating the importance of the sound-speed fluctuations relative to the wind-velocity fluctuations.

Let us examine the general behavior of equation (3.89) for various values of the parameter A. First we consider $|A| \ll 1$, which corresponds to nearly neutral stratification and a predominant role for wind gradients. Such conditions are typical of a windy day or night with cloud cover. In these conditions, $c_* \to 0$, and thus $P_t c_{\text{eff}} \simeq u_* \cos\psi$ and $\phi_m(\xi) \simeq 1$ in equation (3.89). It is useful to introduce the length scale $\Lambda = c_0 u_* / k_v g = (2A/P_t)L_o$, which indicates the height at which the contributions from the wind gradient and the adiabatic lapse rate have a comparable magnitude. Values for Λ range from 20 m to 200 m for a typical range of friction velocities from 0.05 m/s to 0.5 m/s. When $z \lesssim \Lambda |\cos\psi|$, the gradient is approximately proportional to $u_* \cos\psi / k_v z$ and hence the profile is logarithmic, with a sign dependent upon the propagation direction. Well above this height, adiabatic lapse dominates, such that the gradient is $-g(\gamma - 1)/2c_0$. The profile $c_{\text{eff}}(z)$ is thus linear.

Next we consider $A \ll -1$, which corresponds to unstable atmospheric stratification, as typically occurs on a sunny afternoon with low mean wind. The effective sound speed gradient is dominated by the temperature gradient, and is negative at all heights and for all propagation directions. For heights such that $|\xi|$ is small, the gradient is approximately $P_t c_* / k_v z$, and hence the height dependence of $c_{\text{eff}}(z)$ is logarithmic. As $|\xi|$ is increased, the first term on the right-hand side of equation (3.89) weakens, and eventually the second term, associated with the adiabatic lapse rate, dominates. This transition occurs at $z \approx A\Lambda$. Above this height the adiabatic lapse dominates.

Finally, when $A \gg 1$, stratification is stable, as typically occurs on a clear night with calm wind. Near the ground, the gradient is positive for all propagation directions. For heights such that $\xi \lesssim 1$, the gradient is approximately $P_t c_* / k_v z$, and hence the height dependence of $c_{\text{eff}}(z)$ is logarithmic. When $\xi \gg 1$, according to equation (2.47), $\phi_{h,m} \propto z$. Hence both terms in equation (3.89) become height-independent and $c_{\text{eff}}(z)$ is linear. The sign of the gradient depends on the relative strength of the two terms, which depends on variables such as c_*. Dominance of the first term, i.e., a positive gradient, would correspond to a surface-based *temperature inversion*, which according to MOST persists to arbitrarily large z. However, it should be kept in mind that application of MOST is tenuous when $\xi \gtrsim 1$. Inversion layers suppress turbulent mixing so that gradients at the surface decouple from those at higher altitudes. A surface-based inversion will almost always give way to a nearly adiabatic layer above.

The previous discussion of the dependence of the profiles on A is summarized pictorially in figure 3.8. Example ray traces for neutral ($A = 0$), unstable ($A = -2$), and stable ($A = 1$) conditions are shown in figures 3.9–3.11, respectively.

A simple scheme was proposed in reference [428] which partitions the propagation conditions into five refractive categories based on the A value: (1) $A < -2$, *completely upward refraction*; (2) $-2 < A < -0.5$, *predominantly*

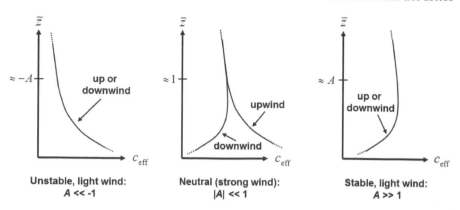

Unstable, light wind:
A << -1

Neutral (strong wind):
|A| << 1

Stable, light wind:
A >> 1

FIGURE 3.8

Qualitative behavior of the effective sound-speed profile for different regimes of the parameter $A = P_t c_* / u_*$, which indicates the relative contributions from sound speed and wind fluctuations to the effective sound-speed gradient. The profiles are shown as a function of the normalized height $\bar{z} = z/\Lambda$, where Λ is described in the text.

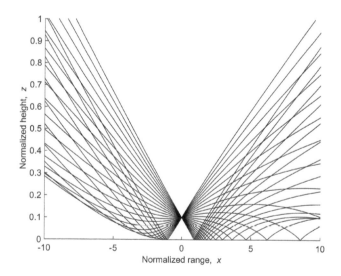

FIGURE 3.9

Ray trace for neutral atmospheric stratification, $A = 0$. Shown is a vertical plane in the upwind and downwind directions (left and right, respectively), with the source at $\bar{z} = 0.1$ and $\bar{x} = 0$. The coordinates have been normalized by the length scale Λ, which is described in the text. Rays are launched at $0.5°$ increments between $-6°$ and $6°$.

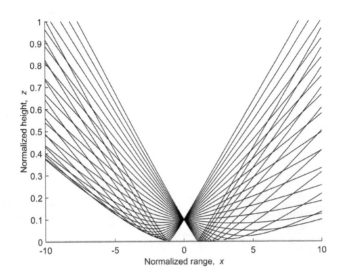

FIGURE 3.10
Same as figure 3.9, except for $A = -2$.

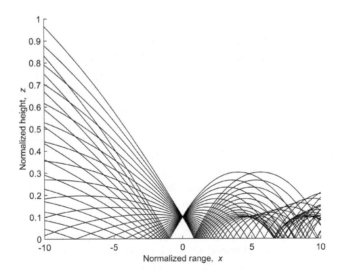

FIGURE 3.11
Same as figure 3.9, except for $A = 1$.

Geometrical acoustics

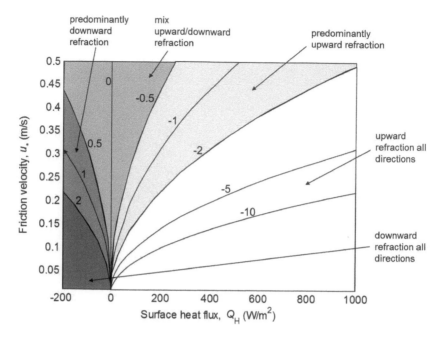

FIGURE 3.12

Contour plot of the parameter A as a function of the friction velocity u_* and surface heat flux Q_H. Shown also are five different categories for propagation conditions based on the range of A.

upward refraction; (3) $-0.5 < A < 0.5$, *wind-controlled refraction (roughly half downward, half upward);* (4) $0.5 < A < 2$, *predominantly downward refraction;* and (5) $2 < A$, *completely downward refraction.* A contour plot of the dependence of A on the atmospheric forcing parameters, namely the friction velocity u_* and sensible heat flux Q_H, is shown in figure 3.12. The five refraction categories are overlaid on the figure.

Besides the scheme mentioned above based on the value of A, many schemes have been proposed for categorizing the impact of atmospheric conditions on sound propagation, i.e., whether refraction is upward or downward, and whether it is weak or strong. Some of the most widely used schemes for refractive categories have been based on heuristic ASL classes familiar elsewhere in micrometeorology, such as in turbulent diffusion modeling. For example, Marsh [240] employed Pasquill's stability classes for atmospheric diffusion (based on wind speed and solar radiation) to partition sound propagation conditions into six refractive classes. Heimann and Salomons [159] approximated the MOST profiles with log-linear profiles and then developed refractive classes based on the log-linear profile parameters. In effect, their scheme amounts to modeling the functions $\phi_{h,m}(\xi)$ as an expansion with a constant plus a term linear in ξ. For stable conditions, this is consistent with

TABLE 3.1
Suggested ranges for "refraction strength" categories based on the effective sound-speed scale, c_{eff}^*.

Range	Description
$c_{\text{eff}}^* < -0.5\,\text{m/s}$	very strong upward refraction
$-0.5\,\text{m/s} < c_{\text{eff}}^* < -0.3\,\text{m/s}$	strong upward refraction
$-0.3\,\text{m/s} < c_{\text{eff}}^* < -0.1\,\text{m/s}$	moderate upward refraction
$-0.1\,\text{m/s} < c_{\text{eff}}^* < 0.1\,\text{m/s}$	weak refraction
$0.1\,\text{m/s} < c_{\text{eff}}^* < 0.3\,\text{m/s}$	moderate downward refraction
$0.3\,\text{m/s} < c_{\text{eff}}^* < 0.5\,\text{m/s}$	strong downward refraction
$0.5\,\text{m/s} < c_{\text{eff}}^*$	very strong downward refraction

TABLE 3.2
Suggested ranges for "profile shape" categories based on the inverse Obukhov length, L_o^{-1}.

Range	Description
$-0.5\,\text{m}^{-1} < L_o^{-1} < -0.1\,\text{m}^{-1}$	strongly convective
$-0.1\,\text{m}^{-1} < L_o^{-1} < -0.02\,\text{m}^{-1}$	moderately convective
$-0.02\,\text{m}^{-1} < L_o^{-1} < -0.005\,\text{m}^{-1}$	weakly convective
$-0.005\,\text{m}^{-1} < L_o^{-1} < 0.005\,\text{m}^{-1}$	neutral
$0.005\,\text{m}^{-1} < L_o^{-1} < 0.02\,\text{m}^{-1}$	weakly stable
$0.02\,\text{m}^{-1} < L_o^{-1} < 0.1\,\text{m}^{-1}$	moderately stable

equation (2.47). For unstable conditions, however, equation (2.47) behaves somewhat differently. This discrepancy may not be important if the primary interest is for low altitudes such that $|\xi| \lesssim 1$.

In general, a parameter involving $\cos \psi$ must be introduced into the classification scheme if we wish to consider dependence on the propagation direction. Based on equation (3.89), a good candidate for such a scheme would involve c_{eff}^* and L_o, which are the two parameters in this equation that explicitly vary with the atmospheric conditions and propagation direction. In reference [428], a scheme based on seven categories for c_{eff}^*, and six for L_o^{-1}, was developed. The former were associated with the strength of the gradient, and the latter with the profile shape. The suggested ranges are shown in tables 3.1 and 3.2. Of course, fewer or more categories could be employed, depending on the desired fidelity. Although there are $42 = 6 \times 7$ mathematical combinations of categories, in actuality not all combinations are physically possible; for example, strong downward refraction cannot occur in a strongly convective atmosphere.

3.5.5 Sound-pressure amplitude

The ray path $\mathbf{R} = \mathbf{R}(\zeta, \nu, \tau)$ in a stratified moving medium can be determined analytically (e.g., equation (3.79)). Therefore, the cross-sectional area of the infinitesimally narrow ray tube can be calculated analytically:

$$d\sigma = \mathbf{n} \cdot \mathbf{s} |(\partial \mathbf{R}/\partial \zeta) \times (\partial \mathbf{R}/\partial \nu)| d\zeta d\nu.$$

If $d\sigma$ is known, the sound-pressure amplitude p_1 can be readily obtained from equation (3.62). This approach has been used for numerical calculations of the sound pressure $p = \exp(ik_0 \Psi) p_1$ in the ocean [120, 121, 363, 364] and the atmosphere [325, 326, 155, 378, 397]. In most of these references, p is calculated to an accuracy of v/c.

For sound propagation in the ocean and in the near-ground atmosphere, not only the terms of order v^2/c^2, but also those of order v/c can be neglected in calculating the amplitude p_1 of the sound pressure, i.e., p_1 can be determined by the following equation:

$$\frac{p_1^2 |(\partial \mathbf{R}/\partial \zeta) \times (\partial \mathbf{R}/\partial \nu)|}{2\varrho c} = \text{constant}.$$

On the other hand, in section 3.6 it is shown that the terms of the order of v^2/c^2 can contribute significantly to the phase of a sound field even for small ranges from a source.

Let us derive an analytical expression for p_1 in a stratified moving medium. Consider the monochromatic quasi-plane wave $p(\mathbf{R}, t) = p_1(z) \exp(i\boldsymbol{\kappa} \cdot \mathbf{r} - i\omega t)$. (Note that in the stratified medium, the solutions of the linearized equations of fluid dynamics can depend on $\mathbf{r}$ and t as $\exp(i\boldsymbol{\kappa} \cdot \mathbf{r} - i\omega t)$.) Since both p_1 and the ambient quantities depend only on z, the law of acoustic energy conservation (3.26) takes the form

$$\frac{d}{dz} \left[E_1 \left(\frac{cq}{k} + v_z \right) \right] = 0.$$

Solving this equation, we obtain the amplitude of the quasi-plane wave:

$$p_1 = \left[\frac{\varrho(q_0 + k_0 v_{z,0}/c_0)}{\varrho_0(q + k v_z/c)} \right]^{1/2} \frac{kc}{k_0 c_0} p_{1,0}. \tag{3.91}$$

Here, the subscript 0 indicates that a quantity is evaluated at a fixed level z_0, e.g., $p_{1,0} = p_1(z_0)$.

3.6 Approximate equations for the eikonal and sound ray path

The inequalities $v/c_0 \ll 1$ and $|\tilde{c}/c_0| \ll 1$ are always valid in both the atmosphere near the ground and the ocean. Here, $\tilde{c} = c - c_0$ is the deviation of the sound speed from its reference value c_0. In this subsection, these inequalities are used to simplify the equations for the sound ray and the eikonal. The equations obtained elucidate the effect of medium motion on sound propagation. They have applications in the analysis of the effects of atmospheric refraction on remote sensing of the wind velocity with sodars [295] and in source localization with elevated acoustic sensor arrays [298]. In what follows, we shall consider the case when $v_z = 0$.

3.6.1 Large elevation angles

In a similar manner to reference [284], we assume first that the modulus of the elevation angle α of the sound ray is greater than $\varepsilon^{1/2}$, where $\varepsilon = \max(|\tilde{c}/c_0|, v/c_0)$ is a small parameter. In equation (3.79), the function χ is given by $\chi = \pm(\omega^2 - \kappa^2 c_0^2 - 2\omega\boldsymbol{\kappa} \cdot \mathbf{v} - 2\tilde{c}c_0\kappa^2)^{1/2}$ to an accuracy of ε. This function can be expanded into a series in ε if the following inequality is valid:

$$\omega^2 - \kappa^2 c_0^2 \gg 2|\boldsymbol{\kappa} \cdot \mathbf{v}\omega + \kappa^2 c_0\tilde{c}|.$$

Dividing both sides of this inequality by ω^2, we express it in the form

$$1 - \frac{\kappa^2}{k_0^2} \gg 2\frac{\kappa}{k_0}\left|\frac{\mathbf{e} \cdot \mathbf{v}_\perp}{c_0} + \frac{\kappa\tilde{c}}{k_0 c_0}\right|. \tag{3.92}$$

If this inequality is valid, $\chi \neq 0$ and, hence, there are no turning points. This inequality is valid if κ is not too close to k_0, i.e., the modulus of the elevation angle α is not too small. For $\pi/2 - |\alpha| \gg v/c_0$, it follows from equation (3.74) that $\kappa = k_0(\cos\alpha - \mathbf{e} \cdot \mathbf{v}_\perp/c_0 - \tilde{c}\cos\alpha/c_0 + o(\varepsilon))$. (If $\pi/2 - |\alpha| \ll v/c_0$, $\kappa \ll k_0$ and the inequality (3.92) is valid.) Substituting this value of κ into the inequality (3.92) and solving it for α to an accuracy of $\varepsilon^{1/2}$ yields

$$|\alpha| \gg (2|\mathbf{e} \cdot \mathbf{v}_\perp/c_0 + \tilde{c}/c_0|)^{1/2} \sim O(\varepsilon^{1/2}). \tag{3.93}$$

Assuming that this inequality (or the equivalent inequality (3.92)) is valid, we expand χ in equation (3.79) into a series in ε. From equation (3.79), one can obtain the desired approximate equation for a sound ray,

$$\mathbf{r} = \mathbf{r}_1 + |z - z_1|\left[\frac{\boldsymbol{\kappa}}{(k_0^2 - \kappa^2)^{1/2}} + \frac{k_0\mathbf{M}}{(k_0^2 - \kappa^2)^{1/2}} + \frac{k_0\boldsymbol{\kappa}(\boldsymbol{\kappa} \cdot \mathbf{M})}{(k_0^2 - \kappa^2)^{3/2}} + \frac{k_0^2 b\boldsymbol{\kappa}}{(k_0^2 - \kappa^2)^{3/2}}\right]. \tag{3.94}$$

Here, the source is located at $\mathbf{R}_1 = (\mathbf{r}_1, z_1)$ and the receiver at $\mathbf{R} = (\mathbf{r}, z)$. The vector $\mathbf{M}$ and parameter b are proportional to the integrated profiles of velocity and the variations in the sound speed:

$$\mathbf{M} = \frac{1}{z - z_1}\int_{z_1}^{z}\frac{\mathbf{v}_\perp(z')}{c_0}\,dz', \quad b = \frac{1}{z - z_1}\int_{z_1}^{z}\frac{\tilde{c}(z')}{c_0}\,dz'.$$

Equation (3.94) is valid to an accuracy of ε. It follows from this equation that the vertical profiles $\mathbf{M}(z)$ and $b(z)$ cause a relatively small deviation of the ray path from line-of-sight sound propagation. It also follows from this equation that the ray path is a three-dimensional curve so that the approximation of the effective sound speed introduced in section 1.2.2 cannot be used to obtain the ray path even to order ε. If $\kappa \ll k_0 M$, the second term in square brackets on the right-hand side of equation (3.94) is much greater than the other terms. In this case, the ray path is close to the z-axis.

Equation (3.94) can be solved for κ. Neglecting the terms of order ε in the resulting equation, we obtain

$$\kappa = (\mathbf{r} - \mathbf{r}_1)k_0/R_0 + O(\varepsilon).$$

Here, $R_0 = |\mathbf{R} - \mathbf{R}_1|$ is the distance between the source and receiver. The second, third, and fourth terms in square brackets in equation (3.94) are of the order of ε. Therefore, in these terms, we can set $\kappa = (\mathbf{r} - \mathbf{r}_1)k_0/R_0$ to the same accuracy as that to which equation (3.94) is valid. Then, solving equation (3.94) for κ, we have

$$\kappa = (k_0/R_0)\left[(1 - b)(\mathbf{r} - \mathbf{r}_1) - R_0\mathbf{M}\right]. \tag{3.95}$$

To an accuracy of ε, this equation determines the ray parameters κ/k_0 of a sound ray arriving at the observation point $(\mathbf{r}, z)$.

Now let us derive the approximate equation for the eikonal Ψ. Expanding the integrand on the right-hand side of equation (3.81) in a series in ε and retaining terms of the order of ε^0 and ε^1 yields

$$\Psi = \Psi_1 + \frac{|z - z_1|k_0}{(k_0^2 - \kappa^2)^{1/2}}\left[1 + \frac{\kappa^2\boldsymbol{\kappa} \cdot \mathbf{M}}{(k_0^2 - \kappa^2)k_0} + \frac{(2\kappa^2 - k_0^2)b}{(k_0^2 - \kappa^2)}\right]. \tag{3.96}$$

This equation determines the eikonal Ψ along the ray with the ray parameters κ/k_0. Substituting the value of κ given by equation (3.95) into equation (3.96) and retaining the terms of order ε^0 and ε^1 in the resulting equation, we obtain the explicit formula for $\Psi(\mathbf{r}, z)$ as a function of $\mathbf{r}$ and z:

$$\Psi(\mathbf{r}, z) = \Psi_1 + (1 - b)R_0 - \mathbf{M} \cdot (\mathbf{r} - \mathbf{r}_1). \tag{3.97}$$

It follows from this formula that the eikonal depends only on the projection of $\mathbf{v}_\perp$ on the vertical plane containing the source and receiver. With the previously assumed accuracy, equation (3.97) can be written in the equivalent form

$$\Psi(\mathbf{r}, z) = \Psi_1 + \frac{R_0}{1 + b + \mathbf{s}_0 \cdot \mathbf{M}}. \tag{3.98}$$

Here, $\mathbf{s}_0 = (\mathbf{R} - \mathbf{R}_1) / |\mathbf{R} - \mathbf{R}_1|$ is the unit vector in the direction from the source to the receiver. When deriving this formula for Ψ, we took into account that $\mathbf{M} \cdot (\mathbf{r} - \mathbf{r}_1)/R_0 = \mathbf{M} \cdot (\mathbf{R} - \mathbf{R}_1) / |\mathbf{R} - \mathbf{R}_1|$. In equation (3.98), the sound speed and medium velocity can be combined into the effective sound speed $c_{\text{eff}}(z') = c_0 + \tilde{c}(z') + \mathbf{s}_0 \cdot \mathbf{v}_\perp(z') = c(z') + \mathbf{s}_0 \cdot \mathbf{v}_\perp(z')$. Neglecting terms of order $v_\perp^2/c_0^2$ (which were already neglected in derivation of equation (3.98)), $\mathbf{s}_0$ can be replaced with the unit vector $\mathbf{s}(z')$ tangential to the sound ray path from the source to the receiver. In this case, the effective sound speed $c_{\text{eff}}(z') = c(z') + \mathbf{s}(z') \cdot \mathbf{v}_\perp(z')$, which can be introduced in equation (3.98), coincides with that defined by equation (1.2).

If $|\mathbf{r} - \mathbf{r}_1| \lesssim R_0 M$, the direction of sound propagation is close to the vertical. Strictly speaking, in this case, the last term on the right-hand side of equation (3.97) should be ignored. Let us derive an approximate equation for the eikonal Ψ which is valid for the case $|\mathbf{r} - \mathbf{r}_1| \lesssim R_0 M$ and has a wider range of applicability than equation (3.97). To this end, we substitute the value of the vector $\boldsymbol{\kappa}$ from equation (3.95) into equation (3.96). We also take into account the fact that $\kappa/k_0 \approx |(\mathbf{r} - \mathbf{r}_1)/R_0 - \mathbf{M}| \sim v/c_0$. These relationships can be obtained from equation (3.69) if $|\mathbf{r} - \mathbf{r}_1| \lesssim R_0 M$. Retaining the terms of the order of $|\tilde{c}/c_0|$ and $(v/c_0)^2$ in equation (3.96), we obtain the desired formula for the eikonal Ψ:

$$\Psi(\mathbf{r}, z) = \Psi_1 + |z - z_1|(1 - b) + \frac{(\mathbf{r} - \mathbf{r}_1 - |z - z_1|\mathbf{M})^2}{2|z - z_1|}. \tag{3.99}$$

When deriving equation (3.96), we omitted the term proportional to $(v/c_0)^2$. It can be shown that this term is proportional to $\kappa^2(\boldsymbol{\kappa} \cdot \mathbf{v}_\perp)^2/k_0^4$ and, since $\kappa/k_0 \sim v/c_0$, it does not contribute to the eikonal Ψ given by equation (3.99). It follows from the latter equation that the eikonal depends not only on the projection of the vector $\mathbf{v}_\perp$ on the vertical plane containing the source and receiver but also on the component of $\mathbf{v}_\perp$ perpendicular to this plane. It can be shown that with this approximation, equation (3.99) coincides with equation (3.97), except for the term $|z - z_1|M^2/2$ which is not present in equation (3.97).

Let us estimate the distance R_0 from the source to the receiver at which the eikonal Ψ and phase $\Phi = k_0\Psi$ can be calculated to an accuracy of v/c. For simplicity, we shall consider a point source in a homogeneous medium moving uniformly parallel to the x-axis. The sound field of this source is given by equation (4.25). It follows from this equation that the term of order $(v/c)^2$ in the phase of the sound field is given by $\Phi_2 = k_0 R(1 + \cos^2 \alpha)(v/c)^2/2$, where α is the angle between the direction from the source to the receiver and the x-axis. The term Φ_2 can be ignored if $\Phi_2 \ll \pi/2$. We assume that this inequality is valid if $\Phi_2 \leq \pi/8$. Substituting the value of Φ_2 into the latter inequality and setting $\alpha = 0$, we obtain

$$R_0 \leq \frac{\lambda c^2}{16 v^2}. \tag{3.100}$$

This inequality determines the distance R_0 from the source to the receiver at which the terms of order $(v/c)^2$ in the phase can be ignored. Let us evaluate the distance R_0 for sound propagation in the atmosphere and ocean. For sound propagation in the atmosphere, setting $v = 5$ m/s and $c = 340$ m/s in equation (3.100) yields $R_0 \leq 289\lambda$. If the acoustic frequency $f = \omega/2\pi = 500$ Hz, the distance $R_0 \leq 196.5$ m. For sound propagation in the ocean, setting $v = 0.5$ m/s and $c = 1500$ m/s, we obtain $R_0 \leq 5.6 \times 10^5 \lambda$. If $f = 1$ kHz, we have $R_0 \leq 840$ km.

3.6.2 Small elevation angles

We now derive the approximate equations for a sound ray and eikonal in the case for which the inequalities (3.92) and (3.93) are not valid, i.e., $1 - \kappa/k_0 \lesssim \varepsilon$ and $|\alpha| \lesssim (2\varepsilon)^{1/2}$. It follows from the latter inequality that the modulus of the sound ray elevation angle is small. In equation (3.79), it is worthwhile expressing χ in the form $\chi = \pm\omega(\chi_1\chi_2)^{1/2}$, where $\chi_1 = (1 - \kappa/k_0 - \kappa\tilde{c}/k_0 c_0 - \mathbf{e} \cdot \mathbf{v}_\perp \kappa/k_0 c_0)$ and $\chi_2 = (1 + \kappa/k_0 + \kappa\tilde{c}/k_0 c_0 - \mathbf{e} \cdot \mathbf{v}_\perp \kappa/k_0 c_0)$. If the inequality $1 - \kappa/k_0 \lesssim \varepsilon$ is valid, $\chi_1 = (1 - \kappa/k_0 - \tilde{c}/c_0 - \mathbf{e} \cdot \mathbf{v}_\perp/c_0)(1 + O(\varepsilon))$ and $\chi_2 = 2 + O(\varepsilon)$.

We substitute the value of χ into equations (3.79) and (3.81) and expand the integrands of these equations in series in ε. These series contain terms of the order of $\varepsilon^{-1/2}$, $\varepsilon^{1/2}$, $\varepsilon^{3/2}$, etc. Retaining terms of order $\varepsilon^{-1/2}$ (inclusion of other terms leads to rather complicated equations) yields the desired approximate formulas for the sound ray and eikonal

$$\mathbf{r} = \mathbf{r}_1 \pm \mathbf{e} \int_{z_1}^z \frac{1}{[2(1 - \kappa/k_0 - \tilde{c}(z')/c_0 - \mathbf{e} \cdot \mathbf{v}_\perp(z')/c_0)]^{1/2}} \, dz', \qquad (3.101)$$

$$\Psi = \Psi_1 \pm \int_{z_1}^z \frac{1}{[2(1 - \kappa/k_0 - \tilde{c}(z')/c_0 - \mathbf{e} \cdot \mathbf{v}_\perp(z')/c_0)]^{1/2}} \, dz'. \qquad (3.102)$$

It follows from the former equation that the ray path is located in the vertical plane containing the source and receiver. Since the integrals in equations (3.101) and (3.102) are the same, Ψ and $\mathbf{r}$ are related by $\Psi = \Psi_1 + |\mathbf{r} - \mathbf{r}_1|$. In these equations, similarly to equation (3.98), the sound speed and medium velocity can be combined into the effective sound speed $c_{\text{eff}}(z') = c(z') + \mathbf{s}(z') \cdot \mathbf{v}_\perp(z')$.

Equations (3.101) and (3.102) are valid if there are no turning points of the sound ray. If there is a turning point, the integral $\int_{z_1}^z dz'$ in these equations should be replaced with $\left(\int_{z_1}^{z_t} + \int_z^{z_t} \right) dz'$.

To illustrate the difference between the ray paths and between the phase changes along these paths, calculated on the basis of the exact equations (3.79) and (3.81) and the approximate ones (3.101) and (3.102), let us consider the linear profile $\mathbf{v} = \mathbf{B}c_0 z$. A point source and receiver are located in the center ($\mathbf{r}_1 = 0$, $z_1 = 0$) of the coordinate system and at the point $(\mathbf{d}, 0)$, respectively, and the sound speed c_0 is constant. First consider the ray path, calculated on the basis of the exact equation (3.79), which passes through the source and receiver and has a turning point at the level $z_t > 0$. For this ray path, equations (3.84)–(3.86) determine the elevation angle α between the direction of the sound ray near the source and the horizontal plane (figure 3.3), the level z_t of the turning point, and the phase change Φ_d along the path. In these equations, the quantity $Bd/2 = v_t d/(2c_0 z_t) \approx \alpha$ is the small parameter, if $v/c \ll 1$. Assuming that the vectors **e** and **B** are collinear, from equations (3.84)–(3.86) we obtain equations for α, Φ_d, and z_t to an accuracy of $(Bd/2)^3$:

$$\alpha = (Bd/2)\left[1 - (3/4)(Bd/2)^2\right], \quad \Phi_d = k_0 d\left[1 - (5/12)(Bd/2)^2\right],$$

$$z_t = v_t/c_0 B = (Bd^2/8)\left[1 - (41/48)(Bd/2)^2\right]. \tag{3.103}$$

For the approximate ray path and eikonal, the analogous quantities α_0, $\Phi_{d,0}$, and $z_{t,0}$ can be calculated from equations (3.101) and (3.102) and equation (3.80), which relates κ/k_0 and α_0:

$$\alpha_0 = Bd/2, \quad \Phi_{d,0} = k_0 d, \quad z_{t,0} = Bd^2/8. \tag{3.104}$$

Equations (3.103) and (3.104) allow us to obtain the differences between the elevation angles α, the levels z_t, and the phase increments Φ_d for the actual and approximate ray paths. Obviously, the eikonal can be calculated on the basis of the approximate equation (3.102) if $|\Phi_{d,0} - \Phi_d| \ll \pi/2$. We assume that this inequality is valid if $|\Phi_{d,0} - \Phi_d| \leq \pi/8$. Substituting the values of Φ_d and $\Phi_{d,0}$ given by equations (3.103) and (3.104) into the latter inequality and introducing the vertical gradient $v' = Bc_0$ of the velocity yields the following limitation on the ray skip distance

$$d \leq \left[(3/5)(c_0/v')^2\lambda\right]^{1/3}. \tag{3.105}$$

Let us estimate d for sound propagation in the atmosphere and ocean. For sound propagation in the atmosphere near the ground, setting $v' = 0.05$ s^{-1} and $c = 340$ m/s in equation (3.105) yields $d_m \leq 302.7\lambda_m^{1/3}$, where d_m and λ_m are the values of d and λ in meters. If $f = 500$ Hz, we have $d \leq 266.2$ m. For sound propagation in the ocean, setting $v' = 2.5 \times 10^{-3}$ s^{-1} and $c = 1500$ m/s yields $d_m \leq 6.0 \times 10^3\lambda_m^{1/3}$. If $f = 1$ kHz, the ray skip distance $d \leq 6.9$ km.

3.7   Acoustic tomography of the atmosphere

Acoustic tomography of the atmosphere enables one to reconstruct the temperature $T(\mathbf{R}, t)$ and wind velocity $\mathbf{v}(\mathbf{R}, t)$ fields within a tomographic volume or area and to monitor their evolution in time. Using various schemes of acoustic tomography, these fields have been reconstructed in different regions of the atmosphere: in the atmospheric surface layer (ASL), in the atmospheric boundary layer up to the heights of a few hundreds meters [79], and in the upper atmosphere up to the heights of about 100 km [85, 103, 213]. In this section, acoustic travel-time tomography of the ASL is briefly reviewed.

3.7.1   Forward and inverse problems in acoustic travel-time tomography

The idea of acoustic tomography of the atmosphere is similar to that in medicine where ultrasound waves or X-rays probe a particular organ of a human body to produce an image of that organ. In the case of acoustic travel-time tomography of the ASL, sound waves are used for measuring the travel times $t_{tr,i}$ of sound propagation between different pairs of sources (speakers) and receivers (microphones) which are usually located a few meters above the ground. Here, the subscript $i = 1, 2, ...N$ indicates a particular sound propagation path and N is the number of such paths. Using equations (3.58) and (3.43), the travel times $t_{tr,i}$ can be expressed in terms of the sound speed $c(\mathbf{R}, t)$ and wind velocity $\mathbf{v}(\mathbf{R}, t)$ within the tomographic area

$$t_{tr,i} = \int_{L_i} \frac{1}{u}\, dl = \int_{L_i} \frac{1}{(c^2 + 2c\mathbf{n}_i \cdot \mathbf{v} + v^2)^{1/2}}\, dl. \tag{3.106}$$

Here, the integration is over the ith sound propagation path and $\mathbf{n}_i$ is the unit vector normal to the wavefront of a sound wave propagating along the path.

Equation (3.106) is a nonlinear expression relating $t_{tr,i}$ with $c(\mathbf{R}, t)$ and $\mathbf{v}(\mathbf{R}, t)$ that significantly complicates solution of the inverse problem. In this equation, the integration can be performed along the straight line connecting the source and receiver if the propagation path L_i is less than a few hundred meters. Furthermore, it is worthwhile to express the sound speed and wind velocity fields as the following sums: $c(\mathbf{R},t) = c_0(t) + \tilde{c}(\mathbf{R},t)$ and $\mathbf{v}(\mathbf{R},t) = \mathbf{v}_0(t) + \tilde{\mathbf{v}}(\mathbf{R},t)$. Here, $c_0(t)$ and $\mathbf{v}_0(t)$ are the mean values (averaged over the tomographic area) of the sound speed and wind velocity, and $\tilde{c}(\mathbf{R},t)$ and $\tilde{\mathbf{v}}(\mathbf{R},t)$ are their fluctuating components. In equation (3.106), we introduce a small parameter $\varepsilon = \max\left(|\tilde{c}/c_0|, \tilde{v}/c_0\right)$. Omitting terms of order ε^2, this equation takes the form [391]

$$t_{tr,i} = \frac{L_i}{c_0}\left(1 - \frac{\mathbf{s}_i \cdot \mathbf{v}_0}{c_0}\right) - \frac{1}{c_0}\int_{L_i}\left(\frac{\tilde{c}}{c_0} + \frac{\mathbf{s}_i \cdot \tilde{\mathbf{v}}}{c_0}\right) dl. \tag{3.107}$$

Here, the integration is along the straight line from the source to the receiver, L_i is the length of this straight line, and $\mathbf{s}_i$ is the unit vector in the direction of the ith propagation path.

Equation (3.107) determines the *forward problem* in acoustic travel-time tomography of the ASL. It expresses the travel times $t_{tr,i}$ in terms of the sound speed $c(\mathbf{R},t)$, wind velocity $\mathbf{v}(\mathbf{R},t)$, the propagation paths L_i, and the unit vectors $\mathbf{s}_i$. The values of L_i and $\mathbf{s}_i$ can be determined if the transducer coordinates are known.

The *inverse problem* in acoustic travel-time tomography is to reconstruct the sound speed and wind velocity fields inside a tomographic area. In other words, one has to solve equation (3.107) for $c(\mathbf{R},t) = c_0(t) + \tilde{c}(\mathbf{R},t)$ and $\mathbf{v}(\mathbf{R},t) = \mathbf{v}_0(t) + \tilde{\mathbf{v}}(\mathbf{R},t)$ given the values of $t_{tr,i}$, L_i, and s_i. Different inverse algorithms can be used to solve this inverse problem. If the sound speed c is reconstructed, equation (1.1) enables calculation of the *acoustic virtual temperature*

$$T_{\text{vir}} = (1 + 0.511q)\, T, \tag{3.108}$$

where q is specific humidity. The thermodynamic temperature T can be obtained from this equation provided that specific humidity is known. Note that other meteorological devices such as an ultrasonic anemometer/thermometer and a radio acoustic sounding system (RASS) enable measurements of the acoustic virtual temperature T_{vir}. The temperature and wind velocity fields reconstructed in acoustic tomography of the atmosphere are important in many practical applications such as boundary layer meteorology, validation of large eddy simulation (LES), theories of turbulence, and wave propagation through a turbulent atmosphere.

Some velocity fields do not contribute to the travel time t_{tr} of sound propagation between a speaker and a microphone. Apparently such fields cannot be reconstructed in acoustic travel-time tomography [185]. Examples of these fields are given in references [45, 183]. However, the fields presented in these references are relatively simple and do not resemble atmospheric turbulence. It is an open question whether such velocity fields are present in the ASL and to what extent they affect acoustic travel-time tomography.

3.7.2 Arrays for acoustic travel-time tomography

The first array for acoustic tomography of the ASL was built at the Pennsylvania State University at the beginning of the 1990s [433]. In the array, three speakers and five microphones were located a few meters above the ground along the perimeter of a square with a side length 200 m. Sound signals were frequency-modulated acoustic pulses. By cross-correlating the transmitted signals with those recorded by microphones, the travel times $t_{tr,i}$ of sound propagation between speakers and microphones were determined. The total number of travel times was $N = 3 \times 5 = 15$. Then, the stochastic inversion algorithm was used to reconstruct the temperature T and wind velocity $\mathbf{v}$ fields from the measured travel times. An overview of initial efforts in acoustic travel-time tomography of the atmosphere is presented in reference [435].

In the mid-1990s, a portable array for acoustic tomography of the ASL was built at the Institute of Meteorology, the University of Leipzig, Germany. In this array, speakers and microphones were mounted on tripods 2 m above the ground. The size of the array varied and was of the order of several hundred meters. The number of speakers and microphones also varied and could be as large as 8 and 12, respectively. Speakers transmitted two short bursts of a harmonic signal separated by about 20 ms; the frequency range of the transmitted signal was around 1 kHz. Since the mid-1990s, this tomography array has been used in several experimental campaigns, e.g., references [9, 10, 320, 400, 441, 442]. Reconstruction of the temperature and wind velocity fields from the measured travel times $t_{tr,i}$ was performed using algebraic reconstruction. Recently, a Kalman filter has also been used in reconstruction [205].

Scientists from the University of Leipzig have also built a scalable, indoor prototype of their outdoor acoustic tomography array [22, 170]. The horizontal size of the indoor array varied from 1 m to 20 m. The transmitted signals were pseudorandom noise. The indoor array was built to test acoustic tomography in controlled environments and to compare different algorithms for reconstruction of the temperature and velocity fields. It can also be used for studies of turbulence in wind channels and over heated surfaces [22].

An array for acoustic tomography of the ASL has recently been built at the Boulder Atmospheric Observatory (BAO) which is a premier meteorological site, located in Erie, Colorado, with many instruments for measuring parameters of the atmospheric boundary layer. The array consists of eight towers, each of height 9.1 m, located along the perimeter of a square with the side length 80 m. Speakers and microphones can be installed on the towers at different levels; currently they are situated at a height of about 8 m thus enabling 2D horizontal slice tomography. Microphones are installed on five towers, while the remaining three towers carry speakers. Figure 3.13(a) depicts a schematic of the BAO acoustic tomography array in the horizontal plane (x, y). The speakers and microphones of the array are connected via underground cables with a central command and data acquisition computer, which is located inside a small modular building. Sound signals transmitted by speakers are similar to those used by German scientists in the outdoor tomography experiments.

Reconstruction of the temperature and wind velocity fields from the travel times $t_{tr,i}$ measured with the BAO tomography array is performed using the recently developed time-dependent stochastic inversion (TDSI) algorithm [391, 392]. The main idea of the TDSI is to assume that the temperature and wind velocity fluctuations are random fields both in space and time with known spatial-temporal correlation functions. To take the full advantage of this idea, the travel times $t_{tr,i}$ are measured repeatedly with a relatively short time interval between the measurements and then used in reconstruction of the T and $\mathbf{v}$ fields. Figures 3.13(b)–(d) show reconstruction of these fields in the tomography experiment carried out at the BAO on 9 July, 2008 [393]. Cold and warm temperature eddies, and slow and fast velocity eddies, are clearly seen in the figure. On the temperature plot, the arrows indicate the direction of the mean wind velocity. The TDSI algorithm has also been used for reconstruction of temperature and velocity fields in acoustic tomography experiments carried out in Germany [394].

A somewhat different scheme of acoustic tomography of the ASL was pro-

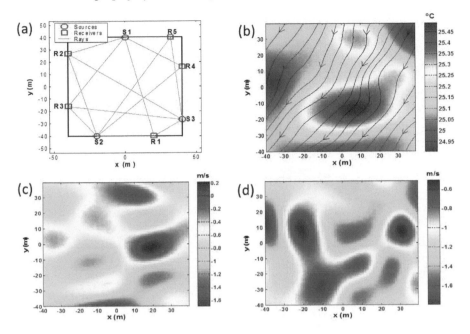

FIGURE 3.13
(a) Schematic of the acoustic tomography array at the Boulder Atmospheric
Observatory (BAO) in the horizontal plane (x, y). Black lines mark the perimeter of the tomographic area. Green lines are sound propagation paths (rays)
from speakers (sources) to microphones (receivers). (b) Temperature and (c),
(d) two components of the wind velocity fields reconstructed in the acoustic
tomography experiment at the BAO on 9 July, 2008 [393].

posed in reference [361] and experimentally implemented in [360]. In this
scheme, an array of microphones located near the ground recorded calling
birds. The positions of birds and time of acoustic calls were unknown. By
measuring the time differences between an acoustic call arrival at different
microphones and then using the inverse algorithm developed, it was possible
to locate a calling bird and to reconstruct the T and $\mathbf{v}$ fields inside the volume
between the bird and the microphones.

4

Wave theory of sound propagation in a stratified moving medium

For many applications, the atmosphere and ocean can be considered as stratified moving media. Geometrical acoustics enables one to calculate the ray paths and sound pressure in a stratified moving medium (sections 3.4–3.6). However, ray theory is not valid near caustics or if the sound wavelength λ is greater than the vertical scale l of the variations in the density, sound speed, or the medium velocity.

The wave theory of sound propagation in a stratified moving medium, considered in this chapter, makes it possible to overcome these difficulties. Section 4.1 presents the starting equations of wave theory. Acoustic sources are introduced in section 4.2. Propagation of a plane wave and the sound field due to a point source are considered in sections 4.3 and 4.4. In these sections, the effects of the medium motion on the sound-pressure field are also analyzed. The high-frequency approximation for the sound field due to a point source located above an impedance surface is studied in sections 4.5–4.6. Section 4.6 also deals with the calculation of the discrete spectrum of the sound field of a point source for any ratio λ/l.

4.1 Starting equations of wave theory

4.1.1 Starting equations

Let $p(\mathbf{r}, z, t)$ be the sound-pressure field of a wave propagating in a stratified moving medium. Here, t is the time, $\mathbf{r} = (x, y)$ are the horizontal coordinates, and z is the vertical coordinate. In section 2.3, calculation of the sound pressure p was reduced to the solution of equation (2.60) for the function $\tilde{p}(\boldsymbol{\kappa}, z, \omega)$, proportional to the spatial-temporal spectrum of p, and, then, to the calculation of the Fourier integral (2.59) over the horizontal components $\boldsymbol{\kappa}$ of the wave vector and the sound frequency ω. Equation (2.60) is an exact consequence of the set of linearized equations of fluid dynamics (2.32)–(2.37) and, hence, describes the propagation of both acoustic and internal gravity waves. In this chapter, we shall consider only acoustic waves with frequencies $\omega \gg \max(\omega_g, N)$, where N is the Brunt–Väisälä frequency and ω_g, defined in section 2.3.2, is of the order of 10^{-1} s^{-1} and 3×10^{-2} s^{-1} in the atmosphere and ocean, respectively. If this inequality is valid, equations (2.59) and (2.60) can be reduced to equations (2.63) and (2.64), which are repeated here:

$$p(\mathbf{r}, z, t) = \left(\frac{\varrho}{\varrho_1}\right)^{1/2} \int_{-\infty}^{\infty} \int_{-\infty}^{\infty} \left(1 - \frac{\boldsymbol{\kappa} \cdot \mathbf{v}}{\omega}\right) \tilde{p}(\boldsymbol{\kappa}, z, \omega) e^{i\boldsymbol{\kappa} \cdot \mathbf{r} - i\omega t} \, d\omega \, d^2\kappa, \quad (4.1)$$

$$\frac{d^2 \tilde{p}}{dz^2} + \left[q^2 - \frac{\boldsymbol{\kappa} \cdot \mathbf{v}''}{\omega - \boldsymbol{\kappa} \cdot \mathbf{v}} - 2 \left(\frac{\boldsymbol{\kappa} \cdot \mathbf{v}'}{\omega - \boldsymbol{\kappa} \cdot \mathbf{v}}\right)^2 + \frac{\varrho'}{\varrho} \frac{\boldsymbol{\kappa} \cdot \mathbf{v}'}{\omega - \boldsymbol{\kappa} \cdot \mathbf{v}} \right.$$

$$\left. + \frac{\varrho''}{2\varrho} - \frac{3}{4}\left(\frac{\varrho'}{\varrho}\right)^2 \right] \tilde{p} = i\omega(\varrho\varrho_1)^{1/2}\hat{Q}. \quad (4.2)$$

Here, $\varrho(z)$ is the density, $\mathbf{v}(z)$ is the horizontal velocity of the medium motion, a prime denotes the derivative of these ambient quantities with respect to z; $\varrho_1 = \varrho(z_1)$, where z_1 is a reference level; $\hat{Q}(\boldsymbol{\kappa}, z, \omega)$ is the spatial-temporal spectrum of a mass source, defined by equation (2.61); and q is the high-frequency approximation for the vertical wavenumber, given by

$$q(z) = \left[\frac{(\omega - \boldsymbol{\kappa} \cdot \mathbf{v}(z))^2}{c^2(z)} - \kappa^2\right]^{1/2}. \quad (4.3)$$

The Fourier integral (4.1) and the second-order ordinary differential equation (4.2) (or their modifications) are the starting equations of the wave theory of sound propagation in a stratified moving medium. It is worthwhile using these equations to calculate the sound pressure p, because various well-developed methods exist for solving ordinary differential equations and calculating Fourier integrals. Note that wave theory is valid for any ratio of λ to l and that equation (4.2) can be considered as a one-dimensional Helmholtz equation.

In many respects equation (4.2) is analogous to the equations used for investigating the stability of flows [33], reflection of a plane wave from inhomogeneous supersonic flows [131, 206], and sound propagation in ducts with mean flow [270, 371]. But the equations used in these studies differ from equation (4.2): in these equations, the direction of the vector $\mathbf{v}$ does not depend on z; the density ϱ is usually constant; and when deriving these equations, it is assumed that the ambient pressure P is constant. Usually, in the atmosphere and ocean, these assumptions are not valid. Furthermore, while the problems listed above have been treated as one-dimensional [131, 206] or two-dimensional [270, 371], sound propagation in the atmosphere and ocean is a three-dimensional problem and, hence, has certain important distinctions.

4.1.2   Early developments in wave theory

Probably, the paper [318] by Pridmore-Brown, published in 1962, was the first paper on the wave theory of sound propagation in a stratified moving medium. In this work, equation (2.79) was derived (see section 2.3) for the case of sound propagation in the atmosphere, and the discrete spectrum of the sound field of a point source located above the impedance ground was calculated approximately.

Subsequently, there have been several developments based on wave theory by Pierce [311], Tatarskii [375], Chunchuzov [77, 78], and Ostashev [281, 282]. In reference [311], the discrete spectrum of the sound field of a point source in a stratified moving atmosphere was calculated more accurately than that in reference [318]. In reference [375], the calculation of the sound pressure in a stratified moving atmosphere (which was considered as an ideal gas with the equation of state (2.39)) was reduced to solving equations (2.77) and (2.78) for the spatial-temporal spectra of the sound pressure and the vertical component of the particle velocity and, then, to calculating the Fourier integral (2.66). Moreover, a high-frequency approximation for the sound field of an acoustic antenna emitting in the vertical direction was obtained.

The sound field due to a point source located on hard ground, in an atmosphere with exponential profiles of temperature and wind velocity, has been calculated [77, 78]. The calculations start from equation (2.89) for the quasi-potential, the solution of which is represented in the form of the Fourier integral (2.90) with the integrand $\tilde{p}$ satisfying equation (2.91). It was shown in Chapter 2 of this book that the integral (2.90) and equation (2.91) are approximations to the integral (4.1) and equation (4.2), respectively. Furthermore, in these references the effective sound speed and effective density (see next subsection) were introduced implicitly. The equation for the quasi-potential was also used in reference [145] to obtain the low-frequency sound field of a point source in the atmosphere with wind stratification. Sound propagation in the nocturnal boundary layer has recently been investigated [399, 398] using the effective sound speed approximation. In particular, the "quite height" located a few meters above the ground was revealed theoretically and experimentally.

Equations (4.1) and (4.2) of wave theory have been obtained in references [281, 282]. In these references, the sound fields of a quasi-plane wave and point source were studied in the high-frequency approximation. Wave theory has been used for calculating the sound field of a point source in the ocean [149], for determining the coefficient of wave reflection by the upper atmosphere [212], for investigating the propagation of sound waves (including impulse propagation) near the ground [443], for studies of waveguide sound propagation in the atmosphere [285], and for calculating the high-frequency approximation for the sound field of a point source located above a porous ground surface [223]. In references [47]–[48], wave theory has also been used for calculating the fields of plane or quasi-plane waves in a layered moving medium, for determining the reflection coefficients of these waves by a moving layer, and for calculations of the discrete spectrum of the sound field due to a point source.

A fast-field program (FFP), developed originally for computations of a sound field in a motionless medium, was generalized to sound propagation in a stratified moving medium [224, 272, 323, 411]. This program is based on numerical solutions of equation (4.2) and its various modifications, or on a numerical solution of the set of equations (2.77) and (2.78), which is equivalent to equation (4.2) for acoustic waves.

4.1.3 Effective sound speed and density

If $\mathbf{v} = 0$, equations (4.1)–(4.3) coincide with those widely used for calculating the sound pressure in a motionless medium [49, 194]. Many approaches and methods known in the acoustics of a motionless medium are used for studying sound propagation in a moving medium in this chapter. Nevertheless, calculations of the sound pressure in a stratified moving medium cannot be reduced rigorously to similar calculations for a motionless medium. This can, however, be done approximately if the effective sound speed c_{eff} and the effective density ϱ_{eff} are introduced:

$$c_{\mathrm{eff}}(z) = c(z) + \mathbf{s}_0 \cdot \mathbf{v}(z), \quad \varrho_{\mathrm{eff}}(z) = \varrho(z) - 2\varrho_0 \mathbf{s}_0 \cdot \mathbf{v}(z)/c_0. \tag{4.4}$$

Here, c_0 and ϱ_0 are the reference values of c and ϱ, and $\mathbf{s}_0 = (\mathbf{R} - \mathbf{R}_1)/|\mathbf{R} - \mathbf{R}_1|$ is the unit vector in the direction from the source location $\mathbf{R}_1$ to the observation point $\mathbf{R}$. Also, we introduce the parameter $\varepsilon = \max\left(|\tilde{c}/c_0|, |\tilde{\varrho}/\varrho_0|, v/c\right)$, where $\tilde{c} = c - c_0$ and $\tilde{\varrho} = \varrho - \varrho_0$, assume that $\varepsilon \ll 1$, and ignore terms of order ε^2. As mentioned in the paragraph following equation (3.98), the unit vector $\mathbf{s}_0$ can be replaced with the unit vector $\mathbf{s}(z)$ tangential to the ray path from the source to the receiver if terms of order ε^2 are omitted. In this case, c_{eff} given by equation (4.4) coincides with the effective sound speed defined in Chapter 1 (equation (1.2)).

Note that the eikonal of a sound wave depends only on c and $\mathbf{v}$, see equation (3.81). Therefore, it is possible to use one effective function c_{eff} for the approximate formulation of the eikonal as in equation (3.98). The need to introduce the second effective function ϱ_{eff} in the current section is due to the fact that the sound field p depends not only on c and $\mathbf{v}$, but also on ϱ.

The vector $\mathbf{v}$ enters into the Fourier integral (4.1) and differential equation (4.2) only as the scalar product $\boldsymbol{\kappa} \cdot \mathbf{v}$. The integration over the domain where $\kappa \lesssim k_0 = \omega/c_0$ usually makes the main contribution to the integral (4.1). In this domain, the vector $\boldsymbol{\kappa}$ can be expressed in the form $\boldsymbol{\kappa} = k_0(\mathbf{r} - \mathbf{r}_1)/|\mathbf{R} - \mathbf{R}_1| + O(\varepsilon)$. (If $1 - \kappa/k_0 \gg \varepsilon$, this equation for $\boldsymbol{\kappa}$ follows from equation (3.95); if $1 - \kappa/k_0 \lesssim \varepsilon$, this equation follows from the latter inequality and the formula $|\mathbf{r} - \mathbf{r}_1| \approx |\mathbf{R} - \mathbf{R}_1|$ which is valid if the latter inequality is valid.) Using this equation for $\boldsymbol{\kappa}$ and taking into account that $\mathbf{v}$ is a horizontal vector, it can be shown that $\boldsymbol{\kappa} \cdot \mathbf{v} = k_0 \mathbf{s}_0 \cdot \mathbf{v} + O(\varepsilon^2)$. Substituting the value of $\boldsymbol{\kappa} \cdot \mathbf{v}$ into equations (4.1) and (4.2), neglecting terms of order ε^2, and introducing the new function $\tilde{\tilde{p}} = \varrho_1^{-1/2}\tilde{p}$, we obtain

$$p(\mathbf{r}, z, t) = \varrho_{\mathrm{eff}}^{1/2} \int_{-\infty}^{\infty} \int_{-\infty}^{\infty} \tilde{\tilde{p}}(\boldsymbol{\kappa}, z, \omega) e^{i\boldsymbol{\kappa} \cdot \mathbf{r} - i\omega t} \, d\omega \, d^2\kappa, \tag{4.5}$$

$$\frac{d^2 \tilde{p}}{dz^2} + \left[\frac{\omega^2}{c_{\text{eff}}^2} - \kappa^2 + \frac{\varrho_{\text{eff}}''}{2\varrho_{\text{eff}}} \right] \tilde{p} = i\omega \varrho_{\text{eff}}^{1/2} \hat{Q}_{\text{eff}}, \tag{4.6}$$

where $\hat{Q}_{\text{eff}} = (1 + v_R(z)/c_0)\hat{Q}(\kappa, z, \omega)$.

In a motionless layered medium, the integral (4.1) and equation (4.2) (where the term $(\varrho'/\varrho)^2 \lesssim \varepsilon^2$ should be neglected) are reduced to equations (4.5) and (4.6) if c_{eff}, ϱ_{eff}, and Q_{eff} are replaced by c, ϱ, and Q. Therefore, to an accuracy of ε, the introduction of the effective functions c_{eff}, ϱ_{eff}, and $\hat{Q}_{\text{eff}}$ allows us to replace the calculation of p in a stratified moving medium to that in a motionless layered medium. However, such a reduction can lead to significant phase distortion even for small ranges from the source, and also to amplitude distortion if the sound field is a sum of two or more sound waves at the receiver. In this chapter, we calculate p to any accuracy of v/c unless stated otherwise.

4.2   Acoustic sources

The right-hand side of equation (4.2) is proportional to the spatial-temporal spectrum $\hat{Q}$ of a sound source. In this chapter, we shall consider the following sound sources: the point mass source, an acoustic transmitting antenna, and a hypothetical source emitting a plane wave. All of these sources are assumed to be monochromatic. Let us consider these sources and show that the sound fields induced by them can be expressed through solutions of equation (4.2) with $\hat{Q} = 0$.

4.2.1   Point mass source

In the equations for sound waves (e.g., in equations (2.9), (2.58), and (2.60)), the point monochromatic mass source is given by the function

$$Q(\mathbf{R}, t) = \hat{a} \exp(-i\omega_s t)\delta(\mathbf{R} - \mathbf{R}_1). \tag{4.7}$$

Here, $\hat{a}$ is the amplitude, $\mathbf{R} = (\mathbf{r}, z)$ is a position vector in the Cartesian coordinate system, the vector $\mathbf{R}_1 = (\mathbf{r}_1, z_1)$ characterizes the source location, $\delta(\mathbf{R})$ is the delta function, and ω_s is the source frequency. The function $Q(\mathbf{R}, t)$ is related to the temporal spectrum $\hat{Q}(\mathbf{R}, \omega)$ of the source by the inverse Fourier transform similar to equation (2.87). (The difference between the functions $\hat{Q}(\mathbf{R}, \omega)$ and $\hat{Q}(\kappa, z, \omega)$ is explained in the text following that equation.) Using equations (2.87) and (4.7), one obtains $\hat{Q}(\mathbf{R}, \omega) = \hat{a}\delta(\omega - \omega_s)\delta(\mathbf{R} - \mathbf{R}_1)$. In a homogenous motionless medium, equation (2.88) for the temporal spectrum $\hat{p}(\mathbf{R}, \omega)$ of the sound pressure reads:

$$(\Delta + k^2)\hat{p} = i\varrho\omega_s\hat{a}(\omega - \omega_s)\delta(\mathbf{R} - \mathbf{R}_1), \tag{4.8}$$

where we substituted with the value of $\widehat{Q}(\mathbf{R}, \omega)$, and $k = \omega_s/c$ is the sound wavenumber. The solution of this equation is readily obtained. Substituting the resulting value of $\widehat{p}(\mathbf{R}, \omega)$ into equation (2.87) and calculating the integral over the frequency ω, we obtain the sound pressure due to the point mass source in a homogeneous motionless medium,

$$p(\mathbf{R}, t) = -\frac{i\varrho\omega_s\widehat{a}}{4\pi|\mathbf{R} - \mathbf{R}_1|} \exp\left(ik|\mathbf{R} - \mathbf{R}_1| - i\omega_s t\right). \tag{4.9}$$

In this book, calculations of the sound pressure are often done for a *unit-amplitude* point source, for which $\widehat{a} = i4\pi/(\varrho\omega_s)$. For this source, equation (4.9) simplifies

$$p(\mathbf{R}, t) = \frac{1}{|\mathbf{R} - \mathbf{R}_1|} \exp\left(ik|\mathbf{R} - \mathbf{R}_1| - i\omega_s t\right). \tag{4.10}$$

To obtain the sound pressure of a different point source, the corresponding result should be multiplied by $p_{\text{free}}R_0$, where p_{free} is the sound pressure at the distance $R_0 = 1$ m from the source in free space.

Substituting the value of $Q(\mathbf{R}, t)$ from equation (4.7) into equation (2.61), we obtain the spatial-temporal spectrum of the unit-amplitude point source

$$\widehat{Q}(\boldsymbol{\kappa}, z, \omega) = \frac{i\delta(z - z_1)\delta(\omega - \omega_s)}{\pi\omega_s\varrho_1} \exp(-i\boldsymbol{\kappa} \cdot \mathbf{r}_1). \tag{4.11}$$

Since $\widehat{Q}(\boldsymbol{\kappa}, z, \omega)$ is proportional to $\delta(\omega - \omega_s)$, the spatial-temporal spectrum $\tilde{p}(\boldsymbol{\kappa}, z, \omega)$ in equation (4.2) is also proportional to $\delta(\omega - \omega_s)$. This delta function enables calculation of the integral over ω in equation (4.1). As a result, the sound pressure $p(\mathbf{R}, t)$ is proportional to the factor $\exp(-i\omega_s t)$ as should be for a monochromatic source. To simplify notations, in the remainder of the chapter (unless stated otherwise) we omit the factors $\delta(\omega - \omega_s)$ and $\exp(-i\omega_s t)$, and also omit the arguments ω and t in the functions $\tilde{p}$, $\widehat{Q}$, and p. With these notations, equations (4.1) and (4.2) take the form

$$p(\mathbf{r}, z) = \left(\frac{\varrho}{\varrho_1}\right)^{1/2} \int_{-\infty}^{\infty} \left(1 - \frac{\boldsymbol{\kappa} \cdot \mathbf{v}}{\omega}\right) \tilde{p}(\boldsymbol{\kappa}, z)e^{i\boldsymbol{\kappa} \cdot \mathbf{r}} \, d^2\kappa, \tag{4.12}$$

$$\frac{d^2\tilde{p}(\boldsymbol{\kappa}, z)}{dz^2} + \left[q^2 - \frac{\boldsymbol{\kappa} \cdot \mathbf{v}''}{\omega - \boldsymbol{\kappa} \cdot \mathbf{v}} - 2\left(\frac{\boldsymbol{\kappa} \cdot \mathbf{v}'}{\omega - \boldsymbol{\kappa} \cdot \mathbf{v}}\right)^2 + \frac{\varrho'}{\varrho}\frac{\boldsymbol{\kappa} \cdot \mathbf{v}'}{\omega - \boldsymbol{\kappa} \cdot \mathbf{v}}\right.$$
$$\left. + \frac{\varrho''}{2\varrho} - \frac{3}{4}\left(\frac{\varrho'}{\varrho}\right)^2\right]\tilde{p}(\boldsymbol{\kappa}, z) = -\frac{\delta(z - z_1)}{\pi} \exp(-i\boldsymbol{\kappa} \cdot \mathbf{r}_1). \tag{4.13}$$

The right-hand side of the latter equation is pertinent for the unit-amplitude source. The solution of this equation can be expressed as (e.g., reference [49])

$$\tilde{p}(\boldsymbol{\kappa}, z) = \frac{\tilde{p}_1(\boldsymbol{\kappa}, z_>)\,\tilde{p}_2(\boldsymbol{\kappa}, z_<)}{\pi W} \exp(-i\boldsymbol{\kappa} \cdot \mathbf{r}_1). \tag{4.14}$$

Here, the functions $\tilde{p}_1$ and $\tilde{p}_2$ are the solutions of the homogeneous equation (4.13) with the right-hand side equal to zero and satisfy the radiation and initial conditions for $z > z_1$ and $z < z_1$, respectively; $W = \tilde{p}_1 d\tilde{p}_2/dz - \tilde{p}_2 d\tilde{p}_1/dz$ is the Wronskian; and $z_>$ and $z_<$ are the largest and smallest values of z and z_1, i.e., $z_> = \max(z, z_1)$ and $z_< = \min(z, z_1)$.

Substituting the value of $\tilde{p}$ given by equation (4.14) into equation (4.12), we obtain the sound field p of a point source in a stratified moving medium

$$p(\mathbf{r}, z) = \left(\frac{\varrho}{\varrho_1}\right)^{1/2} \int_{-\infty}^{\infty} \left(1 - \frac{\boldsymbol{\kappa} \cdot \mathbf{v}}{\omega}\right) \frac{\tilde{p}_1(\boldsymbol{\kappa}, z_>) \, \tilde{p}_2(\boldsymbol{\kappa}, z_<)}{\pi W} e^{i\boldsymbol{\kappa} \cdot (\mathbf{r} - \mathbf{r}_1)} \, d^2\kappa. \quad (4.15)$$

It follows from this formula that the sound pressure depends on the difference of the vectors $\mathbf{r}$ and $\mathbf{r}_1$. To emphasize that the point source is located at $\mathbf{R}_1 = (\mathbf{r}_1, z_1)$, the sound pressure can also be written as $p(\mathbf{R}; \mathbf{R}_1)$. The sound pressure $p_{\text{ext}}(\mathbf{R})$ due to the extended mass source, characterized by the function $Q(\mathbf{R})$, can be expressed in terms of $p(\mathbf{R}; \mathbf{R}_1)$:

$$p_{\text{ext}}(\mathbf{R}) = \int_{-\infty}^{\infty} \frac{Q(\mathbf{R}_1)}{\hat{a}} p(\mathbf{R}; \mathbf{R}_1) \, d^3 R_1. \quad (4.16)$$

4.2.2 Transmitting antenna

Acoustic transmitting antennas are used in sodars and for directional sound transmission. They consist either of a single speaker/driver in a center of a parabolic reflector or a phased array of speakers. The sound field of an extended or directional source can be recalculated to a plane located near the source. This plane can be considered as the aperture of an equivalent acoustic antenna.

In principle, equation (4.16) allows us to determine the field emitted by an acoustic antenna. But it is more convenient to express this field in terms of the sound pressure on the antenna aperture. Let the antenna aperture be located in the horizontal plane $z_1 = $ constant (this is often the case for sodars) and the sound pressure on it is $p_a(\mathbf{r})$. The spatial spectrum of the sound pressure is then

$$\hat{p}_a(\boldsymbol{\kappa}) = \frac{1}{4\pi^2} \int_{-\infty}^{\infty} p_a(\mathbf{r}) e^{-i\boldsymbol{\kappa} \cdot \mathbf{r}} \, d^2 r.$$

The sound field $p(\mathbf{r}, z)$ emitted by an antenna in a stratified medium is given by equation (4.12), where the function $\tilde{p}(\boldsymbol{\kappa}, z)$ is the solution of the homogeneous equation (4.13) and satisfies the initial condition:

$$\tilde{p}(\boldsymbol{\kappa}, z_1) = \frac{\hat{p}_a(\boldsymbol{\kappa})}{1 - \boldsymbol{\kappa} \cdot \mathbf{v}(z_1)/\omega}. \quad (4.17)$$

In some cases, the sound pressure on the antenna aperture can be regarded as a Gaussian function: $p_a(\mathbf{r}) = p_0 \exp(-r^2/D^2 + i\boldsymbol{\kappa}_0 \cdot \mathbf{r})$. Here, p_0 is the amplitude factor, D is the width of the emitted sound beam, and the vector $\boldsymbol{\kappa}_0$ characterizes the horizontal direction of propagation of this beam. The spatial spectrum on the aperture of the Gaussian antenna is given by

$$\widehat{p}_a(\boldsymbol{\kappa}) = \frac{p_0 D^2}{4\pi} \exp\left[-\frac{(\boldsymbol{\kappa} - \boldsymbol{\kappa}_0)^2 D^2}{4}\right]. \tag{4.18}$$

4.2.3   Plane wave source

A Gaussian antenna with the width $D \to \infty$ can be considered as a hypothetical plane wave source. In this limiting case, the spatial spectrum $\widehat{p}_a(\boldsymbol{\kappa})$, given by equation (4.18), is proportional to the delta function $\delta(\boldsymbol{\kappa} - \boldsymbol{\kappa}_0)$. Therefore, a solution of equation (4.13) can be expressed in the form

$$\tilde{p}(\boldsymbol{\kappa}, z) = \delta(\boldsymbol{\kappa} - \boldsymbol{\kappa}_0)\tilde{p}_{\text{pl}}(\boldsymbol{\kappa}, z), \tag{4.19}$$

where the subscript pl stands for "plane." The function $\tilde{p}_{\text{pl}}(\boldsymbol{\kappa}, z)$ satisfies the homogeneous equation (4.13). Substituting equation (4.19) into equation (4.12) and calculating the integral over $\boldsymbol{\kappa}$, we obtain the sound pressure due to a plane wave source:

$$p(\mathbf{r}, z) = \left(\frac{\varrho}{\varrho_1}\right)^{1/2} \left(1 - \frac{\boldsymbol{\kappa} \cdot \mathbf{v}}{\omega}\right) \tilde{p}_{\text{pl}}(\boldsymbol{\kappa}, z) \exp(i\boldsymbol{\kappa} \cdot \mathbf{r}). \tag{4.20}$$

(In this formula, we replaced $\boldsymbol{\kappa}_0$ with $\boldsymbol{\kappa}$.)

It is worthwhile to study the propagation of the waves described by equation (4.20) because the sound fields of real sources such as the acoustic antenna, the point source, and the extended source can be expressed as a superposition of these waves with different values of the horizontal wave vector $\boldsymbol{\kappa}$. In a stratified medium, a sound wave emitted by a plane wave source is usually affected by refraction and diffraction and becomes a *quasi-plane* wave.

4.3   Sound propagation in a homogeneous flow and reflection by a homogeneous moving layer

A homogeneous moving medium, also termed a homogeneous flow, is an important particular case of a stratified moving medium. The significance of investigating sound propagation in a homogeneous flow arises from the fact that exact formulas for the sound fields of a plane wave and a point source can be obtained. These results enable us to analyze the effect of medium motion on the amplitude and phase of the sound field.

4.3.1 Plane wave

First, we calculate the sound pressure $p(\mathbf{r}, z)$ of a plane wave. The wave theory is not applicable when the vertical component v_z of the velocity is finite. Therefore, we will calculate p starting from equation (2.100). In this equation, we replace the operator $\partial/\partial t$ by $-i\omega$ and omit the dependence of p on time. The solution of equation (2.100) is sought in the form of the Fourier integral (4.12). Substituting equation (4.12) into the homogeneous equation (2.100), we obtain the ordinary second-order differential equation for $\tilde{p}(\boldsymbol{\kappa}, z)$ with respect to z. The coefficients of the latter equation are constant. Solving this equation, substituting the value of $\tilde{p}(\boldsymbol{\kappa}, z)$ into equation (4.12), and taking account of the fact that $\tilde{p}$ is proportional to $\delta(\boldsymbol{\kappa} - \boldsymbol{\kappa}_0)$ for the plane wave source, we obtain the sound pressure of a plane wave in homogeneous flow:

$$p = p_0 \exp(i\mathbf{k} \cdot \mathbf{R}) = p_0 \exp(i\boldsymbol{\kappa} \cdot \mathbf{r} + iqz). \tag{4.21}$$

Here, p_0 is the amplitude of the wave, the horizontal vector $\boldsymbol{\kappa}$ characterizes the direction of plane wave propagation, $\mathbf{k} = (\boldsymbol{\kappa}, q)$ is the wave vector, and $q(z)$ is the vertical wavenumber defined by equation (3.65), which becomes equation (4.3) for $v_z = 0$. We assume that the intensity I_1 of the plane wave does not depend on the properties of the medium or the direction of wave propagation. The intensity I_1 is related to the amplitude p_0 of a sound wave by equation (3.36). (Note that in this equation, the amplitude is denoted as p_1.) Solving equation (3.36) for p_0, we obtain the exact formula for the sound pressure of a plane wave with intensity I_1, propagating in homogeneous flow:

$$p = \left[\frac{2\varrho c I_1}{|\mathbf{n} + \mathbf{v}/c| \, (1 + \mathbf{n} \cdot \mathbf{v}/c)} \right]^{1/2} \exp(i\boldsymbol{\kappa} \cdot \mathbf{r} + iqz), \tag{4.22}$$

where $\mathbf{n}$ is the unit vector normal to the wavefront.

As one would expect, the phase of the plane wave (4.22) coincides with its value calculated from geometrical acoustics (see equation (3.81)). Furthermore, using equation (3.65) for q, it can be shown that the frequency ω of this wave and its wave vector $\mathbf{k} = (\boldsymbol{\kappa}, q)$ are related by the dispersion equation (3.35). If $v_z = 0$, the dispersion equation becomes

$$\omega = kc + \boldsymbol{\kappa} \cdot \mathbf{v}. \tag{4.23}$$

4.3.2 Azimuthal dependence of the sound pressure

In a motionless layered medium, the sound fields p of a point source and of a plane wave do not depend on either the azimuthal (horizontal) direction from the source to the observation point or the azimuthal direction of plane wave propagation. The dependence of p on the azimuthal direction is one of the main effects of medium motion on the sound field.

Equation (4.22) enables us to analyze the dependence of the amplitude $|p|$ of the plane wave on the azimuthal angle ψ between the vectors $\mathbf{v}_\perp$ and $\boldsymbol{\kappa}$.

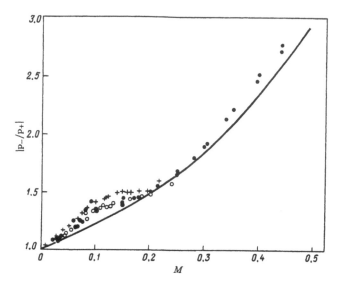

FIGURE 4.1
The ratio $|p_-/p_+|$ of the amplitudes of the sound waves propagating upstream and downstream versus the Mach number M [176]. The crosses and the solid and open points are the experimental data; the solid curve is the prediction from equation (4.24).

We assume that $-\pi < \psi \le \pi$, and $\psi > 0$ if the shortest rotation of vector $\boldsymbol{\kappa}$ until it coincides with vector $\mathbf{v}_\perp$ is anticlockwise. For simplicity, we set $v_z = 0$ in equation (4.22). It follows from equation (3.60) that $1 + \mathbf{n} \cdot \mathbf{v}/c = (1 - \boldsymbol{\kappa} \cdot \mathbf{v}/\omega)^{-1}$. Substituting the latter equality into equation (4.22), it can be shown that $|p(\psi_1, \kappa)| < |p(\psi_2, \kappa)|$ if $|\psi_1| < |\psi_2|$. In other words, the greater the angle ψ, the greater the amplitude of the plane wave. In particular, $|p_+| < |p_-|$, where $|p_+|$ and $|p_-|$ are the amplitudes of the plane waves propagating downstream and upstream, respectively, i.e., when $\mathbf{n} \cdot \mathbf{v} = v$ and $\mathbf{n} \cdot \mathbf{v} = -v$. From equation (4.22), it follows that

$$\left| \frac{p_-}{p_+} \right| = \left| \frac{1 + v/c}{1 - v/c} \right|. \tag{4.24}$$

It can be shown from equation (4.22) that the ratio $|p_-/p_+|$ is still given by equation (4.24) if $v_z \ne 0$.

   The dependence of $|p|$ on the azimuthal angle ψ can be explained with the use of equation (3.36). It follows from this equation that to create the same intensity I_1, the sound pressure downstream (when $\boldsymbol{\kappa} \cdot \mathbf{v} > 0$) must be less than that upstream (when $\boldsymbol{\kappa} \cdot \mathbf{v} < 0$).

   Another approach to the derivation of equation (4.24) has been published [176], in the context of the theoretical and experimental investigation of plane wave propagation in a rectangular duct containing a uniform flow. Experimental data for the dependence of the ratio $|p_-/p_+|$ on the Mach number $M = v/c$ of the flow are shown in figure 4.1, where the solid curve is the theoretical dependence (4.24). The crosses and the solid and open points are experimental data corresponding to various frequencies of the sound waves or to different geometries of the source and receiver. The figure shows that the experimental data are in a rather good agreement with the theoretical predictions if $M \gtrsim 0.2$. It is noted in reference [176] that this agreement would be better if one took into account the difference in the attenuation of the sound waves propagating downstream and upstream.

4.3.3 Point source

Let us now consider the sound field of a unit-amplitude point source in uniform flow. This field can be obtained from equation (5.28) for the sound pressure of the point source moving with the constant velocity in a homogeneous motionless medium. Indeed, in the coordinate system moving with the source, the source is at rest, while the medium is moving uniformly. Rearranging equation (5.28) for this coordinate system and assuming that the source is located at $\mathbf{R}_1 = 0$, we obtain the exact expression for the sound pressure of a point source in a homogeneous flow

$$p(\mathbf{R}) = \sum_{j=1}^{2} \exp\left[\frac{ikR}{(-1)^j(1 - M^2\sin^2\alpha)^{1/2} + M\cos\alpha}\right]$$
$$\left[\frac{(1 - M^2\sin^2\alpha)^{1/2} - (-1)^j M\cos\alpha}{R(1 - M^2\sin^2\alpha)(1 - M^2)} - \frac{iM\cos\alpha}{kR^2(1 - M^2\sin^2\alpha)^{3/2}}\right]. \quad (4.25)$$

Here, $k = \omega/c$ is the sound wavenumber, α is the angle between the vectors $\mathbf{R}$ and $\mathbf{v}$, and $0 \leq \alpha \leq \pi$. The term on the right-hand side of equation (4.25), corresponding to the index $j = 1$, should be omitted if $M < 1$. Therefore, if $M < 1$, there is only one sound wave reaching the observation point. On the other hand, for $M > 1$ and $\alpha < \alpha_M$, where $\alpha_M = \arcsin(1/M)$ is the Mach angle, there are two waves reaching the observation point, corresponding to the indices $j = 1$ and 2. Finally, if $M > 1$ and $\alpha > \alpha_M$, there is no wave reaching the point of observation, so that $p = 0$.

Let us explain qualitatively the distribution of the sound pressure p due to a point source in a homogeneous flow. The directions of propagation of the plane waves entering into the spectral representation (4.12) of such a source coincide with the directions of the group velocities of these waves $\mathbf{u} = \mathbf{v} + c\mathbf{n}$ (see equation (3.28)). Suppose that the beginning of the vector $\mathbf{v}$ is at the source, and the beginning of the vector $c\mathbf{n}$ is at the end of the vector $\mathbf{v}$ (figure 4.2). In this case, the beginning of the vector $\mathbf{u}$ is at the source, and the end of this vector coincides with the end of the vector $c\mathbf{n}$. For a point source, the direction of the vector $\mathbf{n}$ can be arbitrary. Therefore, the ends of all vectors

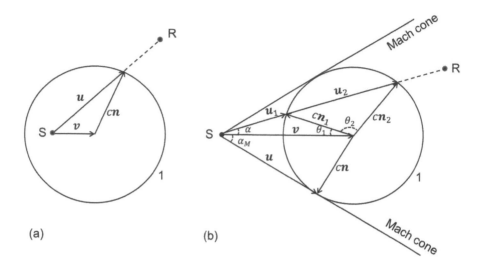

FIGURE 4.2
Group velocities of the sound waves emitted by a point source (S) in (a) subsonic and (b) supersonic homogeneous flows. The receiver is located at point R and the circle 1 is of radius c.

cn and, hence, the ends of all vectors $\mathbf{u}$ are located on a circle of radius c, the center of which is at the end of the vector $\mathbf{v}$.

For $v < c$, it can be seen from figure 4.2(a) that the direction of the vector $\mathbf{u}$ can be arbitrary and, hence, there is one and only one sound wave reaching the observation point. But for $v > c$, figure 4.2(b) shows that the angle α between the vectors $\mathbf{u}$ and $\mathbf{v}$ is always less than the Mach angle α_M; therefore, sound waves cannot reach the observation point located outside the Mach cone when $\alpha > \alpha_M$. There are two plane waves reaching the observation point located inside the Mach cone ($\alpha < \alpha_M$); the group velocities of these, $\mathbf{u}_1$ and $\mathbf{u}_2$, have the same direction but differ in magnitude. The unit vectors $\mathbf{n}_1$ and $\mathbf{n}_2$, normal to the wavefronts of these waves, are also different. It can be shown that the angles θ_1 and θ_2 between the vector $-\mathbf{v}$ and the vectors $\mathbf{n}_1$ and $\mathbf{n}_2$ are given by

$$\theta_{1,2} = \arccos\left[M \sin^2 \alpha \pm (1 - M^2 \sin^2 \alpha)^{1/2} \cos \alpha \right].$$

This qualitative explanation of the formation of the Mach cone in a supersonic flow also enables us to explain the formation of this cone if a source is moving with supersonic speed in a motionless medium. Indeed, in the coordinate system moving with the supersonic flow, the medium is at rest while the source is moving with a velocity $-\mathbf{v}$; therefore, the insonified spatial domain

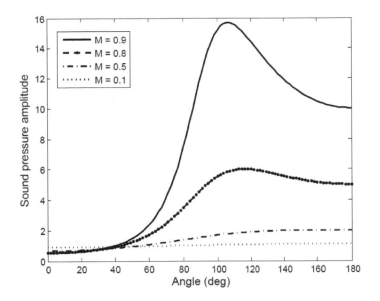

FIGURE 4.3
Dependence of the sound-pressure amplitude $|p|$ (in arbitrary units) due to a point source in subsonic homogeneous flow on the angle α between vectors $\mathbf{R}$ and $\mathbf{v}$.

is located inside the Mach cone depicted in figure 4.2, and there are two sound waves reaching the observation point.

Now let us consider the dependence of the sound-pressure amplitude $|p|$ on the angle α for a point source in subsonic homogeneous flow. If $k_0 R \gg 1$, the second term in the braces in equation (4.25) is much less than the first term and can be ignored. In this case, $|p|$ is given by

$$|p| = \frac{(1 - M^2 \sin^2 \alpha)^{1/2} - M \cos \alpha}{R(1 - M^2 \sin^2 \alpha)(1 - M^2)}. \tag{4.26}$$

Figure 4.3 depicts the dependence of the $|p|$ (in arbitrary units) on the angle α, obtained for $R = 1$ m and different values of M, using equation (4.26). It can be shown that for $M \leq 0.5$, the sound-pressure amplitude increases monotonically as the angle α is increased. But if $M > 0.5$, the function $|p(\alpha)|$ has the maximum at $\alpha = \alpha_M$, where $\cos \alpha_M = -[(1 - M^2)/3M^2]^{1/2}$. If $M \to 1 - 0$, the angle $\alpha_M \to \pi/2 + 0$, and $|p(\alpha_M)| \to \infty$. The dependence of $|p|$ on α, appearing for $M \to 1 - 0$, can be interpreted qualitatively as the "fore-runner" of the Mach cone.

The angle α might not coincide with the azimuthal angle characterizing the horizontal direction from the source to the receiver. However, this coincidence always exists for $\alpha = 0$ and $\alpha = \pi$. In this case, the observation points are located downstream and upstream, respectively. It follows from equation (4.26) that $|p(\alpha = 0)| = |p_+| = 1/(1 + M)R$ and $|p(\alpha = \pi)| = |p_-| = 1/(1 - M)R$. The ratio of the sound-pressure amplitudes $|p_-|$ and $|p_+|$ is given by $|p_-/p_+| = (1 + M)/(1 - M)$, which coincides with equation (4.24) obtained previously for a plane wave. Such a coincidence is explained by the fact that a spherical wave can be considered as a plane wave for large enough ranges from the point source.

Equation (4.25) can also be derived by other approaches. In reference [283], wave theory is used to derive this equation in the case $M < 1$. The sound field of a point source in a homogeneous flow is usually obtained [37, 354, 395] using the equation for the velocity potential φ, which is related to the sound pressure by $p = \varrho(\partial/\partial t + \mathbf{v} \cdot \nabla)\varphi$. Equation (4.25) agrees with equation (1.91) of reference [37] and equation (1.6) of reference [395] if the amplitude factor $(1 - M^2)^{1/2}$ appearing in these references is omitted. This factor is probably due to the non-rigorous introduction of the source in these references.

4.3.4 Reflection of a plane wave by a homogeneous moving layer

In the remainder of this section, we consider the reflection of a plane sound wave from a homogeneous moving layer located between the planes $z = 0$ and $z = d$ (figure 4.4). For $0 < z < d$, the values of the quantities c, ϱ, and $\mathbf{v}$ are denoted by c_2, ϱ_2, and $\mathbf{v}_2$. Suppose that for $z < 0$ and $z > d$ these quantities are constant also, and given by $c_1, \varrho_1, \mathbf{v}_1$ and $c_3, \varrho_3, \mathbf{v}_3$, respectively. The vertical components of the medium velocities $\mathbf{v}_i$, where $i = 1, 2, 3$, are equal to zero. Generally, the vectors $\mathbf{v}_i$ are not collinear so that the medium velocity can vary stepwise at $z = 0$ and $z = d$. Actually, since such a motion of the medium is unstable, the velocity $\mathbf{v}$ varies in a layer of thickness h. However, if the sound wavelength $\lambda \gg h$, we can set $h = 0$ when considering the reflection of the wave at the interface of relative motion.

Suppose that a plane sound wave given by equation (4.21) is incident on the moving layer from the spatial domain $z < 0$. There is also a reflected wave in this spatial domain. Therefore, for $z < 0$, the sound pressure p can be expressed in the form

$$p = p_0 \exp(i\boldsymbol{\kappa} \cdot \mathbf{r} + iq_1 z) + V p_0 \exp(i\boldsymbol{\kappa} \cdot \mathbf{r} - iq_1 z). \tag{4.27}$$

In the moving layer ($0 < z < d$), the sound pressure is the sum of the fields of two plane waves:

$$p = A p_0 \exp(i\boldsymbol{\kappa} \cdot \mathbf{r} + iq_2 z) + B p_0 \exp(i\boldsymbol{\kappa} \cdot \mathbf{r} - iq_2 z). \tag{4.28}$$

For $z > d$, there is only the wave propagating in the positive direction of the z-axis:

$$p = W p_0 \exp(i\boldsymbol{\kappa} \cdot \mathbf{r} + iq_3(z - d)). \tag{4.29}$$

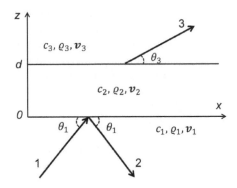

FIGURE 4.4
Reflection of a plane sound wave by a homogeneous moving layer. Straight
lines with arrows indicate (1) incident, (2) reflected, and (3) transmitted
waves.

In these equations, V is the reflection coefficient of a plane wave incident
on the moving layer, W is the transmission coefficient of this wave, A and
B are the amplitudes of the plane waves in the layer, and q_i are the vertical
wavenumbers given by equation (4.3) with c and $\mathbf{v}$ replaced by c_i and $\mathbf{v}_i$.
When deriving equations (4.27)–(4.29), we took into account the fact that the
horizontal component of the wave vector of the plane wave is constant; i.e.,
$\kappa = \text{constant}$. This equality follows from equation (4.20) and the fact that the
moving medium considered here is a particular case of a stratified medium.
From this equality and the dispersion equation (4.23), where $k = (\kappa^2 + q^2)^{1/2}$,
it follows that the vertical wavenumbers of the plane waves in the homogeneous
layers can differ only in their signs (see equations (4.27) and (4.28)).
Without loss of generality, suppose that the wave vector $\mathbf{k}_1 = (\kappa, q_1)$ of
the incident wave is located in the xz-plane. Since $\kappa = \text{constant}$, the wave
vector $\mathbf{k}_1' = (\kappa, -q_1)$ of the wave reflected by the layer, the wave vectors
$\mathbf{k}_2 = (\kappa, q_2)$ and $\mathbf{k}_2' = (\kappa, -q_2)$ of the waves in the layer, and the wave vector
$\mathbf{k}_3 = (\kappa, q_3)$ of the wave transmitted through the layer are all in the xz-plane.
Let θ_i denote the elevation angles between the vectors $\mathbf{k}_i$ and the horizontal
plane. Since the vectors $\mathbf{k}_1$ and $\mathbf{k}_1'$ differ only in the signs of their vertical
components, the elevation angle of the wave reflected by the layer is equal to
the elevation angle θ_1 of the incident wave (see figure 4.4). Substituting the
value of k_i obtained from equation (4.23) into the equality $\cos\theta_i = \kappa/k_i$, we
transform this equality to the form

$$\frac{c_i}{\cos\theta_i} + \mathbf{e} \cdot \mathbf{v}_i = \frac{\omega}{\kappa}, \qquad (4.30)$$

where $\mathbf{e} = \boldsymbol{\kappa}/\kappa$. Equation (4.30) can be considered as the refraction law for
the normal $\mathbf{n}_i(\theta) = \mathbf{k}_i/k_i$ to the wavefront of the plane wave at the interface
between homogeneous moving layers. It can be shown that this law is valid

for an arbitrary number of layers. Equation (4.30) has the same form as the refraction law, equation (3.69), for the wavefront of a sound wave propagating in a stratified moving medium with $v_z = 0$, obtained in geometrical acoustics. Using equation (4.30), the elevation angle θ_3 of the wave transmitted through the layer can be expressed in terms of the angle θ_1:

$$\cos\theta_3 = \frac{c_3 \, \cos\theta_1}{c_1 + \mathbf{e} \cdot (\mathbf{v}_1 - \mathbf{v}_3)\cos\theta_1}.$$

The coefficients V and W can be determined by making use of boundary conditions. Using equations (4.27)–(4.29) and taking account of the fact that the sound pressure p is continuous at $z = 0$ and $z = d$, we obtain

$$1 + V = A + B, \quad A\exp(iq_2 d) + B\exp(-iq_2 d) = W. \tag{4.31}$$

The vertical displacement $\zeta_z(\mathbf{R}, t)$ of the particles of the medium, caused by the propagating sound wave, is also continuous at $z = 0$ and $z = d$ [252, 331]. (Here, we temporarily restore the dependence of the acoustic quantities on time.) Generally, the vertical component $w_z(\mathbf{R}, t)$ of the acoustic particle velocity, given by $w_z = d\zeta_z/dt$, where $d/dt = \partial/\partial t + \mathbf{v} \cdot \nabla$, is discontinuous at $z = 0$ and $z = d$.

Substituting the value of w_z into equation (2.32) and taking into account the fact that $P' = F_z = 0$ for the problem considered, we obtain the following equation:

$$\varrho\frac{d^2\zeta_z}{dt^2} = -\frac{\partial p}{\partial z}.$$

Since p and ζ_z are proportional to $\exp(i\boldsymbol{\kappa} \cdot \mathbf{r} - i\omega t)$ for a plane wave, this equation can be expressed in the form

$$\zeta_z = \frac{1}{q\omega Z}\frac{\partial p}{\partial z},$$

where

$$Z = \frac{\varrho\omega(1 - \boldsymbol{\kappa} \cdot \mathbf{v}/\omega)^2}{q} = \frac{\varrho c}{\sin\theta(1 + \cos\theta\,\mathbf{e} \cdot \mathbf{v}/c)}. \tag{4.32}$$

The quantity Z is proportional to the characteristic acoustic impedance of a medium, ϱc. For $\mathbf{v} = 0$, the value of Z is denoted as $Z_0 = \varrho c/\sin\theta$.

Let us apply the operators $(q_i\omega Z_i)^{-1}\partial/\partial z$, where $i = 1, 2, 3$, to both sides of equations (4.27), (4.28), and (4.29), respectively. Here, Z_i is the value of Z for $c = c_i$, $\varrho = \varrho_i$, $\mathbf{v} = \mathbf{v}_i$, and $\theta = \theta_i$. The left-hand sides of the resulting equations are equal to the vertical displacements ζ_z of the particles of the medium. Taking account of the continuity of ζ_z at $z = 0$ and $z = d$ yields

$$(1 - V)Z_1^{-1} = (A - B)Z_2^{-1}, \quad (Ae^{iq_2 d} - Be^{-iq_2 d})Z_2^{-1} = WZ_3^{-1}. \tag{4.33}$$

Equations (4.31) and (4.33) can be solved for the reflection coefficient V and the transmission coefficient W:

$$V = \frac{(Z_2 + Z_1)(Z_3 - Z_2)e^{iq_2d} + (Z_2 - Z_1)(Z_3 + Z_2)e^{-iq_2d}}{(Z_2 - Z_1)(Z_3 - Z_2)e^{iq_2d} + (Z_2 + Z_1)(Z_3 + Z_2)e^{-iq_2d}}, \tag{4.34}$$

$$W = \frac{4Z_2 Z_3}{(Z_2 - Z_1)(Z_3 - Z_2)e^{iq_2d} + (Z_2 + Z_1)(Z_3 + Z_2)e^{-iq_2d}}. \tag{4.35}$$

It follows from these equations that the dependence of the reflection and transmission coefficients on the thickness of the layer d and the vertical wavenumber $q_2 = \kappa \tan \theta_2$ has an oscillatory character. It has been demonstrated [232] that equations (4.34) and (4.35) can be obtained from the well-known equations for the reflection and transmission coefficients for the case $\mathbf{v} = 0$ (e.g., reference [49]), if in the latter equations the quantity Z_0 and the vertical wavenumber $q_0 = (\omega^2/c^2 - \kappa^2)^{1/2}$ in a motionless medium are replaced with the quantity Z and the vertical wavenumber q in a moving medium. This result is also valid if the plane wave is reflected by an arbitrary number of homogeneous moving layers.

The reflection of a plane wave from a homogeneous moving layer has been treated by many authors [197, 232, 440]. One of the possible practical applications of this problem is to screen the noise of a turbulent jet by a high-temperature gas flow parallel to the jet axis.

For $d = 0$, the problem considered above is reduced to that of plane wave reflection by an interface between moving media, which was first solved by Miles [252] and Ribner [331]. If $d = 0$, it follows from equations (4.34) and (4.35) that

$$V = \frac{Z_3 - Z_1}{Z_3 + Z_1}, \quad W = \frac{2Z_3}{Z_3 + Z_1}. \tag{4.36}$$

It has been shown [331] that the magnitude $|V|$ of the plane wave reflection coefficient can be greater than 1 (in a motionless medium $|V| \leq 1$). The reflection of a spherical wave by a moving medium has also been studied [47, 396].

4.4 High-frequency approximation for the sound field

In section 4.2, the sound fields of a transmitting antenna, a point source, and a plane wave (equations (4.12), (4.15), and (4.20), respectively) are expressed in terms of the functions $\tilde{p}(\kappa, z)$ which are solutions of the homogeneous equation (4.13). In the present section, we obtain high-frequency approximations for these functions and calculate the sound fields of these sources in a stratified moving medium.

Some of the coefficients in equation (4.13) become infinite if $\omega - \boldsymbol{\kappa} \cdot \mathbf{v} \to 0$. The level where the equality $\omega - \boldsymbol{\kappa} \cdot \mathbf{v} = 0$ is valid is called the *critical level.* Such levels can appear, for example, if a plane wave is reflected by an inhomogeneous supersonic flow [47, 131, 206]. Near the critical level there is a resonance interaction between the sound wave and the moving medium. In what follows, we assume always that $\omega \neq \boldsymbol{\kappa} \cdot \mathbf{v}$.

4.4.1   High-frequency approximation

In equation (4.13), q^2 is proportional to ω^2/c_0^2, while the other terms in square brackets are proportional to l, where l is the characteristic scale of variation of $\varrho(z)$, $c(z)$, and $\mathbf{v}(z)$. In the high-frequency approximation, in which $\omega/c_0 \gg 1/l$ and $\lambda \ll l$, these terms can be neglected. This approximation is also called the short-wavelength, quasi-classic, or WKB approximation. With this approximation, the homogeneous equation (4.13) is simplified significantly and takes the form

$$\frac{d^2 \tilde{p}(\boldsymbol{\kappa}, z)}{dz^2} + q^2 \tilde{p}(\boldsymbol{\kappa}, z) = 0. \tag{4.37}$$

The high-frequency asymptotic form of the solution of this equation depends on the function $q^2(z)$. In the remainder of this chapter, we consider profiles $c(z)$ and $\mathbf{v}(z)$ such that $q^2(z) \neq 0$ or the function $q^2(z)$ has only one zero of first order, except where stated otherwise. (The function $q^2(z)$ has a zero of first order at $z = z_t$ if $q^2(z_t) = 0$ and $q^2 \approx (z - z_t)C$ at $z \approx z_t$, where C is a constant.) For such functions $q^2(z)$, it is convenient to use in equation (4.37) a new independent variable $\xi(z)$ and a new unknown function $\Pi(\xi)$, given by

$$\xi \left(\frac{d\xi}{dz} \right)^2 = q^2(z), \quad \tilde{p} = \frac{\Pi}{(d\xi/dz)^{1/2}}. \tag{4.38}$$

In this case equation (4.37) takes the form

$$\frac{d^2 \Pi}{d\xi^2} + \left[\xi + \frac{3}{4} \frac{(d^2\xi/dz^2)^2}{(d\xi/dz)^4} - \frac{1}{2} \frac{d^3\xi/dz^3}{(d\xi/dz)^3} \right] \Pi = 0. \tag{4.39}$$

The first term in square brackets, ξ, is proportional to $\omega^{2/3}$, while the other two terms are proportional to $\omega^{-4/3}$. For high frequencies, the latter terms should be neglected. With this approximation, the solution $\Pi(\xi)$ of equation (4.39) is a linear combination of the Airy functions $\mathrm{Ai}(-\xi)$ and $\mathrm{Bi}(-\xi)$. Substituting the value of Π into equation (4.38), we obtain an approximate solution of the homogeneous equation (4.13) at high frequencies, valid for arbitrary z:

$$\tilde{p}(\boldsymbol{\kappa}, z) = (\xi/q^2)^{1/4} \left[C_1 \mathrm{Ai}(-\xi) + C_2 \mathrm{Bi}(-\xi) \right], \tag{4.40}$$

where C_1 and C_2 are constants. If $\xi \gg 1$, the square of the vertical wavenumber

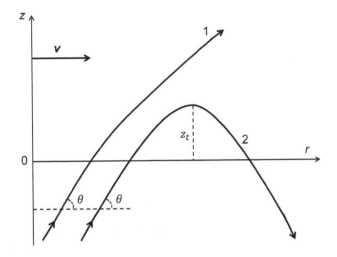

FIGURE 4.5

Ray paths corresponding to quasi-plane sound waves (1) without and (2) with a turning point. The direction of the medium velocity $\mathbf{v}$ coincides with the positive direction of the r-axis and z_t is the height of the turning point.

$q^2(z) > 0$ and the Airy functions $\mathrm{Ai}(-\xi)$ and $\mathrm{Bi}(-\xi)$ can be replaced by their asymptotic forms for large values of the argument. In this case, equation (4.40) becomes

$$\tilde{p}(\kappa, z) = q^{-1/2} \left[C_3 \exp\left(i \int_{z_0}^{z} q(z')\,dz' \right) + C_4 \exp\left(-i \int_{z_0}^{z} q(z')\,dz' \right) \right], \quad (4.41)$$

where C_3 and C_4 are constants and z_0 is a reference level.

4.4.2 Sound field of a quasi-plane wave

Equations (4.40) and (4.41) for $\tilde{p}$, together with equations (4.12), (4.15), and (4.20), allow us to obtain the sound fields due to sources considered in this section. First, we calculate the field of a quasi-plane wave.

Let a plane sound wave be incident on a semi-infinite inhomogeneous moving layer located in the half space $z > 0$ (figure 4.5). In the half space $z < 0$, the medium is homogeneous and at rest: $c = c_1 = $ constant, $\varrho = \varrho_1 = $ constant, and $\mathbf{v} = 0$. For $z < 0$, the sound pressure of the incident wave is given by $p = p_0 \exp(i\boldsymbol{\kappa} \cdot \mathbf{r} + iq_1 z)$, where q_1 is the value of q for $z < 0$. The functions c, ϱ, and $\mathbf{v}$ are continuous at $z = 0$. We consider first the case when $q^2(z) > 0$ for $z > 0$. In this case, the sound wave propagating in the positive direction of the z-axis is determined by equation (4.20). The function $\tilde{p}_{\mathrm{pl}}(\kappa, z)$ in this equation is given by equation (4.41), where $C_4 = 0$ and C_3 is chosen so that the sound pressure is continuous at $z = 0$. Substituting equation (4.41) into equation (4.20), we obtain the high-frequency approximation for the sound field of a quasi-plane wave for $z > 0$:

$$p(\mathbf{r}, z) = \left(\frac{\varrho q_1}{\varrho_1 q}\right)^{1/2} \left(1 - \frac{\boldsymbol{\kappa} \cdot \mathbf{v}}{\omega}\right) p_0 \exp\left(i\boldsymbol{\kappa} \cdot \mathbf{r} + i\int_0^z q\,dz'\right). \qquad (4.42)$$

In this equation, the direction of the vector $\boldsymbol{\kappa}$ coincides with the azimuthal direction of propagation of the plane wave for $z < 0$, and the modulus of $\boldsymbol{\kappa}$ is related to the elevation angle θ (see figure 4.5) of this wave by $\kappa c_1/\omega = \cos\theta$. In equation (4.42), the medium velocity $\mathbf{v}$ is present in the amplitude factor $(1 - \boldsymbol{\kappa} \cdot \mathbf{v}/\omega)$ and in the vertical wavenumber q that appears in the amplitude and phase. Since $q(z)$ can vary significantly even if $v/c \ll 1$, not only the phase but also the amplitude of the quasi-plane wave can depend significantly on $\mathbf{v}$ in both the atmosphere and the ocean. The dependence of $|p|$ on the azimuthal direction of wave propagation has been studied elsewhere [284].

The high-frequency approximation of wave theory and geometrical acoustics are two methods for describing a sound field, which supplement each other. Therefore, as one would expect, the vertical wavenumber, phase, and amplitude of the quasi-plane wave given by equation (4.42) agree with their values calculated in Chapter 3 (see equations (3.65), (3.81), and (3.91), respectively). It can be shown that the frequency ω and the wave vector $\mathbf{k} = (\boldsymbol{\kappa}, q)$ of this wave are related by the dispersion equation (4.23), where $\mathbf{k}$, c, and $\mathbf{v}$ depend on z. Equation (4.23) agrees with the dispersion equation (3.35) obtained using the geometrical acoustics approximation. The phase and group velocities $\mathbf{U}$ and $\mathbf{u}$ of the quasi-plane wave (4.42), the unit vectors $\mathbf{n} = \mathbf{k}/k$ and $\mathbf{s} = \mathbf{u}/u$, etc. are determined by the equations presented in Chapter 3. This wave corresponds to a sound ray determined by the vector $\mathbf{s}(z)$.

4.4.3 Reflection of a quasi-plane wave by an inhomogeneous moving layer

Now we analyze the case in which the vertical wavenumber $q(z)$ of the quasi-plane sound wave becomes zero at a level $z = z_t > 0$. In this case, the sound ray corresponding to this wave has a turning point at the level z_t. Therefore, for $z < 0$, there is a wave reflected by the inhomogeneous layer in addition to the wave incident on it (see figure 4.5). Geometrical acoustics does not allow us to calculate the sound field of the reflected wave and the sound field for $z > 0$. These fields can be obtained in the high-frequency approximation.

If $z < 0$, the sound field is a sum of the fields of the incident and reflected waves,

$$p = p_0 \exp(i\boldsymbol{\kappa} \cdot \mathbf{r} + iq_1 z) + V p_0 \exp(i\boldsymbol{\kappa} \cdot \mathbf{r} - iq_1 z). \qquad (4.43)$$

Here, V is the reflection coefficient of a plane wave from an inhomogeneous moving layer.

For $z > 0$ the sound field p is given by equation (4.20), where $\tilde{p}_{\mathrm{pl}}(\boldsymbol{\kappa}, z)$

is determined by equation (4.40). For the problem considered, in the latter equation, the dimensionless vertical coordinate $\xi > 0$ if $0 \leq z \leq z_t$, and $\xi < 0$ if $z > z_t$. Using these inequalities and equation (4.38) yields:

$$\xi(z) = \left(\frac{3}{2} \int_z^{z_t} q \, dz' \right)^{2/3} \quad \text{for } 0 \leq z \leq z_t,$$

$$\xi(z) = -\left(\frac{3}{2} \int_{z_t}^z (-q^2)^{1/2} \, dz' \right)^{2/3} \quad \text{for } z \geq z_t.$$

It can be shown that for these values of $\xi(z)$, the function $\tilde{p}_{\text{pl}}(\boldsymbol{\kappa}, z)$ given by equation (4.40) satisfies the radiation condition at $z \to \infty$ only if $C_2 = 0$. Substituting $\tilde{p}_{\text{pl}}(\boldsymbol{\kappa}, z)$ into equation (4.20), we obtain the sound field for $z > 0$ [281]

$$p = 2K \left(\frac{\pi \varrho q_1}{\varrho_1 q} \right)^{1/2} \xi^{1/4} \left(1 - \frac{\boldsymbol{\kappa} \cdot \mathbf{v}}{\omega} \right) \text{Ai}(-\xi) p_0 \exp(i\boldsymbol{\kappa} \cdot \mathbf{r}), \tag{4.44}$$

where K is a new constant.

The coefficients V and K in equations (4.43) and (4.44) can be obtained using the continuity conditions for p and ζ_z at $z = 0$, which are equivalent to the continuity conditions for p and $\partial p / \partial z$. We present the values of V and K for the case $\xi(z = 0) \gg 1$, when the Airy function $\text{Ai}(-\xi)$ in equation (4.44) can be replaced by its asymptotic form:

$$V = \exp\left(2i \int_0^{z_t} q \, dz - \frac{i\pi}{2} \right), \quad K = \exp\left(i \int_0^{z_t} q \, dz - \frac{i\pi}{4} \right).$$

It can be shown from equation (4.44) that the sound field p decays exponentially for $z > z_t$.

4.4.4 Two-dimensional method of stationary phase

In what follows, we shall often make use of the two-dimensional method of stationary phase for calculating the integrals over $\boldsymbol{\kappa}$ in equations such as (4.12) and (4.15). In this subsection, we present the basic equations of the method. Let

$$p(\mathbf{r}, z) = \int_{-\infty}^{\infty} U(\boldsymbol{\kappa}', z) \exp(im\Phi(\boldsymbol{\kappa}', \mathbf{r}, z)) \, d^2\kappa',$$

where m is a large positive number, U and Φ are functions, and Φ has only one stationary point $\boldsymbol{\kappa}$, defined by the equation $\partial \Phi / \partial \boldsymbol{\kappa} = 0$. For $m \to \infty$, the two-dimensional integral can be expressed in terms of an asymptotic series with respect to powers of $1/m$. Retaining only the first term of this series yields

$$p = \frac{2\pi}{m |J|^{1/2}} U(\boldsymbol{\kappa}, z)(1 + O(m^{-1})) \exp\left(im\Phi(\boldsymbol{\kappa}, \mathbf{r}, z) + i\pi s / 4 \right). \tag{4.45}$$

Here, $J = \det[\partial^2 \Phi / \partial \kappa_i \partial \kappa_j]$; $i, j = 1$ or 2; $\boldsymbol{\kappa} = (\kappa_1, \kappa_2)$; and $s = \operatorname{sgn}[\partial^2 \Phi / \partial \kappa_i \partial \kappa_j] = \nu_+ - \nu_-$ is the signature, where ν_+ and ν_- are the numbers of positive and negative eigenvalues of the matrix $[\partial^2 \Phi / \partial \kappa_i \partial \kappa_j]$, respectively.

4.4.5   Sound field due to a point source

Next we calculate the sound field of a unit-amplitude point source. Suppose that the source is located at the point $\mathbf{R}_1 = (\mathbf{r}_1 = 0, z_1)$ and there are no turning points of the sound rays (i.e., $q^2 > 0$) in the spatial domain where the sound field is considered. The sound pressure p of this point source is determined from equation (4.15). The functions $\tilde{p}_1$ and $\tilde{p}_2$ in this equation are given by equation (4.41), where the constants C_3 and C_4 are chosen so that $\tilde{p}_1$ corresponds to the wave propagating in the positive direction of the z-axis and $\tilde{p}_2$ corresponds to the wave propagating in the opposite direction. The functions $\tilde{p}_1$ and $\tilde{p}_2$ satisfying these conditions are

$$\tilde{p}_1 = q^{-1/2} \exp\left(i \int_{z_1}^{z} q \, dz' \right), \quad \tilde{p}_2 = q^{-1/2} \exp\left(-i \int_{z_1}^{z} q \, dz' \right).$$

Substituting the values of $\tilde{p}_1$ and $\tilde{p}_2$ into equation (4.15) and calculating their Wronskian W yields

$$p = \frac{i}{2\pi} \left(\frac{\varrho}{\varrho_1} \right)^{1/2} \int_{-\infty}^{\infty} \frac{(1 - \boldsymbol{\kappa}' \cdot \mathbf{v}/\omega)}{(q q_1)^{1/2}} \exp\left(i\boldsymbol{\kappa}' \cdot \mathbf{r} + i \int_{z_<}^{z_>} q \, dz \right) d^2 \kappa', \quad (4.46)$$

where $q_1 = q(z_1)$.

In equation (4.46), the phase Φ of the integrand is given by $\Phi = \boldsymbol{\kappa} \cdot \mathbf{r} + \int_{z_<}^{z_>} q \, dz$. The stationary phase point is determined from the equation $\partial \Phi / \partial \boldsymbol{\kappa} = 0$. Substituting the value of Φ into this equation yields

$$\mathbf{r} = \int_{z_<}^{z_>} \frac{\boldsymbol{\kappa} + (\omega - \boldsymbol{\kappa} \cdot \mathbf{v})\mathbf{v}/c^2}{q(\boldsymbol{\kappa})} \, dz'. \tag{4.47}$$

The solution of this equation for $\boldsymbol{\kappa}$ determines the stationary phase point. It follows from section 3.5 that equation (4.47) is the equation for the sound ray path from the source to the observation point $(\mathbf{r}, z)$. In this equation, the direction and magnitude of the vector $\boldsymbol{\kappa}$ (strictly speaking, of the vector $\boldsymbol{\kappa}/k_0$) are the ray parameters and determine the direction of the ray at the source.

The phase of the integrand in equation (4.46) is expressed in the form $\Phi = \Phi_a \Phi / \Phi_a$, where $\Phi_a = \Phi(\boldsymbol{\kappa}, \mathbf{r}, z)$ is the phase change along the sound ray (4.47) from the source to the observation point. Assuming that the ray path length $L \gg \lambda$, we have $\Phi_a \sim L/\lambda \gg 1$. Therefore, Φ_a can be chosen as a large parameter m in equation (4.45). Calculating the integral in equation (4.46) by the two-dimensional method of stationary phase yields

$$p = i \left(\frac{\varrho}{\varrho_1} \right)^{1/2} \frac{(1 - \boldsymbol{\kappa} \cdot \mathbf{v}/\omega)}{(q q_1 |J|)^{1/2}} \exp\left(i\boldsymbol{\kappa} \cdot \mathbf{r} + i \int_{z_<}^{z_>} q \, dz' + i\pi s/4 \right). \tag{4.48}$$

It can be shown that $[\partial^2 \Phi / \partial \kappa_i \partial \kappa_j] = [-\partial r_i / \partial \kappa_j]$, where r_1, r_2 are the components of the vector $\mathbf{r}$ defined by equation (4.47). Therefore, in equation (4.48), the quantities J and s are given by

$$J = \det \left[-\frac{\partial r_i}{\partial \kappa_j} \right], \quad s = \operatorname{sgn} \left[-\frac{\partial r_i}{\partial \kappa_j} \right]. \tag{4.49}$$

It follows from this equation that J is the Jacobian of the transformation from $\boldsymbol{\kappa}$ to $\mathbf{r}(\boldsymbol{\kappa})$. We substitute the values of the components of the vector $\mathbf{r}$ into the equation for J. After tedious but straightforward transformations, we obtain

$$J = \omega^2 \left[\int_{z_<}^{z_>} \frac{dz}{q} \int_{z_<}^{z_>} \frac{dz}{c^2 q^3} + \int_{z_<}^{z_>} \frac{dz}{c^2 q^3} \int_{z_<}^{z_>} \frac{\kappa^2 u^2}{c^2 q^3} dz - \left(\int_{z_<}^{z_>} \frac{\kappa u}{c^2 q^3} dz \right)^2 \right], \tag{4.50}$$

where $u = [v^2 - (\boldsymbol{\kappa} \cdot \mathbf{v} / \kappa)^2]^{1/2}$ is the magnitude of the component of $\mathbf{v}$ perpendicular to the vector $\boldsymbol{\kappa}$. It follows from the Koschi-Bunjakovskii inequality that the second term on the right-hand side of equation (4.50) is greater than or equal to the third term so that $J > 0$.

Let us calculate the signature s of the matrix $[-\partial r_i / \partial \kappa_j]$. The eigenvalues ν of this matrix are determined by the equation

$$\det[\partial r_i / \partial \kappa_j + \nu \delta_{ij}] = 0,$$

where $\delta_{ij} = 1$ if $i = j$ and $\delta_{ij} = 0$ if $i \neq j$. Solving this equation for ν yields

$$\nu = \frac{1}{2} \left\{ - \left(\frac{\partial r_1}{\partial \kappa_1} + \frac{\partial r_2}{\partial \kappa_2} \right) \pm \left[\left(\frac{\partial r_1}{\partial \kappa_1} - \frac{\partial r_2}{\partial \kappa_2} \right)^2 + 4 \left(\frac{\partial r_1}{\partial \kappa_2} \right)^2 \right]^{1/2} \right\}. \tag{4.51}$$

In deriving this equation, we used the equality $\partial r_1 / \partial \kappa_2 = \partial r_2 / \partial \kappa_1$. This equality can be proved using equation (4.47). From this equation, it can also be shown that:

$$\frac{\partial r_1}{\partial \kappa_1} + \frac{\partial r_2}{\partial \kappa_2} = \int_{z_<}^{z_>} \frac{\omega^2 + q^2 c^2 + \kappa^2 u^2}{c^2 q^3} dz > 0. \tag{4.52}$$

Starting from equation (4.51) and using the inequalities (4.52) and $J > 0$, it is easy to show that the eigenvalues ν are negative. Therefore, the signature $s = -2$.

Substituting the value of s into equation (4.48), we obtain the high-frequency approximation for the sound field of a point source in a stratified moving medium [282]

$$p = \left(\frac{\varrho}{\varrho_1} \right)^{1/2} \frac{(1 - \boldsymbol{\kappa} \cdot \mathbf{v} / \omega)}{(q q_1 J)^{1/2}} \exp \left(i \boldsymbol{\kappa} \cdot \mathbf{r} + i \int_{z_<}^{z_>} q \, dz \right). \tag{4.53}$$

In this equation, one of the vectors $\boldsymbol{\kappa}$ and $\mathbf{r}$ can be expressed in terms of the other vector by making use of equation (4.47) for the ray path. For instance, if the value of $\mathbf{r}$ given by equation (4.47) is substituted into equation (4.53), the latter equation gives the sound pressure along a fixed ray. On the other hand, if equation (4.47) is solved for $\boldsymbol{\kappa}$ and the value of this vector is substituted into equation (4.53), the resulting equation determines the sound pressure at the fixed point $(\mathbf{r}, z)$.

The approximate analytical solution (3.95) of the sound ray equation was obtained in section 3.6. When deriving equation (3.95), only terms of order $\varepsilon = \max(|\tilde{c}/c_0|, v/c_0) \ll 1$ were retained, and it was assumed that the modulus of the elevation angle was greater than $\varepsilon^{1/2}$. Here, $\tilde{c} = c - c_0$ is the deviation of the sound speed c from its reference value c_0. The approximate solution of equation (4.47) for $\boldsymbol{\kappa}$ is also given by equation (3.95) if, in the latter equation, b and $\mathbf{M}$ are replaced by

$$b_1 = \frac{1}{z_> - z_<} \int_{z_<}^{z_>} \frac{\tilde{c}}{c_0}\, dz \quad \text{and} \quad \mathbf{M}_1 = \frac{1}{z_> - z_<} \int_{z_<}^{z_>} \frac{\mathbf{v}}{c_0}\, dz,$$

respectively. Substituting the resulting value of $\boldsymbol{\kappa}$ into equation (4.53) and retaining terms of order ε, we obtain the sound field at a fixed point $(\mathbf{r}, z)$:

$$p(\mathbf{r}, z) = \left\{ 1 - \frac{\mathbf{r} \cdot \mathbf{v}}{R_1 c_0} + \frac{(\mathbf{r} \cdot F_1\left[\mathbf{v}/c_0\right] + R_1 F_1\left[\tilde{c}/c_0\right]) R_1}{(z - z_1)^2} \right\}$$

$$\times \left(\frac{\varrho}{\varrho_1}\right)^{1/2} \frac{1}{R_1} \exp\left[ik_0(R_1 - \mathbf{r} \cdot \mathbf{M}_1 - R_1 b_1)\right]. \quad (4.54)$$

Here, $R_1 = (r^2 + (z-z_1)^2)^{1/2}$ is the distance from the source to the observation point, and $k_0 = \omega/c_0$. The functional $F_1[f]$ of a vector or scalar function f is defined as

$$F_1[f] = \frac{f(z_>) + f(z_<)}{2} - \frac{1}{z_> - z_<} \int_{z_<}^{z_>} f\, dz. \quad (4.55)$$

If $|\varrho - \varrho_1| \ll \varrho_1$, the effective sound speed c_{eff} and the effective density ϱ_{eff} can be introduced in equation (4.54). This is in agreement with the results obtained in section 4.1.

4.4.6   Azimuthal dependence of the sound-pressure amplitude

Equation (4.54) allows us to analyze the dependence of the sound-pressure amplitude $|p|$ on the azimuthal direction to the observation point, which is given by the angle ψ between the vectors $\mathbf{r}$ and $\mathbf{v}(z_1)$. We assume that $-\pi < \psi \leq \pi$, and the angle $\psi > 0$ if the shortest rotation of the vector $\mathbf{r}$ until it coincides with the vector $\mathbf{v}(z_1)$ is anticlockwise. The values of the functionals F_1 in equation (4.54) depend on the profiles $c(z)$ and $\mathbf{v}(z)$. It can be shown from equation (4.55) that $F_1 = 0$ if $c =$ constant and $\mathbf{v} =$ constant, or if $c(z)$ and $\mathbf{v}(z)$ depend linearly on z. For $F_1 = 0$, we obtain from equation (4.54)

$$|p(r, \psi, z)| = \left(\frac{\varrho(z)}{\varrho_1} \right)^{1/2} \frac{1}{R_1} \left[1 - \frac{rv(z)\cos\psi}{R_1 c_0} \right]. \tag{4.56}$$

It follows from this equation that the sound-pressure amplitude increases monotonically as $|\psi|$ is increased.

The functional F_1 is not equal to zero for the square-law profiles $c(z) = c_0 + \mu(z/z_c)^2$ and $\mathbf{v}(z) = \mathbf{v}_0(z/z_v)^2$, where μ, z_c, $\mathbf{v}_0$, and z_v are parameters. Assuming that the source is located at $z_1 = 0$, from equations (4.54) and (4.55) we obtain the sound-pressure amplitude:

$$|p(r, \psi, z)| = \left(\frac{\varrho(z)}{\varrho_1} \right)^{1/2} \frac{1}{R_1} \left[1 + \frac{rv_0 \cos\psi(r - 5z^2)}{6c_0 z_v^2 R_1} + \frac{\mu R_1^2}{6c_0 z_c^2} \right]. \tag{4.57}$$

It follows from this formula that the dependence of $|p|$ on ψ is different for $r^2 < 5z^2$ and $r^2 > 5z^2$. These conditions correspond to $\alpha < 24.1°$ and $\alpha > 24.1°$, respectively, where α is the angle between the direction to the observation point and the horizontal plane. If $|\psi|$ increases, $|p|$ decreases for the first condition and increases for the second one.

Medium motion results in a variation of the sound-pressure amplitude $|p|$ from its value $|p_0|$ for $\mathbf{v} = 0$. This variation is determined by equation (4.57) and depicted qualitatively in figure 4.6, where the angle $\alpha_0 = 24.1°$ and the direction of the vector $\mathbf{v} = (v_x, 0)$ coincides with the x-axis. The sign "+" in the regions D_1 and D_3 means that $|p| > |p_0|$ for observation points located in these regions. The sign "−" in the regions D_2 and D_4 means that $|p| < |p_0|$. Figure 4.6 depicts two azimuthal directions, corresponding to $\psi = 0$ and $\psi = \pi$. It follows from equation (4.57) that the distribution of the sound-pressure amplitude $|p|$ is the same for other azimuthal directions.

Consider two observation points located at the same horizontal and vertical distances r and z, respectively, from the source, where the first point is located downstream ($|\psi| < \pi/2$) and the second is located upstream ($|\psi| > \pi/2$). In figure 4.6, these points are situated in the regions D_1 and D_4, or D_2 and D_3. We denote the sound-pressure amplitudes for $|\psi| < \pi/2$ and $|\psi| > \pi/2$ by $|p_+|$ and $|p_-|$, respectively. It can be shown from equation (4.57) that $|p_+| > |p_0| > |p_-|$ if the observation points are located in regions D_1 and D_4. On the other hand, $|p_+| < |p_0| < |p_-|$ if they are located in regions D_2 and D_3. Therefore, figure 4.6 also allows us to determine which of the two amplitudes $|p_+|$ or $|p_-|$ is greater.

Let us explain the distribution of the sound-pressure amplitude $|p|$ shown in figure 4.6. If the angle α between the direction to the observation point and the horizontal plane is sufficiently small (in the figure, $\alpha < \alpha_0$), the variation of $|p|$ is caused mainly by refraction of sound in the stratified moving medium. This results in an enhancement of the amplitude in the direction of the vector $\mathbf{v}$ and to its reduction in the opposite direction. On the other

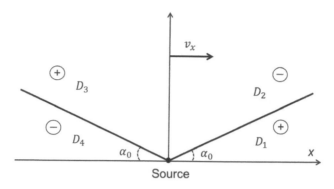

FIGURE 4.6
Distribution of sound-pressure amplitude in the xz-plane for square-law profile $v_x(z)$.

hand, if $\alpha > \alpha_0$, refraction is less pronounced (the projection of $\mathbf{v}$ on the direction of wave propagation is reduced) and does not lead to a significant variation of the sound intensity I_1. However, in a moving medium, to an accuracy of v/c, the sound-pressure amplitude and the intensity are related by the relationship $|p|^2 = 2\varrho c I_1 (1 - 2\boldsymbol{\kappa} \cdot \mathbf{v}/\omega)$, which can be obtained from equation (3.36). It follows from this relationship that for the same intensity I_1, the sound amplitude downstream ($\boldsymbol{\kappa} \cdot \mathbf{v} > 0$) is less than that upstream ($\boldsymbol{\kappa} \cdot \mathbf{v} < 0$). Therefore, $|p_+| < |p_0| < |p_-|$.

Now we show that the distribution of $|p|$ depicted in figure 4.6 is valid not only for square-law profiles $v(z)$ but also for arbitrary monotonically increasing profiles. Of course, the spatial configuration of the regions D_{1-4} depends on the form of the profile $v(z)$, but the signs "+" and "−" remain the same in these regions. If the angle α is sufficiently small, the vertical wavenumber $q(z)$ varies significantly as a function of z and the value of q downstream is less than that upstream. Since p is proportional to $q^{-1/2}$ (see equation (4.53)), the inequality $|p_+| > |p_-|$ is valid if α is small. On the other hand, assuming $z_1 = 0$ and $z > 0$, and setting $\tilde{c} = 0$ for simplicity, we obtain from equation (4.54)

$$|p| = \left(\frac{\varrho}{\varrho_1}\right)^{1/2} \frac{1}{R_1} \left\{ 1 - \frac{(z^2 - r^2)\mathbf{r} \cdot \mathbf{v}}{2z^2 R c_0} - \frac{Rr}{2z^2} \cdot \left[2\mathbf{M}_1(z) - \frac{\mathbf{v}(z=0)}{c_0} \right] \right\}.$$

For monotonic profiles $v(z)$, the sum of the terms in the square brackets of this equation is always greater than zero. Therefore, the inequality $|p_-| > |p_+|$ is valid at least for $z \geq r$ (i.e., for $\alpha \geq 45°$).

Consider a numerical example showing the order of magnitude of these quantities in the near-ground atmosphere. Let $c = $ constant, $v = v_0' z$, and $v_0' = 0.4$ s^{-1}. Assume also that there are two receivers, one located downwind ($\psi = 0$) and one upwind ($\psi = \pi$) at a height $z = 45$ m and horizontal distance $r = 150$ m from the source, which is situated on the ground. In this case, it follows from equation (4.56) that $(|p_-| - |p_+|)/|p_+| = 0.11$. Thus, the difference between $|p_-|$ and $|p_+|$ is small. It follows from equation (4.54) that this difference is of order v/c for arbitrary profiles $v(z)$. Nevertheless, the azimuthal dependence of the sound-pressure amplitude presented above is of interest, because $|p|$ does not depend on ψ in a stratified motionless medium. The difference between $|p_-|$ and $|p_+|$ can be significant if v/c is large enough (see figure 4.1).

4.4.7 Sound field of a transmitting antenna

To conclude this section, we calculate the sound field emitted by an acoustic antenna in a stratified moving medium. Suppose that the aperture of the transmitting antenna is located in the horizontal plane $z_1 = $ constant and that $q^2 > 0$ in the space domain where the sound field is considered. In this case, the high-frequency asymptotic form of the solution of equation (4.37) is given by

$$\tilde{p}(\boldsymbol{\kappa}, z) = Cq^{-1/2} \exp\left(i \int_{z_<}^{z_>} q \, dz\right), \tag{4.58}$$

where $z_> = \max(z, z_1)$, $z_< = \min(z, z_1)$, and C is a constant. At the antenna aperture, this equation becomes $\tilde{p}(\boldsymbol{\kappa}, z_1) = Cq_1^{-1/2}$. The function $\tilde{p}(\boldsymbol{\kappa}, z_1)$ was also expressed in terms of the spatial spectrum $\widehat{p}_a(\boldsymbol{\kappa})$ of the sound pressure at the antenna aperture, see equation (4.17). Eliminating $\tilde{p}(\boldsymbol{\kappa}, z_1)$ between these equations, we express C in terms of $\widehat{p}_a(\boldsymbol{\kappa})$. Substituting equation (4.58) into equation (4.12) and calculating the integral over $\boldsymbol{\kappa}$ by the two-dimensional method of stationary phase, we obtain the sound field emitted by the acoustic antenna in a stratified moving medium:

$$p(\mathbf{r}, z) = 2\pi \left(\frac{\varrho q_1}{\varrho_1 q J}\right)^{1/2} \frac{\omega - \boldsymbol{\kappa} \cdot \mathbf{v}}{\omega - \boldsymbol{\kappa} \cdot \mathbf{v}_1} \widehat{p}_a(\boldsymbol{\kappa}) \exp\left(i\boldsymbol{\kappa} \cdot \mathbf{r} + i \int_{z_<}^{z_>} q \, dz - \frac{i\pi}{2}\right). \tag{4.59}$$

In this equation, the vectors $\boldsymbol{\kappa}$ and $\mathbf{r}$ are related by equation (4.47) for the sound ray, and J is determined by equation (4.50).

4.5 High-frequency sound field of a point source above an impedance surface in a stratified medium

In this section, using the high-frequency approximation, we shall calculate the sound field of a unit-amplitude point source located above an impedance surface in a stratified moving medium. We follow mainly the method given elsewhere [287].

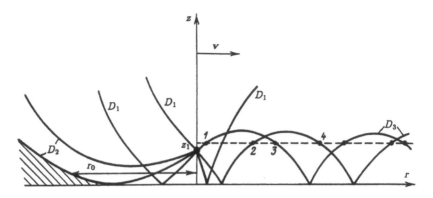

FIGURE 4.7
Sound ray paths in a moving medium. The corresponding region D_i (figure 4.8) is indicated beside each ray. The source is located at the point $(z_1, 0)$, the dotted line is at the level z of the observation point, and r_0 is the horizontal distance to the shadow zone (shaded). The arrow near the top indicates the direction of the medium velocity $\mathbf{v}$.

4.5.1   Regions of integration

Let the point source be located at the point $(\mathbf{r}_1 = 0, z_1)$ above an impedance surface coinciding with the plane $z = 0$ (figure 4.7). We shall consider profiles $c(z)$ and $\mathbf{v}(z)$ for which the square of the vertical wavenumber $q^2(z)$ has no more than one zero of first order at the height z_t. In this case, the sound field is given by the integral (4.15), where the functions $\tilde{p}_1$ and $\tilde{p}_2$ are determined by equation (4.40). In the latter equation, the constants $C_{1,2}$ are different for $\tilde{p}_1$ and $\tilde{p}_2$. These constants and the dimensionless vertical coordinate $\xi(z)$ depend on the form of the function $q^2(z)$ and need to be determined. Therefore, we should study the dependence of q^2 on z before determining $\xi(z)$ and $C_{1,2}$.

We first consider for what profiles of $c(z)$ and $\mathbf{v}(z)$ the function $q^2(z)$ has no more than one zero. To this end, $q^2(z)$ is expressed in the form: $q^2 = (\omega - \boldsymbol{\kappa} \cdot \mathbf{v} - \kappa c)(\omega - \boldsymbol{\kappa} \cdot \mathbf{v} + \kappa c)/c$. For $v \leq c$, only the first factor in this formula can become zero. It is convenient to express this factor as $\omega - \kappa w$, where $w(\varphi, z) = \mathbf{v} \cdot \boldsymbol{\kappa}/\kappa + c = v \cos(\varphi - \psi) + c$. Here, φ and $\psi(z)$ are the angles between the vectors $\boldsymbol{\kappa}$ and $\mathbf{v}(z)$ and the x-axis, respectively.

Suppose that the profiles $c(z)$ and $\mathbf{v}(z)$ are such that the function $w(\varphi, z)$ and, hence, the function $\omega - \kappa w$ are monotonic functions of z for a fixed value of φ. In this case, $q^2(z)$ has no more than one zero. In the atmosphere and ocean, the function $w(\varphi, z)$ can be considered as a bounded function of z: $w_1(\varphi) \leq w(\varphi, z) \leq w_2(\varphi)$. If these assumptions about $w(\varphi, z)$ are valid, there are four regions D_{1-4} in the $\kappa_x \kappa_y$-plane (figure 4.8), between each of

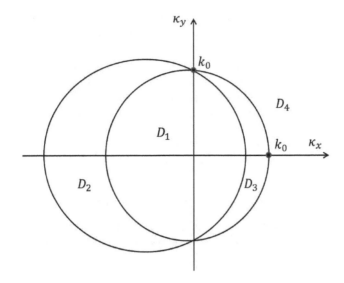

FIGURE 4.8
Example of relative positions of the regions D_{1-4} in the $\kappa_x \kappa_y$-plane.

which the dependence of q^2 on z differs qualitatively. (Here, $\kappa_x = \kappa \cos \varphi$ and $\kappa_y = \kappa \sin \varphi$ are the components of the vector $\boldsymbol{\kappa}$.)

Region D_1 is determined by the inequality $\kappa < \omega/w_2(\varphi)$. In D_1, the function $q^2(z) > 0$.

Region D_2 is determined by the inequalities $\omega/w_2(\varphi) < \kappa < \omega/w_1(\varphi)$ and $\partial w/\partial z < 0$. In D_2, the function $q^2(z) < 0$ if $0 \le z \le z_t$, and $q^2(z) > 0$ if $z > z_t$.

Region D_3 is determined by the inequalities $\omega/w_2(\varphi) < \kappa < \omega/w_1(\varphi)$ and $\partial w/\partial z > 0$. In D_3, the function $q^2(z) > 0$ if $0 \le z \le z_t$, and $q^2(z) < 0$ if $z > z_t$.

Finally, region D_4 is determined by the inequality $\kappa > \omega/w_1(\varphi)$. In D_4, the function $q^2(z) < 0$.

Note that neither the region D_2 nor the region D_3 appears for certain relationships between the gradients of $c(z)$ and $v(z)$. An example of relative positions of regions D_{1-4} in the $\kappa_x \kappa_y$-plane is shown in figure 4.8 for a stratification of $c(z)$ and $\mathbf{v}(z)$ similar to that considered in the numerical example in section 4.5.5. The only difference between these stratifications is that the gradient of the profile $v(z)$ used to draw figure 4.8 is ten times greater than that in section 4.5.5. Such an increase of the gradient is necessary for a clear representation of the regions D_2 and D_3, the area of which, otherwise, is much less than that of region D_1.

The components κ_x and κ_y of the vector $\boldsymbol{\kappa}$ can be treated as ray parameters determining a fixed sound ray (see sections 3.5 and 4.4). This ray corresponds to a quasi-plane wave with the vertical wavenumber $q(\boldsymbol{\kappa}, z)$. Therefore, using the dependence of $q^2(\kappa, z)$ on z considered above, it is possible to determine a qualitative picture of the sound ray paths emitted by the source, as in figure 4.7.

If the values of κ_x and κ_y are in the region D_1, the corresponding sound rays bend upwards. Some of these rays are reflected first by the plane $z = 0$. The sound rays corresponding to κ_x and κ_y in region D_2 are also bent upwards, but in addition have turning points at heights z_t. If κ_x and κ_y are in region D_3, the rays are ducted in the waveguide bounded by the plane $z = 0$ and the level of the ray turning points. Finally, κ_x and κ_y in region D_4 correspond to inhomogeneous waves which decay exponentially as $|z - z_1|$ increases.

4.5.2 Integral representation of the sound field

Thus, we have identified four regions D_{1-4} between which the dependences of q^2 on z differ qualitatively. This allows one to determine the function $\xi(z)$ and the constants $C_{1,2}$ in equation (4.40). The function $\xi(z)$ is a solution of equation (4.38) and takes different forms in the different regions:

$$D_1 : \xi = \left(\frac{3}{2} \int_0^z q\, dz' + B_1 \right)^{2/3} ;$$

$$D_2 : \xi = \left[\begin{array}{ll} \left(\dfrac{3}{2} \displaystyle\int_{z_t}^z q\, dz' \right)^{2/3} & \text{for} \quad z \geq z_t, \\[4mm] -\left(\dfrac{3}{2} \displaystyle\int_z^{z_t} (-q^2)^{1/2}\, dz' \right)^{2/3} & \text{for} \quad 0 \leq z \leq z_t; \end{array} \right.$$

$$D_3 : \xi = \left[\begin{array}{ll} -\left(\dfrac{3}{2} \displaystyle\int_{z_t}^z (-q^2)^{1/2}\, dz' \right)^{2/3} & \text{for} \quad z > z_t, \\[4mm] \left(\dfrac{3}{2} \displaystyle\int_z^{z_t} q\, dz' \right)^{2/3} & \text{for} \quad 0 \leq z \leq z_t; \end{array} \right.$$

$$D_4 : \xi = -\left(\frac{3}{2} \int_0^z (-q^2)^{1/2}\, dz' + B_2 \right)^{2/3} . \tag{4.60}$$

Here, $B_{1,2}$ are positive constants much greater than 1.

The constants $C_{1,2}$ in the function $\tilde{p}_1(\kappa, z)$ can be obtained by making use of the radiation condition for $z \to \infty$. If the components κ_x and κ_y of the vector κ are in regions $D_{1,2}$, $\tilde{p}_1$ must correspond to a propagating wave for $z \to \infty$. If these components are in regions $D_{3,4}$, the function $\tilde{p}_1$ must correspond to a wave decaying exponentially for $z \to \infty$. Replacing the functions $\mathrm{Ai}(-\xi)$ and $\mathrm{Bi}(-\xi)$ in equation (4.40) by their asymptotic forms for large values of $|\xi|$, it can be shown that these radiation conditions are valid if $C_1 = i$ and $C_2 = 1$ in $D_{1,2}$, and $C_1 = 1$ and $C_2 = 0$ in $D_{3,4}$.

The constants $C_{1,2}$ in the function $\tilde{p}_2(\kappa, z)$ can be obtained by making use of the boundary condition at $z = 0$. The function $\tilde{p}_2$ satisfies the same boundary condition as the function $\tilde{p}$ given by equation (4.14). Let us find the boundary condition for $\tilde{p}$. To this end, consider the known boundary condition for the sound pressure p of a plane wave incident obliquely on an impedance surface located at $z = 0$:

$$\frac{\partial p}{\partial z} = -ik_0\beta(\kappa)p, \quad \beta = \left(1 - \frac{\kappa^2}{k_0^2}\right)^{1/2}\frac{1 - \mathcal{R}_p(\kappa)}{1 + \mathcal{R}_p(\kappa)}. \tag{4.61}$$

Here, $k_0 = \omega/c_0$, $c_0 = c(z = 0)$, and β is the normalized admittance of the surface (see the text following equation (10.21)). In equation (4.61), $\mathcal{R}_p$ is the plane wave reflection coefficient of the impedance surface, depending on the angle of incidence and hence on κ. For an acoustically hard surface $\mathcal{R}_p = 1$ and $\beta = 0$, while for an acoustically soft surface $\mathcal{R}_p = -1$ and $\beta = \infty$. The admittance β does not depend on κ for a locally reacting surface. Note that in many cases, the ground can be considered as a locally reacting surface [16, 74].

If the integral operator $\int d^2\kappa'$ is omitted on the right-hand side of equation (4.12), which is equivalent to equation (4.15) and also describes the point source sound field, one obtains equation (4.20) for the sound pressure p of the plane waves that appear in the spectral representation of the field. Equation (4.20) contains the function $\tilde{p}$. Substituting equation (4.20) into equation (4.61), neglecting terms proportional to $v' \sim v/l$ and $\varrho' \sim \varrho/l$ in comparison with those proportional to $d\tilde{p}/dz \sim k_0\tilde{p}$, and assuming that $\mathbf{v}(z = 0) = 0$, we obtain the boundary condition at $z = 0$:

$$\frac{d\tilde{p}}{dz} = -ik_0\beta\tilde{p}. \tag{4.62}$$

The function $\tilde{p}_2$ satisfies the same boundary condition. Substituting $\tilde{p}_2$ given by equation (4.40) into the boundary condition (4.62), one determines the constants $C_{1,2}$; these have the same form in all of the regions D_{1-4}:

$$C_1 = \frac{K}{E}, \quad C_2 = -1. \tag{4.63}$$

Here,

$$K = \xi_0'\text{Bi}'(-\xi_0) - ik_0\beta\text{Bi}(-\xi_0), \quad E = \xi_0'\text{Ai}'(-\xi_0) - ik_0\beta\text{Ai}(-\xi_0), \tag{4.64}$$

where Bi' and Ai' are the derivatives of the Airy functions, $\xi_0 = \xi(z = 0)$ and $\xi_0' = (d\xi/dz)_{(z=0)}$.

Thus, we have obtained the constants $C_{1,2}$ for the functions $\tilde{p}_1$ and $\tilde{p}_2$. Substituting the values of $\tilde{p}_1$ and $\tilde{p}_2$ into equation (4.15) and calculating the Wronskian W, one obtains the high-frequency sound field of a point source located above an impedance surface in a stratified moving medium:

$$p(\mathbf{r}, z) = \left(\frac{\varrho}{\varrho_1}\right)^{1/2}\int_{-\infty}^{\infty}\left(1 - \frac{\boldsymbol{\kappa}\cdot\mathbf{v}}{\omega}\right)\tilde{p}(\boldsymbol{\kappa}, z)e^{i\boldsymbol{\kappa}\cdot\mathbf{r}}\,d^2\kappa, \tag{4.65}$$

where

$$
\tilde{p} = \left(\frac{\xi\xi_1}{q^2 q_1^2}\right)^{1/4} \left\{ \frac{[\text{Bi}(-\xi_>) + i\text{Ai}(-\xi_>)][K\text{Ai}(-\xi_<) - E\text{Bi}(-\xi_<)]}{K + iE} \right.
$$
$$
\left. \times \Theta(D_1, D_2) - \text{Ai}(-\xi_>)\frac{K\text{Ai}(-\xi_<) - E\text{Bi}(-\xi_<)}{E}\Theta(D_3, D_4) \right\}. \quad (4.66)
$$

Here, $\xi_1 = \xi(z_1)$; $\xi_> = \xi(z_>)$; $\xi_< = \xi(z_<)$; and the function $\Theta(D_i, D_j) = 1$ if the components κ_x, κ_y of the vector κ are in the region D_i or D_j, and $\Theta(D_i, D_j) = 0$ if κ_x, κ_y are not in these regions. Equation (4.66) is valid for arbitrary z.

A calculation of the Fourier integral in equation (4.65) can be done by different methods, including numerical integration. In the remainder of this section, this integral is calculated by making use of the two-dimensional method of stationary phase. This will allow us to express $p(\mathbf{r}, z)$ as the sum of the sound fields corresponding to the sound rays of different kinds.

4.5.3   Asymptotic form of the integrand

Assume that the levels of the source and the receiver are far away from the level z_t of the ray turning point, which in turn is not too close to the surface. In this case, it follows from equation (4.60) that $|\xi| \gg 1$, $|\xi_1| \gg 1$, and $|\xi_0| \gg 1$. If these inequalities are valid, the Airy functions Ai and Bi in equation (4.66) for $\tilde{p}$ can be replaced by their asymptotic forms for large values of the argument. These asymptotic forms depend on the signs of ξ, ξ_1, and ξ_0, which are different in different regions D_{1-4}.

If the components κ_x and κ_y of the vector κ are in region D_1, the inequalities $\xi > 0$, $\xi_1 > 0$, and $\xi_0 > 0$ are valid. Starting from equation (4.66) and taking these inequalities into account, one obtains the approximate expression for $\tilde{p}(\kappa, z)$ in region D_1, valid for arbitrary z and z_1:

$$
\tilde{p} = \frac{i}{2\pi(qq_1)^{1/2}}\left\{ \exp\left(i\int_{z_<}^{z_>} q\,dz\right) + \mathcal{R}_p \exp\left[i\left(\int_0^z + \int_0^{z_1}\right)q\,dz'\right] \right\}. \quad (4.67)
$$

The two terms in braces correspond to a sound wave, which does not interact with the surface and has no turning points, and a sound wave reflected by the impedance surface.

If the components κ_x and κ_y of the vector κ are in the region D_2, the paths of the sound rays corresponding to these values of κ are located in the spatial region where $z > z_t$ and $z_1 > z_t$ (see figure 4.7). It follows from equation (4.60) that $\xi > 0$ and $\xi_1 > 0$ in this spatial region. Moreover, $\xi_0 < 0$. Starting from equation (4.66) and taking these inequalities into account, one obtains the approximate expression for the function $\tilde{p}(\kappa, z)$ in region D_2, for $z, z_1 > z_t$:

$$
\tilde{p} = \frac{i}{2\pi(qq_1)^{1/2}}\left\{ \exp\left(i\int_{z_<}^{z_>} q\,dz\right) + \exp\left[i\left(\int_{z_t}^z + \int_{z_t}^{z_1}\right)q\,dz' - \frac{i\pi}{2}\right] \right\}. \quad (4.68)
$$

The two terms in square brackets correspond to a sound wave, which does not interact with the surface and has no turning points, and a sound wave with a turning point. The term $-\pi/2$ in the phase of the latter wave is due to the turning point.

If the components κ_x and κ_y of the vector $\boldsymbol{\kappa}$ are in region D_3, the paths of the corresponding sound rays are located in the spatial region where $z < z_t$ and $z_1 < z_t$ (see figure 4.7). In this spatial region, $\xi > 0$, $\xi_1 > 0$, and $\xi_0 > 0$. If $\xi_0 \gg 1$, the function E^{-1} in equation (4.66) is given by

$$
E^{-1} = -\frac{2\pi^{1/2}\xi_0^{1/4}}{q_0 + k_0\beta}\left[1 - \mathcal{R}_p \exp\left(2i\int_0^{z_t} q\,dz - \frac{i\pi}{2}\right)\right]^{-1}
$$

$$
\times \exp\left(i\int_0^{z_t} q\,dz + \frac{i\pi}{4}\right),
$$

where $q_0 = q(z=0) = (k_0^2 - \kappa^2)^{1/2}$. The expression in square brackets can be written in the equivalent form

$$
\sum_{m=0}^{\infty} \mathcal{R}_p^m \exp\left(2mi\int_0^{z_t} q\,dz - \frac{im\pi}{2}\right).
$$

Substituting the value of E^{-1} into equation (4.66), one obtains the function $\tilde{p}(\boldsymbol{\kappa}, z)$ in region D_3, for $z, z_1 < z_t$:

$$
\tilde{p} = \frac{i}{2\pi(qq_1)^{1/2}}\left\{\exp\left(i\int_{z_<}^{z_>} q\,dz\right) + \mathcal{R}_p \exp\left[i\left(\int_0^z + \int_0^{z_1}\right)q\,dz'\right]\right.
$$

$$
+ \exp\left[i\left(\int_{z_1}^{z_t} + \int_z^{z_t}\right)q\,dz' - \frac{i\pi}{2}\right] + \mathcal{R}_p \exp\left[i\left(2\int_0^{z_t} - \int_{z_<}^{z_>}\right)q\,dz\right.
$$

$$
\left.\left.- \frac{i\pi}{2}\right]\right\}\sum_{m=0}^{\infty} \mathcal{R}_p^m \exp\left(2mi\int_0^{z_t} q\,dz - \frac{im\pi}{2}\right). \quad (4.69)
$$

Consider the physical meaning of the terms in this expression. In figure 4.7, the points 1, 2, 3, 4, etc. denote the intersections of the ray paths in the waveguide with the level z of the receiver. If $m = 0$, there are four terms on the right-hand side of equation (4.69), which correspond to receiver positions at points 1, 2, 3, 4 and describe the first cycle of ray paths in the waveguide. The next four terms in equation (4.69), corresponding to $m = 1$, describe the second cycle of the ray paths, etc.

Finally, it follows from equation (4.60) that $\xi, \xi_0, \xi_1 < 0$ in region D_4. Therefore, for arbitrary z and z_1, the function $\tilde{p}(\boldsymbol{\kappa}, z)$ takes the form

$$
\tilde{p} = \frac{1}{2\pi(q^2 q_1^2)^{1/4}}\left\{\exp\left(-\int_{z_<}^{z_>}(-q^2)^{1/2}\,dz'\right)\right.
$$

$$
\left.+ \mathcal{R}_p \exp\left[-\left(\int_0^z + \int_0^{z_1}\right)(-q^2)^{1/2}\,dz'\right]\right\}. \quad (4.70)
$$

4.5.4 Sound fields of different type

Equations (4.67)-(4.70) give approximate expressions for the function $\tilde{p}$ if $|\xi|, |\xi_1|, |\xi_0| \gg 1$. We substitute these expressions into equation (4.65) and combine the integrals over D_{1-4}, which have the same integrands. As a result, the sound field p is expressed as the sum of five fields:

$$p = \sum_{i=1}^{5} p_i. \tag{4.71}$$

Here, the p_1 corresponds to the sound wave which does not interact with the surface and has no turning points:

$$p_1 = \frac{i}{2\pi} \left(\frac{\varrho}{\varrho_1} \right)^{1/2} \int_{D_1 \cup D_2 \cup D_3} \frac{(1 - \boldsymbol{\kappa} \cdot \mathbf{v}/\omega)}{(qq_1)^{1/2}} \exp\left(i\boldsymbol{\kappa} \cdot \mathbf{r} + i \int_{z_<}^{z_>} q \, dz \right) d^2\kappa, \tag{4.72}$$

where the integral is taken over the union of the regions D_1, D_2, and D_3. As one should expect, the p_1 coincides with the sound field given by equation (4.46) obtained above for the case when there are no turning points of the sound rays.

The p_2 corresponds to the sound wave reflected by the surface:

$$p_2 = \frac{i}{2\pi} \left(\frac{\varrho}{\varrho_1} \right)^{1/2} \int_{D_1 \cup D_3} \frac{(1 - \boldsymbol{\kappa} \cdot \mathbf{v}/\omega)}{(qq_1)^{1/2}} R_p$$
$$\times \exp\left[i\boldsymbol{\kappa} \cdot \mathbf{r} + i \left(\int_0^z + \int_0^{z_1} \right) q \, dz' \right] d^2\kappa. \tag{4.73}$$

The p_3 corresponds to the sound wave with a turning point:

$$p_3 = \frac{i}{2\pi} \left(\frac{\varrho}{\varrho_1} \right)^{1/2} \int_{D_2 \cup D_3} \frac{(1 - \boldsymbol{\kappa} \cdot \mathbf{v}/\omega)}{(qq_1)^{1/2}}$$
$$\times \exp\left[i\boldsymbol{\kappa} \cdot \mathbf{r} \pm i \left(\int_{z_t}^z + \int_{z_t}^{z_1} \right) q \, dz' - \frac{i\pi}{2} \right] d^2\kappa. \tag{4.74}$$

Hereafter, the upper sign is chosen if $\boldsymbol{\kappa}$ is in D_2, and the lower sign is chosen if $\boldsymbol{\kappa}$ is in D_3.

The p_4 describes the sound field in the waveguide:

$$p_4 = \frac{i}{2\pi} \left(\frac{\varrho}{\varrho_1} \right)^{1/2} \sum_{n=1}^{4} \sum_{m=0}^{\infty} \int_{D_3} \frac{(1 - \boldsymbol{\kappa} \cdot \mathbf{v}/\omega)}{(qq_1)^{1/2}} R_p^{m+\alpha_n} e^{i\Phi_{nm}} \, d^2\kappa,$$

$$\Phi_{nm} = \boldsymbol{\kappa} \cdot \mathbf{r} + \left(\int_{l_n} + 2m \int_0^{z_t} \right) q \, dz - (m + \gamma_n)\pi/2. \tag{4.75}$$

Here, $\alpha_n = 0$ if $n = 1, 2$; $\alpha_n = 1$ if $n = 3, 4$; $\gamma_n = 0$ if $n = 1, 3$; and $\gamma_n = 1$ if $n = 2, 4$. The symbol $\int_{l_n}$ takes the form

$$\int_{z_<}^{z_>} \qquad \text{for } n = 1,$$

$$\left(\int_z^{z_t} + \int_{z_1}^{z_t} \right) \qquad \text{for } n = 2,$$

$$\left(\int_0^z + \int_0^{z_1} \right) \qquad \text{for } n = 3,$$

$$\left(2 \int_0^{z_t} - \int_{z_<}^{z_>} \right) \qquad \text{for } n = 4.$$

The first three terms in equation (4.75) (for $m = 0$ and $n = 1, 2, 3$) should be omitted because they have already been presented in the sound fields p_1, p_2, and p_3.

Finally, the p_5 corresponds to inhomogeneous sound waves:

$$p_5 = \frac{1}{2\pi} \left(\frac{\varrho}{\varrho_1} \right)^{1/2} \int_{D_4} \frac{(1 - \boldsymbol{\kappa} \cdot \mathbf{v}/\omega)}{(q^2 q_1^2)^{1/4}} \left\{ \exp \left(i\boldsymbol{\kappa} \cdot \mathbf{r} - \int_{z_<}^{z_>} (-q^2)^{1/2} \, dz \right) \right.$$
$$\left. + \mathcal{R}_p \exp \left[i\boldsymbol{\kappa} \cdot \mathbf{r} - \left(\int_0^z + \int_0^{z_1} \right) (-q^2)^{1/2} \, dz' \right] \right\} d^2\kappa. \quad (4.76)$$

When calculating the function $\tilde{p}$ in the $\kappa_x\kappa_y$-plane, it was worthwhile to separate out the four regions D_{1-4}, because this function is given by different expressions in these regions. However, when calculating the sound field p, it is useful to express it as the sum given by equation (4.71), because the p_i are sound fields corresponding to rays of particular kinds.

The integrals over $\boldsymbol{\kappa}$ in equations (4.72)–(4.76) can be calculated by using the two-dimensional method of stationary phase. Since equation (4.72) for sound field p_1 coincides with equation (4.46), one finds that p_1 is given by equation (4.53), where J is determined by equation (4.50) and $\boldsymbol{\kappa}$ and $\mathbf{r}$ are related by the ray equation (4.47).

The integral over $\boldsymbol{\kappa}$ in equation (4.73) is calculated in an analogous way to that in equation (4.46). It can be shown (see reference [284]) that the sound field p_2 is still given by equations (4.47)–(4.53) if the integral operator $\int_{z_<}^{z_>} dz'$ is replaced by $\left(\int_0^z + \int_0^{z_1} \right) dz'$ in these equations:

$$p_2 = \left(\frac{\varrho}{\varrho_1} \right)^{1/2} \frac{(1 - \boldsymbol{\kappa} \cdot \mathbf{v}/\omega)\mathcal{R}_p}{(q q_1 J_2)^{1/2}} \exp \left[i\boldsymbol{\kappa} \cdot \mathbf{r} + i \left(\int_0^z + \int_0^{z_1} \right) q \, dz' \right],$$

$$\mathbf{r} = \left(\int_0^z + \int_0^{z_1} \right) \frac{\boldsymbol{\kappa} + (\omega - \boldsymbol{\kappa} \cdot \mathbf{v})\mathbf{v}/c^2}{q} \, dz',$$

$$J_2 = \omega^2 \left\{ \left[\left(\int_0^z + \int_0^{z_1} \right) \frac{dz'}{c^2 q^3} \right] \left[\left(\int_0^z + \int_0^{z_1} \right) \left(1 + \frac{\kappa^2 u^2}{c^2 q^2} \right) \frac{dz'}{q} \right] \right.$$
$$\left. - \left[\left(\int_0^z + \int_0^{z_1} \right) \frac{\kappa u}{c^2 q^3} \, dz' \right]^2 \right\}. \quad (4.77)$$

Here, $J_2 > 0$.

Let $\varepsilon = \max \left(|\tilde{c}/c_0|, v/c \right) \ll 1$ and let the modulus of the elevation angle of the sound ray be greater than $\varepsilon^{1/2}$. The equation of the sound ray $\mathbf{r} = \mathbf{r}(\kappa, z)$, reflected by the surface and determined by equation (4.77), can be solved for κ by the approach considered in section 3.6 for approximate solution of the ray equation (3.79). As a result, κ is given by equation (3.95), where R, b, $\mathbf{M}$ are replaced by

$$R_2 = \sqrt{r^2 + (z + z_1)^2}, \quad b_2 = \frac{1}{z + z_1} \left(\int_0^z + \int_0^{z_1} \right) \frac{\tilde{c}}{c_0} \, dz',$$

$$\mathbf{M}_2 = \frac{1}{z + z_1} \left(\int_0^z + \int_0^{z_1} \right) \frac{\mathbf{v}}{c_0} \, dz',$$

respectively. Substituting the resulting value of κ into equation (4.77) and retaining terms of order ε, we obtain the sound field p_2, at a fixed spatial point:

$$p_2(\mathbf{r}, z) = \left\{ 1 - \frac{\mathbf{r} \cdot \mathbf{v}}{R_2 c_0} + \frac{(\mathbf{r} \cdot F_2[\mathbf{v}/c_0] + R_2 F_2[\tilde{c}/c_0]) R_2}{(z + z_1)^2} \right\}$$
$$\times \left(\frac{\varrho}{\varrho_1} \right)^{1/2} \frac{R_p}{R_2} \exp \left[i k_0 (R_2 - \mathbf{r} \cdot \mathbf{M}_2 - R_2 b_2) \right], \quad (4.78)$$

where

$$F_2[f] = \frac{f(z) + f(z_1)}{2} - \frac{1}{z + z_1} \left(\int_0^z + \int_0^{z_1} \right) f \, dz'.$$

If $|\varrho - \varrho_1| \ll \varrho_1$, the effective functions c_{eff} and ϱ_{eff} can be introduced in equation (4.78).

The calculation of the integral over κ in equation (4.74) somewhat differs from that in equation (4.46). The phase Φ of the integrand in the former equation is given by the expression in the exponent. The point of stationary phase, κ, is determined by the equality $\partial \Phi / \partial \kappa = 0$. Substituting the value of Φ into this equality yields the equation of the sound ray with the turning point z_t:

$$\mathbf{r} = \pm \left(\int_{z_t}^z + \int_{z_t}^{z_1} \right) \frac{\kappa + (\omega - \kappa \cdot \mathbf{v}) \mathbf{v}/c^2}{q} \, dz'. \quad (4.79)$$

Calculating the integral in equation (4.74) using the two-dimensional method of stationary phase and retaining only the first term of the asymptotic series with respect to the small parameter λ/L, where L is the path length of the sound ray (4.79), one obtains the sound field corresponding to the ray with a turning point

$$p_3 = \left(\frac{\varrho}{\varrho_1}\right)^{1/2} \frac{(1 - \boldsymbol{\kappa} \cdot \mathbf{v}/\omega)}{(qq_1|J_3|)^{1/2}} \exp\left[i\boldsymbol{\kappa} \cdot \mathbf{r} \pm i\left(\int_{z_t}^{z} + \int_{z_t}^{z_1}\right) q\, dz' + i\pi s_3/4\right].$$

(4.80)

Here, J_3 and s_3 are expressed in terms of $\partial r_i/\partial \kappa_j$ by equation (4.49), where the vector $\mathbf{r} = (r_1, r_2)$ is determined by the ray equation (4.79). Since $q(z_t) = 0$, the calculation of $\partial r_i/\partial \kappa_j$ using equation (4.79) yields singular terms. Therefore, the right-hand side of equation (4.79) should be integrated by parts first. Taking into account the fact that $1/q = [2/(k^2)']dq/dz$, where $k = (\omega - \boldsymbol{\kappa} \cdot \mathbf{v})/c$ is the wavenumber and $(k^2)' = dk^2/dz$ (it is assumed that $(k^2)' \neq 0$), one obtains

$$\mathbf{r} = \pm\frac{2(\boldsymbol{\kappa} + k\mathbf{v}/c)q}{(k^2)'}\left(\Big|_z + \Big|_{z_1}\right) \mp 2\left(\int_{z_t}^{z} + \int_{z_t}^{z_1}\right) q\, \frac{d}{dz'}\left(\frac{\boldsymbol{\kappa} + k\mathbf{v}/c}{(k^2)'}\right) dz'.$$

(4.81)

Hereafter, any function f written in front of the symbol $\left(\big|_z + \big|_{z_1}\right)$ is interpreted as the sum $f(z) + f(z_1)$.

Applying the operator $\partial/\partial \kappa_j$ to both sides of equation (4.81) yields

$$\frac{\partial r_i}{\partial \kappa_j} = \mp\frac{2(\kappa_i + kv_i/c)(\kappa_j + kv_j/c)}{(k^2)'q}\left(\Big|_z + \Big|_{z_1}\right) \pm \left(\int_{z_t}^{z} + \int_{z_t}^{z_1}\right)$$
$$\times \left\{\delta_{ij} - \frac{v_i v_j}{c^2} + 2\frac{d}{dz'}\left[\frac{(\kappa_i + kv_i/c)(\kappa_j + kv_j/c)}{(k^2)'}\right]\right\}\frac{dz'}{q}, \quad (4.82)$$

where v_1 and v_2 are the components of the vector $\mathbf{v}$. Substituting $\partial r_i/\partial \kappa_j$ into equations (4.49) and (4.51), one determines J_3 and s_3. The expression for J_3 is rather complicated and is not presented here. If only the terms of order v/c are retained in this expression, it simplifies to:

$$J_3 = \left[\left(\int_{z_t}^{z} + \int_{z_t}^{z_1}\right)\frac{dz'}{q}\right]\left\{\left(\int_{z_t}^{z} + \int_{z_t}^{z_1}\right)\left[1 + 2\frac{d}{dz'}\left(\frac{\kappa^2 + 2k\boldsymbol{\kappa} \cdot \mathbf{v}/c}{(k^2)'}\right)\right]\right.$$
$$\left.\times \frac{dz'}{q} - \frac{2(\kappa^2 + 2k\boldsymbol{\kappa} \cdot \mathbf{v}/c)}{(k^2)'q}\left(\Big|_z + \Big|_{z_1}\right)\right\} + O(v^2/c^2). \quad (4.83)$$

The value of the signature s_3 depends on the profiles $c(z)$ and $\mathbf{v}(z)$. Equations (4.79)–(4.83) and (4.49) determine the sound field p_3.

The integral over $\boldsymbol{\kappa}$ in equation (4.75) for p_4 is calculated in a similar manner to that in equation (4.74). As a result, one obtains the sound field p_4:

$$p_4 = \left(\frac{\varrho}{\varrho_1}\right)^{1/2} \sum_{n=1}^{4} \sum_{m=0}^{\infty} \frac{(1 - \boldsymbol{\kappa} \cdot \mathbf{v}/\omega)}{(qq_1|J_4|)^{1/2}}\mathcal{R}_p^{m+\alpha_n} e^{i\Phi_{nm} + is_4\pi/4},$$

$$\mathbf{r} = \left(\int_{l_n} + 2m \int_0^{z_t} \right) \frac{\boldsymbol{\kappa} + (\omega - \boldsymbol{\kappa} \cdot \mathbf{v})\mathbf{v}/c^2}{q} \, dz',$$

$$\frac{\partial r_i}{\partial \kappa_j} = -\frac{2(\kappa_i + kv_i/c)(\kappa_j + kv_j/c)}{(k^2)'q} \left(\Big|_{l_n} - 2m \Big|_{z=0} \right) + \left(\int_{l_n} + 2m \int_0^{z_t} \right)$$
$$\times \left\{ \delta_{ij} - \frac{v_i v_j}{c^2} + 2\frac{d}{dz} \left[\frac{(\kappa_i + kv_i/c)(\kappa_j + kv_j/c)}{(k^2)'} \right] \right\} \frac{dz}{q},$$

$$J_4 = \left[\left(\int_{l_n} + 2m \int_0^{z_t} \right) \frac{dz}{q} \right] \left\{ \left(\int_{l_n} + 2m \int_0^{z_t} \right) \left[2\frac{d}{dz} \left(\frac{\kappa^2 + 2k\boldsymbol{\kappa} \cdot \mathbf{v}/c}{(k^2)'} \right) \right. \right.$$
$$\left. \left. + 1 \right] \frac{dz}{q} - \frac{2(\kappa^2 + 2k\boldsymbol{\kappa} \cdot \mathbf{v}/c)}{(k^2)'q} \left(\Big|_{l_n} - 2m \Big|_{z=0} \right) \right\} + O(v^2/c^2). \quad (4.84)$$

These equations are analogous to equations (4.80), (4.79), (4.82), and (4.83), respectively. In equation (4.84), the symbol $|_{l_n}$ denotes the limits of the integral $\int_{l_n} dz$. For example, a function f written in front of the symbol $(|_{l_1} - 2m|_{z=0})$ is interpreted as $f(z_>) - f(z_<) - 2mf(z = 0)$. The quantities J_4 and s_4 should be calculated by using equation (4.49), where $\partial r_i/\partial \kappa_j$ is determined from equation (4.84). The p_4 is the sum of the sound fields corresponding to the sound rays in the waveguide.

Finally, consider the sound field p_5 of the inhomogeneous waves. It follows from equation (4.76) that p_5 is exponentially small in comparison with p_{1-4} if $|z - z_1|k_0 \gg 1$. If $|z - z_1|k_0 \lesssim 1$, the integrand in equation (4.76) has no points of stationary phase. Evaluating the integral in this equation yields $p_5 \sim 1/[R(k_0 R)^{1/2}]$, where $R = [r^2 + (z - z_1)^2]^{1/2}$ is the distance between the source and receiver. The sound field p_5 can be neglected in comparison with the p_{1-4} if $k_0 R \gg 1$.

Thus, using the high-frequency approximation, the sound field p of a point source located above an impedance surface in a stratified moving medium is expressed as the sum of the fields p_{1-4} corresponding to rays of the particular kinds. To calculate the sound field at an observation point, the rays arriving at this point should be determined first, and the sound fields corresponding to these rays should then be summed.

4.5.5   Numerical example

The equations obtained above will now be used to study the effects of wind stratification and impedance ground on the interference of the direct wave from the source to the receiver and the wave reflected from the ground. The source and receiver are located at heights $z_1 = 100$ m and $z = 1.6$ m, respectively, the sound frequency is $f = \omega/2\pi = 1.5$ kHz, $c = c_0 = 330$ m/s, the wind velocity $v(z) = v_0' z$ increases linearly with height up to 150 m and then it is constant, and the horizontal range r between the source and receiver is less than 650 m. In this case, the sound field p is the sum of the fields p_1 and p_2, that is, the direct wave and the wave reflected from the ground. We shall analyze the dependence of the sound-pressure amplitude $|p|$ on the azimuthal angle ψ between the direction to the observation point and the direction of wind velocity, for different values of the horizontal range r.

First, consider sound propagation above an acoustically hard ground with $\beta = 0$. (In this case, the reflection coefficient $\mathcal{R}_p$ of a sound wave from the ground is equal to 1.) The dependence of $|p|R/2$ on ψ, calculated using equations (4.54) and (4.78), is shown in figure 4.9(a) for four values of r. The amplitude $|p|$ depends slightly on ψ if r is small (in this case, $r = 50$ m). For large enough horizontal ranges ($r = 75$, 250, and 650 m), the effect of the wind stratification on the interference of the direct and reflected waves becomes significant. This leads to the appearance of maxima and minima in the dependence of $|p|$ on ψ; $|p|R/2 = 1$ at maxima and $|p|R/2 = 0$ at minima. Note that the maximum of the sound pressure might not occur downwind.

For the case of an impedance ground with $\beta^{-1} = 2.6 + i2.2$, the analogous dependence of $|p|R/2$ on ψ is shown in figure 4.9(b). This value of β represents grass-covered ground. The interference maxima and minima in figure 4.9(b) are not as pronounced as those in figure 4.9(a) and their values are different for different r and ψ. This dependence of $|p|$ on ψ is due to the fact that the modulus of the reflection coefficient $|\mathcal{R}_p|$ is less than 1 and depends on the angle of incidence α of a sound wave on the ground. The angle α, in turn, depends on the horizontal range r, resulting in different values of the interference maxima and minima at $r = 75$, 250, and 650 m. Moreover, the angle α depends on the azimuthal angle ψ because of the medium motion. This results in different values of maxima and minima at $r = 650$ m.

4.6 Discrete spectrum of a sound field

In this section, an asymptotic form for the discrete spectrum of a point source field is obtained for the propagation ranges greater than the sound wavelength [285]. The resulting equations are then used for calculating the sound field in a waveguide located near the ground.

4.6.1 Asymptotic form for the discrete spectrum

The sound field of a point source is given by equation (4.15) or by equation (4.12). These equations are valid for arbitrary profiles $c(z)$, $\varrho(z)$, and $\mathbf{v}(z)$. In equation (4.12), it is worthwhile introducing the angle $\psi(z)$ between the vectors $\mathbf{v}(z)$ and $\mathbf{r}$, and the angle φ between the vectors $\boldsymbol{\kappa}$ and $\mathbf{r}$. Hereafter, it is assumed that $-\pi \le \psi \le \pi$ and $-\pi/2 \le \varphi \le 3\pi/2$.

The integration over $\boldsymbol{\kappa}$ in equation (4.12) is replaced by an integration over

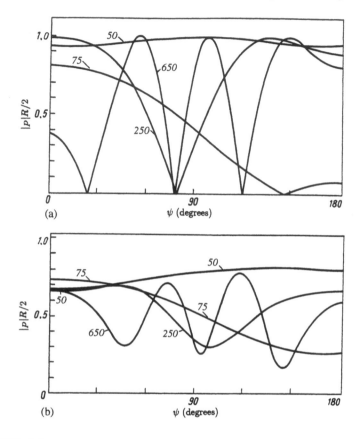

FIGURE 4.9

Function $|p|R/2$, proportional to sound-pressure amplitude, versus azimuthal angle ψ for different values of the horizontal range r. The numbers beside the curves indicate the value of r in meters. The calculations are for (a) acoustically hard ground and (b) impedance ground.

κ and φ. The integral over φ is expressed as the sum of two integrals with the limits of integration $-\pi/2 \leq \varphi \leq \pi/2$ and $\pi/2 \leq \varphi \leq 3\pi/2$, respectively. As a result, the sound field p is expressed as the sum of two fields:

$$p = p^{(1)} + p^{(2)},$$

where $p^{(1)}$ and $p^{(2)}$ are given by

$$p^{(1)} = \int_{-\pi/2}^{\pi/2} d\varphi \int_0^\infty \kappa \Lambda(\kappa, \varphi, z) \tilde{p}(\kappa, \varphi, z) e^{i\kappa r \cos \varphi} \, d\kappa, \tag{4.85}$$

$$p^{(2)} = \int_{\pi/2}^{3\pi/2} d\varphi \int_0^\infty \kappa \Lambda(\kappa, \varphi, z) \tilde{p}(\kappa, \varphi, z) e^{i\kappa r \cos \varphi} \, d\kappa. \tag{4.86}$$

Here, $\Lambda = (\varrho/\varrho_1)^{1/2}[1 - \kappa v \cos(\varphi - \psi)/\omega]$.

Consider the integral over κ in equation (4.85). The poles of the integrand coincide with poles of the function $\tilde{p}$ and, hence, can be determined from the equality $W = 0$ (see equation (4.14)). If infinitesimal absorption is present in the medium, these poles are shifted from the path of integration into the first or fourth quadrant of the complex plane κ, where $\kappa = \kappa' + i\kappa''$. The path of integration over κ in equation (4.85) is then completed by the part L_1 of an infinitely large circle and by the imaginary half-axis Γ_1 (figure 4.10). Since $\cos\varphi > 0$ in equation (4.85), the integral along L_1 is equal to zero. Using the theorem of residues, one obtains the sound field $p^{(1)}$:

$$p^{(1)} = \int_{-\pi/2}^{\pi/2} \left[2\pi i \sum_n \kappa_n \Lambda_n \operatorname{Res} \tilde{p}_n e^{i\kappa_n r \cos\varphi} \right.$$
$$\left. - \left(\int_{i\infty}^{0} + \int_{B_1} \right) \kappa \Lambda \tilde{p} e^{i\kappa r \cos\varphi} \, d\kappa \right] d\varphi. \quad (4.87)$$

In this expression, κ_n are the poles of the function $\tilde{p}$ in the first quadrant, the sum is taken over all of these poles, $\Lambda_n = \Lambda(\kappa = \kappa_n)$, $\operatorname{Res} \tilde{p}_n$ are the residues of the function $\tilde{p}$ at the poles κ_n, and B_1 denotes possible cuts of this function.

The integral over κ in equation (4.86) can be calculated in an analogous manner. Since $\cos\varphi < 0$ in equation (4.86), the path of integration should be completed in the fourth quadrant of the complex plane (see figure 4.10). In this case, the integral along the part L_2 of the infinitely large circle is equal to zero. As a result, the sound field $p^{(2)}$ is given by

$$p^{(2)} = -\int_{\pi/2}^{3\pi/2} \left[2\pi i \sum_m \kappa_m \Lambda_m \operatorname{Res} \tilde{p}_m e^{i\kappa_m r \cos\varphi} \right.$$
$$\left. + \left(\int_{-i\infty}^{0} + \int_{B_2} \right) \kappa \Lambda \tilde{p} e^{i\kappa r \cos\varphi} \, d\kappa \right] d\varphi. \quad (4.88)$$

Here, the sum is taken over all of the poles κ_m of the function $\tilde{p}$ in the fourth quadrant, B_2 denotes possible cuts of this function, and $\Lambda_m = \Lambda(\kappa = \kappa_m)$.

Consider the integrals in equations (4.87) and (4.88). It can be shown that the integrals along the imaginary half-axis Γ_1 and Γ_2 (figure 4.10) do not contribute to the asymptotic form of the discrete spectrum for relatively large range when $\kappa'_{n,m} r \gg 1$. The integrals along the cuts B_1 and B_2 make a contribution only to the continuous spectrum of the sound field. Uniting the remainder terms in equations (4.87) and (4.88), corresponding to the residues of the function $\tilde{p}$ at the poles κ_n and κ_m, one obtains

$$p = \sum_n p_n + \sum_m p_m = 2\pi i \sum_n \int_{-\pi/2}^{\pi/2} \kappa_n \Lambda_n \operatorname{Res} \tilde{p}_n \, e^{i\kappa_n r \cos\varphi} \, d\varphi$$
$$- 2\pi i \sum_m \int_{\pi/2}^{3\pi/2} \kappa_m \Lambda_m \operatorname{Res} \tilde{p}_m \, e^{i\kappa_m r \cos\varphi} \, d\varphi. \quad (4.89)$$

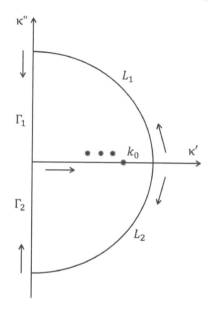

FIGURE 4.10
Paths of integration in the complex plane. L_1 and L_2 are parts of the infinitely large circle. Arrows indicate the direction of path-tracing. Dots show the location of the poles κ_n of the function $\tilde{p}$ for the case of waveguide sound propagation near the ground.

Let us consider the poles κ_i, where the index i takes the values of the indices n and m. The sound fields p_i (i.e., the fields p_n and p_m in equation (4.89)), corresponding to the residues of the function $\tilde{p}$ at these poles, will be called the asymptotic forms of the normal waves (modes) and the sum of p_i will be called the asymptotic form of the discrete spectrum. The discrete spectrum of the sound field and the normal modes are very convenient for describing waveguide propagation.

Since κ_i, Λ_i, and $\mathrm{Res}\,\tilde{p}_i$ depend on the angle φ in a moving medium, the integrals over φ in equation (4.89) cannot be calculated exactly. These integrals can be evaluated approximately by making use of the method of steepest descent provided that $|\kappa_i|r \gg 1$. To this end, the arguments of the exponents in equation (4.89) are denoted by $f_i = i\Phi_i$, where $\Phi_i = \kappa_i r \cos\varphi$ are complex phase functions. The saddle points φ_i are determined by the equalities $\partial f_i/\partial\varphi = 0$ and are different for different modes. Substituting f_i into these equalities, one obtains the equations for φ_i:

$$\tan\varphi_i = \frac{1}{\kappa_i}\frac{\partial\kappa_i}{\partial\varphi_i}. \tag{4.90}$$

Calculating the integrals in equation (4.89) by using the method of steepest descent and retaining only the first term of the asymptotic series in the small parameter $1/(|\kappa_i|r)$, one obtains the desired expression for the asymptotic form of the discrete spectrum:

$$p = i(2\pi)^{3/2} \sum_i (-1)^{\nu_i} \kappa_i \Lambda_i J_i^{-1/2} \operatorname{Res} \tilde{p}_i \exp\left[i\kappa_i r \cos\varphi_i + \frac{i}{2}\left(\frac{\pi}{2} - s_i\right)\right].$$

(4.91)

In this expression, the sum is taken over all of the poles κ_i of the function $\tilde{p}$ in the first and fourth quadrants and $\nu_i = \nu_n = 0$ in the first quadrant and $\nu_i = \nu_m = 1$ in the fourth quadrant. Furthermore, J_i and s_i are given by

$$J_i = \left|\frac{\partial^2 \Phi_i}{\partial\varphi_i^2}\right|, \quad s_i = \arg\frac{\partial^2 \Phi_i}{\partial\varphi_i^2},$$

(4.92)

where

$$\frac{\partial^2 \Phi_i}{\partial\varphi_i^2} = -r\left[\cos\varphi_i\left(\kappa_i - \frac{\partial^2 \kappa_i}{\partial\varphi_i^2}\right) + 2\sin\varphi_i \frac{\partial\kappa_i}{\partial\varphi_i}\right].$$

(4.93)

Since $\operatorname{Im}\kappa_n > 0$ and $\cos\varphi_n > 0$ in the first quadrant, and $\operatorname{Im}\kappa_m < 0$ and $\cos\varphi_m < 0$ in the fourth quadrant, the normal modes of the discrete spectrum (4.91) are attenuated exponentially if r is increased. Using equation (4.14) (where we set $\mathbf{r}_1 = 0$ without loss of generality), the residues of the function $\tilde{p}$ in equation (4.91) can be expressed in terms of the solutions $\tilde{p}_1$ and $\tilde{p}_2$ of the homogeneous equation (4.13) and their Wronskian W:

$$\operatorname{Res}\tilde{p} = \frac{\tilde{p}_1 \tilde{p}_2}{\pi \partial W/\partial\kappa}.$$

(4.94)

In a motionless medium, the poles κ_n and κ_m do not depend on φ. In this case, it can be shown from equations (4.89) and (4.90) that $\varphi_n = 0$, and $\varphi_m = \pi$. It follows from these equalities that the wavefronts of the normal modes p_n and p_m propagate in the directions away from the source and towards it, respectively. In a motionless stratified medium, the normal modes p_m do not satisfy the radiation condition and, therefore, should be eliminated from the discrete spectrum (4.91). But in a moving medium, the normal modes p_m can satisfy the radiation condition in some cases (e.g., see [377]) and hence should be retained in equation (4.91).

4.6.2 Asymptotic form for normal modes

Now we go back to the problem considered in section 4.5, namely the sound field of a point source located above an impedance surface in a stratified moving medium. The asymptotic forms of the normal modes in the waveguide located near the surface will be obtained by using equations derived in the preceding subsection.

The function $\tilde{p}(\kappa, z)$ for this problem has been calculated in the high-frequency approximation (see equation (4.66)). First, let us determine the location of the poles of this function in the $\kappa_x \kappa_y$-plane, where κ_x and κ_y are the components of the vector κ. It can be shown from equation (4.60) that in equation (4.66) the ratio ξ/q^2 cannot be zero. Therefore, it follows from equation (4.66) that the poles of the function $\tilde{p}$ are determined from the equality $K + iE = 0$ if κ_x, κ_y are in the regions $D_{1,2}$ (see figure 4.8), and these poles are determined from the equality $E = 0$ if κ_x, κ_y are in the regions $D_{3,4}$. It can be shown that the poles determined by the equality $K + iE = 0$ correspond to creeping waves, which are eliminated from further consideration.

The normal modes in the waveguide correspond to the poles of the function $\tilde{p}(\kappa, z)$ located in region D_3. These poles are determined from the equality $E = 0$. Substituting the value of E given by equation (4.64) into this equality, calculating the derivative ξ_0' in region D_3, and taking into account the fact that $\mathbf{v}(z = 0) = 0$ yields

$$(k_0^2 - \kappa^2)^{1/2} \mathrm{Ai}'(-\xi_0) + ik_0 \xi_0^{1/2} \beta(\kappa) \mathrm{Ai}(-\xi_0) = 0. \tag{4.95}$$

Here, $k_0 = \omega/c_0$ is a reference wavenumber, where $c_0 = c(z = 0)$, and

$$\xi_0 = \left(\frac{3}{2} \int_0^{z_t} q\,dz \right)^{2/3} = \left\{ \frac{3}{2} \int_0^{z_t} \left[\frac{([\omega - \kappa v \cos(\varphi - \psi)]^2}{c^2} - \kappa^2 \right]^{1/2} dz \right\}^{2/3}. \tag{4.96}$$

The left-hand side of equation (4.95) becomes zero only for certain values of $\kappa = \kappa_i$, which are the poles of the function $\tilde{p}(\kappa, \varphi, z)$ in the complex plane $\kappa = \kappa' + i\kappa''$ (see figure 4.10). It follows from equations (4.95) and (4.96) that κ_i depend on φ in a moving medium.

Substituting the function $\tilde{p}$ given by equation (4.66) into equation (4.91) and calculating the residues of this function, one obtains the sound field of a point source in the waveguide located near the surface:

$$p = 2i(2\pi)^{1/2} \sum_i (-1)^{\nu_i} \kappa_i \Lambda_i J_i^{-1/2} \left[\left(\frac{\xi \xi_1}{q^2 q_1^2} \right)^{1/4} \mathrm{Ai}(-\xi)\mathrm{Ai}(-\xi_1)F \right]_i$$
$$\times \exp\left[i\kappa_i r \cos\varphi_i + \frac{i}{2}(\pi/2 - s_i) \right]. \tag{4.97}$$

In this expression, the arguments κ and φ of the functions in the braces are equal to κ_i and φ_i, respectively, and the function F is given by

$$F = \frac{q_0^2}{\xi_0^{1/2} \mathrm{Ai}^2(-\xi_0)} \left[\left(q_0^2 + \frac{ik_0 \beta q_0}{2\xi_0^{3/2}} - k_0^2 \beta^2 \right) d - ik_0 \left(q_0 \frac{\partial \beta}{\partial \kappa} + \frac{\beta \kappa}{q_0} \right) \right]^{-1}. \tag{4.98}$$

Here, d is given by

$$d = \int_0^{z_t} \frac{\kappa + (\omega - \kappa v \cos(\varphi - \psi))v \cos(\varphi - \psi)/c^2}{q} \, dz.$$

The terms of the series in equation (4.97) are the asymptotic forms of the normal modes for $|\kappa_i| r \gg 1$. It follows from equations (4.95)–(4.98) that in a moving medium the amplitudes and phases of these normal modes depend on the angle ψ between the vectors $\mathbf{v}$ and $\mathbf{r}$. The number of normal modes in the waveguide also depends on the angle ψ. Note that equation (4.97) is valid both for $z \leq z_t$ ($z_1 \leq z_t$) and for $z \geq z_t$ ($z_1 \geq z_t$).

In equation (4.97), the poles κ_i and the saddle points φ_i are determined from equations (4.95) and (4.90), respectively. These equations cannot be solved exactly. We shall obtain approximate solutions, which will allow us to analyze the dependence of κ_i and φ_i on the parameters of the problem considered.

Expanding the function $q^2(z)$ into a Taylor series near the point $z = 0$ and retaining the first two terms of the series yields

$$q^2(z) = k_0^2 - \kappa^2 - zH^{-3}. \tag{4.99}$$

Here,

$$H(a, \varphi) = \left[2k_0^2 \left(\frac{\kappa v_0' \cos(\varphi - \psi_0)}{\omega} + \frac{c_0'}{c_0} \right) \right]^{-1/3}$$

is the effective height of the waveguide and

$$v_0' = v'\big|_{z=0}, \quad c_0' = c'\big|_{z=0}, \quad \psi_0 = \psi\big|_{z=0}.$$

Suppose that the function $q^2(z)$ can be approximated by expression (4.99) up to the level z_t of the turning point and that $\beta \ll 1$ or $\beta \gg 1$. Then, equation (4.95), where ξ_0 depends on $q(z)$ (see equation (4.96)), can be solved for κ_i. For $\beta \ll 1$, one obtains (see [285] for more detail)

$$\kappa_i = k_0 \left[1 - \frac{y_i'}{2k_0^2 H^2(k_0, \varphi)} + \frac{i\beta(\kappa_i^0)}{2y_i' k_0 H(k_0, \varphi)} \right]. \tag{4.100}$$

If $\beta \gg 1$, the κ_i are given by

$$\kappa_i = k_0 \left[1 - \frac{y_i}{2k_0^2 H^2(k_0, \varphi)} + \frac{i}{2k_0^3 H^3(k_0, \varphi)\beta(\kappa_i^\infty)} \right]. \tag{4.101}$$

Here, $-y_i$ and $-y_i'$ are successive zeros of the Airy function Ai and its derivative Ai$'$, where $i = 1, 2, 3, \ldots$. The first two values of y_i and y_i' are $y_1 = 2.34$, $y_2 = 4.09$ and $y_1' = 1.02$, $y_2' = 3.25$. The quantities $\kappa_i^0 = k_0[1 - y_i'/(2k_0^2 H^2(k_0, \varphi))]$ and $\kappa_i^\infty = k_0[1 - y_i/(2k_0^2 H^2(k_0, \varphi))]$ are the values of κ_i for $\beta = 0$ and $\beta = \infty$, respectively. Equations (4.100) and (4.101) are valid if $\kappa_i' \sim k_0$ and $\kappa_i'' \ll k_0$.

It follows from equations (4.100)–(4.101) that the normal modes in the waveguide located near the ground are attenuated exponentially if r is increased. This is explained by the interaction of the normal modes with the impedance ground. Since the effective height H of the waveguide depends on the angle ψ_0 between the vectors $\mathbf{v}$ and $\mathbf{r}$, the real (κ_i') and imaginary (κ_i'') parts of the poles κ_i also depend on this angle. Moreover, $\kappa_i'' > 0$ since $H > 0$. Therefore, the poles κ_i of the function $\tilde{p}$ are located in the first quadrant near the point $\kappa = k_0$. These poles are shown schematically in figure 4.10. In what follows, only the normal modes corresponding to the poles in the first quadrant will be considered and therefore the index i will be replaced by n.

When determining the values of the saddle points φ_n, we assume also that $\beta \ll 1$ or $\beta \gg 1$. However, the linear dependence of the function q^2 on z (see equation (4.99)) is not required now. (In this case, the poles κ_n are determined from equation (4.95), rather than by equations (4.100) and (4.101).) Note that for $\beta \ll 1$ and $\beta \gg 1$, the expressions for the function F are simplified significantly and take the forms

$$F = \frac{1}{\xi_0^{1/2} d \mathrm{Ai}^2(-\xi_0)} \quad \text{and} \quad F = \frac{1}{\xi_0^{1/2} d (\mathrm{Ai}'(-\xi_0))^2},$$

respectively.

Using equations (4.90) and (4.95), the following equation for φ_n can be derived [285]

$$\sin \varphi_n \int_0^{z_{t,n}} \frac{\kappa_n}{q_n} \, dz = - \int_0^{z_{t,n}} \frac{(\omega - \kappa_n v \cos(\varphi_n - \psi)) v \sin \psi}{q_n c^2} \, dz. \qquad (4.102)$$

Here, q_n and $z_{t,n}$ are the values of q and z_t for $\kappa = \kappa_n$ and $\varphi = \varphi_n$. It follows from equation (4.102) that $\varphi_n \sim -(v/c) \sin \psi$. Therefore, $\cos \varphi_n$ in equations (4.97) and (4.98) can be considered to be equal to 1 to an accuracy of v/c. (The angle φ_n itself can be set to zero only if ψ differs from $\pm\pi/2$.) Equation (4.93) can be expressed in the form

$$\frac{\partial^2 \Phi_n}{\partial \varphi_n^2} = -\kappa_n r \left[1 \pm O \left(\frac{v}{c} \right) \right]. \qquad (4.103)$$

It follows from this formula and equation (4.92) that $s_n = -\pi$ in equation (4.97).

The normal modes representation (4.97) of the sound field in the waveguide is convenient for relatively large horizontal ranges r from the source, when there are many sound rays arriving at the observation point. For relatively short ranges, there are only few sound rays arriving at the observation point, and it is worthwhile to express the sound field as the sum of fields corresponding to these rays (equation (4.84)).

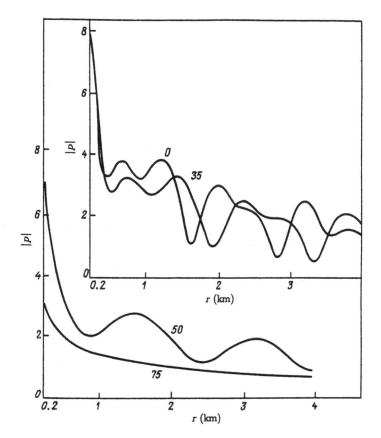

FIGURE 4.11
Sound-pressure amplitude $|p|$ (in arbitrary units) in atmospheric waveguide
versus horizontal range r for different values of the azimuthal angle ψ. Numbers beside the curves are the values of ψ in degrees.

4.6.3 Numerical example

Using the equations obtained above, we shall analyze the effect of temperature and wind velocity stratification on sound propagation in an atmospheric
waveguide located near the ground. The source and receiver are situated on
acoustically hard ground ($\beta = 0$), and the sound frequency $f = 30$ Hz. The
wind velocity $v(z)$ and temperature $T(z)$ are linear functions of z in the interval $0 \leq z \leq z_2 = 100$ m and are constant for $z > z_2$. Moreover, $v(z = 0) = 0$,
$v(z_2) = 10$ m/s, $T(z = 0) = 273$ K, and $T(z_2) = 270$ K (in a dry atmosphere,
the temperature T is related to the sound speed c by equation (2.42)).

Figure 4.11 depicts the sound-pressure amplitude $|p|$ versus the horizontal
range r for different values of the angle ψ between the vectors **v** and **r**. If
$\psi = 0°$, there are three normal modes in the waveguide. The interference of

these modes results in spatial oscillations of $|p|$. If $\psi = 35°$, there are still three normal modes in the waveguide, but now their interference pattern is stretched more along the r-axis compared with that for $\psi = 0°$. This results from the fact that the effective height H of the waveguide depends on the angle ψ. If $\psi = 50°$, there are two normal modes in the waveguide; this increases the period of spatial oscillations of $|p|$ compared with the cases $\psi = 0°$ and $\psi = 35°$. Finally, if $\psi = 75°$, the waveguide contains only one normal mode, the amplitude of which decreases as $r^{-1/2}$.

4.6.4 Additional comments about the discrete spectrum

For $\mathbf{v} = 0$, the approach considered here is known in the literature (e.g., reference [356]). The components p_n of the discrete spectrum given by equation (4.89) or (4.91) satisfy the radiation condition and are asymptotic forms of solutions of the wave equation in a motionless medium for $r \gg 1/|\kappa_n|$.

Usually, in oceanic acoustics, the discrete spectrum is obtained by another approach (e.g., reference [91]). Replacing the integration over $\boldsymbol{\kappa}$ in equation (4.12) by an integration over κ and φ yields

$$p = \int_0^\infty d\kappa \int_0^{2\pi} \kappa \Lambda(\kappa, \varphi, z) \tilde{p}(\kappa, \varphi, z) e^{i\kappa r \cos \varphi} \, d\varphi. \tag{4.104}$$

If $\mathbf{v} = 0$, the functions Λ and $\tilde{p}$ do not depend on the azimuthal angle φ, so the integral over φ can be calculated exactly. In the resulting equation, the path of integration over κ can be extended from $-\infty$ to ∞. Then, the path can be completed by an infinitely large semicircle in the upper half-plane of the complex values of κ. Normal modes, which we denote by Π_n, correspond to residues of the function $\tilde{p}$ in this half-plane. The normal modes Π_n satisfy the wave equation and, for $r \gg 1/|\kappa_n|$, coincide asymptotically with the normal modes p_n of the discrete spectrum (4.91) in a motionless stratified medium.

The discrete spectrum of the sound field of a point source in a stratified moving medium has usually been obtained to an accuracy of v/c [77, 78, 149, 281, 318, 443]. Commonly, the integral over φ in equation (4.104) has been calculated by using the method of stationary phase (or the method of steepest descent). The integral over κ is then calculated by using the approach described in the preceding paragraph. This results in equations (4.91)–(4.93) for the discrete spectrum, where $\cos \varphi_n = 1$ and $\partial^2 \Phi_n / \partial \varphi_n^2 = -\kappa_n r$.

When discussing the discrete spectrum (4.97), it was noted that $\cos \varphi_n = 1$ to an accuracy of v/c and $\partial^2 \Phi_n / \partial \varphi_n^2 = -\kappa_n r + O(v/c)$. Therefore, the approach for calculating the discrete spectrum in references [77, 78, 149, 281, 318, 443] allows one to determine the phases of normal modes to the accuracy used in these references. This approach does not enable to account for terms of order v/c in the amplitudes of normal modes. However, these terms can be ignored to a good accuracy. The absence of these terms is due to the fact that poles of the function $\tilde{p}$ are located near the path of integration over φ. Therefore, the method of stationary phase, used in the above papers for calculating the integral over φ, strictly speaking, cannot be used. It is more important, however, that in these references the phases of the normal modes are calculated only to an accuracy of v/c. This can result in a significant distortion of the sound field if it is a sum of two or more normal modes. Therefore, an alternative approach has been developed [285], which allows determination of the asymptotic form of the discrete spectrum with an arbitrary accuracy in terms of v/c (see equations (4.90)–(4.93)). In many respects, this approach is similar to one proposed earlier [311].

A rigorous approach for obtaining the discrete spectrum of the sound field in a stratified moving medium has been developed elsewhere [47]. The integral over κ in equation (4.104) is calculated first by using the approach similar to that in reference [91]. Then, the integral over φ is calculated by using the method of stationary phase. The normal modes obtained in [47], which we denote by P_n, satisfy the wave equation for the case of a motionless medium. It can be shown that in both a motionless medium and a moving one, the modes P_n and p_n coincide asymptotically for $r \gg 1/|\kappa_n|$, and differ only for $r \lesssim 1/|\kappa_n|$. In particular, equations (15.65)–(15.68) for the modes P_n for $r \gg 1/|\kappa_n|$, obtained in [47], coincide with equations (4.89)–(4.93) for the modes p_n.

5

Moving sound sources and receivers

Sound is generated by moving sources such as cars, trains, airplanes, helicopters, rockets, ships, and submarines. Microphones can be installed on moving platforms such as military vehicles, cars, helicopters, and unattended aerial vehicles (UAVs). Since the sound field produced by a source moving with a constant velocity in a motionless medium is analogous to the field produced by the same source at rest in a uniformly moving medium, moving sources are often considered in the context of acoustics of moving media [37, 144, 264].

In this chapter, the effects of the source and receiver motions on the transmitted and received acoustic signals in an inhomogeneous moving medium are analyzed. The sound field of a point source moving with an arbitrary velocity in a homogeneous motionless medium is studied in section 5.1. In section 5.2, the Galilean transformation is presented and the sound field of a source moving in an inhomogeneous moving medium is analyzed. Sound aberration and the change in propagation direction of a sound wave emitted by a moving source are studied in section 5.3. A general case of the Doppler effect, when both a source and receiver are moving at different velocities in an inhomogeneous moving medium, is considered in section 5.4.

5.1 Sound field for a moving source in a homogeneous motionless medium

In this section, the sound-pressure field $p(\mathbf{R}, t)$ of a point mass source moving with an arbitrary velocity $\mathbf{u}(t)$ in a homogeneous motionless medium is calculated and analyzed. Here, $\mathbf{R} = (x, y, z)$ are the Cartesian coordinates and t is the time. This problem has been studied in the literature (e.g., references [37, 144, 230, 264]).

5.1.1 Exact formula for the sound field

Consider a point source moving along the path

$$\mathbf{R}_u(t) = \int_0^t \mathbf{u}(t') \, dt' \tag{5.1}$$

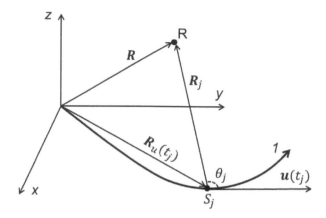

FIGURE 5.1
Point source moving with arbitrary velocity. 1 is the source path (the arrow indicates the direction of source motion), R is the receiver position, and S_j is the source location at the time t_j of sound emission.

as shown in figure 5.1. Here, the vector $\mathbf{R}_u(t)$ determines the source position at time t; the source is located at the point $\mathbf{R} = 0$ when $t = 0$. We assume that the density ϱ and the sound speed c of the medium are constant, that the medium velocity $\mathbf{v} = 0$, and that there are no forces $\mathbf{F}$ acting on the medium. In this case equation (2.100) becomes

$$\frac{1}{c^2}\frac{\partial^2 p}{\partial t^2} - \Delta p = \varrho\frac{\partial Q}{\partial t}. \tag{5.2}$$

Here, $\Delta = \partial^2/\partial x^2 + \partial^2/\partial y^2 + \partial^2/\partial z^2$ and the function $Q(\mathbf{R}, t) = a(t)\delta(\mathbf{R} - \mathbf{R}_u(t))$ characterizes the moving point mass source, where δ is the delta-function and $a(t)$ determines the dependence of Q on time in the coordinate system moving with the source. The sound-pressure field $p(\mathbf{R}, t)$ of the source is a solution of equation (5.2).

Now we introduce the potential $\varphi(\mathbf{R}, t)$ of the acoustic particle velocity, which is related to the sound pressure by $p = \varrho\partial\varphi/\partial t$. It follows from equation (5.2) that φ satisfies the wave equation

$$\frac{1}{c^2}\frac{\partial^2 \varphi}{\partial t^2} - \Delta\varphi = Q. \tag{5.3}$$

Writing down the well-known solution of this equation in integral form and replacing φ by p, one obtains the expression for the sound field of the moving source:

$$p = \frac{\varrho}{4\pi}\frac{\partial}{\partial t}\int dt' \int \frac{Q(\mathbf{R}', t')}{|\mathbf{R} - \mathbf{R}'|}\delta\left(|\mathbf{R} - \mathbf{R}'|/c - t + t'\right) d^3R'. \tag{5.4}$$

The integrals over t' and $\mathbf{R}'$ in this expression can be evaluated. To this end, we substitute $Q = a(t)\delta(\mathbf{R}' - \mathbf{R}_u(t'))$ into equation (5.4) and calculate the integral over $\mathbf{R}'$. The result is

$$p = \frac{\varrho}{4\pi} \frac{\partial}{\partial t} \int \frac{a(t')}{|\mathbf{R} - \mathbf{R}_u(t')|} \delta\left(|\mathbf{R} - \mathbf{R}_u(t')|/c - t + t'\right) dt'. \tag{5.5}$$

When calculating the integral over t', we use the formula

$$\int Y(t')\delta(X(t')) \, dt' = \sum_j Y(t_j)|\partial X/\partial t'|_{t'=t_j}^{-1},$$

where $Y(t)$ and $X(t)$ are functions and t_j is the jth root of the equation $X(t') = 0$. In equation (5.5) the function $|\mathbf{R} - \mathbf{R}_u(t')|/c - t + t'$ plays the role of the function $X(t')$. Calculating the derivative of the former function with respect to t', one obtains:

$$p(\mathbf{R}, t) = \frac{\varrho}{4\pi} \frac{\partial}{\partial t} \sum_j \frac{a(t_j)}{||\mathbf{R} - \mathbf{R}_u(t_j)| - (\mathbf{R} - \mathbf{R}_u(t_j)) \cdot \mathbf{u}(t_j)/c|}. \tag{5.6}$$

Here, the sum is taken over all solutions t_j of the equation $|\mathbf{R} - \mathbf{R}_u(t')|/c - t + t' = 0$. Substituting the value of $\mathbf{R}_u$ into this equation yields the equation for determining t_j:

$$(t - t_j)c = R_j, \tag{5.7}$$

where R_j is the magnitude of the vector $\mathbf{R}_j$ given by:

$$\mathbf{R}_j = \mathbf{R} - \mathbf{R}_u(t_j) = \mathbf{R} - \int_0^{t_j} \mathbf{u}(t') \, dt'. \tag{5.8}$$

Consider the physical meaning of t_j. To this end, we refer to figure 5.1, where the vector $\mathbf{R}$ characterizes the receiver position, $\mathbf{R}_u(t_j)$ determines the source position at the time t_j, and $\mathbf{R}_j$ characterizes the relative position of the source and receiver at this time. It follows from equation (5.7) that $t_j = t - R_j/c$, where R_j/c is the time of sound propagation from the source to the receiver. Therefore, t_j is the time of emission of the sound wave which arrives at the observation point $\mathbf{R}$ at time t.

Equation (5.7) has no more than one solution if $u < c$. To demonstrate this, consider the function $Z(t_j) = (t_j - t)c - R_j$. Since the derivative $\partial Z(t_j)/\partial t_j = c - \mathbf{R}_j \cdot \mathbf{u}(t_j)/R_j > 0$, the function $Z(t_j)$ is a monotonically increasing function. Therefore, the equation $Z(t_j) = 0$ and its equivalent, equation (5.7), have no more than one solution.

Equation (5.6) can be written in the following form:

$$p(\mathbf{R}, t) = \frac{\varrho}{4\pi} \frac{\partial}{\partial t} \sum_j \frac{a(t - R_j/c)}{|R_j - \mathbf{R}_j \cdot \mathbf{M}_j|}, \tag{5.9}$$

where $\mathbf{M}_j = \mathbf{u}(t_j)/c$ is the Mach number. Using equations (5.7) and (5.8), one can calculate the derivatives with respect to t appearing in equation (5.9):

$$\frac{\partial R_j}{\partial t} = -\frac{cM_j \cos \theta_j}{1 - M_j \cos \theta_j}, \tag{5.10}$$

$$\frac{\partial t_j}{\partial t} = \frac{1}{1 - M_j \cos \theta_j}, \tag{5.11}$$

$$\frac{\partial}{\partial t} \frac{1}{|R_j - \mathbf{R}_j \cdot \mathbf{M}_j|} = \frac{cM_j(\cos \theta_j - M_j) + \mathbf{R}_j \cdot \mathbf{M}'_j}{R_j^2 |1 - M_j \cos \theta_j|^3}. \tag{5.12}$$

Here, $\mathbf{M}'_j = \partial M(t_j)/\partial t_j$ and θ_j is the angle between the vectors $\mathbf{R}_j$ and $\mathbf{u}(t_j)$ (see figure 5.1). Using these equations, we obtain the sound field of a point mass source moving in a homogeneous motionless medium:

$$p(\mathbf{R}, t) = \frac{\varrho}{4\pi} \sum_j \left[\frac{c \, M_j(\cos \theta_j - M_j)a(t - R_j/c)}{R_j^2 \, |1 - M_j \cos \theta_j|^3} \right.$$

$$\left. + \frac{M'_j \cos \theta'_j a(t - R_j/c)}{R_j |1 - M_j \cos \theta_j|^3} + \frac{\nu_j a'(t - R_j/c)}{R_j (1 - M_j \cos \theta_j)^2} \right]. \tag{5.13}$$

Here, $a'(t) = \partial a/\partial t$, $\nu_j = \operatorname{sgn}(1 - M_j \cos \theta_j)$, and θ'_j is the angle between the vectors $\mathbf{R}_j$ and $\mathbf{M}'_j$.

It follows from equation (5.13) that the sound field $p(\mathbf{R}, t)$ is determined by the values of the functions $a(t_j)$ and $a'(t_j)$ at the time of sound wave emission t_j. It follows also from this equation that the sound-pressure amplitude $|p| \to \infty$ if $1 - M_j \cos \theta_j \to 0$, which could be the case for a supersonic source. In this case, linear acoustics is not valid and equation (5.13) does not describe the sound field of the moving source. (If the source is moving with a constant velocity, the equality $1 - M_j \cos \theta_j = 0$ is valid on the Mach cone; see figure 5.2 below.)

The first term in braces in equation (5.13) is proportional to $1/R_j^2$ and can be ignored at some distance from the source. The second and third terms are proportional to $1/R_j$ and describe the far field. The second term exists only if the source is moving with an acceleration. The fields corresponding to the second and third terms in equation (5.13) have radiation patterns determined by the functions $M'_j \cos \theta'_j/|1 - M_j \cos \theta_j|^3$ and $1/(1 - M_j \cos \theta_j)^2$, respectively. These functions result in an increase in the pressure amplitude $|p|$ in the direction of source motion, i.e., for $\theta < \pi/2$, and a decrease in $|p|$ in the opposite direction, i.e., for $\pi/2 < \theta < \pi$.

For a monochromatic source, the function $a(t)$ is given by

$$a(t) = \hat{a} \exp(-i\omega_1 t), \tag{5.14}$$

where $\hat{a}$ is the amplitude and ω_1 the angular frequency. In what follows, we consider a unit-amplitude point source (section 4.2.1), for which $\widehat{a} = i4\pi/\varrho\omega_1$. Substituting $a(t)$ into equation (5.13) yields

$$p = \sum_j \left[\nu_j + i\frac{cM_j(\cos\theta_j - M_j) + \mathbf{R}_j \cdot \mathbf{M}'_j}{R_j\omega_1|1 - M_j\cos\theta_j|} \right] \frac{\exp(-i\omega_1(t - R_j/c))}{R_j(1 - M_j\cos\theta_j)^2}. \quad (5.15)$$

In this formula, $\mathbf{R}_j$, $\mathbf{M}_j$, $\mathbf{M}'_j$, and θ_j depend on t. Therefore, the sound field emitted by a moving monochromatic source appears as non-monochromatic for an observer at rest with respect to the coordinate system (x, y, z). This sound field can be considered as approximately monochromatic if in one period of oscillation there is no significant change in amplitude and the phase depends almost linearly on t.

5.1.2 Approximate formula for the sound field

Let us obtain conditions under which the sound field given by (5.15) can be considered as monochromatic. We shall also obtain an approximate expression for the sound pressure $p(\mathbf{R}, t)$ under such conditions. Note that for the particular case $\mathbf{u} = \text{constant}$, these conditions are given in [144].

The phase and amplitude of the sound field given by equation (5.15) are expressed as series in $t - t_0$ near a fixed time t_0. It is convenient to recast the amplitude factor $R_j^{-1}(1 - M_j\cos\theta_j)^{-2}$ on the right-hand side of this equation as $R_j/(R_j - \mathbf{R}_j \cdot \mathbf{M}_j)^2$. Expanding the numerator and denominator of the latter expression as the Taylor series and using equation (5.12) yields

$$R_j(t) = R_j(t_0) \left[1 - (t - t_0) \left(\frac{cM_j\cos\theta_j}{R_j(1 - M_j\cos\theta_j)} \right)_{t=t_0} + \ldots \right], \quad (5.16)$$

$$\frac{1}{(R_j(t) - \mathbf{R}_j(t) \cdot \mathbf{M}_j(t))^2} = \frac{1}{(R_j - \mathbf{R}_j \cdot \mathbf{M}_j)^2_{t=t_0}}$$
$$\times \left[1 + 2(t - t_0) \left(\frac{cM_j(\cos\theta_j - M_j) + \mathbf{R}_j \cdot \mathbf{M}'_j}{R_j(1 - M_j\cos\theta_j)^2} \right)_{t=t_0} + \ldots \right]. \quad (5.17)$$

The factors R_j and $(R_j - \mathbf{R}_j \cdot \mathbf{M}_j)^{-2}$ can be considered to be independent of t, if the second terms in the square brackets in equations (5.16) and (5.17) are much less than 1 for a time interval $|t - t_0| \lesssim 2\pi/\omega$. Here, ω is the angular acoustic frequency measured by an observer at rest, given by equation (5.23) derived below. The second terms in the square brackets in equations (5.16) and (5.17) are much less than 1 if the following inequalities are valid:

$$M'_j \ll f_1 \left| \frac{1 - M_j\cos\theta_j}{2\cos\theta'_j} \right|, \quad (5.18)$$

$$R_j \gg \lambda_1 M_j\max\left[|\cos\theta_j|, \left| \frac{2(M_j - \cos\theta_j)}{1 - M_j\cos\theta_j} \right| \right]. \quad (5.19)$$

Here, $f_1 = \omega_1/2\pi$ and $\lambda_1 = c/f_1$ is the wavelength.

It can be shown that the second term in the square brackets in equation (5.15) can be ignored in comparison with the first one if the inequalities (5.18) and (5.19) are valid. Thus, if these inequalities are valid and the time interval $|t - t_0| \lesssim 2\pi/\omega = 2\pi(1 - M_j \cos\theta_j)/\omega_1$, the complex amplitude A of the sound field determined by equation (5.15) does not depend on t and is given by

$$A = \left[\frac{\nu_j}{R_j(1 - M_j \cos\theta_j)^2} \right]_{t=t_0}. \tag{5.20}$$

The phase of this sound field is given by $\Phi(t) = (R_j/c - t)\omega_1$. Expanding $\Phi(t)$ into a Taylor series near $t = t_0$ and taking into account the fact that $\partial R_j/\partial t$ is given by equation (5.10) and that $\partial^2 R_j/\partial t^2 = (c^2 M_j^2 \sin^2\theta_j/R_j - cM_j' \cos\theta_j')/(1 - M_j \cos\theta_j)^3$, one obtains

$$\Phi(t) = \left(\frac{R_j(t_0)}{c} - t_0 \right)\omega_1 - \frac{\omega_1(t - t_0)}{(1 - M_j \cos\theta_j)_{t=t_0}}$$
$$\times \left[1 - \frac{(t - t_0)}{2} \left(\frac{cM_j^2 \sin^2\theta_j/R_j - M_j' \cos\theta_j'}{(1 - M_j \cos\theta_j)^2} \right) + \dots \right]_{t=t_0}. \tag{5.21}$$

The frequency of the sound field measured by an observer at rest is determined by $\omega = -\partial\Phi/\partial t$. Therefore, the sound field given by equation (5.15) can be considered as monochromatic if the second term in the square brackets in equation (5.21) is much less than 1 for $|t - t_0| \lesssim 2\pi/\omega$. This condition is fulfilled if the inequality (5.18) and the inequality

$$R_j \gg \lambda_1 \frac{M_j^2 \sin^2\theta_j}{2|1 - M_j \cos\theta_j|} \tag{5.22}$$

are valid. In this case, the frequency measured by an observer at rest is obtained from equation (5.21):

$$\omega = \frac{\omega_1}{1 - M_j \cos\theta_j}. \tag{5.23}$$

This formula expresses the Doppler effect for an arbitrary moving source and relates the frequency ω_1 in the coordinate system moving with source to the frequency ω measured by an observer at rest. Equation (5.23) coincides with the well-known formula for the Doppler effect for the case of a source moving uniformly in a straight line. Thus, if the sound field of an arbitrary moving source is approximately monochromatic for an observer at rest, the frequency ω of this field is the same as that for the case $\mathbf{u} = $ constant. In particular, ω depends only on the source speed $u(t_j)$ and the angle θ_j, and does not depend on the source acceleration and its trajectory $R_u(t)$.

If the inequalities (5.18), (5.19), and (5.22) are valid, the approximate expression for the sound field of the moving source is obtained with equations (5.20) and (5.21):

$$p = \sum_j \left[\frac{\nu_j}{R_j(1 - M_j \cos\theta_j)^2} \right]_{t=t_0} \exp\{i\omega_1 [R_j(t_0)/c - t_0] - i\omega(t - t_0)\}.$$

$$(5.24)$$

The amplitude and frequency of this sound-pressure field are approximately independent of t for a few periods of oscillations. However, they can vary significantly over longer time intervals.

Consider the inequalities (5.18), (5.19), and (5.22) in detail. Inequalities (5.19) and (5.22) impose a limitation on the distance R_j between the source and observation point. They are valid if the observation point is not too close to the line where $1 - M_j \cos\theta_j = 0$ and if $R_j \gg \lambda_1 M_j$. Inequality (5.18) imposes a limitation on the source acceleration $u'_j = cM'_j$. For sound propagation in the atmosphere, where $c = 330$ m/s, this inequality takes the form $u'_j \ll 165\widetilde{A}f_{1\mathrm{Hz}}$ m/s^2, where $f_{1\mathrm{Hz}}$ is the value of the source frequency in Hz and $\widetilde{A} = |(1 - M_j \cos\theta_j)/\cos\theta'_j|$. The latter inequality is valid for almost all sources moving in the atmosphere, provided the observation point is not too close to the line where $1 - M_j \cos\theta_j = 0$.

5.1.3 Source moving with constant velocity

In the case $\mathbf{u} = $ constant, explicit expressions can be derived for the factors R_j and $\cos\theta_j$ which appear in the equation for the sound field $p(\mathbf{R}, t)$. To this end, we replace the right-hand side of equation (5.7) with $R_j = |\mathbf{R}_j - \mathbf{u}t_j|$, square both sides of the resulting equation, and obtain a quadratic equation for t_j. Solving this equation and taking into account the fact that $R_j = (t - t_j)c$, we find the explicit expression for the distance R_j:

$$R_j = R_t \frac{(-1)^j (1 - M^2 \sin^2\alpha)^{1/2} - M\cos\alpha}{1 - M^2}.$$

$$(5.25)$$

Here, $j = 1, 2$, $M = u/c$, and R_t is the magnitude of the vector $\mathbf{R}_t$ characterizing the relative position of the source and receiver at time t (see figure 5.2), which is given by

$$\mathbf{R}_t = \mathbf{R} - \mathbf{u}t.$$

$$(5.26)$$

In equation (5.25), α is the angle between the vectors $\mathbf{R}_t$ and $-\mathbf{u}$.

For $\mathbf{u} = $ constant, $\mathbf{R}_j = \mathbf{R} - \mathbf{u}t_j = \mathbf{R}_t + R_j\mathbf{u}/c$, and therefore

$$\cos\theta_j = \frac{\mathbf{R}_j \cdot \mathbf{u}}{R_j u} = \frac{\mathbf{R}_t \cdot \mathbf{u}}{R_j u} + M.$$

Substituting the value of R_j into this equation, one obtains the expression for

$\cos\theta_j$:

$$\cos\theta_j = M\sin^2\alpha - (-1)^j(1 - M^2\sin^2\alpha)^{1/2}\cos\alpha. \tag{5.27}$$

Using this expression, it can be shown that $\nu_j = \text{sgn}(1 - M\cos\theta_j) = (-1)^j$. Substituting the values of R_j, $\cos\theta_j$, and ν_j into equation (5.15) yields the expression for the sound field

$$p = \sum_{j=1}^{2} \left[\frac{(1 - M^2\sin^2\alpha)^{1/2} - (-1)^j M\cos\alpha}{(1 - M^2)(1 - M^2\sin^2\alpha)R_t} - \frac{icM\cos\alpha}{\omega_1(1 - M^2\sin^2\alpha)^{3/2}R_t^2} \right]$$

$$\times \exp\left\{ -i\omega_1\left[t - \frac{R_t}{[(-1)^j(1 - M^2\sin^2\alpha)^{1/2} + M\cos\alpha]c} \right] \right\}. \tag{5.28}$$

Equations (5.25)–(5.28) allow one to analyze the sound field of a source moving uniformly in a straight line. First, consider the case $M < 1$. In this case, it follows from equation (5.25) that $R_1 < 0$, and $R_2 > 0$. Since R_j is a distance and must be greater than 0, the solution $t_1 = t - R_1/c$ of equation (5.7) is extraneous and should be omitted. Consequently, for $M < 1$, there is only one sound wave, emitted at time $t_2 = t - R_2/c$ that arrives at an observation point at time t. The sound field p at the observation point is determined by equation (5.28), in which only the terms corresponding to the index $j = 2$ should be retained. If $R_t \gg c/\omega_1$, the second term in square brackets in this equation is much less than the first one and can be omitted. In this case, the sound-pressure amplitude of the moving source is given by

$$|p| = \frac{(1 - M^2\sin^2\alpha)^{1/2} - M\cos\alpha}{(1 - M^2)(1 - M^2\sin^2\alpha)R_t}. \tag{5.29}$$

Comparison of this formula to equation (4.26) reveals that the dependence of $|p|$ on α for a source moving in a motionless medium is, as one would expect, exactly the same as for a source at rest in a uniformly moving medium. The latter dependence is shown in figure 4.3.

Consider now the case $M > 1$, i.e., when the source is moving with supersonic speed. In this case, the sound field of the source is qualitatively different for $\alpha < \alpha_M$ and $\alpha > \alpha_M$, where $\alpha_M = \arcsin(1/M)$ is the Mach angle. The vertex of the Mach cone coincides with the source and its apex angle is equal to $2\alpha_M$ (figure 5.2). The Mach cone moves with the source and at any moment divides space into two domains. If $\alpha > \alpha_M$, so that the observation point is outside the Mach cone, it follows from equation (5.25) that the values of R_j are complex. Therefore, for $\alpha > \alpha_M$ there are no real solutions t_j of equation (5.7) and $p = 0$. If $\alpha < \alpha_M$, so that the observation point is located inside the Mach cone, equation (5.25) yields two distances $R_1 > 0$ and $R_2 > 0$. In this case, there are two sound waves arriving simultaneously at the observation point, emitted at $t_1 = t - R_1/c$ and $t_2 = t - R_2/c$, respectively (see figure 5.2). The sound field p, given by equation (5.28), is the sum of these

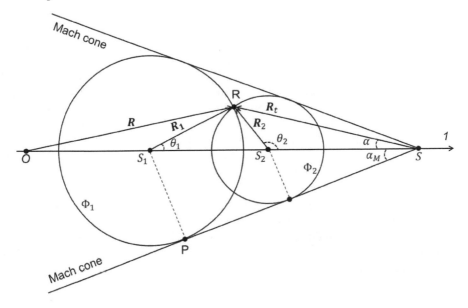

FIGURE 5.2
Point source moving with constant supersonic speed. O is the center of the coordinate system, 1 is the source trajectory, and R is the receiver position. S_1, S_2, and S are the source positions at the instants t_1, t_2, and t, respectively. Φ_1 and Φ_2 are the wavefronts of the sound waves at time t, emitted at times t_1 and t_2, respectively.

waves. The frequencies of these waves, determined from equations (5.23) and (5.27), are different. It can be shown from equations (5.25) and (5.27) that $t_1 < t_2 < t$, $\theta_1 < \pi/2$, and θ_2 can be greater or smaller than $\pi/2$ but always $\theta_2 > \theta_1$. Note that if the observation point is located on the Mach cone, it follows from equation (5.27) that $M \cos \theta_j = 1$.

A qualitative explanation of the appearance of the Mach cone is presented in section 4.2 by considering the sound field due to a point source in supersonic flow. Usually the appearance of the Mach cone is explained in a different manner. At time t the wavefront Φ_1 of the sound wave emitted by the source at time t_1 is located at a distance $S_1 P = (t - t_1)c$ from the point S_1 of this wave emission (see figure 5.2). The tangent to the wavefront, passing through the point S, makes an angle α with the source trajectory. This angle is determined by the formula $\sin \alpha = S_1 P / S_1 S$. Since $S_1 S = (t - t_1)u$, the angle $\alpha = \arcsin(1/M) = \alpha_M$ coincides with the Mach angle. Since α_M does not depend on t_1, the wavefronts of the sound waves emitted at any time $t' < t$ are tangential to the cone, the vertex of which is at the point S and the apex angle of which is equal to $2\alpha_M$. Thus, the Mach cone is the envelope of these wavefronts. It follows from this consideration that $p = 0$ outside the Mach cone.

5.1.4 Other developments for moving sources

Finally we point out published extensions of the problem considered above. The sound field of a monochromatic point mass source moving with a subsonic speed above an impedance surface has been calculated [133, 56]. The sound field of a distributed mass source moving with a variable subsonic speed has been considered [173]. The combined effect of Doppler shift and sound attenuation in an absorbing medium on the field of a broadband point mass source moving uniformly with a subsonic speed has been investigated [234]. Note that moving sound sources can include not only the mass sources considered above but also forces and turbulent stresses acting on the medium [225, 230].

In underwater acoustics, the interest has been to detect, localize, and track a moving source by using matched-field, matched-mode, and matched-ray processing methods (e.g., reference [69]). The sound field of a point source moving in an oceanic waveguide has been calculated [172, 227]. The sound field in an oceanic waveguide due to a source moving uniformly in the atmosphere has been considered [193].

5.2 Sound field for a moving source in an inhomogeneous moving medium

In the previous section, the sound field of the point source moving in a homogeneous motionless medium was studied. In this section, the analysis is extended to the case of various acoustic sources moving in an inhomogeneous moving medium. Explicit expressions for the sound fields of these sources moving in a stratified medium are obtained.

5.2.1 Galilean transformation

Consider two inertial Cartesian coordinate systems K and K_1 with the radius vectors $\mathbf{R} = (x, y, z)$ and $\mathbf{R}_1 = (x_1, y_1, z_1)$, respectively (figure 5.3). The time in each system is denoted as t and t_1. Suppose that the coordinate system K_1 is moving with a constant velocity $\mathbf{u}$ with respect to the coordinate system K. In this case, the coordinates $\mathbf{R}$ and $\mathbf{R}_1$ and the times t and t_1 are related by the Galilean transformation:

$$\mathbf{R} = \mathbf{R}_1 + \mathbf{u}t, \quad t = t_1. \tag{5.30}$$

The coordinate system K_1 coincides with the coordinate system K at $t = 0$. The rule for the addition of velocities is

$$\mathbf{v} = \mathbf{v}_1 + \mathbf{u}, \tag{5.31}$$

where $\mathbf{v}$ and $\mathbf{v}_1$ are the velocities of a material point in the coordinate systems K and K_1, respectively. This rule is a consequence of equation (5.30).

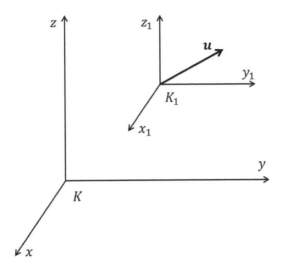

FIGURE 5.3
Inertial coordinate systems K and K_1. The coordinate system K_1 is moving with the velocity $\mathbf{u}$ with respect to the coordinate system K.

Let us show that the equations of fluid dynamics equations (2.3)–(2.7) are invariant under the Galilean transformation (5.30). In the coordinate system K, the Euler equation (2.3) has the form

$$\left[\frac{\partial}{\partial t} + \mathbf{v}(\mathbf{R}, t) \cdot \frac{\partial}{\partial \mathbf{R}}\right] \mathbf{v}(\mathbf{R}, t) + \frac{1}{\varrho(\mathbf{R}, t)} \frac{\partial P(\mathbf{R}, t)}{\partial \mathbf{R}} - \mathbf{g} = \mathbf{F}(\mathbf{R}, t). \qquad (5.32)$$

Here, P is the pressure, ϱ is the density, $\mathbf{v}$ is the medium velocity, the vector $\mathbf{F}$ characterizes the forces acting on the medium, and $\mathbf{g}$ is the vector of the acceleration due to gravity. Using the Galilean transformation and taking into account the fact that

$$\frac{\partial}{\partial t}\mathbf{v}(\mathbf{R}, t) = \left(\frac{\partial}{\partial t} - \mathbf{u} \cdot \frac{\partial}{\partial \mathbf{R}_1}\right) \mathbf{v}(\mathbf{R}_1, t),$$

one can recast equation (5.32) in the coordinate system K_1

$$\left[\frac{\partial}{\partial t} + (\mathbf{v}(\mathbf{R}_1, t) - \mathbf{u}) \cdot \frac{\partial}{\partial \mathbf{R}_1}\right] (\mathbf{v}(\mathbf{R}_1, t) - \mathbf{u}) + \frac{1}{\varrho(\mathbf{R}_1, t)} \frac{\partial P(\mathbf{R}_1, t)}{\partial \mathbf{R}_1} - \mathbf{g}$$
$$= \mathbf{F}(\mathbf{R}_1, t). \qquad (5.33)$$

Taking into account that $\mathbf{v}_1 = \mathbf{v} - \mathbf{u}$ and comparing equations (5.32) and (5.33) reveals that the Euler equation in the moving coordinate system K_1 has the same form as in the coordinate system K. It can be shown analogously that the other equations of fluid dynamics and the equations for sound waves in moving media are invariant under the Galilean transformation.

We now obtain forms of the fluid dynamic equations (2.3)–(2.7) that are valid in a non-inertial coordinate system. Suppose that the coordinate system K_1 is moving with a time-dependent velocity $\mathbf{u}(t)$ with respect to the inertial coordinate system K (see figure 5.3). In this case, the relationships between the coordinates $\mathbf{R}$ and $\mathbf{R}_1$ and the times t and t_1 are given by

$$\mathbf{R} = \mathbf{R}_1 + \int_0^{t_1} \mathbf{u}(t')\, dt', \quad t = t_1. \tag{5.34}$$

Using this transformation, it can be shown that the Euler equation (2.3) in the coordinate system K_1 takes the form of equation (5.33), in which $\mathbf{g}$ is replaced by

$$\mathbf{g}_u = \mathbf{g} - \frac{\partial \mathbf{u}}{\partial t}. \tag{5.35}$$

The appearance of the acceleration $\partial \mathbf{u}/\partial t$ in equation (5.33) is due to the inertial force. Similarly, it can be shown that the equations of fluid dynamics (2.3)-(2.7), in the non-inertial coordinate system, retain their form if $\mathbf{g}$ is replaced by $\mathbf{g}_u$.

5.2.2   Sound field of a moving source

Now we proceed to calculate the sound field $p(\mathbf{R}, t)$ due to a source moving with a constant velocity $\mathbf{u}$ with respect to the coordinate system K. Figure 5.3 applies to this situation if the coordinate systems K_1 and K are connected with the source and receiver, respectively. In the coordinate system K, the sound speed c, the density ϱ, and the medium velocity $\mathbf{v}$ depend on $\mathbf{R}$ and t.

For $\mathbf{u} = 0$, the sound field p of the source is determined by the linearized fluid dynamic equations (2.8)-(2.12). Therefore, in the coordinate system K, this field can be represented in the form

$$p(\mathbf{R}, t) = f\left[\mathbf{R},\, t;\, c(\mathbf{R}, t),\, \varrho(\mathbf{R}, t),\, \mathbf{v}(\mathbf{R}, t)\right],$$

where f is a functional depending on the source.

For $\mathbf{u} \neq 0$, we first consider the sound field $p(\mathbf{R}_1, t)$ in the coordinate system K_1 moving with the source. In this coordinate system the source is at rest and the medium velocity is equal to $\mathbf{v} - \mathbf{u}$; therefore

$$p(\mathbf{R}_1, t) = f\left[\mathbf{R}_1,\, t;\, c(\mathbf{R}_1, t),\, \varrho(\mathbf{R}_1, t),\, \mathbf{v}(\mathbf{R}_1, t) - \mathbf{u}\right].$$

Transforming from the moving coordinate system K_1 to the coordinate system K (figure 5.3), one obtains the desired expression for the sound field of the moving source:

$$p(\mathbf{R}, t) = f\left[\mathbf{R} - \mathbf{u}t,\, t;\, c(\mathbf{R} - \mathbf{u}t, t),\, \varrho(\mathbf{R} - \mathbf{u}t, t),\, \mathbf{v}(\mathbf{R} - \mathbf{u}t, t) - \mathbf{u}\right]. \tag{5.36}$$

Thus, the calculation of the sound field of a source moving with a constant velocity is reduced to that of the same source at rest in an inhomogeneous moving medium. The sound field of a source moving with an arbitrary velocity $\mathbf{u}(t)$ can be expressed analogously in terms of the field of a motionless source. In doing this, one must take into account the fact that in the non-inertial coordinate system moving with the source $\mathbf{g}$ should be replaced by $\mathbf{g}_u$ in the equations of fluid dynamics.

This approach for calculating the sound field of a moving source is useful if the field of the same source at rest in an inhomogeneous moving medium is known. Analytical expressions for the sound fields due to sources at rest have been obtained for the case of a stratified moving medium in which the vertical component v_z of the medium velocity is equal to zero. For such a stratified medium and a source moving with a constant velocity $\mathbf{u}_\perp$ in the horizontal plane, equation (5.36) is simplified. Indeed, in this case, the Galilean transformation (5.30) takes the form

$$\mathbf{r} = \mathbf{r}_1 + \mathbf{u}_\perp t, \quad z = z_1, \quad t = t_1,$$

where $\mathbf{r} = (x, y)$ and $\mathbf{r}_1 = (x_1, y_1)$ are the horizontal coordinates in the coordinate systems K and K_1, respectively. Using these relations in equation (5.36) and taking into account the fact that c, ϱ, and the horizontal component $\mathbf{v}_\perp$ of the medium velocity depend only on z, one obtains the sound field of the source moving in a stratified medium:

$$p(\mathbf{r}, z, t) = f\left[\mathbf{r} - \mathbf{u}_\perp t, z, t; c(z), \varrho(z), \mathbf{v}_\perp(z) - \mathbf{u}_\perp\right]. \tag{5.37}$$

Expressions for the sound pressure p induced by different motionless sources in a homogeneous or stratified moving medium are given in sections 4.3 and 4.4. These expressions allow one to determine the explicit form of the functionals $f[\mathbf{R}, t; c, \varrho, \mathbf{v}]$ for the sources considered in these sections. The sound fields of these sources when they are moving with velocity $\mathbf{u}$ can then be calculated by using equations (5.36) and (5.37). In the remainder of this section, we shall present expressions for the sound fields induced by the following monochromatic moving sources: a plane wave source, a point source, and a transmitting antenna.

5.2.3 Plane wave source

First, consider the sound field induced by a plane wave source moving with a constant velocity $\mathbf{u} = (\mathbf{u}_\perp, u_z)$ in a homogeneous flow $\mathbf{v} = (\mathbf{v}_\perp, v_z)$. The Gaussian antenna with a horizontal aperture and a beam width $D \to \infty$ (section 4.2) will be used to represent the plane wave source. The equations obtained below are also valid for Gaussian antennas with non-horizontal apertures, if $\mathbf{v}_\perp$, $\mathbf{u}_\perp$ and v_z, u_z are treated as the projections of the medium and source velocities on the aperture plane and on the direction perpendicular to it.

For a plane wave source, the functional f is given by

$$f[\mathbf{R}, t; c, \varrho, \mathbf{v}] = p\exp(-i\omega_1 t). \tag{5.38}$$

Here, ω_1 is the angular frequency of the source and p is determined by equation (4.21). Substituting the value of f into equation (5.36), one obtains the sound field of the plane wave source moving in a homogeneous flow:

$$p(\mathbf{r}, z, t) = p_0 \exp(i\boldsymbol{\kappa} \cdot \mathbf{r} + iqz - i\omega t). \tag{5.39}$$

Here, p_0 is the sound-pressure amplitude, and $\boldsymbol{\kappa}$ and q are the horizontal and vertical components of the wave vector $\mathbf{k} = (\boldsymbol{\kappa}, q)$. The vector $\boldsymbol{\kappa}$ determines the direction of propagation of the emitted wave. The angular frequency ω and the vertical wavenumber q of the plane wave (5.39) are given by

$$\omega = \omega_1 + \mathbf{k} \cdot \mathbf{u} \tag{5.40}$$

and

$$
\begin{aligned}
q &= \frac{\pm c \left[\sigma_1^2 - (c^2 - (v_z - u_z)^2 \kappa^2\right]^{1/2} - (v_z - u_z)\sigma_1}{c^2 - (v_z - u_z)^2} \\
&= \frac{\pm c \left[\sigma^2 - (c^2 - v_z^2)\kappa^2\right]^{1/2} - v_z\sigma}{c^2 - v_z^2}.
\end{aligned} \tag{5.41}
$$

In this equation, $\sigma_1 = \omega_1 - \boldsymbol{\kappa} \cdot (\mathbf{v}_\perp - \mathbf{u}_\perp)$, $\sigma = \omega - \boldsymbol{\kappa} \cdot \mathbf{v}_\perp$, and the upper and lower signs correspond to wave propagation in the positive and negative directions of the z-axis, respectively.

Equation (5.40) relates the frequency ω_1 of a plane wave emitted by a moving source and the frequency ω measured by an observer at rest, i.e., it describes the Doppler effect. It follows from this equation that the frequency $\omega = \omega_1 + \boldsymbol{\kappa} \cdot \mathbf{u}_\perp$, measured by the observer at rest, does not depend on the medium velocity $\mathbf{v}$ if $u_z = 0$. But ω depends on $\mathbf{v}$ if $u_z \neq 0$. For $\mathbf{v} = 0$, equation (5.40) coincides with the equation (5.23) for the Doppler effect. Indeed, if $\mathbf{v} = 0$, it can be shown from equation (5.41) that $k = (\kappa^2 + q^2)^{1/2} = \omega/c$. Therefore, $\mathbf{k} \cdot \mathbf{u} = \mathbf{s} \cdot \mathbf{u}\omega/c$, where $\mathbf{s} = \mathbf{k}/k$ is the unit vector in the direction of wave propagation. Substituting the value of $\mathbf{k} \cdot \mathbf{u}$ into equation (5.40), one obtains $\omega = \omega_1/(1 - \mathbf{s} \cdot \mathbf{u}/c)$. The latter equation coincides with equation (5.23) since $\mathbf{s} \cdot \mathbf{u}/c = M\cos\theta$, where θ is the angle between the direction of plane wave propagation and the direction of source motion.

5.2.4 Point source and acoustic antenna

Consider a unit-amplitude point source at rest in a stratified moving medium with $v_z = 0$. For this source, the functional f is given by

$$f\left[\mathbf{r}, z, t; c(z), \varrho(z), \mathbf{v}_\perp(z)\right] = p\exp(-i\omega_1 t),$$

where p is determined by equations (4.53), (4.50), and (4.47). We recall that these equations are valid in the high-frequency approximation and far away from turning points. Substituting the value of f into equation (5.37), one obtains the sound field of a point source moving in the horizontal plane $z = z_0$ with constant velocity $\mathbf{u}_\perp$ in a stratified moving medium

$$p(\mathbf{r}, z, t) = \left(\frac{\varrho}{\varrho_0}\right)^{1/2} \frac{(\omega - \boldsymbol{\kappa} \cdot \mathbf{v}_\perp)}{\omega_1(q_0 q J)^{1/2}} \exp\left(i\boldsymbol{\kappa} \cdot \mathbf{r} + i \int_{z_<}^{z_>} q \, dz - i\omega t\right). \quad (5.42)$$

Here, $\varrho_0 = \varrho(z_0)$, $q_0 = q(z_0)$, $z_> = \max(z, z_0)$, $z_< = \min(z, z_0)$; ω and q are determined by equations (5.40) and (5.41), respectively, where $v_z = u_z = 0$. The Jacobian J in equation (5.42) is given by

$$J = \omega_1^2 \left[\int_{z_<}^{z_>} \frac{dz}{c^2 q^3} \int_{z_<}^{z_>} \left(1 + \frac{\kappa^2 \tilde{v}_\perp^2}{c^2 q^2}\right) \frac{dz}{q} - \left(\int_{z_<}^{z_>} \frac{\kappa \tilde{v}_\perp}{c^2 q^3} dz\right)^2\right], \quad (5.43)$$

where $\tilde{v}_\perp = [(\mathbf{v}_\perp - \mathbf{u}_\perp)^2 - (\boldsymbol{\kappa} \cdot (\mathbf{v}_\perp - \mathbf{u}_\perp))^2/\kappa^2]^{1/2}$. The vectors $\boldsymbol{\kappa}$ and $\mathbf{r}$ in equation (5.42) are related by

$$\mathbf{r} - \mathbf{u}_\perp t = \int_{z_<}^{z_>} \frac{\boldsymbol{\kappa} + [\omega_1 - \boldsymbol{\kappa} \cdot (\mathbf{v}_\perp - \mathbf{u}_\perp)] (\mathbf{v}_\perp - \mathbf{u}_\perp)/c^2}{q} \, dz. \quad (5.44)$$

This equation can be transformed to an equation for a sound ray in a stratified moving medium. To this end, the vector $\mathbf{u}_\perp t$ is expressed as

$$\mathbf{u}_\perp t = (t - t_j)\mathbf{u}_\perp + t_j \mathbf{u}_\perp. \quad (5.45)$$

Here, t_j is the time of emission of the sound wave that arrives at the observation point at time t. In equation (5.45), the vector $t_j \mathbf{u}_\perp = \mathbf{r}_j$ determines the position of the source in the horizontal plane at time t_j, and $t - t_j$ is the time of sound propagation from the point of emission $(\mathbf{r}_j, z_0)$ to the observation point $(\mathbf{r}, z)$. It follows from equation (3.58) that $t - t_j = (\Psi - \Psi_1)/c_0$, where Ψ and Ψ_1 are the eikonals at the emission and observation points, respectively. For a stratified medium, the difference $\Psi - \Psi_1$ is equal to the integral in the first line of equation (3.81), where $v_z = 0$, ω is determined by equation (5.40), and z and z_1 are replaced by $z_<$ and $z_>$, respectively. Substituting $\mathbf{u}_\perp t = \mathbf{r}_j + (\Psi - \Psi_1)\mathbf{u}_\perp/c_0$ into equation (5.44) yields

$$\mathbf{r} = \mathbf{r}_j + \int_{z_<}^{z_>} \frac{\boldsymbol{\kappa} + (\omega - \boldsymbol{\kappa} \cdot \mathbf{v}_\perp)\mathbf{v}_\perp/c^2}{q} \, dz. \quad (5.46)$$

This is the equation for a sound ray $\mathbf{r} = \mathbf{r}(z, \boldsymbol{\kappa})$ in a stratified moving medium. Equation (5.46) relates the two vectors $\mathbf{r}$ and $\boldsymbol{\kappa}$, appearing in equation (5.42) for the sound field of a moving source. Thus, these equations determine the sound field of a moving point source.

Using the approach described above and equations (4.59), (4.50), (4.47), and (5.37), one can derive an expression for the high-frequency sound field of the transmitting antenna moving with a constant velocity in a stratified medium. This expression coincides with the right-hand side of equation (5.42) if the term $1/(\omega_1 q_0^{1/2})$ is replaced by $-2i\pi q_0^{1/2}\widehat{p}_a(\boldsymbol{\kappa})/(\omega - \boldsymbol{\kappa}\cdot\mathbf{v}_0)$, where $\mathbf{v}_0 = \mathbf{v}(z_0)$ and $\widehat{p}_a(\boldsymbol{\kappa})$ is the spatial spectrum of the sound pressure at the antenna aperture.

It follows from equations (5.44) and (5.46) that the vector $\boldsymbol{\kappa}$ depends on t. Therefore, as the source moves, quasi-plane waves with different values of $\boldsymbol{\kappa}$ arrive at the observation point $(\mathbf{r}, z)$. For an observer at rest, the sound field of a moving source, given by equation (5.42), can be considered as monochromatic only under certain conditions analogous to the inequalities (5.19) and (5.22). As with these inequalities, one can expect that the conditions are valid if the distance between the source and receiver $R \gg \lambda_1$.

Assuming that these conditions are valid, let us determine the angular frequency $\widetilde{\omega}$ of the sound field given by equation (5.42). By definition $\widetilde{\omega} = -\partial\Phi/\partial t$, where the phase Φ is determined by the argument of the exponential function in equation (5.42). Therefore,

$$\widetilde{\omega} = \omega + t\,\mathbf{u}_\perp \cdot \frac{\partial\boldsymbol{\kappa}}{\partial t} - \mathbf{r}\cdot\frac{\partial\boldsymbol{\kappa}}{\partial t} - \int_{z_<}^{z_>} \frac{\partial q}{\partial t}\,dz. \tag{5.47}$$

It can be shown that

$$\frac{\partial q}{\partial t} = -\frac{\partial\boldsymbol{\kappa}}{\partial t}\cdot\frac{\boldsymbol{\kappa} + (\omega - \boldsymbol{\kappa}\cdot\mathbf{v}_\perp)(\mathbf{v}_\perp - \mathbf{u}_\perp)/c^2}{q}.$$

Substituting the value of $\partial q/\partial t$ into equation (5.47) and replacing the integral over z by the left-hand side of equation (5.44), one obtains $\widetilde{\omega} = \omega$. Thus for an observer at rest, the frequency of the sound field emitted by a moving point source (or an acoustic antenna) is given by $\omega = \omega_1 + \boldsymbol{\kappa}\cdot\mathbf{u}_\perp$ and coincides with the frequency of the quasi-plane wave arriving at the observation point. For $\mathbf{v} = 0$, the formula $\omega = \omega_1 + \boldsymbol{\kappa}\cdot\mathbf{u}_\perp$ can be transformed to equation (5.23) for the Doppler effect, as in the case of the plane wave.

Finally, let us consider the plane wave given by equation (5.39) and the quasi-plane waves in the spectral representations of the sound fields of a point source and transmitting antenna, which are determined by the exponential function in equation (5.42). When plane and quasi-plane waves are emitted by moving sources, the frequency and, as will be shown in section 5.3, the initial direction of propagation of these waves are changed. However, it follows from equations (5.39)–(5.42) and (5.46) that when these waves propagate in the medium, their velocities, dispersion equation, ray paths, and phase changes along the paths coincide with the analogous quantities for waves emitted by a source at rest and with frequency ω.

5.3 Sound aberration and the change in propagation direction of a sound wave emitted by a moving source

The phenomenon of light aberration is well known in optics and astronomy. It describes the change in direction of propagation of an electromagnetic wave in transforming from a moving coordinate system to one at rest. The equations for light aberration also determine the change in direction of propagation of a plane electromagnetic wave emitted by a source moving with a constant velocity $\mathbf{u}$ in a vacuum, measured in the coordinate system K, which is at rest (see figure 5.3). Indeed, in the coordinate system K_1 moving with the source, the direction of propagation of this wave does not depend on $\mathbf{u}$, since there is no light ether. Therefore, the equations describing light aberration can be used to determine the direction of wave propagation in the coordinate system K at rest.

In analogy to light aberration, we refer to the change in propagation direction of a sound wave, as results from transforming the moving coordinate system K_1 to the motionless system K, as *sound aberration*. Sound aberration is an effect which is different from the change in direction of propagation of a plane sound wave emitted by a moving source, since sound waves propagate in a medium which can be considered as "acoustic ether." Sound aberration should be taken into account in some practical applications such as localization of sound sources on the ground with moving microphone arrays.

In this section, sound aberration and the change in direction of propagation of a sound wave emitted by the source moving with a constant velocity $\mathbf{u}$ are considered. These effects differ from the analogous effects in optics and were considered in reference [289].

5.3.1 Sound aberration

Suppose that the coordinate system K_1 is moving with the velocity $\mathbf{u}$ with respect to the coordinate system K, which is at rest (figure 5.3). In these coordinate systems, the radius vectors are $\mathbf{R}$ and $\mathbf{R}_1$, respectively. In the coordinate system K, the medium velocity $\mathbf{v}$ is constant.

Consider also a plane sound wave. The sound field of this wave in coordinate system K is given by

$$p(\mathbf{R}, t) = p_0 \exp(i\mathbf{k} \cdot \mathbf{R} - i\omega t). \tag{5.48}$$

Using equation (5.30) which relates the vectors $\mathbf{R}$ and $\mathbf{R}_1$, the sound field of this plane wave in coordinate system K_1 is recast as

$$p(\mathbf{R}_1, t) = p_0 \exp\left[i\mathbf{k} \cdot \mathbf{R}_1 - i(\omega - \mathbf{k} \cdot \mathbf{u})t\right]. \tag{5.49}$$

It follows from equations (5.48) and (5.49) that the unit vector $\mathbf{n} = \mathbf{k}/k$ normal to the wavefront of the plane wave is the same in both of these coordinate systems. On the other hand, the unit vectors $\mathbf{s}$ and $\mathbf{s}_1$ in the directions of the group velocities of the plane wave in these coordinate systems are different. These vectors characterize the directions of propagation of the plane wave. We refer to the difference in the vectors $\mathbf{s}$ and $\mathbf{s}_1$ as sound aberration.

The unit vector $\mathbf{s}$ is determined by equation (3.44) $\mathbf{s} = (c\mathbf{n} + \mathbf{v})/|c\mathbf{n} + \mathbf{v}|$. Since the medium velocity in the coordinate system K_1 is equal to $\mathbf{v} - \mathbf{u}$, the unit vector $\mathbf{s}_1 = (c\mathbf{n} + \mathbf{v} - \mathbf{u})/|c\mathbf{n} + \mathbf{v} - \mathbf{u}|$. The equation for $\mathbf{s}$ can be solved for $\mathbf{n}$, yielding equation (3.45). Substituting the vector $\mathbf{n}$ into the expression for $\mathbf{s}_1$, one obtains:

$$\mathbf{s}_1 = \frac{\mathbf{s}\,\eta - \mathbf{u}/c}{(\eta^2 - 2\eta\,\mathbf{s}\cdot\mathbf{u}/c + u^2/c^2)^{1/2}}, \tag{5.50}$$

where η depends only on $\mathbf{s}$ and $\mathbf{v}/c$ and is given by

$$\eta = \left\{ 1 - \frac{v^2}{c^2} + 2\frac{\mathbf{s}\cdot\mathbf{v}}{c}\left[1 - \frac{v^2}{c^2} + \left(\frac{\mathbf{s}\cdot\mathbf{v}}{c}\right)^2 \right]^{1/2} + 2\left(\frac{\mathbf{s}\cdot\mathbf{v}}{c}\right)^2 \right\}^{1/2}. \tag{5.51}$$

Equation (5.50) is the desired formula for sound aberration which expresses the unit vector $\mathbf{s}_1$ in terms of the unit vector $\mathbf{s}$ and the velocity $\mathbf{u}$ of the coordinate system K_1.

It follows from equation (5.50) that vectors $\mathbf{s}_1$, $\mathbf{s}$, and $\mathbf{u}$ are located in the same plane. This equation simplifies for the case $\mathbf{v} = 0$ for which $\eta = 1$. Introducing the angles γ and γ_1 between the vectors $\mathbf{s}$ and $\mathbf{u}$ and between the vectors $\mathbf{s}_1$ and $\mathbf{u}$, respectively, and starting from equation (5.50), one obtains the following formula for the sound aberration in a motionless medium:

$$\cos\gamma_1 = \frac{\cos\gamma - u/c}{(1 - 2\cos\gamma\,u/c + u^2/c^2)^{1/2}}. \tag{5.52}$$

This formula differs from the well-known formula for the light aberration [216]

$$\cos\gamma_1 = \frac{\cos\gamma - u/c}{1 - \cos\gamma\,u/c} \tag{5.53}$$

by terms of the order of u^2/c^2.

Sound aberration should be addressed when acoustic sensor arrays on moving platforms, such as helicopters and UAVs, are used to detect and localize sound sources on the ground. Consider a small domain D centered on the receiver moving with the velocity $\mathbf{u}$ with respect to the sound source located on the ground. In this domain, the sound speed c and wind velocity $\mathbf{v}$ can be considered as constant. The analysis of sound aberration applies to this situation if the coordinate systems K and K_1 shown in figure 5.3 are connected with the source and receiver, respectively. Equation (5.50) relates the unit vector $\mathbf{s}_1$ in propagation direction of the sound signal measured by the moving microphone array with the unit vector $\mathbf{s}$ in the actual direction to the source in the domain D. Beyond this domain, refraction corrections to source localization in a stratified moving atmosphere might also be important [298].

5.3.2 Change in propagation direction of a plane wave emitted by a moving source

Now let us consider the change in direction of propagation of a sound wave emitted by a source moving with a constant velocity $\mathbf{u} = (\mathbf{u}_\perp, u_z)$ with respect to the coordinate system K, which is at rest (figure 5.3). Suppose that $c =$ constant and $\mathbf{v} = (\mathbf{v}_\perp, v_z) =$ constant in this coordinate system and that the source emits a plane sound wave with amplitude $|p| = 1$. The sound field of this wave in the coordinate system K is given by equation (5.39). If $\mathbf{u} = \mathbf{v} = 0$, it follows from equation (5.39) that the source emits the plane wave

$$p(\mathbf{r}, z, t) = p_0 \exp(i\boldsymbol{\kappa} \cdot \mathbf{r} + iq_1 z - i\omega_1 t), \tag{5.54}$$

where $q_1 = (\omega_1^2/c^2 - \kappa^2)^{1/2}$. This wave propagates in the direction of the unit vector $\mathbf{n}_1 = (\boldsymbol{\kappa}, q_1)/(\kappa^2 + q_1^2)^{1/2}$. If $\mathbf{u} \neq 0$ and/or $\mathbf{v} \neq 0$, the direction of propagation of the emitted wave is changed and, in the coordinate system K, is given by the unit vector $\mathbf{s}$ in the direction of the group velocity of the plane wave determined by equation (5.39).

The difference between the vectors $\mathbf{n}_1$ and $\mathbf{s}$ constitutes the change in the direction of propagation of the sound wave emitted by the moving source in comparison with the source at rest. This effect is different from the sound aberration, since the vector $\mathbf{n}_1$ does not coincide with the vector $\mathbf{s}_1$, which characterizes the direction of wave propagation in the coordinate system K_1 moving with the source. The difference between $\mathbf{n}_1$ and $\mathbf{s}_1$ is due to the fact that the medium velocity is not zero in the coordinate system K_1.

It is worthwhile to express the vectors $\mathbf{n}_1$ and $\mathbf{s}$ in the forms $\mathbf{n}_1 = (\mathbf{e}\cos\theta, \sin\theta)$ and $\mathbf{s} = (\mathbf{m}\cos\alpha, \sin\alpha)$. Here, $\mathbf{e}$ and $\mathbf{m}$ are the unit vectors characterizing the azimuthal (horizontal) directions of the vectors $\mathbf{n}_1$ and $\mathbf{s}$, and θ and α are the elevation angles, i.e., the angles between the vectors $\mathbf{n}_1$ and $\mathbf{s}$ and the horizontal plane. Note that the geometry of the vectors $\mathbf{n}_1$ and $\mathbf{s}$ is the same as for the vectors $\mathbf{n}$ and $\mathbf{s}$ shown in figure 3.2. (The physical meanings of the vectors $\mathbf{s}$, $\mathbf{n}_1$, and $\mathbf{n}$ are different between this chapter and Chapter 3.) The vector $\mathbf{n}_1$ can also be expressed in the form $\mathbf{n}_1 = (\mathbf{e}\cos\theta, \sin\theta) = (\boldsymbol{\kappa}, q_1)c/\omega_1$. Using this formula, one obtains the horizontal component of the wave vector as $\boldsymbol{\kappa} = (\omega_1\mathbf{e}/c)\cos\theta$. Assuming that $q^2 \geq 0$ and substituting the vector $\boldsymbol{\kappa}$ and the value of q given by equation (5.41) into the equation $\mathbf{n} = (\boldsymbol{\kappa}, q)/(\kappa^2 + q^2)^{1/2}$ yields

$$\mathbf{n} = \left(\mathbf{e}\cos\theta/\chi, \left(1 - \cos^2\theta/\chi^2\right)^{1/2}\right). \tag{5.55}$$

Here,

$$\chi = \frac{1 + \mathbf{e}\cdot\mathbf{w}_\perp\cos\theta \pm w_z\left[(1 + \mathbf{e}\cdot\mathbf{w}_\perp\cos\theta)^2 - \left(1 - w_z^2\right)\cos^2\theta\right]^{1/2}}{1 - w_z^2}, \tag{5.56}$$

where $\mathbf{w}_\perp = (\mathbf{u}_\perp - \mathbf{v}_\perp)/c$ and $w_z = (u_z - v_z)/c$. Using equation (5.55) and the equality $\mathbf{s} = (c\mathbf{n} + \mathbf{v})/|c\mathbf{n} + \mathbf{v}|$, one obtains:

$$\mathbf{s} = \frac{\left(\mathbf{e}\cos\theta + \chi\mathbf{v}_\perp/c,\, (\chi^2 - \cos^2\theta)^{1/2} + \chi v_z/c\right)}{\left[\chi^2\left(1 + v^2/c^2\right) + 2\chi\left(\mathbf{e}\cdot\mathbf{v}_\perp\cos\theta + v_z\,(\chi^2 - \cos^2\theta)^{1/2}\right)/c\right]^{1/2}}. \tag{5.57}$$

This equation determines the vector $\mathbf{s}$ in terms of the vector $\mathbf{n}_1 = (\mathbf{e}\cos\theta, \sin\theta)$ and, hence, describes the change in direction of propagation of a plane wave emitted by a moving source. Equation (5.57) is valid for $q^2 \geq 0$; if this inequality is fulfilled, $\chi^2 \geq \cos^2\theta$.

If $q^2 < 0$ and $w_z = 0$, the moving source emits an inhomogeneous plane wave which propagates in the horizontal plane ($\alpha = 0$) in the direction of the vector $\mathbf{e}\cos\theta + \chi\mathbf{v}_\perp/c$ and is attenuated exponentially in the direction of the z-axis. If $w_z \neq 0$ and q is a complex quantity, the angle α between the direction of propagation of this inhomogeneous plane wave and the horizontal plane is not zero. This latter case is not considered here.

It follows from equation (5.57) that motion of the source and medium result in a variation of both the elevation angle of the emitted wave and its azimuthal direction of propagation, which can be characterized by the angle ψ between the vectors $\mathbf{e}$ and $\mathbf{m}$. We shall assume that $-\pi < \psi \leq \pi$, and $\psi > 0$ if the shortest rotation of the vector $\mathbf{e}$ until it coincides with the vector $\mathbf{m}$ is anticlockwise. Using equation (5.57) and the formula $\mathbf{s} = (\mathbf{m}\cos\alpha, \sin\alpha)$, one obtains the angles α and ψ:

$$\cos\alpha = \left[\frac{\cos^2\theta + 2\cos\theta\,\chi\mathbf{e}\cdot\mathbf{v}_\perp/c + \chi^2 v_\perp^2/c^2}{\chi^2\left(1 + v^2/c^2\right) + 2\chi\left(\mathbf{e}\cdot\mathbf{v}_\perp\cos\theta + (\chi^2 - \cos^2\theta)^{1/2}\,v_z\right)/c}\right]^{1/2}, \tag{5.58}$$

$$\cos\psi = \frac{\cos\theta + \chi\mathbf{e}\cdot\mathbf{v}_\perp/c}{[\cos^2\theta + 2\cos\theta\,\chi\mathbf{e}\cdot\mathbf{v}_\perp/c + \chi^2 v_\perp^2/c^2]^{1/2}}, \tag{5.59}$$

$$\sin\psi = \frac{\chi\mathbf{e}_\perp\cdot\mathbf{v}_\perp/c}{[\cos^2\theta + 2\cos\theta\,\chi\mathbf{e}\cdot\mathbf{v}_\perp/c + \chi^2 v_\perp^2/c^2]^{1/2}}. \tag{5.60}$$

Here, the unit vector $\mathbf{e}_\perp$ is located in the horizontal plane perpendicular to the vector $\mathbf{e}$. (The relative position of these vectors is shown in figure 3.2.) Equations (5.58)–(5.60) determine the variations in the elevation angle of the plane wave emitted by the moving source and in its azimuthal direction of propagation. These equations are simplified significantly if $\mathbf{v} = 0$. In this case, it follows from equations (5.59) and (5.60) that $\psi = 0$; moreover, from equations (5.56) and (5.58) one obtains

$$\cos\alpha = \frac{\left(1 - \beta_z^2\right)\cos\theta}{1 + \beta_\perp\cos\theta + \beta_z\left[(1 + \beta_\perp\cos\theta)^2 - (1 - \beta_z^2)\cos^2\theta\right]^{1/2}}. \tag{5.61}$$

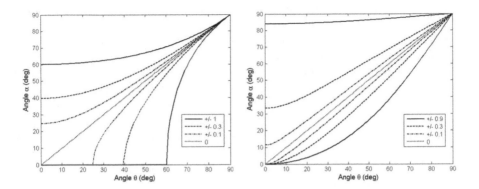

FIGURE 5.4

Dependence of the angle α on the angle θ, calculated using equation (5.61) for different values of β_z and $\beta_\perp$. (Left) $\beta_z = 0$ and different values of $\beta_\perp$. The curves located above and below the dotted straight line ($\beta_\perp = 0$) correspond to positive and negative values of $\beta_\perp$. (Right) $\beta_\perp = 0$ and different values of β_z. The curves located above and below the dotted straight line ($\beta_z = 0$) correspond to positive and negative values of β_z.

Here, $\beta_\perp = \mathbf{e} \cdot \mathbf{u}_\perp/c$ and $\beta_z = \pm|u_z/c|$, where the upper and lower signs correspond to wave propagation in the direction of the source motion and in the opposite direction, respectively.

The dependence of the angle α on the angle θ, calculated using equation (5.61), is depicted in figure 5.4 for different values of $\beta_\perp$ and β_z. This figure shows that the difference between the angles α and θ may be of the order of tens of degrees even for small values of $\beta_\perp$ and β_z of order 0.1–0.3. For $\beta_\perp < 0$, the curves $\alpha = \alpha(\theta)$ in the left plot intersect the x-axis at the points $\theta_0 = \arccos[1/(1 - \beta_\perp)]$. If $\theta < \theta_0$, the quantity $q^2 < 0$ and, hence, $\alpha = 0$ in accordance with the discussion of equation (5.57).

5.3.3 Quasi-plane wave, point source, and acoustic antenna

When deriving equations (5.57)–(5.61) it was assumed that the plane sound wave propagates in a homogeneous flow. This wave becomes a quasi-plane wave in a stratified moving medium where the characteristic scales of variation in ϱ, c, and $\mathbf{v}$ are much greater than the sound wavelength λ. In such a medium, the values of ϱ, c, and $\mathbf{v}$ can be considered as constants for vertical ranges of a few wavelengths from the plane wave source. Therefore, for these ranges, the change in direction of propagation of the wave emitted by the moving source is determined by equations (5.57)–(5.61). For larger ranges from the source, direction of propagation of the quasi-plane wave changes due to refraction which is considered in detail in Chapter 3.

The change in direction of propagation of a plane or quasi-plane sound wave emitted by a moving source is of interest because it enables determination of the change in direction of propagation of the plane or quasi-plane waves in the spectral representation of the sound field of a real moving source such as a point source or a transmitting antenna.

For a real moving source, this effect leads to the result that a different quasi-plane wave arrives at the observation point than that which arrives when the source is motionless. The amplitudes of the quasi-plane waves in the spectral representation of the source field are different. For a moving point source, therefore, the effect results in a small variation of the sound-pressure amplitude at the observation point. However, if the moving source has a narrow radiation pattern, the change in direction of propagation of the quasi-plane waves can become greater than the radiation pattern. In this case, the motion of the source results in a significant variation of the insonified spatial domain. Therefore, this effect must be taken into account when orientating moving transmitting antennas with narrow radiation patterns.

Antennas with narrow radiation patterns are used in atmospheric and oceanic acoustics. Such transmitting antennas are installed, for example, on helicopters and small airplanes to give audible warnings, for purposes such as directing people away from the sites of forest fires and natural disasters, police work, and military applications. Equation (5.61) and figure 5.4 show that the biggest variation of the radiation pattern due to aircraft motion occurs for large values of aircraft speed and small values of the elevation angle θ of the quasi-plane waves in the spectral representation of the antenna's field (with the exception of the case $\beta_\perp = 0$ and $\beta_z < 0$). Consider an aircraft speed $u_\perp = \mathbf{e} \cdot \mathbf{u}_\perp = 200$ km/h, corresponding to $\beta_\perp = 0.17$ and $u_z = 0$, i.e., the direction of aircraft motion is parallel to the antenna aperture. In this case, the difference between the angles α and θ may be greater than 30° (see figure 5.4, left plot). For a ship traveling at speed $\mathbf{u}_\perp = \mathbf{e} \cdot \mathbf{u}_\perp = 50$ km/h ($\beta_\perp \approx 0.01$) in the ocean, the change in direction of propagation of quasi-plane waves may be 8°.

5.4 Doppler effect in an inhomogeneous moving medium

The Doppler effect is one of the main manifestations of source motion. In this section, the classical equation for the Doppler effect in a homogeneous medium at rest is considered first. This equation is then generalized to the case of an inhomogeneous moving medium.

5.4.1 Homogeneous medium at rest

Equation (5.23) for the Doppler effect in a homogeneous medium at rest was derived in sections 5.1 and 5.2 using different approaches and for various

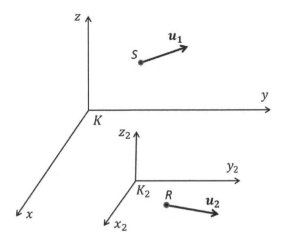

FIGURE 5.5
Source (S) and receiver (R) moving with velocities $\mathbf{u}_1$ and $\mathbf{u}_2$ with respect to the coordinate system K. The coordinate system K_2 is moving with the receiver.

sources. It was assumed that the source was moving with respect to the coordinate system K while the receiver was at rest in this coordinate system.

Now we generalize equation (5.23) to the case in which both the source and the receiver are moving with respect to the coordinate system K (see figure 5.5). The medium velocity $\mathbf{v}$ in this coordinate system is constant and the source and receiver velocities are $\mathbf{u}_1$ and $\mathbf{u}_2$, respectively.

Assume that the source emits a plane sound wave with frequency ω_1. The sound pressure p in coordinate system K, induced by the moving source, is determined by equation (5.39). Transforming equation (5.39) to the coordinate system K_2 moving with the receiver (see figure 5.5), one obtains the expression for the sound pressure

$$p(\mathbf{R}_2, t) = p_0 \exp(i\mathbf{k} \cdot \mathbf{R}_2 - i\omega_2 t). \tag{5.62}$$

Here, $\mathbf{R}_2 = (x_2, y_2, z_2)$ is the radius vector in coordinate system K_2; $\mathbf{k} = (\kappa, q)$ is the wave vector of the plane wave, which takes the same form in all inertial coordinate systems, including K and K_2; $\omega_2 = \omega - \mathbf{k} \cdot \mathbf{u}_2$ is the frequency measured by the moving receiver; and $\omega = \omega_1 + \mathbf{k} \cdot \mathbf{u}_1$ is the frequency of the wave in the coordinate system K (see equations (5.39) and (5.40)). Elimination of the frequency ω between the last two equations yields

$$\omega_2 = \omega_1 + \mathbf{k} \cdot (\mathbf{u}_1 - \mathbf{u}_2). \tag{5.63}$$

This equation has been obtained by using the assumption that the medium velocity $\mathbf{v} = \text{constant}$ in the coordinate system K. In the remaining part of this subsection, we assume that $\mathbf{v} = 0$; we will return to the case $\mathbf{v} = \text{constant}$ in the next subsection.

If $\mathbf{v} = 0$, then $\omega = kc = \omega_1 + \mathbf{k} \cdot \mathbf{u}_1$. Also we set $\mathbf{k} = \mathbf{n}k$, where $\mathbf{n}$ is the unit vector normal to the wavefront of the sound wave emitted by the moving source and arriving at the receiver. For $\mathbf{v} = 0$, the vector $\mathbf{n}$ coincides with the unit vector $\mathbf{s}$, the direction of propagation of this wave in the coordinate system K, i.e., it coincides with the unit vector in the direction from the source to the receiver. From the last equation for ω, we obtained the value of the wavenumber $k = \omega_1/(c - \mathbf{n} \cdot \mathbf{u}_1)$. Substituting this value into equation (5.63) yields

$$\omega_2 = \omega_1 \frac{1 - \mathbf{n} \cdot \mathbf{u}_2/c}{1 - \mathbf{n} \cdot \mathbf{u}_1/c}. \tag{5.64}$$

This equation describes the Doppler effect for the case when the source and receiver are moving with the velocities $\mathbf{u}_1$ and $\mathbf{u}_2$ with respect to the coordinate system K, in which the medium velocity $\mathbf{v} = 0$ and c is constant. Equation (5.64) is well known in the literature (e.g., reference [217, §68]).

For $\mathbf{u}_2 = 0$, equation (5.64) becomes equation (5.23) and describes the Doppler effect for the case when the source is moving and the receiver is at rest. On the other hand, for $\mathbf{u}_1 = 0$, equation (5.64) describes the Doppler effect for the case when the source is at rest while the receiver is moving. Note that the Doppler effect for $\mathbf{u}_1 \neq 0$, $\mathbf{u}_2 = 0$ differs from that for $\mathbf{u}_1 = 0$, $\mathbf{u}_2 \neq 0$.

5.4.2   Homogeneous flow

Now we generalize equation (5.64) to the case of an inhomogeneous moving medium using a method published elsewhere [289].

Assume first that in the coordinate system K, the sound speed c and the medium velocity $\mathbf{v}$ are constant. In this case, equation (5.63) determines a relationship between the frequencies ω_1 and ω_2 of the transmitted and received sound waves, respectively. In the coordinate system K_2 moving with the receiver, the medium velocity is given by $\mathbf{v} - \mathbf{u}_2$. Therefore, the dispersion equation (3.35) takes the form $\omega_2 = kc + \mathbf{k} \cdot (\mathbf{v} - \mathbf{u}_2)$ in this coordinate system. Solving this equation for the vector $\mathbf{k}$ yields $\mathbf{k} = \omega_2 \mathbf{n}/[c + \mathbf{n} \cdot (\mathbf{v} - \mathbf{u}_2)]$. Substituting $\mathbf{k}$ into equation (5.63), one obtains

$$\omega_2 = \omega_1 \frac{1 + \mathbf{n} \cdot (\mathbf{v} - \mathbf{u}_2)/c}{1 + \mathbf{n} \cdot (\mathbf{v} - \mathbf{u}_1)/c}. \tag{5.65}$$

This equation describes the Doppler effect in a homogeneous flow where c and $\mathbf{v}$ are constant.

Consider the inertial coordinate system K_w moving with a constant velocity $\mathbf{w}$ with respect to the coordinate system K. In the coordinate system K_w, the velocities of the source, receiver, and medium are given by $\mathbf{u}_1 - \mathbf{w}_1$, $\mathbf{u}_2 - \mathbf{w}$, and $\mathbf{v} - \mathbf{w}$, respectively. These vectors must replace the vectors $\mathbf{u}_1$, $\mathbf{u}_2$, and **v** in equation (5.65) if the Doppler effect is considered in the coordinate system K_w. Taking account of the fact that the unit vector **n** has the same form in all inertial coordinate systems, one obtains that in the coordinate system K_w the Doppler effect is still described by equation (5.65). Thus, as one would expect, the frequency measured by the receiver does not depend on the choice of the inertial coordinate system with respect to which the velocities of the source, receiver, and medium are determined. It should be emphasized that equation (5.65) contains the unit vector **n**, which is invariant under the Galilean transformation, rather than the unit vector **s** in the direction of wave propagation (i.e., in the direction from the source to the receiver). The latter vector is not such an invariant, because of sound aberration. The vectors **n** and **s** coincide only in the case **v** = 0.

5.4.3 Inhomogeneous moving medium

Now let us derive the equation for the Doppler effect for the case in which the source and receiver are moving in an inhomogeneous moving medium. The geometry of the problem is still given by figure 5.5. Suppose that in the coordinate system K, the characteristic scale l of the variations in $\varrho(\mathbf{R})$, $c(\mathbf{R})$, and $\mathbf{v}(\mathbf{R})$ is much greater than the sound wavelength λ. Consider also a quasi-plane sound wave emitted by the source and arriving at the receiver. The frequency ω of this wave, measured in the coordinate system K, does not change when this wave propagates through the inhomogeneous moving medium.

Let us express the frequency ω in terms of the frequencies ω_1 and ω_2 of this wave, measured in the coordinate systems moving with the source and receiver, respectively. To this end, consider two domains D_1 and D_2 centered on the source and receiver. The scales of these domains are greater than λ but smaller than l. In these domains, c and $\mathbf{v}$ can be treated as constant and equal to c_1 and $\mathbf{v}_1$, and c_2 and $\mathbf{v}_2$, respectively. Taking account of the fact that in domain D_1 the source frequency ω_1 and the frequency ω of the emitted wave are related by equation (5.65) in which $\mathbf{u}_2 = 0$, one obtains

$$\omega = \omega_1 \frac{1 + \mathbf{n}_1 \cdot \mathbf{v}_1/c_1}{1 + \mathbf{n}_1 \cdot (\mathbf{v}_1 - \mathbf{u}_1)/c_1}. \tag{5.66}$$

Here, $\mathbf{n}_1$ is the unit vector normal to the wavefront of the emitted wave in domain D_1.

In domain D_2, the frequency ω of this wave and the frequency ω_2 measured by the moving receiver are also related by equation (5.65), where now $\mathbf{u}_1 = 0$. Therefore,

$$\omega_2 = \omega \frac{1 + \mathbf{n}_2 \cdot (\mathbf{v}_2 - \mathbf{u}_2)/c_2}{1 + \mathbf{n}_2 \cdot \mathbf{v}_2/c_2}, \tag{5.67}$$

where $\mathbf{n}_2$ is the unit vector normal to the wavefront of the wave at the receiver.

Eliminating ω between equations (5.66) and (5.67), one obtains [289, 290]

$$\omega_2 = \omega_1 \frac{1 + \mathbf{n}_1 \cdot \mathbf{v}_1 / c_1}{1 + \mathbf{n}_1 \cdot (\mathbf{v}_1 - \mathbf{u}_1)/c_1} \frac{1 + \mathbf{n}_2 \cdot (\mathbf{v}_2 - \mathbf{u}_2)/c_2}{1 + \mathbf{n}_2 \cdot \mathbf{v}_2/c_2}. \tag{5.68}$$

This formula describes the Doppler effect in the most general case considered so far, namely, when the source and receiver are moving with different velocities $\mathbf{u}_1$ and $\mathbf{u}_2$ in an inhomogeneous moving medium.

To find the vectors $\mathbf{n}_1$ and $\mathbf{n}_2$ in equation (5.68), one must find a ray path in an inhomogeneous moving medium connecting the source and receiver. Using equation (3.45), the vectors $\mathbf{n}_1$ and $\mathbf{n}_2$ can be expressed in terms of the unit vectors $\mathbf{s}_1$ and $\mathbf{s}_2$, which characterize (in the coordinate system K) the directions of propagation of the emitted wave and the wave at the receiver. In this case, equation (5.68) can be expressed in the form

$$\omega_2 = \omega_1 \frac{1 - \mathbf{m}_2 \cdot \mathbf{u}_2 / c_2}{1 - \mathbf{m}_1 \cdot \mathbf{u}_1 / c_1}. \tag{5.69}$$

Here, the vectors $\mathbf{m}_i = c_i \mathbf{n}_i / (c_i + \mathbf{n}_i \cdot \mathbf{v}_i)$ are defined by

$$\mathbf{m}_i = \frac{\mathbf{s}_i \left\{ \left[1 + (\mathbf{s}_i \cdot \mathbf{v}_i / c_i)^2 - v_i^2 / c_i^2 \right] + \mathbf{s}_i \cdot \mathbf{v}_i / c_i \right\} - \mathbf{v}_i / c_i}{1 + (\mathbf{s}_i \cdot \mathbf{v}_i / c_i)^2 - v_i^2 / c_i^2 + [1 + (\mathbf{s}_i \cdot \mathbf{v}_i / c_i)^2 - v_i^2 / c_i^2]^{1/2} \, \mathbf{s}_i \cdot \mathbf{v}_i / c_i}, \tag{5.70}$$

where $i = 1, 2$. Equation (5.69) has the same form as equation (5.64) for the Doppler effect in a motionless medium. However, the vectors $\mathbf{m}_i$ and $\mathbf{n}$ in equation (5.69) and equation (5.64) have different meaning.

It follows from equations (5.69) and (5.70) that $\omega_2 \neq \omega_1$ if $\mathbf{s}_1 \cdot \mathbf{u}_1 = \mathbf{s}_2 \cdot \mathbf{u}_2 = 0$, i.e., if the directions of sound wave propagation near the source and receiver are perpendicular to their velocities. In other words, there is a transverse Doppler effect in a moving medium, as mentioned elsewhere [39]. However there is no transverse Doppler effect in a motionless medium. This result follows from equation (5.64) and the equalities $\mathbf{s}_1 = \mathbf{s}_2 = \mathbf{n}$, which are valid in a motionless medium.

Equation (5.68) for the Doppler effect in an inhomogeneous moving medium is used, for example, to determine a correct formula for calculating the wind velocity in bistatic and monostatic acoustic sounding of the atmosphere with sodars [295].

Part II

Sound propagation and scattering in random moving media

In Part II of the book, we analyze sound propagation and scattering in media with random inhomogeneities in the sound speed, density, and medium velocity. The results are pertinent to atmospheric acoustics, for which sound waves are affected by fluctuations in temperature, humidity, and wind velocity. The temperature and humidity fluctuations produce fluctuations in the sound speed and density, which appear in equations for the sound field. In the ocean, sound waves are primarily affected by the temperature-induced sound speed fluctuations [117]. Nevertheless, the sound field can also be influenced by relatively weak random inhomogeneities in the density [72], salinity, and current velocity. The effects of current velocity fluctuations on the statistical moments of the sound field have been revealed by several experiments [30, 50, 98, 116, 436]. The density fluctuations in a fluid ocean bottom significantly affect the sound wave reflected by the bottom. Sound waves propagating through turbulent gases and liquids are affected by fluctuations in the sound speed and medium velocity.

Random inhomogeneities in a medium scatter sound waves, cause fluctuations in amplitude and phase, and diminish the coherence. In this part of the book, we will study these phenomena for different geometries: scattering of sound at arbitrary angles, line-of-sight propagation, and multipath propagation with refraction and/or reflections from the boundaries. The methods developed in theories of electromagnetic, seismic, and sound wave propagation in random media will be employed. Reviews of these theories can be found in general books describing waves in random media [72, 178, 339, 374], electromagnetic wave propagation in a turbulent atmosphere [6, 153, 334, 405, 406], sound propagation in a fluctuating ocean [117], and seismic wave propagation in the heterogeneous Earth [349]. Many methods and results presented in these books are applicable to sound propagation in a random medium with sound speed fluctuations. What distinguishes this book from the others is its systematic treatment of sound propagation through a medium with fluctuations not only in the sound speed, but also in the density and medium velocity.

In this part of the book, we consider propagation and scattering of monochromatic sound waves. Broadband sound propagation in a medium with spatial-temporal fluctuations in temperature and velocity is studied in reference [304].

6

Random inhomogeneities in a moving medium and scattering of sound

One of the main goals of analytical studies of wave propagation in a random medium is to relate the statistical characteristics of a propagating wave to the statistics of the random inhomogeneities occurring in the medium. Therefore, we begin Part II of this book with the statistical description of fluctuations of temperature, velocity, and the concentration of a component dissolved in the medium (section 6.1). Examples of the latter are fluctuations of the water vapor in the atmosphere and salinity in the ocean. For isotropic turbulence, the random inhomogeneities in the medium are often described with von Kármán, Kolmogorov, and Gaussian spectra (section 6.2). Parameters of these spectra pertinent to the atmospheric surface layer and the upper mixed layer in the ocean are presented.

Equations for sound waves in random media contain the sound speed and density fluctuations. In section 6.3, these fluctuations are expressed in terms of the fluctuations in temperature and the concentration of the dissolved component, which are most often measured experimentally. In section 6.4, sound scattering by the random inhomogeneities in a medium is investigated. Using the single scattering approximation, we derive a formula for the sound scattering cross section, which is important in many applications such as acoustic sounding of the atmosphere with sodars.

6.1 Statistical description of random inhomogeneities

Many media have random inhomogeneities such as fluctuations in the ambient temperature and velocity. It is difficult or even impossible to describe these inhomogeneities and fluctuations with the use of deterministic functions. Therefore, a statistical analysis is employed for such description.

This section presents a statistical characterization of the fluctuations in the temperature, velocity, and concentration of a component dissolved in the medium. A detailed description of random fields can be found in a number of books, e.g., references [339, 374]. Here, we briefly overview the results which will be needed for analysis of the statistical moments of the sound field propagating in a random moving medium. The terms "random inhomogeneities," "fluctuations," "random fields" will often be used as synonyms.

6.1.1   Random inhomogeneities in moving media

Random inhomogeneities in a moving medium can be caused by a variety of factors. In the atmosphere, the fluctuations in temperature, humidity, and wind velocity on time scales less than an hour are caused primarily by turbulence and internal gravity waves. Turbulence is most significant in the atmospheric boundary layer (ABL) which extends from the ground to the height of about 0.5–2 km. In the ABL, turbulence is caused by the wind shear near the ground and at the top of the ABL, and by buoyancy forces due to solar heating of the ground. On a clear night, when buoyancy forces are suppressed, intermittent turbulence and internal gravity waves are common sources of fluctuations. Three-dimensional spectra of temperature and wind velocity fluctuations due to the internal gravity waves in a stably stratified nighttime ABL have been suggested [292]. Internal gravity waves can also cause significant fluctuations in the temperature and wind velocity in the upper atmosphere, e.g., at the heights of about 40–50 km above the ground.

In the ocean, the primary cause of temperature fluctuations are internal gravity waves. In the upper layer of the ocean, turbulent mixing also results in fluctuations of temperature and velocity. Strong tidal currents in shallow and narrow channels can generate turbulent fluctuations in temperature, salinity, and velocity [99]. Salinity fluctuations can be important near river mouths and melting ice.

Turbulence occurs in many natural and man-made flows at high Reynolds numbers. Examples include intense currents and wind shear, air rising over heated grids, flow over vehicles such as cars, aircraft, and ships, and a jet exhausting from a nozzle. These turbulent flows, while they may initially appear quite different, can be described within a common framework of spatial scales and subranges. Turbulence causes the formation of eddies of different length scales l, which are in the range $\eta \lesssim l \lesssim L$. Here, L is the characteristic length scale of the largest, most energetic, and least dissipative eddies in a turbulent flow. In the wave propagation literature, L is termed the *outer scale* of turbulence; we will follow this terminology. In the ABL, L can be as large as several km. The *inner scale* of turbulence, η, is characteristic of the smallest, most dissipative, and least energetic eddies; it is also called the Kolmogorov microscale. In atmospheric turbulence, η is of order 1 to 10 mm, which is many orders of magnitude smaller than L. The outer and inner scales mark the boundary of the three primary subranges of turbulence: the energy-containing (or source), where $l \sim L$, the inertial $\eta \ll l \ll L$, and the dissipation $l \sim \eta$.

The largest eddies in a turbulent flow (with scales of order L) derive energy from the mean flow. These eddies break down by inertial forces, with essentially no energy loss, into progressively smaller eddies. This process continues down to the dissipation subrange, where viscous forces become significant.

Turbulent eddies in the energy subrange depend on the dynamics and boundary conditions of a particular flow. As a result, they are usually statistically anisotropic (dependent on orientation) and inhomogeneous (dependent on position). Statistical characterization of this subrange is not universal, depending on the details of the mean velocity and density profiles. On the other hand, the spectrum in the inertial subrange can be regarded as universal and given by the Kolmogorov spectrum.

6.1.2 Statistically homogeneous random fields

Let $\widetilde{T}(\mathbf{R})$ indicate the temperature fluctuations, $\widetilde{C}(\mathbf{R})$ the fluctuations in the concentration of the component dissolved in the medium, and $\widetilde{\mathbf{v}}(\mathbf{R})$ the velocity fluctuations. In the atmosphere, $\widetilde{C}$ corresponds to humidity fluctuations; in the ocean, it describes salinity fluctuations. The random fields $\widetilde{T}(\mathbf{R})$, $\widetilde{C}(\mathbf{R})$, and $\widetilde{\mathbf{v}}(\mathbf{R})$ are functions of the spatial coordinates $\mathbf{R} = (x, y, z)$. The random velocity field can be written as $\widetilde{\mathbf{v}} = (\widetilde{v}_1, \widetilde{v}_3, \widetilde{v}_3)$, where $\widetilde{v}_i(\mathbf{R})$, $i = 1, 2, 3$, indicates the velocity components along the x, y, and z axes.

The mean values of the random fields $\widetilde{T}(\mathbf{R})$, $\widetilde{C}(\mathbf{R})$, and $\widetilde{v}_i(\mathbf{R})$ are, by definition, zero:

$$\langle \widetilde{T}(\mathbf{R}) \rangle = 0, \quad \langle \widetilde{C}(\mathbf{R}) \rangle = 0, \quad \langle \widetilde{v}_i(\mathbf{R}) \rangle = 0. \tag{6.1}$$

Here, the angle brackets indicate the averaging over an ensemble of realizations of the corresponding random fields. We shall also assume that the scalar random fields, $\widetilde{T}$ and $\widetilde{C}$, do not correlate with the components $\widetilde{v}_i$ of the velocity fluctuations

$$\left\langle \widetilde{T}(\mathbf{R}_1)\widetilde{v}_i(\mathbf{R}_2) \right\rangle = 0, \quad \left\langle \widetilde{C}(\mathbf{R}_1)\widetilde{v}_i(\mathbf{R}_2) \right\rangle = 0. \tag{6.2}$$

These conditions are valid rigorously for isotropic turbulence and can be used approximately in the inertial subrange. However, generally they are not fulfilled in the energy-containing subrange. The scalar random fields $\widetilde{T}$ and $\widetilde{C}$ might be correlated even in the inertial subrange as is the case for the temperature and humidity fluctuations in the atmospheric boundary layer [111]. Therefore, we will account for the cross correlation between the scalar random fields $\widetilde{T}$ and $\widetilde{C}$.

For statistical description of the random fields $\widetilde{T}$, $\widetilde{C}$, and $\widetilde{v}_i$ and the cross correlation of $\widetilde{C}$ and $\widetilde{T}$, we shall use the correlation functions B_T, B_C, B_{ij}, and B_{CT} of the corresponding random fields, and their three-dimensional spectral densities (spectra) Φ_T, Φ_C, Φ_{ij}, and Φ_{CT}. For statistically homogeneous turbulence, these correlation functions (for a particular field or set of velocity components) depend only on the difference $\mathbf{R} = \mathbf{R}_1 - \mathbf{R}_2$ between two points of observation and are defined as follows:

$$B_T(\mathbf{R}_1 - \mathbf{R}_2) = \langle \widetilde{T}(\mathbf{R}_1)\widetilde{T}(\mathbf{R}_2) \rangle = \int \Phi_T(\boldsymbol{\kappa})e^{i\boldsymbol{\kappa}\cdot(\mathbf{R}_1-\mathbf{R}_2)}\, d^3\kappa, \tag{6.3}$$

$$B_C(\mathbf{R}_1 - \mathbf{R}_2) = \langle \widetilde{C}(\mathbf{R}_1)\widetilde{C}(\mathbf{R}_2) \rangle = \int \Phi_C(\boldsymbol{\kappa}) e^{i\boldsymbol{\kappa} \cdot (\mathbf{R}_1 - \mathbf{R}_2)} \, d^3\kappa, \qquad (6.4)$$

$$B_{CT}(\mathbf{R}_1 - \mathbf{R}_2) = \langle \widetilde{C}(\mathbf{R}_1)\widetilde{T}(\mathbf{R}_2) \rangle = \int \Phi_{CT}(\boldsymbol{\kappa}) e^{i\boldsymbol{\kappa} \cdot (\mathbf{R}_1 - \mathbf{R}_2)} \, d^3\kappa, \qquad (6.5)$$

$$B_{ij}(\mathbf{R}_1 - \mathbf{R}_2) = \langle \widetilde{v}_i(\mathbf{R}_1)\widetilde{v}_j(\mathbf{R}_2) \rangle = \int \Phi_{ij}(\boldsymbol{\kappa}) e^{i\boldsymbol{\kappa} \cdot (\mathbf{R}_1 - \mathbf{R}_2)} \, d^3\kappa. \qquad (6.6)$$

Here, $\boldsymbol{\kappa}$ is the turbulence wave vector, $i, j = 1, 2, 3$, and B_{ij} and Φ_{ij} are the components of the correlation and spectral tensors, respectively. In this chapter, if the limits of integration are not indicated, they are assumed to be from $-\infty$ to ∞.

For the Kolmogorov spectrum of turbulence, which will be considered in section 6.2, the correlation functions are singular at the origin. However, the structure functions D_T, D_C, D_{CT}, and D_{ij} remain mathematically well behaved and are used instead. These functions are defined as:

$$D_T(\mathbf{R}_1 - \mathbf{R}_2) = \langle [\widetilde{T}(\mathbf{R}_1) - \widetilde{T}(\mathbf{R}_2)]^2 \rangle$$
$$= 2\int \Phi_T(\boldsymbol{\kappa}) \left[1 - \cos(\boldsymbol{\kappa} \cdot (\mathbf{R}_1 - \mathbf{R}_2))\right] d^3\kappa, \qquad (6.7)$$

$$D_C(\mathbf{R}_1 - \mathbf{R}_2) = \langle [\widetilde{C}(\mathbf{R}_1) - \widetilde{C}(\mathbf{R}_2)]^2 \rangle$$
$$= 2\int \Phi_C(\boldsymbol{\kappa}) \left[1 - \cos(\boldsymbol{\kappa} \cdot (\mathbf{R}_1 - \mathbf{R}_2))\right] d^3\kappa, \qquad (6.8)$$

$$D_{CT}(\mathbf{R}_1 - \mathbf{R}_2) = \langle [\widetilde{C}(\mathbf{R}_1) - \widetilde{C}(\mathbf{R}_2)][\widetilde{T}(\mathbf{R}_1) - \widetilde{T}(\mathbf{R}_2)] \rangle$$
$$= 2\int \Phi_{CT}(\boldsymbol{\kappa}) \left[1 - \cos(\boldsymbol{\kappa} \cdot (\mathbf{R}_1 - \mathbf{R}_2))\right] d^3\kappa, \qquad (6.9)$$

$$D_{ij}(\mathbf{R}_1 - \mathbf{R}_2) = \langle [\widetilde{v}_i(\mathbf{R}_1) - \widetilde{v}_i(\mathbf{R}_2)][\widetilde{v}_j(\mathbf{R}_1) - \widetilde{v}_j(\mathbf{R}_2)] \rangle$$
$$= 2\int \Phi_{ij}(\boldsymbol{\kappa}) \left[1 - \cos(\boldsymbol{\kappa} \cdot (\mathbf{R}_1 - \mathbf{R}_2))\right] d^3\kappa. \qquad (6.10)$$

Here, the D_{ij} are the components of the structure-function tensor. For homogeneous turbulence, the structure functions also depend only on the difference $\mathbf{R} = \mathbf{R}_1 - \mathbf{R}_2$ between the observation points.

If the correlation function $B(\mathbf{R})$ of a random field exists, it can be expressed in terms of the structure function:

$$B(\mathbf{R}) = \frac{1}{2}\left[D(\mathbf{R} \to \infty) - D(\mathbf{R})\right]. \qquad (6.11)$$

Similarly, the structure function $D(\mathbf{R})$ can be expressed in terms of a correlation function:

$$D(\mathbf{R}) = 2[B(0) - B(\mathbf{R})]. \qquad (6.12)$$

These relationships between the correlation and structure functions are applicable for any random field, e.g., $\widetilde{T}$, $\widetilde{C}$, or $\widetilde{v}_i$.

6.1.3 Statistically homogeneous and isotropic random fields

Although random fields typically exhibit anisotropies, particularly at the larger scales, it can nonetheless sometimes be useful to model the fields as isotropic. For statistically homogeneous and isotropic scalar fields $\widetilde{T}$ and $\widetilde{C}$ and their cross correlation, the correlation and structure functions $B_T(R)$, $B_C(R)$, $B_{CT}(R)$ and $D_T(R)$, $D_C(R)$, $D_{CT}(R)$ depend only on the *magnitude* of the vector $\mathbf{R}$. Furthermore, the three-dimensional spectra $\Phi_T(\kappa)$, $\Phi_C(\kappa)$, and $\Phi_{CT}(\kappa)$ are functions of the modulus of the vector κ.

On the other hand, for a statistically homogeneous and isotropic vector field $\widetilde{\mathbf{v}}$, the correlation and structure tensors, B_{ij} and D_{ij}, depend on the vector $\mathbf{R}$ and are given by [25, 374]

$$B_{ij}(\mathbf{R}) = (\delta_{ij} - m_i m_j)B_\perp(R) + m_i m_j B_{||}(R), \tag{6.13}$$

$$D_{ij}(\mathbf{R}) = (\delta_{ij} - m_i m_j)D_\perp(R) + m_i m_j D_{||}(R). \tag{6.14}$$

Moreover, the components of the tensor $\Phi_{ij}(\kappa)$ are functions of the vector κ. In equations (6.13) and (6.14), δ_{ij} is the Kronecker symbol (equal to 1 when $i = j$ and 0 otherwise) and m_i are components of the unit vector $\mathbf{m} = \mathbf{R}/R = (m_1, m_2, m_3)$. Furthermore, $B_{||}(R) = \langle \widetilde{v}_{||}(\mathbf{R}' + \mathbf{R})\widetilde{v}_{||}(\mathbf{R}') \rangle$ and $B_\perp(R) = \langle \widetilde{v}_\perp(\mathbf{R}' + \mathbf{R})\widetilde{v}_\perp(\mathbf{R}') \rangle$ are the longitudinal and transverse correlation functions, where $\widetilde{v}_{||}(\mathbf{R})$ and $\widetilde{v}_\perp(\mathbf{R})$ are the components of the vector $\widetilde{\mathbf{v}}(\mathbf{R})$ in the direction of the vector $\mathbf{R}$ and in the direction perpendicular to $\mathbf{R}$; $D_{||}(R) = \langle [\widetilde{v}_{||}(\mathbf{R}' + \mathbf{R}) - \widetilde{v}_{||}(\mathbf{R}')]^2 \rangle$ and $D_\perp(R) = \langle [\widetilde{v}_\perp(\mathbf{R}' + \mathbf{R}) - \widetilde{v}_\perp(\mathbf{R}')]^2 \rangle$ are the longitudinal and transverse structure functions. Note that the functions $B_{||}(R)$, $B_\perp(R)$ and $D_{||}(R)$, $D_\perp(R)$ depend only on the modulus of the vector $\mathbf{R}$.

If the homogeneous and isotropic vector random field $\widetilde{\mathbf{v}}$ is also solenoidal ($\nabla \cdot \widetilde{\mathbf{v}} = 0$), then [25, 374]

$$B_\perp(R) = \frac{1}{2R}\frac{d}{dR}\left[R^2 B_{||}(R)\right], \tag{6.15}$$

$$D_\perp(R) = \frac{1}{2R}\frac{d}{dR}\left[R^2 D_{||}(R)\right], \tag{6.16}$$

$$\Phi_{ij}(\kappa) = \left(\delta_{ij} - \frac{\kappa_i \kappa_j}{\kappa^2}\right)\frac{E(\kappa)}{4\pi\kappa^2}. \tag{6.17}$$

Here, $E(\kappa)$ is the energy spectrum of turbulence, such that

$$\int_0^\infty E(\kappa)\,d\kappa \tag{6.18}$$

is the turbulent kinetic energy per unit mass. The integral in equation (6.18) is equal to $(3/2)\sigma_v^2$, where σ_v^2 is the variance of a single velocity component.

Equations (6.13)–(6.17) allow one to express all components of the tensors B_{ij}, D_{ij}, and Φ_{ij} in terms of one function $B_{||}(R)$, $D_{||}(R)$, or $E(\kappa)$. Note that equation (6.2) is rigorously valid [374] if the random fields $\widetilde{T}$, $\widetilde{C}$, and $\widetilde{\mathbf{v}}$ are statistically homogeneous and isotropic and the field $\widetilde{\mathbf{v}}$ is solenoidal.

For the considered case of homogeneous and isotropic turbulence, the correlation functions $B_T(R)$ and $B_{\shortparallel}(R)$ can be expressed in terms of the spectra $\Phi_T(\kappa)$ and $E(\kappa)$ [167, 374]:

$$B_T(R) = \frac{4\pi}{R} \int_0^\infty \Phi_T(\kappa) \sin(\kappa R) \kappa \, d\kappa, \tag{6.19}$$

$$B_{\shortparallel}(R) = \int_0^\infty d\kappa \, \cos(\kappa R) \int_\kappa^\infty \frac{\kappa_1^2 - \kappa^2}{\kappa_1^3} E(\kappa_1) \, d\kappa_1. \tag{6.20}$$

The formula for $B_T(R)$ is a consequence of equation (6.3). The correlation functions $B_C(R)$ and $B_{CT}(R)$ are expressed in terms of their spectra $\Phi_C(\kappa)$ and $\Phi_{CT}(\kappa)$ with equation (6.19) if in this equation, the subscript T is replaced with C and CT, respectively. To derive equation (6.20), one substitutes B_{ij} from equation (6.13) and Φ_{ij} from equation (6.17) into equation (6.6) and replaces $B_\perp$ with $B_{\shortparallel}$ by using equation (6.15). The resulting equation contains only two functions, $B_{\shortparallel}(R)$ and $E(\kappa)$, and it can be recast as equation (6.20).

As we will see later, the statistical moments of a sound field propagating in a random medium depend on the three-dimensional spectra of random inhomogeneities $\Phi_T(\boldsymbol{\kappa})$, $\Phi_{CT}(\boldsymbol{\kappa})$, $\Phi_C(\boldsymbol{\kappa})$, and $\Phi_{ij}(\boldsymbol{\kappa})$. Devices that provide observations of the atmosphere at a single point (such as ultrasonic and hot-wire anemometers and thermometers) can actually be used to infer the three-dimensional spectra. First, a time series is recorded, which can then be transformed to a frequency spectrum. Next, the frequency spectrum can be converted to a one-dimensional spatial spectrum using Taylor's frozen turbulence hypothesis. Lastly, we evoke the assumptions of homogeneity and isotropy to relate the 1D spectrum to the 3D spectrum. For example, in equation (6.6), setting $\mathbf{R} = \mathbf{R}_1 - \mathbf{R}_2 = (x, y = 0, z = 0)$, $\boldsymbol{\kappa} = (\kappa_1, \kappa_2, \kappa_3)$, and $i = j = 1$ yields

$$B_{11}(x, 0, 0) \equiv B_{\shortparallel}(x) = \int \Theta_{11}(\kappa_1) e^{i\kappa_1 x} \, d\kappa_1. \tag{6.21}$$

Here, $B_{11}(x)$ is the longitudinal correlation function of velocity fluctuations and $\Theta_{11}(\kappa_1)$ is the one-dimensional longitudinal spectrum, related to the three-dimensional spectrum by the formula

$$\Theta_{11}(\kappa_1) = \int \int \Phi_{11}(\kappa_1, \kappa_2, \kappa_3) \, d\kappa_2 \, d\kappa_3 \tag{6.22}$$

Similarly, setting $\mathbf{R} = \mathbf{R}_1 - \mathbf{R}_2 = (x, y = 0, z = 0)$, $\boldsymbol{\kappa} = (\kappa_1, \kappa_2, \kappa_3)$, and $i = j = 2$ in equation (6.6), we obtain the transverse (lateral) correlation function of velocity fluctuations

$$B_{22}(x, 0, 0) \equiv B_\perp(x) = \int \Theta_{22}(\kappa_1) e^{i\kappa_1 x} \, d\kappa_1. \tag{6.23}$$

The one-dimensional transverse spectrum $\Theta_{22}(\kappa_1)$ is related to three-dimensional spectrum $\Phi_{22}(\kappa_1, \kappa_2, \kappa_3)$ by equation (6.22) if in this equation, the subscripts 11 are replaced with the subscripts 22.

Substituting Φ_{11} from equation (6.17) into equation (6.22), we express the one-dimensional longitudinal spectrum $\Theta_{11}(\kappa_1)$ in terms of the energy spectrum

$$\Theta_{11}(\kappa_1) = \frac{1}{2} \int_{\kappa_1}^{\infty} \left(1 - \frac{\kappa_1^2}{\kappa^2}\right) \frac{E(\kappa)}{\kappa} \, d\kappa. \tag{6.24}$$

Substitution of this formula into equation (6.21) results in equation (6.20), as it should. The one-dimensional transverse spectrum can be expressed in terms of the longitudinal spectrum as follows [25]

$$\Theta_{22}(\kappa_1) = \frac{1}{2} \left[\Theta_{11}(\kappa_1) - \kappa_1 \frac{d\Theta_{11}(\kappa_1)}{d\kappa_1}\right]. \tag{6.25}$$

6.2 Spectra of turbulence

As an example of a turbulent flow, consider atmospheric turbulence which spans a very broad range of spatial scales, from millimeters to kilometers. No one model accurately describes the entire turbulence spectrum for different meteorological regimes of the ABL. In the energy-containing subrange, spectra of atmospheric turbulence are inherently inhomogeneous and anisotropic. For example, the variance for turbulent velocity fluctuations in the direction of the mean wind is several times larger than the variance for vertical velocity fluctuations. An example of inhomogeneity is the dependence of the variances on the height above the ground. Similar manifestations of anisotropy and inhomogeneity are evident in the outer scales of atmospheric turbulence.

Models for anisotropic spectra and correlation functions of atmospheric turbulence have been presented elsewhere [152, 207, 414]. Kristensen et al. [211] developed an anisotropic spectral model for turbulence and applied it to turbulence originating from various instability mechanisms, including the surface-layer shear. Mann [237] obtained a velocity spectral tensor using rapid distortion theory, which hypothesizes that the shear distortions to an eddy can be calculated from linearized Navier–Stokes equations.

In the energy-containing subrange, spectra of other turbulent flows generally are also inhomogeneous and anisotropic and differ from those in the ABL. However, for turbulent flows with a well-developed inertial subrange, the Kolmogorov spectrum can be regarded as universally applicable within that subrange.

Kolmogorov, von Kármán, and Gaussian spectra are often used to describe the spectra of isotropic turbulence. The Kolmogorov spectrum is a realistic spectrum of turbulence in the inertial subrange $\eta \ll l \ll L$, where l is the scale of turbulent eddies, and η and L are the inner and outer scales of turbulence. The von Kármán spectrum coincides with the Kolmogorov spectrum in the inertial subrange and accounts approximately for the eddies in the energy subrange, where $l \sim L$. The Gaussian spectrum, as typically used, accounts approximately only for the largest eddies in a medium [418]. The von Kármán and Gaussian models—with parameters correctly chosen—can be a reasonable approximation for the energy subrange in some situations, because they give roughly the correct qualitative behavior of the spectra. But the limitations of these models must be kept in mind. For example, they are usually applied assuming homogeneous and isotropic turbulence when, in reality, the energy subrange is inherently inhomogeneous and anisotropic. Some of these limitations can be relaxed, e.g., by allowing the parameters of the spectra to depend on the height above the ground (section 6.2.4).

6.2.1   Von Kármán spectrum

The von Kármán spectrum of temperature fluctuations is given by

$$\Phi_T^{vK}(\kappa) = \frac{\Gamma(11/6)}{\pi^{3/2}\Gamma(1/3)} \frac{\sigma_T^2 L_T^3}{(1 + \kappa^2 L_T^2)^{11/6}}. \tag{6.26}$$

Here, the superscript vK indicates that the function corresponds to the von Kármán spectrum, Γ is the gamma function, σ_T^2 is the variance of the temperature fluctuations, and L_T is the von Kármán length scale, which is representative of the outer scale of temperature fluctuations. The numerical coefficient appearing on the right-hand side of equation (6.26) ensures that the variance of the temperature fluctuations is equal to σ_T^2. The normalized von Kármán temperature spectrum, $\Phi_T^{vK}(\kappa)/\sigma_T^2 L_T^3$, is plotted in figure 6.1 versus κL_T. For large values of κL_T, the spectrum has a power-law dependence $\Phi_T^{vK}(\kappa) \sim \kappa^{-11/3}$. For small values of κL_T, the spectrum is nearly constant.

Similarly to equation (6.26), formulas for $\Phi_C^{vK}(\kappa)$ and $\Phi_{CT}^{vK}(\kappa)$ are given by:

$$\Phi_C^{vK}(\kappa) = \frac{\Gamma(11/6)}{\pi^{3/2}\Gamma(1/3)} \frac{\sigma_C^2 L_C^3}{(1 + \kappa^2 L_C^2)^{11/6}}, \tag{6.27}$$

$$\Phi_{CT}^{vK}(\kappa) = \frac{\Gamma(11/6)}{\pi^{3/2}\Gamma(1/3)} \frac{\sigma_{CT} L_{CT}^3}{(1 + \kappa^2 L_{CT}^2)^{11/6}}. \tag{6.28}$$

Here, σ_C^2 is the variance of the fluctuations in the dissolved component in the medium, σ_{CT} is the covariance between $\widetilde{C}$ and $\widetilde{T}$, and L_T and L_{CT} are the corresponding length scales.

The von Kármán energy spectrum is given by [167]

$$E^{vK}(\kappa) = \frac{55\Gamma(5/6)}{9\pi^{1/2}\Gamma(1/3)} \frac{\sigma_v^2 \kappa^4 L_v^5}{(1 + \kappa^2 L_v^2)^{17/6}}. \tag{6.29}$$

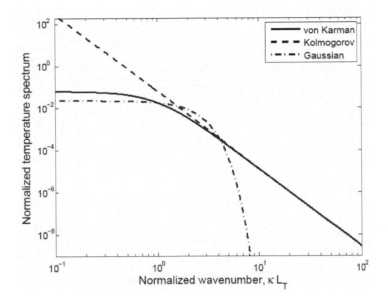

FIGURE 6.1
Normalized von Kármán (solid line), Kolmogorov (dashed line), and Gaussian (dash-dotted line) spectra of temperature fluctuations versus the normalized turbulence wavenumber.

Here, σ_v^2 is the variance of one of the velocity components and L_v is the von Kármán length scale of velocity fluctuations. Figure 6.2 depicts the normalized von Kármán energy spectrum, $E^{vK}(\kappa)/\sigma_v^2 L_v$, versus κL_v . The spectrum has a power-law dependence $E^{vK}(\kappa) \sim \kappa^{-5/3}$ for large values of κL_v. For small values of κL_v, the spectrum has the dependence $E^{vK}(\kappa) \sim \kappa^4$. The maximum of the spectrum occurs at $\kappa L_v = 1.55$ which is in between these two regimes.

The von Kármán spectra are sometimes presented with the high-wavenumber cutoff factor $\exp(-\kappa^2 \eta^2)$, which accounts for the dissipation subrange. We have omitted such a factor from equations (6.26)–(6.29) since the sound wavelength is usually much greater than the inner scale of turbulence η and turbulent eddies in the dissipation subrange have less energy than those in the inertial subrange; as a result sound scattering by inhomogeneities with scales of the order of η is negligible.

Substituting $\Phi_T^{vK}(\kappa)$ from equation (6.26) into equation (6.19) and calculating the integral over κ, one obtains the correlation function for the von Kármán spectrum of the temperature fluctuations:

$$B_T^{vK}(R) = \frac{2^{2/3}}{\Gamma(1/3)} \sigma_T^2 (R/L_T)^{1/3} K_{1/3}(R/L_T), \tag{6.30}$$

where $K_{1/3}$ is the modified Bessel function. It can be shown from this equation that $B_T^{vK}(0) = \sigma_T^2$, as it should. The equations for $B_C^{vK}(R)$ and $B_{CT}^{vK}(R)$ are

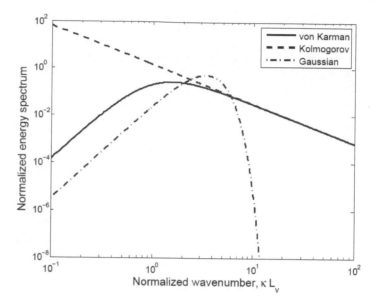

FIGURE 6.2
Normalized von Kármán (solid line), Kolmogorov (dashed line), and Gaussian (dash-dotted line) energy spectra of turbulence versus the normalized turbulence wavenumber.

given by equation (6.30) if σ_T^2 is replaced by σ_C^2 and σ_{CT}, and L_T is replaced by L_C and L_{CT}, respectively.

Similarly, substituting $E^{vK}(\kappa)$ from equation (6.29) into equation (6.20), we obtain $B_\|^{vK}(R)$:

$$B_\|^{vK}(R) = \frac{2^{2/3}}{\Gamma(1/3)}\sigma_v^2 (R/L_v)^{1/3} K_{1/3}(R/L_v). \tag{6.31}$$

It follows from this equation that $B_v^{vK}(0) = \sigma_v^2$. Note that the correlation functions $B_T^{vK}(R)$ and $B_\|^{vK}(R)$ have a similar form.

Using equation (6.12), we obtain formulas for the structure functions of temperature and longitudinal velocity fluctuations:

$$D_T^{vK}(R) = 2\sigma_T^2 \left[1 - \frac{2^{2/3}}{\Gamma(1/3)}(R/L_T)^{1/3}K_{1/3}(R/L_T)\right], \tag{6.32}$$

$$D_\|^{vK}(R) = 2\sigma_v^2 \left[1 - \frac{2^{2/3}}{\Gamma(1/3)}(R/L_v)^{1/3}K_{1/3}(R/L_v)\right]. \tag{6.33}$$

The normalized structure function of temperature fluctuations, $D_T^{vK}(R)/\sigma_T^2$, is plotted in figure 6.3 versus R/L_T. For small values of R/L_T, the structure function has a power-law dependence $D_T^{vK}(R) \sim R^{2/3}$. For large val-

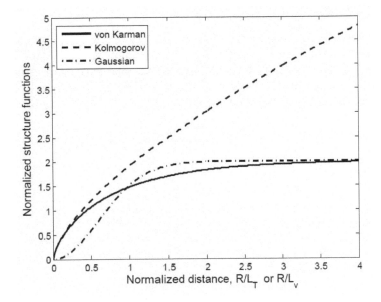

FIGURE 6.3

Normalized structure functions of temperature and longitudinal velocity fluctuations for von Kármán (solid line), Kolmogorov (dashed line), and Gaussian (dash-dotted line) spectra. The horizontal axis represents R/L_T for the case of temperature fluctuations and R/L_v for velocity fluctuations.

ues of R/L_T, the structure function is nearly constant. It follows from equations (6.32) and (6.33) that the normalized structure functions $D_T^{vK}(R)/\sigma_T^2$ and $D_{\parallel}^{vK}(R)/\sigma_v^2$ coincide if $L_T = L_v$. Therefore, the normalized structure function of longitudinal velocity fluctuations, $D_{\parallel}^{vK}(R)/\sigma_v^2$, is also given by the solid line in figure 6.3 if the x-axis represents the values of R/L_v.

The von Kármán spectrum provides a fairly good approximation to the spectrum of turbulence in jet- or grid-generated turbulent flows and over heated grids. If $\kappa \gg 1/L$ (where L is L_T, L_v, L_C, or L_{CT}), this spectrum coincides with the Kolmogorov spectrum. The von Kármán spectrum of temperature fluctuations is used both in the theory of turbulence [167] and in the theory of waves in random media [339]. In atmospheric acoustics, turbulence in the ABL is often modeled with the von Kármán spectra. The relatively simple formulas for the spectra enable analytical calculations of some statistical moments of sound fields. Using the Monin–Obukhov similarity theory (MOST), realistic, height-dependent parameterizations of the von Kármán spectra have been obtained (section 6.2.4).

6.2.2   Kolmogorov spectrum

In the inertial subrange, where $1/L \ll \kappa \ll 1/\eta$, the Kolmogorov spectrum applies. For this spectrum, the three-dimensional spectral densities of the temperature and velocity fluctuations and the fluctuations in the concentration of the component dissolved in the medium can be obtained from equations (6.26)–(6.29) in the limit $\kappa \gg 1/L$:

$$\Phi_T^K(\kappa) = \mathcal{Q}\frac{C_T^2}{\kappa^{11/3}}, \quad \Phi_C^K(\kappa) = \mathcal{Q}\frac{C_C^2}{\kappa^{11/3}}, \quad \Phi_{CT}^K(\kappa) = \mathcal{Q}\frac{C_{CT}}{\kappa^{11/3}},$$

$$E^K(\kappa) = \frac{22\pi\mathcal{Q}}{3}\frac{C_v^2}{\kappa^{5/3}}. \tag{6.34}$$

Here, the superscript K indicates the Kolmogorov spectrum, $\mathcal{Q}$ is a numerical coefficient, given by

$$\mathcal{Q} = \frac{5}{18\pi\Gamma(1/3)} \approx 0.0330, \tag{6.35}$$

and C_T^2, C_C^2, C_{CT}, and C_v^2 are the structure-function parameters of the corresponding random fields:

$$C_T^2 = \frac{3\Gamma(5/6)}{\pi^{1/2}}\frac{\sigma_T^2}{L_T^{2/3}}, \quad C_C^2 = \frac{3\Gamma(5/6)}{\pi^{1/2}}\frac{\sigma_C^2}{L_C^{2/3}}, \quad C_{CT} = \frac{3\Gamma(5/6)}{\pi^{1/2}}\frac{\sigma_{CT}}{L_{CT}^{2/3}},$$

$$C_v^2 = \frac{3\Gamma(5/6)}{\pi^{1/2}}\frac{\sigma_v^2}{L_v^{2/3}}. \tag{6.36}$$

These formulas relate the structure-function parameters, which characterize the intensity of turbulence in the inertial subrange, with the variances and the length scales appearing in the von Kármán spectra. It is convenient to express the structure-function parameter C_{CT} in the form:

$$C_{CT} = \varrho_{CT}C_C C_T. \tag{6.37}$$

Here, ϱ_{CT} is the correlation coefficient of the random fields $\widetilde{C}$ and $\widetilde{T}$.

The normalized Kolmogorov temperature spectrum, $\Phi_T^K(\kappa)/\sigma_T^2 L_T^3$, is plotted in figure 6.1 versus κL_T. In the inertial subrange (large values of κL_T), the spectrum coincides with the von Kármán temperature spectrum. Figure 6.2 depicts the normalized Kolmogorov energy spectrum, $E^K(\kappa)/\sigma_v^2 L_v$. Similarly to the temperature fluctuations, the spectrum coincides with the von Kármán energy spectrum for large values of κL_v.

Using equations (6.7)–(6.10) and (6.34), one obtains the structure functions for the considered random fields

$$D_T^K(R) = C_T^2 R^{2/3}, \quad D_C^K(R) = C_C^2 R^{2/3}, \quad D_{CT}^K(R) = RC_C C_T R^{2/3},$$

$$D_{RR}^K(R) = C_v^2 R^{2/3}. \tag{6.38}$$

When deriving a formula for $D_{RR}^K(R)$, the spectral tensor $\Phi_{ij}^K(\boldsymbol{\kappa})$ was first calculated with equations (6.34) and (6.17). Then, using equation (6.10), the structure tensor $D_{ij}^K(\mathbf{R})$ was obtained. Finally, equations (6.14) and (6.16) were used to derive a formula for $D_{RR}^K(R)$. The structure functions given by equations (6.38) can also be obtained from equations (6.32), (6.33), and similar equations for $D_C^{vK}(R)$ and $D_{CT}^{vK}(R)$ if in all these equations, the modified Bessel functions $K_{1/3}$ are expanded in the Taylor series for small values of $\mathbf{R}$ and the first two terms are kept.

The normalized structure function of temperature fluctuations, $D_T^K(R)/\sigma_T^2$, is plotted in figure 6.3 versus R/L_T. In the inertial subrange (small values of R/L_T), it coincides with the normalized structure function for the von Kármán spectrum, $D_T^{vK}(R)/\sigma_T^2$. In the energy subrange, where $R/L_T \gtrsim 1$, these structure functions deviate remarkably. Using equations (6.36) and (6.38), it can be shown that $D_T^K(R)/\sigma_T^2 = D_{RR}^K(R)/\sigma_v^2$ if $L_T = L_v$. Therefore, the dependence of the normalized structure function of the longitudinal velocity fluctuations for the Kolmogorov spectrum, $D_{RR}^K(R)/\sigma_v^2$, on R/L_v is also given by the dashed line in figure 6.3.

6.2.3 Gaussian spectrum

The Gaussian correlation function of temperature fluctuations is given by

$$B_T^G(R) = \sigma_T^2 \exp\left(-R^2/\mathcal{L}_T^2\right). \tag{6.39}$$

Here, the superscript G indicates that the function corresponds to the Gaussian spectrum, σ_T^2 is the variance of the temperature fluctuations, and $\mathcal{L}_T$ is the Gaussian length scale. We assume that the variance σ_T^2 in equation (6.39) is the same as that for the von Kármán spectrum; as will be discussed later in this section, the relationship between the length scales in the two spectral models depends on further modeling assumptions. The expressions for $B_C^G(R)$ and $B_{CT}^G(R)$ are given by equation (6.39) if in that equation, σ_T^2 is replaced by σ_C^2 and σ_{CT}, respectively, and $\mathcal{L}_T$ is replaced by $\mathcal{L}_C$ and $\mathcal{L}_{CT}$.

Since for an isotropic and solenoidal vector random field $\tilde{\mathbf{v}}$ the correlation functions $B_{ij}(\mathbf{R})$ are anisotropic, we ascribe the Gaussian function to the longitudinal correlation function:

$$B_{\shortparallel}^G(R) = \sigma_v^2 \exp(-R^2/\mathcal{L}_v^2). \tag{6.40}$$

Here, σ_v^2 is the variance of the medium velocity component $\tilde{v}_{\shortparallel}$, which is the same as that for the von Kármán spectrum, and $\mathcal{L}_v$ is the Gaussian length scale. Using equations (6.13), (6.15), and (6.40), one can calculate the correlation functions $B_{ij}(\mathbf{R})$. For example, for $i = 1$ and $j = 1$, we have

$$B_{11}^G(\mathbf{R}) = \sigma_v^2 \left(1 - \frac{y^2 + z^2}{\mathcal{L}_v^2}\right) \exp(-R^2/\mathcal{L}_v^2), \tag{6.41}$$

where $\mathbf{R} = (x, y, z)$ and the variance of the medium velocity component $\tilde{v}_1$

is equal to σ_v^2. The total variance of the isotropic vector random field $\widetilde{\mathbf{v}}$ is given by $\langle \widetilde{\mathbf{v}}^2 \rangle = 3\langle \widetilde{v}_1^2 \rangle = 3\sigma_v^2$. Note that, although equation (6.41) is derived for isotropic turbulence, the correlation function does not depend only on the magnitude of $\mathbf{R}$.

Substituting equation (6.39) into equation (6.12), one obtains the structure function of the temperature fluctuations

$$D_T^G(R) = 2\sigma_T^2 \left[1 - e^{-R^2/\mathcal{L}_T^2}\right]. \tag{6.42}$$

Similarly, substituting equation (6.40) into equation (6.12), we obtain a formula for the structure function of the longitudinal velocity fluctuations:

$$D_{\shortparallel}^G(R) = 2\sigma_v^2 \left[1 - e^{-R^2/\mathcal{L}_v^2}\right]. \tag{6.43}$$

The three-dimensional spectrum of temperature fluctuations is obtained with equations (6.3) and (6.39)

$$\Phi_T^G(\kappa) = \frac{\sigma_T^2 \mathcal{L}_T^3}{8\pi^{3/2}} \exp\left(-\kappa^2 \mathcal{L}_T^2/4\right). \tag{6.44}$$

The expressions for $\Phi_C^G(\kappa)$ and $\Phi_{CT}^G(\kappa)$ are given by equation (6.44), if in that equation σ_T^2 is replaced by σ_C^2 and σ_{CT}, respectively, and $\mathcal{L}_T$ is replaced by $\mathcal{L}_C$ and $\mathcal{L}_{CT}$. Making use of equations (6.6), (6.17), and (6.41), the energy spectrum can be obtained as

$$E^G(\kappa) = \frac{\sigma_v^2 \kappa^4 \mathcal{L}_v^5}{8\pi^{1/2}} \exp\left(-\kappa^2 \mathcal{L}_v^2/4\right). \tag{6.45}$$

It should always be kept in mind that the Gaussian spectral model does not approximate well the spectra of high Reynolds number turbulent flows, which possess a broad range of scales of random inhomogeneities. However, with properly chosen values for $\mathcal{L}_T$, $\mathcal{L}_v$, $\mathcal{L}_C$, and $\mathcal{L}_{CT}$, this spectrum can be used for calculations of the statistical moments of sound fields when they are primarily affected by inhomogeneities in the energy subrange. In this case, the Gaussian spectrum is very convenient since it often leads to analytical expressions for the statistical moments. For example, it is used in the time-dependent stochastic inversion algorithm [391] for calculations of the correlation functions of the travel-time fluctuations along sound propagation paths in acoustic tomography of the atmosphere (section 3.7). This is justified by the fact these fluctuations are caused primarily by the largest inhomogeneities in the atmosphere. The Gaussian spectrum was also used in early studies of sound propagation in the ocean [72] and atmosphere [86, 88].

In order to perform a meaningful comparison between the Gaussian spectra of temperature and velocity fluctuations and the corresponding von Kármán spectra, relationships between the various models' length scales are needed.

For this purpose, we shall assume that the integral length scales of the temperature fluctuations of the two models are the same,

$$\int_0^\infty \frac{B_T^{vK}(R)}{B_T^{vK}(0)}\, dR = \int_0^\infty \frac{B_T^G(R)}{B_T^G(0)}\, dR, \tag{6.46}$$

as are the longitudinal integral length scales of the velocity fluctuations:

$$\int_0^\infty \frac{B_{\|}^{vK}(R)}{B_{\|}^{vK}(0)}\, dR = \int_0^\infty \frac{B_{\|}^G(R)}{B_{\|}^G(0)}\, dR. \tag{6.47}$$

Substituting $B_T^{vK}(R)$ from equation (6.30) and $B_T^G(R)$ from equation (6.39) into equation (6.46) and calculating the integrals over R yields the relationship between $\mathcal{L}_T$ and L_T:

$$\mathcal{L}_T = \frac{2\Gamma(5/6)}{\Gamma(1/3)} L_T \approx 0.843 L_T. \tag{6.48}$$

A similar relationship is obtained when substituting $B_{\|}^{vK}(R)$ given by equation (6.31) and $B_{\|}^G(R)$ given by equation (6.40) into equation (6.47) and calculating the integrals over R:

$$\mathcal{L}_v = \frac{2\Gamma(5/6)}{\Gamma(1/3)} L_v \approx 0.843 L_v. \tag{6.49}$$

These formulas allow one to express length scales, appropriate to the Gaussian spectra of temperature and velocity fluctuations, in terms of the length scales appropriate to the von Kármán spectra of these random fields.

To compare the Gaussian spectra with the von Kármán and Kolmogorov spectra, in equations (6.42)–(6.45) $\mathcal{L}_T$ and $\mathcal{L}_v$ are replaced with L_T and L_v, respectively. The normalized Gaussian spectrum of temperature fluctuations, $\Phi_T^G(\kappa)/\sigma_T^2 L_T^3$, is shown in figure 6.1 versus κL_T. For small values of κL_T, the spectrum is nearly constant, which is similar to the behavior of the von Kármán spectrum. For large values of κL_T, the Gaussian spectrum deviates significantly from both the von Kármán and Kolmogorov spectra. Figure 6.2 depicts the normalized Gaussian energy spectrum, $E^G(\kappa)/\sigma_v^2 L_v$. The spectrum has the same power-law dependence, $E^G(\kappa) \sim \kappa^4$, as the von Kármán spectrum for small values of κL_v; for large values of κL_v, the Gaussian spectrum again deviates from the others. Finally, figure 6.3 shows the normalized structure functions of temperature and velocity fluctuations for the Gaussian spectrum, $D_T^G(R)/\sigma_T^2$ and $D_{\|}^G(R)/\sigma_v^2$, versus R/L_T and R/L_v, respectively. These structure functions coincide with those for the von Kármán spectrum for large values of R (i.e., in the energy subrange). On the other hand, for small values of R, the Gaussian structure functions (since they lack a realistic inertial subrange) deviate significantly from the von Kármán and Kolmogorov structure functions.

6.2.4   Parameters of spectra in the atmosphere and ocean

In this subsection, we present the parameters of the von Kármán and Kolmogorov spectra in the atmospheric surface layer (ASL) and in the ocean.

The temperature spectrum in the ASL can be parameterized using the Monin–Obukhov similarity theory (MOST), as was described and applied to the vertical profiles in section 2.2.3. In MOST, quantities are first non-dimensionalized using the height z, the friction velocity u_*, and the surface-layer temperature scale $T_* = -Q_H/(\varrho_0 c_P u_*)$ (where Q_H is the surface sensible heat flux, ϱ_0 the air density, and c_P the specific heat at constant pressure). The resulting normalized quantities are then functions of z/L_o, where $L_o = -u_*^3 T_s \varrho_0 c_P/(g \kappa_v Q_H)$ is the Obukhov length, T_s the surface temperature, g gravitational acceleration, and $\kappa_v = 0.40$ von Kármán constant.

Thus, according to MOST, the temperature variance σ_T^2 and the length scale L_T in the von Kármán spectrum, when normalized by T_*^2 and z, respectively, become functions of z/L_o. On the basis of previous experimental observations of the ASL, the following expressions were obtained in reference [299]:

$$\frac{\sigma_T^2(z)}{T_*^2} = \frac{4.0}{[1 + 10(-z/L_o)]^{2/3}},\tag{6.50}$$

and

$$\frac{L_T(z)}{z} = 2.0\frac{1 + 7.0(-z/L_o)}{1 + 10(-z/L_o)}.\tag{6.51}$$

Equations (6.50) and (6.51) are valid in the limiting cases of a surface layer with purely shear-generated turbulence ($-z/L_o \to 0$, in which case the dependence on g/T_s vanishes) and in free-convection turbulence ($-z/L_o \to \infty$, in which case the dependence on u_* vanishes). Ranges of the values for u_* and Q_H in the ASL were presented in section 2.2.3.

Equations (6.50) and (6.51) should similarly apply to σ_C^2, L_C, σ_{CT}, and L_{CT}, with appropriate normalizations.

Application of MOST to the velocity fluctuations would imply that σ_v^2/u_*^2 and L_v/z are functions of z/L_o. However, experimental evidence has shown conclusively (see, for example, reference [186]) that the spectra of the *horizontal* velocity fluctuations in the ASL do *not* obey MOST in unstable conditions. Although the spectra behave as expected when shear dominates, when buoyancy production dominates, the free-convection limit is not observed, as would be consistent with MOST. The underlying cause is the presence of large eddies spanning the entire depth of the ABL in convective conditions, as illustrated in figure 6.4. These large eddies have little impact on *vertical* velocity in the ASL, since they are blocked by the surface. As a result, heat and momentum fluxes, and thus the vertical profiles and temperature spectrum, are relatively unaffected by them, and MOST applies to these quantities as discussed previously.

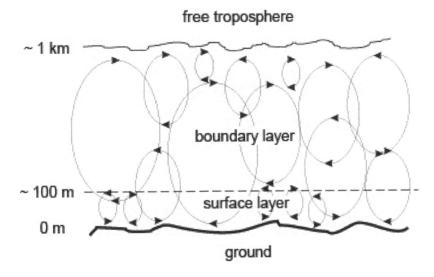

FIGURE 6.4
Schematic of large eddies in the atmospheric boundary layer (ABL) for convective conditions.

Instead of obeying free-convection similarity in the limit $-z/L_o \to \infty$, the statistics of the horizontal velocity fluctuations in the ASL are dominated by a different form of similarity, called *mixed-layer* similarity, in which the height z_i of the ABL replaces z [186]. Referring back to the discussion in section 2.2.3, the pertinent velocity scale is thus $w_* = (z_i g Q_H/(\varrho_0 c_P T_s))^{1/3}$.

Considering first the limiting case of shear surface-layer turbulence, the following similarity expressions were derived in reference [415] for the Kármán spectrum parameters:

$$\frac{\sigma_v^2}{u_*^2} = 3.0, \quad \frac{L_v}{z} = 1.8. \tag{6.52}$$

For mixed-layer similarity, the expressions are [415]:

$$\frac{\sigma_v^2}{w_*^2} = 0.35, \quad \frac{L_v}{z_i} = 0.23. \tag{6.53}$$

A general similarity theory combining these two cases might, for example, involve normalizing length scales by z and velocities by u_*, and then setting the result to a function of z/z_i. However, Højstrup [168] proposed a simpler approach to combining the two types of similarity, which has since been widely adopted. This approach is to add together independent spectra, one obeying shear surface-layer similarity, and the other mixed-layer similarity. In effect, such expressions interpolate between the two cases. For example, $E(\kappa)$ can be written with two terms, one calculated using the parameters in

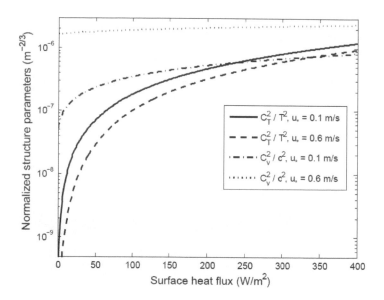

FIGURE 6.5
Normalized structure-function parameters of temperature and wind velocity fluctuations, C_T^2/T^2 and C_v^2/c^2, versus the surface heat flux Q_H for two values of the friction velocity u_*.

equation (6.52), and the other using the parameters in equation (6.53). By implication, other quantities such as variances and structure-function parameters can also be written with two such additive terms.

Equations (6.50)–(6.53) account for realistic vertical inhomogeneity of turbulence, in atmospheric conditions involving turbulent production by both shear and buoyancy instabilities.

The structure-function parameters C_T^2, C_C^2, and C_v^2 characterize the intensity of turbulence in the inertial subrange. The structure-function parameters vary by orders of magnitude for different meteorological regimes of the ASL. The C_T^2 and C_v^2 can be calculated with equations (6.36) and the values of the variances and characteristic lengths given by equations (6.50)–(6.53). Figure 6.5 depicts the normalized structure-function parameters, C_T^2/T^2 and C_v^2/c^2, versus the surface heat flux Q_H for two values of the friction velocity $u_* = 0.1$ m/s and 0.6 m/s, corresponding to light and strong wind. In the calculations, the height above the ground is $z = 20$ m, the height of the ABL $z_i = 1$ km, and the surface temperature $T_s = 18\,°C$. For strong wind, C_v^2/c^2 is always greater than C_T^2/T^2. For light wind, C_T^2/T^2 is greater than C_v^2/c^2 for $Q_H \gtrsim 230$ W/m^2.

Over dry ground the value of the structure-function parameter C_C^2 is given by [188]

$$C_C^2 \sim 3 \times 10^{-9}\,\mathrm{m}^{-2/3} \tag{6.54}$$

and is usually less than C_T^2/T_0^2. But over warm wet ground and swamps, and in marine and coastal atmospheres, C_C^2 may be of the order of or greater than C_T^2/T_0^2. For example, over the warm ocean

$$C_C^2 \sim 2.5 \times 10^{-7}\,\mathrm{m}^{-2/3}. \tag{6.55}$$

In the atmosphere, the coefficient ϱ_{CT} for the cross correlation of $\widetilde{C}$ and $\widetilde{T}$ can be positive or negative, and often $|\varrho_{CT}| \sim 1$.

In the ocean, for small-scale inhomogeneities ($l \lesssim 1$ m), the values of C_T^2/T_0^2 and C_v^2/c_0^2 are in the ranges [261, 46]

$$2 \times 10^{-10}\,\mathrm{m}^{-2/3} \lesssim C_T^2/T_0^2 \lesssim 2 \times 10^{-6}\,\mathrm{m}^{-2/3},$$
$$2 \times 10^{-11}\,\mathrm{m}^{-2/3} \lesssim C_v^2/c_0^2 \lesssim 2 \times 10^{-9}\,\mathrm{m}^{-2/3}. \tag{6.56}$$

The salinity fluctuations are noticeable near melting ice and river mouths.

6.2.5   Inhomogeneous turbulence

When we account for the height dependence of the variances and length scales as in equations (6.50)–(6.52), a random medium becomes statistically inhomogeneous. Here, we present a description of inhomogeneous turbulence.

The correlation function of temperature fluctuations for the inhomogeneous case is given by

$$B_T(\mathbf{R}_1, \mathbf{R}_2) = \langle \widetilde{T}(\mathbf{R}_1)\widetilde{T}(\mathbf{R}_2) \rangle. \tag{6.57}$$

Note that the B_T depends on the coordinates $\mathbf{R}_1$ and $\mathbf{R}_2$ of two points of observation rather than on their difference as for homogenous turbulence, equation (6.3). It is worthwhile introducing the center and difference coordinates

$$\mathbf{R}_c = (\mathbf{R}_1 + \mathbf{R}_2)/2, \quad \mathbf{R}_d = \mathbf{R}_1 - \mathbf{R}_2. \tag{6.58}$$

Then, the correlation function $B_T(\mathbf{R}_1, \mathbf{R}_2)$ can be written as

$$B_T(\mathbf{R}_1, \mathbf{R}_2) = B_T(\mathbf{R}_c + \mathbf{R}_d/2, \mathbf{R}_c - \mathbf{R}_d/2) \equiv \mathcal{B}_T(\mathbf{R}_c, \mathbf{R}_d). \tag{6.59}$$

Next we express the correlation function $\mathcal{B}_T(\mathbf{R}_c, \mathbf{R}_d)$ as a Fourier integral with respect to the difference coordinates $\mathbf{R}_d$:

$$\mathcal{B}_T(\mathbf{R}_c, \mathbf{R}_d) = \int \Phi_T(\mathbf{R}_c; \boldsymbol{\kappa})e^{i\boldsymbol{\kappa}\cdot\mathbf{R}_d}\,d^3\kappa. \tag{6.60}$$

The temperature spectrum $\Phi_T(\mathbf{R}_c; \boldsymbol{\kappa})$ for inhomogeneous turbulence depends on both the turbulence wave vector $\boldsymbol{\kappa}$ and the center coordinates $\mathbf{R}_c$.

The von Kármán, Kolmogorov, and Gaussian spectra of temperature fluctuations with the height-dependent parameters can be written in a form similar to the spectrum $\Phi_T(\mathbf{R}_c; \boldsymbol{\kappa})$. For example, the von Kármán spectrum, equation (6.26), can be recast as

$$\Phi_T^{vK}(z; \kappa) = \frac{\Gamma(11/6)}{\pi^{3/2}\Gamma(1/3)} \frac{\sigma_T^2(z)L_T^3(z)}{(1 + \kappa^2 L_T^2(z))^{11/6}}, \tag{6.61}$$

with the variance and length scale given by equations (6.50) and (6.51). It is evident from this equation that the height z above the ground plays the role of the center coordinates $\mathbf{R}_c$ of two points of observation.

The case of the wind velocity fluctuations can be considered similarly.

6.3   Fluctuations in the sound speed and density

The statistical characteristics of the temperature fluctuations $\widetilde{T}$ and the fluctuations $\widetilde{C}$ of the concentration of a component dissolved in the medium (the water vapor in the atmosphere or salinity in the ocean) were studied in detail in the previous section. Since the equations for the sound field (Chapter 2) contain the sound speed c and density ϱ, in this section the fluctuations in c and ϱ will be expressed in terms of $\widetilde{T}$ and $\widetilde{C}$.

6.3.1   Formulas for fluctuations

We will assume that the medium has only one dissolved component. The sound speed c and density ϱ are functions of the ambient pressure P, temperature T, and concentration C of the dissolved component:

$$c = c(P, T, C), \quad \varrho = \varrho(P, T, C). \tag{6.62}$$

By definition, $C = \varrho_C/\varrho_m$. Here, ϱ_C is the density of a component dissolved in the medium, which has density ϱ_m, so that the total density of the mixture is $\varrho = \varrho_C + \varrho_m$. For instance, in the atmosphere $C = \varrho_w/\varrho_a$, where ϱ_w and ϱ_a are the densities of water vapor and dry air, respectively. In the ocean $C = 10^{-3}s$, where s is the salinity in parts per thousand.

Let

$$c = c_0 + \widetilde{c}, \quad \varrho = \varrho_0 + \widetilde{\varrho}, \tag{6.63}$$

where c_0 and ϱ_0 are the mean values of the sound speed and density, and $\widetilde{c}$ and $\widetilde{\varrho}$ are their fluctuating components. Similarly,

$$T = T_0 + \widetilde{T}, \quad C = C_0 + \widetilde{C}, \tag{6.64}$$

where T_0 and C_0 are the mean values of the temperature and the concentration of the dissolved component, and $\widetilde{T}$ and $\widetilde{C}$ are the fluctuations. In the atmosphere and ocean, the fluctuations $\widetilde{c}$ and $\widetilde{\varrho}$ are caused primarily by temperature fluctuations $\widetilde{T}$ and by fluctuations $\widetilde{C}$ of the concentration of the dissolved component, while pressure fluctuations practically do not affect $\widetilde{c}$ and $\widetilde{\varrho}$. Substituting $T = T_0 + \widetilde{T}$ and $C = C_0 + \widetilde{C}$ into equation (6.62), expanding c and ϱ into the Taylor series in powers of $\widetilde{T}$ and $\widetilde{C}$, and retaining the first two terms, we obtain the following equations for $\widetilde{c}$ and $\widetilde{\varrho}$:

$$\widetilde{c} = c - c_0 = c_0 \left(\beta_c \frac{\widetilde{T}}{2T_0} + \eta_c \frac{\widetilde{C}}{2} \right), \tag{6.65}$$

$$\widetilde{\varrho} = \varrho - \varrho_0 = \varrho_0 \left(\beta_\varrho \frac{\widetilde{T}}{T_0} + \eta_\varrho \widetilde{C} \right). \tag{6.66}$$

(Since $\widetilde{C}$ is a dimensionless quantity, it is not normalized to C_0.) The coefficients β_c, η_c, β_ϱ, and η_ϱ appearing in equations (6.65) and (6.66) are given by

$$\beta_c = \frac{2T_0}{c_0} \frac{\partial c(P_0, T_0, C_0)}{\partial T_0}, \quad \eta_c = \frac{2}{c_0} \frac{\partial c(P_0, T_0, C_0)}{\partial C_0},$$
$$\beta_\varrho = \frac{T_0}{\varrho_0} \frac{\partial \varrho(P_0, T_0, C_0)}{\partial T_0}, \quad \eta_\varrho = \frac{1}{\varrho_0} \frac{\partial \varrho(P_0, T_0, C_0)}{\partial C_0}. \tag{6.67}$$

Here, P_0 is the mean value of the ambient pressure. The values of the coefficients β_c, β_ϱ, η_c, and η_ϱ in the atmosphere and ocean are calculated in the next two subsections.

6.3.2 Values of the coefficients in the atmosphere

Humid air can be considered as a mixture of two ideal gases: dry air and water vapor. According to the equation of state of an ideal gas, the densities of dry air and water are given by

$$\varrho_a = \frac{(P - e)\mu_a}{RT}, \quad \varrho_w = \frac{e\mu_w}{RT}, \tag{6.68}$$

where $R = 8.314$ J/(K mol) is the universal gas constant [27], e is the partial pressure of water vapor, and $\mu_a = 28.97$ g/mol and $\mu_w = 18.02$ g/mol are molecular weights of dry air and water vapor, respectively. Using equations (6.68), one obtains the equation of state for humid air:

$$\varrho = \varrho_a + \varrho_w = \frac{P}{R_a T} \left[1 + \frac{e}{P}(1/\alpha - 1) \right], \tag{6.69}$$

where $R_a = R/\mu_a$ is the gas constant for dry air, and $\alpha = \mu_a/\mu_w$. Making use of equations (6.68), one can also obtain the concentration C of water vapor in the atmosphere

$$C = \frac{\varrho_w}{\varrho_a} = \frac{e/P}{\alpha(1 - e/P)}. \tag{6.70}$$

The value of C depends on the climate regime and may vary from 0.002 for polar regions to 0.02 for tropical marine atmospheres [111]. In the numerical examples below we shall assume that $C = 0.008$. This value is typical for many climate regimes including the California coast, the Central Plains, and mid-latitude marine atmospheres. The concentration C of water vapor is often called the *mixing ratio* in the atmospheric sciences. The concentration C and specific humidity $q = \varrho_w/(\varrho_w + \varrho_a)$ are related by the simple formula $q = C/(1 + C)$. Since in the atmosphere $C \ll 1$, it follows from this formula that $C \cong q$. Another useful approximate relationship between C and e follows from equation (6.70): $C \cong e/\alpha P \cong 0.622 e/P$.

Solving equation (6.70) for e/P and substituting the resulting value into equation (6.69), one obtains the desired equation for the density of humid air,

$$\varrho = \frac{P}{R_a T} \frac{1 + C}{1 + \alpha C}. \tag{6.71}$$

The coefficients β_ϱ and η_ϱ can be calculated using equations (6.67) and (6.71):

$$\beta_\varrho = -1, \quad \eta_\varrho = -\frac{\alpha - 1}{(1 + C)(1 + \alpha C)}. \tag{6.72}$$

The value of η_ϱ depends slightly on C; $\eta_\varrho = -0.605$ if $C = 0.002$, and $\eta_\varrho = -0.577$ if $C = 0.02$. For $C = 0.008$ one obtains $\eta_\varrho = -0.596$. This value of η_ϱ is used in numerical calculations here.

The square of the sound speed is given by $c^2 = (\partial P/\partial \varrho)_{S,C}$, where S is the entropy. This equation can be expressed in the equivalent form:

$$c^2 = (\partial P/\partial \varrho)_{T,C} \frac{(\partial P/\partial \varrho)_{S,C}}{(\partial P/\partial \varrho)_{T,C}}. \tag{6.73}$$

Substituting the thermodynamic equality [27]

$$\frac{c_P}{c_V} = \frac{(\partial P/\partial \varrho)_{S,C}}{(\partial P/\partial \varrho)_{T,C}} \tag{6.74}$$

into equation (6.73) yields

$$c^2 = \gamma (\partial P/\partial \varrho)_{T,C}, \tag{6.75}$$

where $\gamma = c_P/c_V$ is the ratio of the specific heat at constant pressure c_P to the specific heat at constant volume c_V in humid air. In humid air these specific heats are given by [27]

$$c_P = c_{P_a} + C c_{P_w}, \quad c_V = c_{V_a} + C c_{V_w}. \tag{6.76}$$

Here, c_{P_a} and c_{P_w} are the specific heats at constant pressure for dry air and water vapor, respectively, and c_{V_a} and c_{V_w} are the corresponding specific heats at constant volume. The specific heats c_{P_a} and c_{V_a} are related by the well-known equations

$$c_{P_a} - c_{V_a} = R/\mu_a, \quad c_{P_a} = \gamma_a c_{V_a}, \tag{6.77}$$

where γ_a is the ratio of specific heats for dry air. Solving equations (6.77) for c_{P_a} and c_{V_a} yields

$$c_{P_a} = \frac{\gamma_a R}{\mu_a(\gamma_a - 1)}, \quad c_{V_a} = \frac{R}{\mu_a(\gamma_a - 1)}. \tag{6.78}$$

Analogously, one can find the specific heat at constant pressure, c_{P_w}, and the specific heat at constant volume, c_{V_w}, for water vapor:

$$c_{P_w} = \frac{\gamma_w R}{\mu_w(\gamma_w - 1)}, \quad c_{V_w} = \frac{R}{\mu_w(\gamma_w - 1)}, \tag{6.79}$$

where $\gamma_w = c_{P_w}/c_{V_w}$ is the ratio of specific heats for water vapor.

Using equations (6.76), (6.78), and (6.79), one obtains the ratio of specific heats for humid air:

$$\gamma = \frac{c_P}{c_V} = \frac{c_{P_a} + C c_{P_w}}{c_{V_a} + C c_{V_w}} = \gamma_a \frac{1 + \alpha\delta C}{1 + \alpha\nu C}, \tag{6.80}$$

where $\nu = (\gamma_a - 1)/(\gamma_w - 1)$ and $\delta = (1 - 1/\gamma_a)/(1 - 1/\gamma_w)$.

The value of $(\partial P/\partial\varrho)_{T,C}$ can be calculated from equation (6.71)

$$(\partial P/\partial\varrho)_{T,C} = P/\varrho = R_a T \frac{1 + \alpha C}{1 + C}. \tag{6.81}$$

Substituting equations (6.80) and (6.81) into equation (6.75), one obtains the sound speed in humid air:

$$c^2 = \gamma_a R_a T \frac{(1 + \alpha\delta C)(1 + \alpha C)}{(1 + \alpha\nu C)(1 + C)}. \tag{6.82}$$

This result and equations (6.67) allow one to calculate the coefficients β_c and η_c:

$$\beta_c = 1, \quad \eta_c = \frac{\alpha(1 + \delta - \nu) - 1 + 2\alpha(\alpha\delta - \nu)C + (\alpha\delta\nu + \delta - \nu - \delta\nu)\alpha^2 C^2}{(1 + C)(1 + \alpha C)(1 + \alpha\delta C)(1 + \alpha\nu C)}. \tag{6.83}$$

The value of η_c depends on γ_a, γ_w, and C. It follows from equations (6.78) and (6.79) that $\gamma_a = c_{P_a}/(c_{P_a} - R_a)$ and $\gamma_w = c_{P_w}/(c_{P_w} - R_w)$, where $R_w = R/\mu_w$ is the gas constant for water vapor. The values of c_{P_a} and c_{P_w} depend slightly on the temperature; $c_{P_a} = 1.004$ J/(K kg) and $c_{P_w} = 1.861$ J/(K kg) at

$T = 20°$ C [19]. Substituting these values of c_{P_a} and c_{P_w} into the equations for γ_a and γ_w yields: $\gamma_a = 1.400$ and $\gamma_w = 1.330$. Using these values of γ_a and γ_w and equation (6.83), one finds that $\eta_c = 0.508$ if $C = 0.002$, and $\eta_c = 0.487$ if $C = 0.02$. For $C = 0.008$ it follows from equation (6.83) that $\eta_c = 0.501$. This value of η_c is used in the numerical calculations below.

Neglecting terms of the order of C^2, equation (6.82) can be written in the form

$$c^2 = \gamma_a R_a T[1 + (\alpha(1 + \delta - \nu) - 1)C] = \gamma_a R_a T(1 + 0.511C), \qquad (6.84)$$

where the numerical coefficient before C is calculated for $\gamma_a = 1.400$ and $\gamma_w = 1.330$. Equations (6.82) and (6.84) allow one to study the effect of different meteorological conditions (different values of C, γ_a, and γ_w) on c^2 and η_c.

A detailed derivation of equation (6.82) has been given, since various equations for the sound speed c in humid air are used in the literature. For instance, in [188] the square of the sound speed is given as

$$c^2 = \gamma_a R_a T(1 + 0.450C), \qquad (6.85)$$

while in [402] it is determined by the formula

$$c^2 = \gamma_a R_a T(1 + 0.494C). \qquad (6.86)$$

The numerical coefficient of C in equations (6.85) and (6.86) differs from those in equation (6.84). This difference is due to the fact that in [188] the incorrect value $\gamma_w = 1.30$ is used, and in [402] the ratio of specific heats for humid air is given as $\gamma = \gamma_a(1 - e/(10P\gamma_a))$, while it follows from equation (6.80) that $\gamma = \gamma_a(1 - e/(11.79P\gamma_a))$.

6.3.3   Values of the coefficients in the ocean

The values of the coefficients β_c, β_ϱ, η_c, and η_ϱ in the ocean are presented in this subsection.

According to reference [72], in the ocean

$$\beta_\varrho \sim -7.5 \times 10^{-2}, \quad \eta_\varrho \sim -(0.15 \ \text{to} \ 0.2). \qquad (6.87)$$

The sound speed in the ocean is given by [49]

$$c_{m/s} = 1449.2 + 4.6T°C - 0.055\,(T°C)^2$$
$$+ 0.00029\,(T°C)^3 + (1.34 - 0.01\,T°C)(s - 35). \quad (6.88)$$

Here, $c_{m/s}$ denotes the sound speed in meters per second and $T\,°C$ denotes the temperature in degrees Celsius. Using equation (6.88) and taking into account that $C = 10^{-3} \times s$, the coefficients β_c and η_c can be calculated easily. Since the equations for β_c and η_c are rather involved, they are expressed here to an accuracy of 10%:

$$\beta_c = 1.73 - 3.46 \times 10^{-2} T^\circ C + 3.29 \times 10^{-4} (T^\circ C)^2, \tag{6.89}$$

$$\eta_c = 1.79 - 1.33 \times 10^{-2} T^\circ C. \tag{6.90}$$

6.4 Scattering cross section

The sound scattering cross section σ is one of the most important statistical characteristics of a sound field propagating in a random medium. The derivation of the equation for σ for the case of sound scattering by temperature and wind velocity fluctuations in the atmosphere is well known [374]. In this section, a more general equation for σ is derived, which is valid for the case of sound scattering in a random moving medium with an arbitrary equation of state. The equation derived will allow us to study sound scattering by fluctuations in temperature, velocity, and the concentration of the component dissolved in the medium.

6.4.1 Starting equation

We will consider sound scattering of a monochromatic sound wave with the frequency ω. The frequency spectrum $\widehat{p}(\mathbf{R})$ of the sound pressure satisfies the Helmholtz-type equation (2.88), which is repeated here for convenience:

$$[\nabla^2 + k^2(1 + \varepsilon) - (\nabla \ln(\varrho/\varrho_0)) \cdot \nabla - \frac{2i}{\omega} \frac{\partial \widetilde{v}_i}{\partial x_j} \frac{\partial^2}{\partial x_i \partial x_j}$$
$$+ \frac{2ik}{c_0} \widetilde{\mathbf{v}} \cdot \nabla] \widehat{p}(\mathbf{R}) = \varrho(i\omega - \widetilde{\mathbf{v}} \cdot \nabla) \widehat{Q}(\mathbf{R}). \tag{6.91}$$

Here, $\mathbf{R} = (x_1, x_2, x_3) = (x, y, z)$ are the Cartesian coordinates, $\nabla = (\partial/\partial x, \partial/\partial y, \partial/\partial z)$, $k = \omega/c_0$ is the sound wavenumber, and $\widetilde{\mathbf{v}} = (\widetilde{v}_1, \widetilde{v}_2, \widetilde{v}_3)$ represents the velocity fluctuations. The sound speed $c(\mathbf{R})$ and density $\varrho(\mathbf{R})$ are given by equation (6.63), where c_0 and ϱ_0 are their mean values and $\widetilde{c}$ and $\widetilde{\varrho}$ are the fluctuating components; $\varepsilon = c_0^2/c^2 - 1$ is the deviation from 1 of the square of the acoustic refractive index in a motionless medium. In equation (6.91), repeated subscripts are summed from 1 to 3. The function $\widehat{Q}(\mathbf{R})$ describes a mass source; it is assumed that there are no forces $\mathbf{F}$ acting on the medium. To simplify notations, in comparison with Chapter 2, we omit ω in the arguments of the functions $\widehat{p}$ and $\widehat{Q}$.

The range of applicability of equation (6.91) is considered in detail in section 2.4. This equation has a wider range of applicability than those used previously in the theory of sound propagation in a random moving medium.

For instance, unlike Monin's equation, which has been widely used in literature (e.g., references [52, 259, 374]), equation (6.91) enables one to describe
sound propagation and scattering in the turbulent ocean and liquid marine
sediments, and sound scattering by humidity fluctuations in the atmosphere.

To an accuracy of $\widetilde{c}/c_0$ and $\widetilde{\varrho}/\varrho_0$, the functions ε and $\ln(\varrho/\varrho_0)$ appearing
in equation (6.91) can be expressed in the form

$$\varepsilon = -\frac{2\widetilde{c}}{c_0}, \quad \ln\frac{\varrho}{\varrho_0} = \frac{\widetilde{\varrho}}{\varrho_0}. \tag{6.92}$$

The first two terms on the left-hand side of equation (6.91) are the same as
those in the corresponding equation for electromagnetic waves. Light velocity fluctuations $\widetilde{c}$ enter into the latter equation without any derivatives and,
hence, scatter the electromagnetic field like monopoles. In comparison with
electromagnetic wave scattering, a sound wave is scattered not only by sound
speed fluctuations $\widetilde{c}$, but also by density and medium velocity fluctuations,
$\widetilde{\varrho}$ and $\widetilde{\mathbf{v}}$, which are described by the third, fourth, and fifth terms on the
left-hand side of equation (6.91). Furthermore, the radiation patterns due to
sound scattering by $\widetilde{c}$, $\widetilde{\varrho}$, and $\widetilde{\mathbf{v}}$ are different.

Indeed, it follows from equations (6.91) and (6.92) that the sound speed
fluctuations $\widetilde{c}$ enter into equation (6.91) without any derivatives and, as in
electromagnetics, they scatter the sound field like monopoles. On the other
hand, the density fluctuations $\widetilde{\varrho}$ enter into equation (6.91) with derivatives
up to first order and, hence, they scatter the sound field like a combination
of monopoles and dipoles. Finally, the medium velocity fluctuations $\widetilde{\mathbf{v}}$ enter
into equation (6.91) with derivatives up to second order and scatter the sound
field like a combination of monopoles, dipoles, and quadrupoles.

6.4.2   Single-scattered sound field

Let the random inhomogeneities $\widetilde{c}$, $\widetilde{\varrho}$, and $\widetilde{\mathbf{v}}$ be located in the volume V (figure
6.6). The center O of the volume coincides with the center of the coordinate
system. A monochromatic, unit-amplitude point source located at $\mathbf{R}_0$ emits
a spherical sound wave, the dependence of which on the spatial coordinates is
given by

$$p^{(0)}(\mathbf{R} - \mathbf{R}_0) = \frac{\exp(ik|\mathbf{R} - \mathbf{R}_0|)}{|\mathbf{R} - \mathbf{R}_0|}. \tag{6.93}$$

(The unit-amplitude point source was introduced in section 4.2.1.) The re-
ceiver is located outside the scattering volume V. For $\mathbf{R} \neq \mathbf{R}_0$, the frequency
spectrum $\widehat{p}(\mathbf{R}, \mathbf{R}_0)$ of the sound pressure due to the source satisfies equation (6.91) with the right-hand side equal to zero. This equation can be ex-
pressed in the form:

$$(\Delta + k^2)\widehat{p} = -\left[k^2\varepsilon - \left(\nabla\ln\frac{\varrho}{\varrho_0}\right)\nabla - \frac{2i}{\omega}\frac{\partial\widetilde{v}_i}{\partial x_j}\frac{\partial^2}{\partial x_i\partial x_j} + \frac{2ik}{c_0}\widetilde{\mathbf{v}}\cdot\nabla\right]\widehat{p}. \tag{6.94}$$

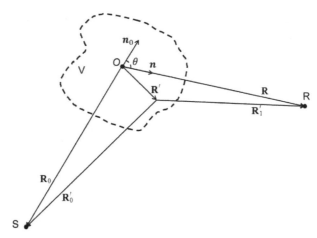

FIGURE 6.6
Geometry of sound scattering. V is the scattering volume, S source, and R receiver.

Here, the functions ε and $\ln(\varrho/\varrho_0)$ can be represented in terms of the fluctuations $\tilde{c}$ and $\tilde{\varrho}$, as in equation (6.92).

Equation (6.94) can be written in the equivalent integral form

$$\hat{p}(\mathbf{R}, \mathbf{R}_0) = p^{(0)}(\mathbf{R} - \mathbf{R}_0) + \frac{1}{4\pi} \int_V \frac{e^{ikR'_1}}{R'_1} \left[k^2 \varepsilon(\mathbf{R}') - \left(\nabla' \ln \frac{\varrho(\mathbf{R}')}{\varrho_0} \right) \cdot \nabla' \right.$$
$$\left. - \frac{2i}{\omega} \frac{\partial \tilde{v}_i(\mathbf{R}')}{\partial x'_j} \frac{\partial^2}{\partial x'_i \partial x'_j} + \frac{2ik^2}{\omega} \tilde{\mathbf{v}}(\mathbf{R}') \cdot \nabla' \right] \hat{p}(\mathbf{R}', \mathbf{R}_0) \, d^3 R'. \quad (6.95)$$

Here, integration is performed over the scattering volume V, the vector $\mathbf{R}' = (x'_1, x'_2, x'_3)$ specifies a point at which the sound wave is scattered, the vector $\mathbf{R}$ specifies the receiver position, $\mathbf{R}'_1 = \mathbf{R} - \mathbf{R}'$ (see figure 6.6), and $\nabla' = \partial/\partial \mathbf{R}'$.

The integral equation (6.95) can be solved by iterations replacing $\hat{p}(\mathbf{R}', \mathbf{R}_0)$ in the right-hand side with its value given by the left-hand side. The second term in the resulting series is the sound field $\hat{p}_s$ singly scattered by the random fields $\tilde{c}$, $\tilde{\varrho}$, and $\tilde{\mathbf{v}}$:

$$\hat{p}_s(\mathbf{R}) = \frac{1}{4\pi} \int_V \frac{e^{ikR'_1}}{R'_1} \left[k^2 \varepsilon(\mathbf{R}') - \left(\nabla' \ln \frac{\varrho(\mathbf{R}')}{\varrho_0} \right) \cdot \nabla' \right.$$
$$\left. - \frac{2i}{\omega} \frac{\partial \tilde{v}_i(\mathbf{R}')}{\partial x'_j} \frac{\partial^2}{\partial x'_i \partial x'_j} + \frac{2ik^2}{\omega} \tilde{\mathbf{v}}(\mathbf{R}') \cdot \nabla' \right] \frac{e^{ikR'_0}}{R'_0} \, d^3 R'. \quad (6.96)$$

Here, the argument $\mathbf{R}_0$ of $\hat{p}_s$ is omitted to simplify notations and $\mathbf{R}'_0 = \mathbf{R}_0 - \mathbf{R}'$ (figure 6.6). The single scattering approximation, which has been used to derive equation (6.96), is also called the *Born approximation*.

Let us assume that $R_0 \gg \max(\lambda, L_V)$, where λ is the wavelength and L_V is the characteristic scale of the scattering volume. Then, the following relationship is valid:

$$\frac{\partial}{\partial x'_i} \frac{\exp(ikR'_0)}{R'_0} \approx ikn_{0,i} \frac{\exp(ikR'_0)}{R_0}. \qquad (6.97)$$

Here, $n_{0,i}$ are the components of the unit vector $\mathbf{n}_0 = -\mathbf{R}_0/R_0$ in the direction of propagation of the emitted wave (figure 6.6). Equation (6.97) allows one to calculate the derivatives of the function $\exp(ikR'_0)/R'_0$ in equation (6.96). Replacing R'_1 in the denominator of the integrand in equation (6.96) by R (this is valid if $R \gg L_V$), one obtains

$$\widehat{p}_s(\mathbf{R}) = \frac{1}{4\pi R R_0} \int_V e^{ik(R'_0+R'_1)} \big[k^2\varepsilon - ik\mathbf{n}_0 \cdot \nabla' \ln(\varrho/\varrho_0)$$
$$+ 2ik(\mathbf{n}_0 \cdot \nabla')(\mathbf{n}_0 \cdot \widetilde{\mathbf{v}}/c_0) - 2k^2 \mathbf{n}_0 \cdot \widetilde{\mathbf{v}}/c_0 \big] \, d^3R'. \qquad (6.98)$$

The integral over $\mathbf{R}'$ of the third term in square brackets in this equation can be expressed in the following form:

$$\int_V e^{ik(R'_0+R'_1)}(\mathbf{n}_0 \cdot \nabla')(\mathbf{n}_0 \cdot \mathbf{v}) \, d^3R' = \int_V \nabla' \cdot \left[e^{ik(R'_0+R'_1)} \mathbf{n}_0(\mathbf{n}_0 \cdot \mathbf{v}) \right] d^3R'$$
$$- \int_V e^{ik(R'_0+R'_1)} ik(\mathbf{n}_0 \cdot \widetilde{\mathbf{v}})\mathbf{n}_0 \cdot (\nabla'(R'_0 + R'_1)) \, d^3R', \qquad (6.99)$$

where the factor $2ik/c_0$ is omitted for simplicity. If $R \gg L_V$ and $R_0 \gg L_V$, the vector $\nabla'(R'_0 + R'_1)$ in equation (6.99) approximately equals $(\mathbf{n}_0 - \mathbf{n})$, where $\mathbf{n} = \mathbf{R}/R$ is the unit vector in the direction of propagation of the scattered wave. Using the Gauss integral theorem, the first integral on the right-hand side of equation (6.99) can be represented as an integral over the surface of the scattering volume. If $L_V \gg \lambda$, this surface integral can be neglected in comparison with the integral from the last term in square brackets in equation (6.98). Analogously, one can eliminate the operator ∇' in the second term in the square brackets in equation (6.98). As a result of these manipulations, one obtains:

$$\widehat{p}_s(\mathbf{R}) = \frac{k^2}{4\pi R R_0} \int_V e^{ik(R'_0+R'_1)}$$
$$\times \big[\varepsilon - (1 - \mathbf{n}_0 \cdot \mathbf{n}) \ln(\varrho/\varrho_0) - 2(\mathbf{n}_0 \cdot \mathbf{n})(\mathbf{n}_0 \cdot \widetilde{\mathbf{v}}/c_0) \big] \, d^3R'. \qquad (6.100)$$

This formula has been derived from equation (6.96) assuming that

$$R, R_0 \gg L_V \gg \lambda. \qquad (6.101)$$

The angle θ between the vectors $\mathbf{n}_0$ and $\mathbf{n}$ (see figure 6.6) is termed the *scattering angle.* Substituting the values of ε and $\ln(\varrho/\varrho_0)$ from equation (6.92) into equation (6.100) and taking into account that $\mathbf{n}_0 \cdot \mathbf{n} = \cos\theta$ yields

$$\widehat{p}_s(\mathbf{R}) = -\frac{k^2}{4\pi RR_0} \int_V e^{ik(R'_0 + R'_1)}$$
$$\times \left[2\widetilde{c}/c_0 + (1 - \cos\theta)\widetilde{\varrho}/\varrho_0 + 2\cos\theta(\mathbf{n}_0 \cdot \widetilde{\mathbf{v}}/c_0)\right] d^3R'. \quad (6.102)$$

This formula expresses the single-scattered field in terms of the random fields $\widetilde{c}$, $\widetilde{\varrho}$, and $\widetilde{\mathbf{v}}$. The coefficients appearing in front of these random fields describe the radiation patterns of the scattering.

Using equations (6.65) and (6.66), the field $\widehat{p}_s$ can be expressed in terms of the random fields $\widetilde{T}$, $\widetilde{C}$, and $\widetilde{\mathbf{v}}$:

$$\widehat{p}_s(\mathbf{R}) = -\frac{k^2}{4\pi RR_0} \int_V e^{ik(R'_0 + R'_1)} \left[\frac{\beta(\theta)\widetilde{T}}{T_0} + \eta(\theta)\widetilde{C} + \frac{2\cos\theta\,\mathbf{n}_0 \cdot \widetilde{\mathbf{v}}}{c_0}\right] d^3R'.$$
$$(6.103)$$

Here,

$$\beta(\theta) = \beta_c + 2\beta_\varrho \sin^2\frac{\theta}{2}, \quad \eta(\theta) = \eta_c + 2\eta_\varrho \sin^2\frac{\theta}{2}, \quad (6.104)$$

are the functions of θ which determine the radiation patterns due to sound scattering by the random fields $\widetilde{T}$ and $\widetilde{C}$. The coefficients β_c, β_ϱ, η_c, and η_ϱ in equation (6.104) were determined in section 6.3.

6.4.3 Acoustic energy flux

The scattering cross section is related closely to the acoustic energy flux (the *intensity*) $\mathbf{I}(\mathbf{R}, t)$ of the scattered field:

$$\mathbf{I} = \operatorname{Re} p \operatorname{Re} \mathbf{w}. \quad (6.105)$$

Here, $p(\mathbf{R}, t) = \widehat{p}_s(\mathbf{R}) \exp(-i\omega t)$ and $\mathbf{w}(\mathbf{R}, t) = \mathbf{w}_s(\mathbf{R}) \exp(-i\omega t)$ are the sound pressure and acoustic particle velocity in the single scattering approximation, and t is time. From equation (2.8), it follows that $\partial\mathbf{w}/\partial t + \varrho^{-1}\nabla p = O(\mu^2)$. Here, μ is of order $\widetilde{c}$, $\widetilde{\varrho}$, and $\widetilde{\mathbf{v}}$; note that p and $\mathbf{w}$ are of order μ. From the last formula, we have $\mathbf{w} = -i\nabla p/(\omega\varrho) + O(\mu^2)$. Substituting this value of $\mathbf{w}$ into equation (6.105) and neglecting terms of order μ^3, one obtains

$$\mathbf{I} = -\frac{i}{4\omega\varrho}(p\nabla p + p^*\nabla p - p\nabla p^* - p^*\nabla p^*).$$

The value of $\mathbf{I}(\mathbf{R}, t)$ averaged over the period of sound oscillations $\mathcal{T} = 2\pi/\omega$ is denoted by $\mathbf{I}_s(\mathbf{R}, t)$:

$$\mathbf{I}_s(\mathbf{R}) = \frac{1}{\mathcal{T}} \int_0^{\mathcal{T}} \mathbf{I}(\mathbf{R}, t)\, dt.$$

Substituting the value of $\mathbf{I}$ into this equation and calculating the integral over time t yields

$$\mathbf{I}_s = -\frac{i}{4\omega\varrho}(\widehat{p}_s^* \nabla \widehat{p}_s - \widehat{p}_s \nabla \widehat{p}_s^*). \tag{6.106}$$

Applying the operator ∇ to both sides of equation (6.103) and taking into account the fact that $R \gg \max(\lambda, L_V)$, one obtains $\nabla\widehat{p}_s = ik\mathbf{n}\widehat{p}_s$. Substitution of $\nabla\widehat{p}_s$ into equation (6.106) yields the energy flux of the single-scattered sound field:

$$\mathbf{I}_s = \frac{\mathbf{n}k\widehat{p}_s\widehat{p}_s^*}{2\omega\varrho}. \tag{6.107}$$

We now substitute the value of $\widehat{p}_s$ from equation (6.103) into equation (6.107). Both sides of the resulting equation are averaged over an ensemble of realizations of the random fields $\widetilde{T}$, $\widetilde{C}$, and $\widetilde{\mathbf{v}}$, which are assumed to be statistically homogeneous. Using equations (6.2), (6.3)–(6.6), and the equality $\mathbf{n}_0 \cdot \widetilde{\mathbf{v}} = n_{0,i}\widetilde{v}_i$, one obtains the formula for the mean value of the acoustic energy flux:

$$\begin{aligned}
\langle \mathbf{I}_s \rangle = \frac{\mathbf{n}k^5}{8\omega\varrho\pi^2 R^2 R_0^2} \int_V \int_V e^{i\psi} \bigg[&\frac{\beta^2(\theta)B_T(\mathbf{R}_1 - \mathbf{R}_2)}{4T_0^2} \\
+ \frac{\beta(\theta)\eta(\theta)B_{CT}(\mathbf{R}_1 - \mathbf{R}_2)}{2T_0} + &\frac{\eta^2(\theta)B_C(\mathbf{R}_1 - \mathbf{R}_2)}{4} \\
+ \frac{\cos^2\theta n_{0,j}n_{0,i}B_{ij}(\mathbf{R}_1 - \mathbf{R}_2)}{c_0^2} \bigg] &\, d^3R_1\, d^3R_2. \tag{6.108}
\end{aligned}$$

Here, $\psi = k(|\mathbf{R} - \mathbf{R}_1| + |\mathbf{R}_0 - \mathbf{R}_1| - |\mathbf{R} - \mathbf{R}_2| - |\mathbf{R}_0 - \mathbf{R}_2|)$.

Let us simplify equation (6.108). First, we expand the function ψ into a Taylor series in the small parameters $R_{1,2}/R_0$ and $R_{1,2}/R$, retaining linear and quadratic terms. Denoting $\mathbf{R}' = \mathbf{R}_1 - \mathbf{R}_2$ and $\mathbf{R}'' = (\mathbf{R}_1 + \mathbf{R}_2)/2$, the function ψ can be expressed in the form

$$\begin{aligned}
\psi = \mathbf{q} \cdot \mathbf{R}' + k[\mathbf{R}' \cdot \mathbf{R}'' - (\mathbf{n} \cdot \mathbf{R}')(\mathbf{n} \cdot \mathbf{R}'')]/R \\
+ k[\mathbf{R}' \cdot \mathbf{R}'' - (\mathbf{n}_0 \cdot \mathbf{R}')(\mathbf{n}_0 \cdot \mathbf{R}'')]/R_0, \tag{6.109}
\end{aligned}$$

where $\mathbf{q} = k(\mathbf{n}_0 - \mathbf{n})$ is called the *scattering vector*. Since R'' is of the order of the scattering volume L_V and R' of order the outer scale of turbulence L, the second and third terms on the right-hand side of equation (6.109) are of order $kL_V L/R$ and $kL_V L/R_0$, respectively. Assuming that $kL_V L \ll R, R_0$, these terms can be neglected. In this case, $\psi = \mathbf{q} \cdot \mathbf{R}'$.

Second, the correlation functions $B_T(\mathbf{R})$, $B_{CT}(\mathbf{R})$, $B_C(\mathbf{R})$, and $B_{ij}(\mathbf{R})$ in equation (6.108) are replaced by their spectral densities $\Phi_T(\boldsymbol{\kappa})$, $\Phi_{CT}(\boldsymbol{\kappa})$, $\Phi_C(\boldsymbol{\kappa})$, and $\Phi_{ij}(\boldsymbol{\kappa})$ by making use of equations (6.3)–(6.6).

Third, the integration over $\mathbf{R}_1$ and $\mathbf{R}_2$ in equation (6.108) is replaced by integration over $\mathbf{R}'$ and $\mathbf{R}''$. Since the integrand does not depend on $\mathbf{R}''$, the integral over $\mathbf{R}''$ simply equals the scattering volume V. On the other hand, the characteristic scale of the integrand with respect to $\mathbf{R}'$ is of order L. Assuming that $L \ll L_V$, the limits of integration over $\mathbf{R}'$ can be extended to infinity. In this case, the integral over $\mathbf{R}'$ is equal to $8\pi^3\delta(\mathbf{q} + \boldsymbol{\kappa})$. The remaining integral over $\boldsymbol{\kappa}$ is calculated exactly. As a result, one obtains the final formula for the mean energy flux of the single-scattered sound field

$$\langle \mathbf{I}_s \rangle = \frac{2\pi k^4 V I_0 \mathbf{n}}{R^2} \left[\frac{\beta^2(\theta)\Phi_T(\mathbf{q})}{4T_0^2} + \frac{\beta(\theta)\eta(\theta)\Phi_{CT}(\mathbf{q})}{2T_0} \right.$$
$$\left. + \frac{\eta^2(\theta)\Phi_C(\mathbf{q})}{4} + \frac{\cos^2\theta n_{0,i}n_{o,j}\Phi_{ij}(\mathbf{q})}{c_0^2} \right]. \quad (6.110)$$

Here, $I_0 = 1/(2\varrho c_0 R_0^2)$ is the intensity of the spherical wave produced by the unit-amplitude source, incident on the scattering volume. Equation (6.110) has been derived assuming that

$$R, R_0 \gg L_V \gg L, \lambda; \quad R, R_0 \gg kL_V L; \quad \sigma_0 L_V \ll 1, \quad (6.111)$$

where

$$\sigma_0 = \oint \sigma(\mathbf{n})\, d\Omega(\mathbf{n}) = \int_0^{2\pi} d\varphi \int_0^{\pi} \sigma(\theta, \varphi) \sin\theta\, d\theta \quad (6.112)$$

is the total scattering cross section. Here, $d\Omega(\mathbf{n})$ is the solid angle in the direction of the unit vector $\mathbf{n}$, σ is determined by equation (6.114) below, and θ and φ are angles in the spherical coordinate system. The last inequality in equation (6.111) represents the range of applicability of the single scattering approximation.

If the inequalities (6.111) are valid, the amplitude of the unperturbed spherical wave $p^{(0)}(\mathbf{R}')$ is nearly constant within the scattering volume and the phase of this wave is approximately equal to $k\mathbf{n}_0 \cdot \mathbf{R}'$. These features of the sound field $p^{(0)}$ were used in the derivation of equation (6.110). Therefore, equation (6.110) remains valid if, within the scattering volume, the amplitude of a sound field $p^{(0)}(\mathbf{R}')$ induced by an arbitrary source is nearly constant, and if the phase is approximately equal to $k\mathbf{n}_0 \cdot \mathbf{R}'$. In this case, I_0 in equation (6.110) is the intensity of the sound field incident on the scattering volume.

6.4.4   Sound scattering cross section

By definition, the scattering cross section per unit volume into a unit solid angle in the direction of the vector $\mathbf{n}$ is given by:

$$\sigma(\mathbf{n} - \mathbf{n}_0) = \frac{\langle I_s \rangle R^2}{I_0 V}. \quad (6.113)$$

Substituting the value of $\langle I_s \rangle$ from equation (6.110) into this formula, one obtains the scattering cross section of a sound wave in a random moving medium:

$$\sigma(\mathbf{n} - \mathbf{n}_0) = 2\pi k^4 \left[\frac{\beta^2(\theta)\Phi_T(\mathbf{q})}{4T_0^2} + \frac{\beta(\theta)\eta(\theta)\Phi_{CT}(\mathbf{q})}{2T_0} \right.$$
$$\left. + \frac{\eta^2(\theta)\Phi_C(\mathbf{q})}{4} + \frac{\cos^2\theta n_{0,i} n_{0,j} \Phi_{ij}(\mathbf{q})}{c_0^2} \right]. \quad (6.114)$$

This equation is also derived in [214], where it is assumed that $\Phi_C = \Phi_{CT} = 0$.

For locally homogeneous and isotropic turbulence, the three-dimensional spectral densities Φ_T, Φ_{CT}, and Φ_C depend only on the magnitude of the vector $\mathbf{q}$, and Φ_{ij} is given by equation (6.17). In this case, taking into account the fact that $n_{0,i} n_{0,j} \Phi_{ij}(\mathbf{q}) = \cot^2(\theta/2) E(q)/(16\pi k^2)$, equation (6.114) becomes

$$\sigma(\theta) = 2\pi k^4 \left[\frac{\beta^2(\theta)\Phi_T(q)}{4T_0^2} + \frac{\beta(\theta)\eta(\theta)\Phi_{CT}(q)}{2T_0} \right.$$
$$\left. + \frac{\eta^2(\theta)\Phi_C(q)}{4} + \frac{\cos^2\theta \cot^2(\theta/2) E(q)}{16\pi k^2 c_0^2} \right]. \quad (6.115)$$

Here, $q = 2k\sin(\theta/2)$ is the magnitude of the scattering vector.

An interesting characteristic of single scattering is its selective dependence on the size of the random inhomogeneities. It follows from equations (6.114) and (6.115) that for a given value of the scattering angle θ, a sound wave is scattered only by random inhomogeneities $\tilde{T}$, $\tilde{C}$, and $\tilde{\mathbf{v}}$ with the spatial scale $\Lambda = 2\pi/q = \lambda/[2\sin(\theta/2)]$. Furthermore, the intensity of the scattered field is proportional to $\Phi_T(2\pi/\Lambda)$, $\Phi_{CT}(2\pi/\Lambda)$, $\Phi_C(2\pi/\Lambda)$, and $E(2\pi/\Lambda)$.

Substitution of Φ_T, Φ_{CT}, Φ_C, and E from equation (6.34) into equation (6.115) yields the scattering cross section, for the case of sound scattering by locally homogeneous and isotropic turbulence as described by the Kolmogorov spectrum:

$$\sigma(\theta) = \frac{5\Gamma(2/3)}{2^{20/3}3^{3/2}\pi} \frac{k^{1/3}}{[\sin(\theta/2)]^{11/3}} \left[\frac{\beta^2(\theta)C_T^2}{T_0^2} + \frac{2\eta(\theta)\beta(\theta)\varrho_{CT}C_C C_T}{T_0} \right.$$
$$\left. + \eta^2(\theta)C_C^2 + \frac{22}{3}\frac{\cos^2\theta \cos^2(\theta/2)C_v^2}{c_0^2} \right]. \quad (6.116)$$

The numerical value of the factor $5\Gamma(2/3)/\left(2^{20/3}3^{3/2}\pi\right)$ in this equation is 4.08×10^{-3}.

For sound scattering by locally homogeneous and isotropic turbulence modeled with von Kármán and Gaussian spectra, the scattering cross section $\sigma(\theta)$ can be readily obtained by substituting the values of Φ_T, Φ_{CT}, Φ_C, and E for these spectra (equations (6.26)–(6.29), (6.44), and (6.45)) into equation (6.115).

6.4.5 Sound scattering in the atmosphere

For sound scattering in the atmosphere, the coefficients β_c, β_ϱ, η_c, and η_ϱ in equation (6.104) are obtained in section 6.3: $\beta_c = -\beta_\varrho = 1$, $\eta_c = 0.501$, and $\eta_\varrho = -0.596$. Substituting these values into equation (6.104) yields $\beta(\theta) = \cos\theta$ and $\eta(\theta) = -0.095 + 0.596\cos\theta$. In this case, equation (6.116) for the sound scattering cross section becomes

$$
\sigma(\theta) = 4.08 \times 10^{-3} \frac{k^{1/3}}{[\sin(\theta/2)]^{11/3}} \left[\frac{\cos^2\theta C_T^2}{T_0^2} \right.
$$
$$
+ \frac{2\cos\theta(-0.095 + 0.596\cos\theta)\varrho_{CT}C_C C_T}{T_0}
$$
$$
\left. + (-0.095 + 0.596\cos\theta)^2 C_C^2 + \frac{22}{3}\frac{\cos^2\theta\cos^2(\theta/2)C_v^2}{c_0^2} \right]. \quad (6.117)
$$

For the case $C_C^2 = 0$, this equation was derived by Monin [259]. Kraichnan [210] had previously derived the formula for $\sigma(\theta)$ assuming $C_T^2 = C_C^2 = 0$. Equation (6.117), with $C_C^2 = 0$, was verified experimentally [187].

The sound scattering cross section $\sigma(\theta)$ plays an important role in applications. It qualitatively describes sound scattering into a refractive shadow zone in outdoor sound propagation. It is also used as a theoretical basis for acoustic sounding of the atmosphere with sodars (SOund Detection And Ranging). In bistatic acoustic sounding, one antenna transmits an acoustic pulse and a second antenna, located at some distance from the transmitting antenna, receives the pulse after it has been scattered by atmospheric turbulence. The intensity of the scattered pulse is proportional to $\sigma(\theta)$. In monostatic sounding, the two antennas coincide, or the same antenna transmits and receives. The height of the scattering volume in acoustic sounding can vary from tens of meters above the ground to the top of the atmospheric boundary layer. Sodars enable remote sensing of inversion layers and vertical profiling of wind velocity, turbulence quantities, and stability classes. The first sodar was designed in 1968 by McAllister [246]. Sodars have since become used worldwide for observations of the atmospheric boundary layer [44, 52, 359].

It follows from equation (6.117) and the typical values of the structure-function parameters C_T^2, C_C^2, C_V^2, and the coefficient ϱ_{CT} (section 6.2.4) that over dry ground, sound scattering is caused mainly by wind and temperature fluctuations. In marine and coastal atmospheres, and over warm, wet ground and swamps, humidity fluctuations and the cross correlation of temperature and humidity fluctuations may contribute significantly to scattering.

The scattering cross section (6.117) can be expressed in the form

$$
\sigma(\theta) = \sigma_T(\theta) + \sigma_{CT}(\theta) + \sigma_C(\theta) + \sigma_v(\theta).
$$

Here, the functions $\sigma_T(\theta)$, $\sigma_{CT}(\theta)$, $\sigma_C(\theta)$, and $\sigma_v(\theta)$ are the contributions to $\sigma(\theta)$ due to sound scattering by temperature, temperature-humidity, humidity, and wind velocity fluctuations, respectively. The functions $\sigma_T(\theta)$, $\sigma_{CT}(\theta)$,

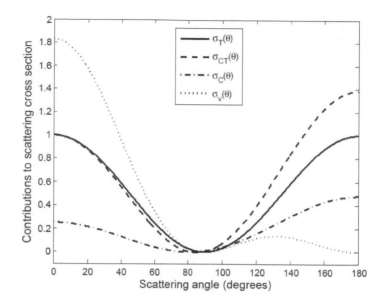

FIGURE 6.7
Normalized contributions to the sound scattering cross section due to fluctuations in temperature ($\sigma_T(\theta)$), temperature-humidity ($\sigma_{CT}(\theta)$), humidity
($\sigma_C(\theta)$), and wind velocity ($\sigma_v(\theta)$) versus the scattering angle θ.

$\sigma_C(\theta)$, and $\sigma_v(\theta)$, normalized by $4.08 \times 10^{-3} k^{1/3} / (\sin(\theta/2))^{11/3}$, are plotted
in figures 6.7 and 6.8 for $C_T^2/T_0^2 = C_C^2 = \varrho_{CT} = 1$, and $C_v^2/c_0^2 = 1/4$. In figure
6.7, the range of the scattering angle θ is $0°$ to $180°$. In figure 6.8, the range
of θ is $75°$ to $100°$, to provide a more detailed representation of the behavior
of the functions near $\theta = 90°$. Note that the numerical value of $1/4$ for C_v^2/c_0^2
is chosen to insure clear distinction between all curves in figures 6.7 and 6.8.
These figures demonstrate that the dependencies of $\sigma_T(\theta)$, $\sigma_{CT}(\theta)$, $\sigma_C(\theta)$, and
$\sigma_v(\theta)$ on θ differ significantly.

Equation (6.117) and figure 6.8 show that, at the scattering angle $\theta = 90°$,
the sound wave is scattered only by humidity fluctuations:

$$\sigma(90°) = \sigma_C(90°) = 1.45 \times 10^{-2}(\eta_c + \eta_\varrho)^2 k^{1/3} C_C^2 = 1.31 \times 10^{-4} k^{1/3} C_C^2.$$

$$(6.118)$$

Measurements of the scattering cross section $\sigma(90°)$ by bistatic acoustic sounding make it possible to retrieve the structure-function parameter C_C^2 of humidity fluctuations. Knowledge of C_C^2 is important in applications; humidity fluctuations lead to fluctuations in the refractive index for both electromagnetic
and sound waves, and play a significant role in boundary layer meteorology.

Monostatic sodars enable measurements of the sound backscattering cross

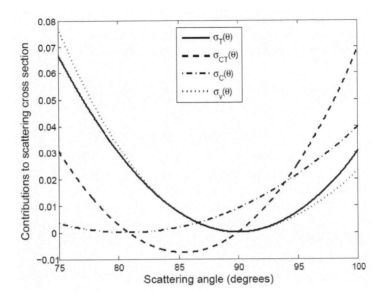

FIGURE 6.8
Same as figure 6.7, but for the scattering angle θ in the range $75° \le \theta \le 100°$.

section $\sigma(180°)$. For $\theta = 180°$, equation (6.117) becomes

$$\sigma(180°) = 4.08 \times 10^{-3}k^{1/3} \left(\frac{C_T^2}{T_0^2} + \frac{1.382\varrho_{CT}C_C C_T}{T_0} + 0.477C_C^2 \right). \quad (6.119)$$

It can be seen from this equation and figure 6.7 that wind velocity fluctuations do not contribute to the backscattering cross section. Over dry ground, $C_C^2 \ll C_T^2/T_0^2$, and, hence, $\sigma(180°) \approx 4.08 \times 10^{-3}k^{1/3}C_T^2/T_0^2$. Therefore, by measuring $\sigma(180°)$, it is possible to retrieve the structure-function parameter of temperature fluctuations, C_T^2, which is of primary importance in boundary layer meteorology. In some climate zones, C_C^2 can be comparable to or even exceed C_T^2/T_0^2 (section 6.2.4). In this case, according to equation (6.119), humidity fluctuations affect the backscattering cross section significantly.

There have been incorrect formulations of sound scattering in a humid atmosphere. For example, the following equation for σ was derived in reference [402]

$$\sigma(\theta) = 0.03 \frac{k^{1/3}}{(\sin(\theta/2))^{11/3}} [0.136 \cos^2 \theta \, 4C_N^2 + \cos^2 \theta \cos^2(\theta/2)C_v^2/c_0^2]. \quad (6.120)$$

Here,

$$C_N^2 = C_T^2/4T_0^2 + \eta_c \varrho_{CT}C_C C_T/2T_0 + \eta_c^2 C_C^2/4 \quad (6.121)$$

is the structure-function parameter of the fluctuations in the refractive index of the motionless atmosphere, $N = c_0/c \approx 1 - \tilde{c}/c_0 = 1 - \tilde{T}/2T_0 - \eta_c \tilde{C}/2$ (see equation (6.65)). The numerical value of the coefficient η_c in equation (6.121) is 0.494, in accordance with equation (6.86). In this case, equation (6.120) can be written as

$$\sigma(\theta) = 0.03 \frac{k^{1/3}}{[\sin(\theta/2)]^{11/3}} \left[0.136 \cos^2\theta \left(C_T^2/T_0^2 + 0.988 \varrho_{CT} C_C C_T/T_0 + \right.\right.$$

$$\left.\left. +0.244 C_C^2 \right) + \cos^2\theta \cos^2(\theta/2) C_v^2/c_0^2 \right]. \quad (6.122)$$

This equation was adopted (e.g., references [52, 57, 58]) for studying the effects of humidity fluctuations on sound scattering in a turbulent atmosphere.

However, equation (6.122) differs significantly from the correct result, equation (6.117). Indeed, not only do the coefficients of $\varrho_{CT} C_C C_T$ and C_C^2 in equations (6.117) and (6.122) differ quantitatively, but they also have different dependencies on the scattering angle θ. In particular, it follows from equation (6.122) that

$$\sigma(90°) = 0,$$

$$\sigma(180°) = 4.08 \times 10^{-3} k^{1/3} (C_T^2/T_0^2 + 0.988 \varrho_{CT} C_C C_T/T_0 + 0.244 C_C^2),$$

while the correct formulas for $\sigma(90°)$ and $\sigma(180°)$ are given by equations (6.118) and (6.119).

The term $C_T^2/4T_0^2$ on the right-hand side of equation (6.121) enters into the well-known equation for σ in the atmosphere, where $C_C^2 = 0$ [259, 374]. In reference [402], when calculating the sound scattering due to humidity fluctuations in a turbulent atmosphere, the term $C_T^2/4T_0^2$ was replaced by C_N^2 in accordance with equation (6.121). However, this substitution is incorrect, because it does not take into account sound scattering by density fluctuations $\tilde{\varrho}$.

6.4.6　Sound scattering in the ocean

For a turbulent ocean, with random inhomogeneities modeled by the Kolmogorov spectrum, the sound scattering cross section is given by equation (6.116). It follows from this equation and the values of the structure-function parameters presented in section 6.2.4 that sound scattering is mainly caused by temperature fluctuations $\tilde{T}$. Sound scattering by velocity fluctuations can be significant, however, in the upper mixed layer of the ocean or in turbulent currents. In equation (6.116), the function $\beta(\theta)$ in front of C_T^2/T_0^2 is a linear combination of the coefficients β_c and β_ϱ which describe the dependencies of the sound speed and density fluctuations on the temperature fluctuations (see equations (6.65) and (6.66)). It can be shown from equations (6.104) and (6.116) that the contribution of the density fluctuations $\tilde{\varrho}$ to the scattering cross section σ increases with increasing scattering angle θ. Taking into account that $\beta_c/\beta_\varrho \sim 20$ (section 6.3), for $\theta = 180°$ this contribution can be about 20% of that due to sound speed fluctuations $\tilde{c}$.

7

Line-of-sight sound propagation in a random moving medium

In this chapter, we consider line-of-sight sound propagation through a random moving medium. This is relevant, for example, to outdoor sound propagation from elevated sources such as helicopters and airplanes to microphones on the ground. It also occurs in propagation from acoustic sources on the ground (e.g., gun and rifle shots, rocket launches, and explosions) to elevated acoustic sensor arrays, such as those suspended below aerostats or installed on helicopters and unattended aerial vehicles (UAVs). If there is only a single, dominant propagation path, the effect of atmospheric refraction on the signal statistics is often relatively small, and such effects can be determined approximately assuming straight line sound propagation. Acoustic tomography of atmospheric turbulence (section 3.7) is based on measurements of the travel times of signals propagating along straight lines between sources and microphones located above the ground.

In ocean acoustics, line-of-sight sound propagation is pertinent to experiments conducted in the upper mixed layer of the ocean or in turbulent tidal channels. In such experiments, the propagation range is usually less than a few kilometers. Straight-line propagation also occurs in laboratory studies of ultrasound propagation in wind tunnels, turbulent jets, and above heated grids. Measurements of the statistical moments of a sound field can be used for the subsequent reconstruction of statistical characteristics of a medium and the mean velocity.

In a random medium, line-of-sight propagation occurs only on average. For one realization of the medium, a sound wave exhibits random refraction and diffraction, such that its propagation path can deviate from a straight line. Theories of line-of-sight propagation in a motionless random medium are well developed in electromagnetics [178, 339, 374], ocean acoustics [72, 117], and seismology [349]. What distinguishes this chapter from most previous texts is that we consider sound scattering by velocity fluctuations and fluctuations in the concentration of a component dissolved in the medium (water vapor in the atmosphere or salinity in the ocean). In the following, we present theoretical formulations for the variances and correlation functions of phase and log-amplitude fluctuations, the mean sound field, and the mutual coherence function. Analytical expressions for these statistical moments are obtained for arbitrary spectra of random inhomogeneities and, then, analyzed for the von Kármán, Kolmogorov, and Gaussian models of turbulence. The presentation is based on the first edition of this book [290] and several papers [300, 301, 302, 422] published thereafter.

7.1　Parabolic equation and Markov approximation in a random moving medium

Analysis of line-of-sight propagation in a random moving medium will be based on the parabolic equation for sound waves, which was derived in Chapter 2. In this section, we formulate the geometry of the problem, specify the parabolic equation for the case of a random medium, and present the correlation function of medium inhomogeneities in the Markov approximation.

7.1.1　Geometry and starting equation

The geometry of sound propagation considered in this chapter is shown in figure 7.1. A monochromatic source with the frequency ω and an arbitrary initial waveform is located in the plane $x = 0$ of the Cartesian coordinate system (x, y, z). The sound-pressure field $p(\mathbf{R}, t)$ due to the source is monitored at range x by one, two, or more receivers (in figure 7.1, two receivers are depicted). Here, $\mathbf{R} = (x, y, z) = (x, \mathbf{r})$ denotes the full spatial coordinates, the vector $\mathbf{r} = (y, z)$ denotes the transverse coordinates. To simplify notations, we omit the dependence of the sound pressure on time t, which is given by the factor $\exp(-i\omega t)$. The receivers are located near the x-axis so that sound propagation close to this axis is considered. In the medium, the sound speed $c(\mathbf{R}) = c_0 + \widetilde{c}(\mathbf{R})$. Here, c_0 is the mean value of the sound speed which does not depend on $\mathbf{R}$, and $\widetilde{c}(\mathbf{R})$ is the sound speed fluctuation which is much smaller than c_0. The mean medium velocity is assumed to be zero; the velocity fluctuation is denoted as $\widetilde{\mathbf{v}}(\mathbf{R})$. The mean values of the fluctuations are zero: $\langle \widetilde{c}(\mathbf{R}) \rangle = 0$ and $\langle \widetilde{\mathbf{v}}(\mathbf{R}) \rangle = 0$. Here, the brackets $\langle \ \rangle$ denote averaging over an ensemble of realizations of fluctuations. In this chapter, we will calculate and analyze the statistical characteristics of the sound field p at the observation points.

Similarly to equation (2.104), the sound pressure $p(\mathbf{R})$ is expressed in the form

$$p(\mathbf{R}) = A(\mathbf{R}) \exp(ikx). \tag{7.1}$$

Here, $A(\mathbf{R})$ is the complex amplitude, $k = 2\pi/\lambda$ the sound wavenumber, and λ is the wavelength. In the parabolic equation approximation, $A(\mathbf{R})$ satisfies equation (2.110) which is recast here as

$$2ik\frac{\partial A}{\partial x} + \nabla_\perp^2 A + k^2 \varepsilon_{\text{mov}} A + \frac{2ik}{c_0}\widetilde{\mathbf{v}}_\perp \cdot \nabla_\perp A = 0. \tag{7.2}$$

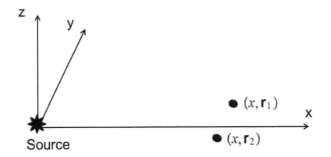

FIGURE 7.1
Schematic of the geometry of the problem. An acoustic source is located in the plane $x = 0$ of the Cartesian coordinate system (x, y, z). Receivers are located near the x-axis at the same range from the source. Sound propagates through a random moving medium.

Here, $\tilde{v}_x(\mathbf{R})$ and $\tilde{\mathbf{v}}_\perp(\mathbf{R})$ are the components of the medium velocity fluctuation $\tilde{\mathbf{v}} = (\tilde{v}_x, \tilde{\mathbf{v}}_\perp)$ in the direction of the x-axis and in the plane perpendicular to it, and the operator $\nabla_\perp = (\partial/\partial y, \partial/\partial z) \equiv \partial/\partial \mathbf{r}$. In equation (7.2), the random field $\varepsilon_{\text{mov}}(\mathbf{R})$ is given by $\varepsilon_{\text{mov}} = \varepsilon - 2\tilde{v}_x/c_0$, where $\varepsilon(\mathbf{R}) = c_0^2/c^2(\mathbf{R}) - 1$. To an accuracy of $\tilde{c}/c_0$, the function $\varepsilon = -2\tilde{c}/c_0$. In this case, ε_{mov} can be expressed in the form:

$$\varepsilon_{\text{mov}}(\mathbf{R}) = -2\frac{\tilde{c}(\mathbf{R}) + \tilde{v}_x(\mathbf{R})}{c_0}. \tag{7.3}$$

Consider the physical meaning of ε_{mov}. The acoustic refractive index in a moving medium is given by $N_{ph} = c_0/(c + \mathbf{n} \cdot \tilde{\mathbf{v}})$; see equation (3.59). Here, $\mathbf{n}$ is a unit vector normal to the wavefront of the sound wave; in the parabolic equation approximation $\mathbf{n} = (1, 0, 0)$ to order ε_{mov}. Neglecting terms of order $\tilde{v}^2/c_0^2$ and $\tilde{c}^2/c_0^2$ yields $N_{ph} = 1 - (\tilde{c} + \tilde{v}_x)/c_0$. The mean value of N_{ph} is 1, and the fluctuation of the acoustic refractive index is given by $\tilde{N}_{ph} = -(\tilde{c} + \tilde{v}_x)/c_0$. Thus, the random field $\varepsilon_{\text{mov}} = 2\tilde{N}_{ph}$ is twice the fluctuation $\tilde{N}_{ph}$ of the acoustic refractive index in a moving medium.

It follows from section 6.3 that for sound propagation in the atmosphere without humidity fluctuations, $\tilde{c}/c_0 = \tilde{T}/(2T_0)$. Here, T_0 and $\tilde{T}$ are the mean temperature and its fluctuation. In this case, equation (7.3) becomes

$$\varepsilon_{\text{mov}}(\mathbf{R}) = -\frac{\tilde{T}(\mathbf{R})}{T_0} - 2\frac{\tilde{v}_x(\mathbf{R})}{c_0}. \tag{7.4}$$

This form of $\varepsilon_{\text{mov}}(\mathbf{R})$ was used in the older theories of sound propagation in a turbulent atmosphere (e.g., references [374, 52, 87, 182, 247]) to adapt the results known for sound scattering by a scalar random field (such as temperature fluctuations $\tilde{T}$) for sound scattering by $\tilde{v}_x$. The argument was that the relative contribution from velocity fluctuations to the statistical moments of a sound field for line-of-sight propagation is the same as that from temperature fluctuations if the structure-function parameter (or the variance) of temperature fluctuations is replaced by that of velocity fluctuations multiplied by a factor 4 (the square of the coefficient 2 appearing in equation (7.4).) This argument is, however, incorrect, since $\widetilde{T}$ is a scalar random field, while $\widetilde{v}_x$ is a component of a vector random field. The importance of this distinction is evident even for isotropic turbulence, for which the correlation function of temperature fluctuations $B_T(R)$ depends only on the magnitude of the vector $\mathbf{R}$, while that of the x-component of velocity fluctuations, $B_{11}(\mathbf{R})$, also depends on the direction of this vector (see section 6.1.3). Due to this difference, the relative contributions from temperature and velocity fluctuations to the statistical moments of a sound field, generally, have different dependencies on parameters of the problem such as the sound wavenumber k, propagation range x, the distance r_d between two points of observation, and the outer scales of turbulence L_T and L_v. This conclusion is illustrated below on formulas for the statistical moments of a sound field.

The ranges of applicability of equation (7.2) are considered in section 2.5. For this equation to be valid, the sound wavelength λ should, strictly speaking, be less than the scale l of random inhomogeneities. In a turbulent medium, there are inhomogeneities with different scales l, which are greater than the inner scale of turbulence η and smaller than the outer scale L (section 6.1.1): $\eta \lesssim l \lesssim L$. Since η is of order 1 mm in the atmosphere and 1 cm in the ocean, the condition $\lambda < \eta$ seems to limit dramatically the range of applicability of equation (7.2). In reality, however, this equation can be used to describe sound propagation if a much lesser restriction on the wavelength λ is imposed:

$$\eta < \lambda \ll L. \tag{7.5}$$

Indeed, the main part of the turbulent energy is in the large-scale inhomogeneities. Therefore, the sound-field fluctuations in the direction of wave propagation are caused primarily by inhomogeneities with scales $l > \lambda$, while inhomogeneities with scales $l < \lambda$ result in relatively weak scattering in all directions which is ignored in the parabolic equation approximation. In Chapter 8, the inequality (7.5) is proved more rigorously using the diagram technique.

In a random medium, the parabolic equation (7.2) can be simplified. The fourth term, proportional to $\widetilde{\mathbf{v}}_\perp$, is significant if $\theta_m > 1/(kl)$ (see section 2.5). Here, θ_m is the maximum angle between the direction of wave propagation and the x-axis. From equation (2.109), the fourth and second terms in equation (7.2) are of the order of $k^2\theta_m A\widetilde{v}_\perp/c_0$ and $k^2\theta_m^2 A$, respectively. The ratio of the mean squares of these terms is given by

$$b = \frac{\langle k^2\theta_m A\widetilde{v}_\perp/c_0\rangle^2}{\left(k^2\theta_m^2 A\right)^2} = \frac{\langle\widetilde{v}_\perp^2\rangle}{c_0^2\theta_m^2}.$$

Since $\theta_m > 1/(kl)$, we have $b < k^2l^2\langle\widetilde{v}_\perp^2\rangle/c_0^2$. The right-hand side of the latter inequality is much less than 1 in the Markov approximation which will be used in theoretical formulations of the statistical moments of a sound field. (The ranges of applicability of this approximation are given by equation (7.23).) Thus, $b \ll 1$ so that the fourth term in equation (7.2) is smaller than the second and can be omitted. As a result, the parabolic equation (7.2) for the complex amplitude simplifies:

$$2ik\frac{\partial A}{\partial x} + \nabla_\perp^2 A + k^2 \varepsilon_{\text{mov}} A = 0. \tag{7.6}$$

The initial condition for this equation is formulated at $x = 0$:

$$A(x = 0, \mathbf{r}) = p(x = 0, \mathbf{r}) \equiv p_0(\mathbf{r}), \tag{7.7}$$

where $p_0(\mathbf{r})$ is the initial sound pressure due to the source. Using the relationship (7.1) between A and p, one obtains the parabolic equation for the sound field:

$$2ik\frac{\partial p}{\partial x} + \nabla_\perp^2 p + 2k^2 \left(1 + \frac{\varepsilon_{\text{mov}}}{2}\right) p = 0. \tag{7.8}$$

This equation, as well as equation (7.6), serves as the starting point of our analysis of the sound-field statistics.

7.1.2 Effective correlation function and effective spectrum

According to equations (7.6) and (7.8) and the approximations underlying their derivation, the fluctuations in the sound field p are caused by the random field ε_{mov}. As a result, the statistical moments of p can be expressed in terms of the statistical characteristics of ε_{mov}. In this subsection, we present a statistical description of this random field.

In the general case of a *statistically* inhomogeneous and anisotropic random medium, the correlation function of ε_{mov} is defined as

$$B_{\text{eff}}(\mathbf{R}_1, \mathbf{R}_2) = \langle \varepsilon_{\text{mov}}(\mathbf{R}_1)\varepsilon_{\text{mov}}(\mathbf{R}_2) \rangle. \tag{7.9}$$

Replacing $\varepsilon_{\text{mov}}(\mathbf{R})$ by the right-hand side of equation (7.3), we have

$$B_{\text{eff}}(\mathbf{R}_1, \mathbf{R}_2) = \frac{4}{c_0^2} \left[B_{\tilde{c}}(\mathbf{R}_1, \mathbf{R}_2) + B_{11}(\mathbf{R}_1, \mathbf{R}_2)\right]. \tag{7.10}$$

Here,

$$B_{\tilde{c}}(\mathbf{R}_1, \mathbf{R}_2) = \langle \tilde{c}(\mathbf{R}_1)\tilde{c}(\mathbf{R}_2) \rangle, \quad B_{11}(\mathbf{R}_1, \mathbf{R}_2) = \langle \tilde{v}_x(\mathbf{R}_1)\tilde{v}_x(\mathbf{R}_2) \rangle, \tag{7.11}$$

are the correlation functions of the sound speed fluctuations $\tilde{c}$ and the fluctuations $\tilde{v}_x \equiv \tilde{v}_1$ of the velocity along the x-axis. The function B_{11} is a component of the correlation tensor B_{ij} similar to that given by equation (6.6) for homogeneous turbulence. As in Chapter 6 (see the text following equation (6.2)), when deriving equation (7.10), we assumed that the scalar random field $\tilde{c}$ does not correlate with components of the vector random field $\tilde{\mathbf{v}}$. The function B_{eff}, which accounts for both the sound speed and velocity fluctuations, is called the *effective* correlation function.

Using equation (6.65), the correlation function of the sound speed fluctuations can be expressed as

$$B_{\tilde{c}}(\mathbf{R}_1, \mathbf{R}_2) = \frac{c_0^2}{4} \left[\frac{\beta_c^2 B_T(\mathbf{R}_1, \mathbf{R}_2)}{T_0^2} + \frac{2\eta_c \beta_c B_{CT}(\mathbf{R}_1, \mathbf{R}_2)}{T_0} + \eta_c^2 B_C(\mathbf{R}_1, \mathbf{R}_2) \right],$$

(7.12)

where B_T and B_C are the correlation functions of the temperature fluctuations $\tilde{T}$ and the fluctuations $\tilde{C}$ in the concentration of the component dissolved in the medium (e.g., the fluctuations in water vapor in the atmosphere or salinity in the ocean), and B_{CT} is the cross correlation between $\tilde{C}$ and $\tilde{T}$. For the case of homogeneous turbulence, these correlation functions are given by equations (6.3)–(6.5). The coefficients η_c and β_c are determined with equation (6.67).

For subsequent analysis, it is helpful to introduce the center $\mathbf{R}_c$ and difference $\mathbf{R}_d$ coordinates

$$\mathbf{R}_c = (\mathbf{R}_1 + \mathbf{R}_2)/2, \quad \mathbf{R}_d = \mathbf{R}_1 - \mathbf{R}_2.$$ (7.13)

The new coordinates can be written as $\mathbf{R}_c = (x_c, \mathbf{r}_c)$ and $\mathbf{R}_d = (x_d, \mathbf{r}_d)$, where

$$x_c = (x_1 + x_2)/2, \quad \mathbf{r}_c = (\mathbf{r}_1 + \mathbf{r}_2)/2,$$ (7.14)

$$x_d = x_1 - x_1, \quad \mathbf{r}_d = \mathbf{r}_1 - \mathbf{r}_2.$$ (7.15)

Taking into account that $\mathbf{R}_1 = \mathbf{R}_c + \mathbf{R}_d/2$ and $\mathbf{R}_2 = \mathbf{R}_c - \mathbf{R}_d/2$, the effective correlation function can be written as

$$B_{\text{eff}}(\mathbf{R}_1, \mathbf{R}_2) = B_{\text{eff}}(\mathbf{R}_c + \mathbf{R}_d/2, \mathbf{R}_c - \mathbf{R}_d/2) \equiv \mathcal{B}_{\text{eff}}(\mathbf{R}_c, \mathbf{R}_d).$$ (7.16)

Similarly, the correlation functions of other random fields, as functions of $\mathbf{R}_c$ and $\mathbf{R}_d$, are denoted as $\mathcal{B}_{\tilde{c}}(\mathbf{R}_c, \mathbf{R}_d)$, $\mathcal{B}_{11}(\mathbf{R}_c, \mathbf{R}_d)$, $\mathcal{B}_T(\mathbf{R}_c, \mathbf{R}_d)$, $\mathcal{B}_{CT}(\mathbf{R}_c, \mathbf{R}_d)$, and $\mathcal{B}_C(\mathbf{R}_c, \mathbf{R}_d)$. The dependence of the correlation functions on the center coordinate $\mathbf{R}_c$ is due to inhomogeneity of the corresponding random fields. In the atmosphere, this inhomogeneity can be caused by the dependence of the variances and length scales of temperature, humidity, and wind velocity fluctuations on the height above the ground (section 6.2.4).

The effective correlation function is expressed as a Fourier integral with respect to the coordinates x_d and $\mathbf{r}_d$:

$$\mathcal{B}_{\text{eff}}(\mathbf{R}_c; x_d, \mathbf{r}_d) = \int \int \Phi_{\text{eff}}(\mathbf{R}_c; \kappa_x, \boldsymbol{\kappa}_\perp) \exp\left(i\kappa_x x_d + i\boldsymbol{\kappa}_\perp \cdot \mathbf{r}_d\right) d\kappa_x \, d^2\kappa_\perp.$$

(7.17)

Here, κ_x and $\boldsymbol{\kappa}_\perp$ are the components of the turbulence wave vector $\boldsymbol{\kappa} = (\kappa_x, \boldsymbol{\kappa}_\perp)$ in the direction of the x-axis and perpendicular to it, and $\Phi_{\mathrm{eff}}(\mathbf{R}_c; \kappa_x, \boldsymbol{\kappa}_\perp)$ is the three-dimensional spectrum of the effective correlation function, which will be termed the *effective* spectrum. In this chapter, if the limits of integration are not provided (as in equation (7.17)), they are assumed to be from $-\infty$ to $+\infty$.

The three-dimensional spectra $\Phi_{\tilde{c}}(\mathbf{R}_c; \kappa_x, \boldsymbol{\kappa}_\perp)$, $\Phi_{11}(\mathbf{R}_c; \kappa_x, \boldsymbol{\kappa}_\perp)$, $\Phi_T(\mathbf{R}_c; \kappa_x, \boldsymbol{\kappa}_\perp)$, $\Phi_{CT}(\mathbf{R}_c; \kappa_x, \boldsymbol{\kappa}_\perp)$, $\Phi_C(\mathbf{R}_c; \kappa_x, \boldsymbol{\kappa}_\perp)$ of the corresponding correlation functions $\mathcal{B}_{\tilde{c}}(\mathbf{R}_c, \mathbf{R}_d)$, $\mathcal{B}_{xx}(\mathbf{R}_c, \mathbf{R}_d)$, $\mathcal{B}_T(\mathbf{R}_c, \mathbf{R}_d)$, $\mathcal{B}_{CT}(\mathbf{R}_c, \mathbf{R}_d)$, $\mathcal{B}_C(\mathbf{R}_c, \mathbf{R}_d)$ are determined by equations similar to equation (7.17).

With these notations and using equation (7.10), the effective spectrum Φ_{eff} is expressed in terms of the spectra of the sound speed and velocity fluctuations

$$\Phi_{\mathrm{eff}}(\mathbf{R}_c; \kappa_x, \boldsymbol{\kappa}_\perp) = \frac{4}{c_0^2} \left[\Phi_{\tilde{c}}(\mathbf{R}_c; \kappa_x, \boldsymbol{\kappa}_\perp) + \Phi_{11}(\mathbf{R}_c; \kappa_x, \boldsymbol{\kappa}_\perp) \right]. \tag{7.18}$$

The spectrum of the sound speed fluctuations can be obtained from equation (7.12):

$$\Phi_{\tilde{c}}(\mathbf{R}_c; \kappa_x, \boldsymbol{\kappa}_\perp) = \frac{c_0^2}{4} \left[\frac{\beta_c^2 \Phi_T(\mathbf{R}_c; \kappa_x, \boldsymbol{\kappa}_\perp)}{T_0^2} \right.$$
$$\left. + \frac{2\eta_c \beta_c \Phi_{CT}(\mathbf{R}_c; \kappa_x, \boldsymbol{\kappa}_\perp)}{T_0} + \eta_c^2 \Phi_C(\mathbf{R}_c; \kappa_x, \boldsymbol{\kappa}_\perp) \right]. \tag{7.19}$$

7.1.3 Markov approximation

Along the x-axis, the correlation between $\varepsilon_{\mathrm{mov}}(x_1, \mathbf{r}_1)$ and $\varepsilon_{\mathrm{mov}}(x_2, \mathbf{r}_2)$ persists if $|x_d| = |x_1 - x_2| \lesssim L$, where L is the outer scale of turbulence. If the propagation range is greater than L, the scale L becomes the smallest parameter along the x-axis in the parabolic equation (7.6) and can be set formally to zero. This results in the Markov approximation, for which the effective correlation function is given by

$$\mathcal{B}_{\mathrm{eff}}(x_c, \mathbf{r}_c; x_d, \mathbf{r}_d) = \delta(x_d) b_{\mathrm{eff}}(x_c, \mathbf{r}_c; \mathbf{r}_d). \tag{7.20}$$

Here, we expressed the vector $\mathbf{R}_c$ as $\mathbf{R}_c = (x_c, \mathbf{r}_c)$, δ is the delta function, and b_{eff} is the *transverse* correlation function. This function is obtained by integrating both sides of equation (7.20) and replacing $\mathcal{B}_{\mathrm{eff}}$ with its value from equation (7.17):

$$b_{\mathrm{eff}}(x_c, \mathbf{r}_c; \mathbf{r}_d) = 2\pi \int \Phi_{\mathrm{eff}}(x_c, \mathbf{r}_c; 0, \boldsymbol{\kappa}_\perp) \exp\left(i\boldsymbol{\kappa}_\perp \cdot \mathbf{r}_d\right) d^2\kappa_\perp. \tag{7.21}$$

Eliminating b_{eff} between the last two equations, we have

$$\mathcal{B}_{\mathrm{eff}}(x_c, \mathbf{r}_c; x_d, \mathbf{r}_d) = 2\pi\delta(x_d) \int \Phi_{\mathrm{eff}}(x_c, \mathbf{r}_c; 0, \boldsymbol{\kappa}_\perp) \exp\left(i\boldsymbol{\kappa}_\perp \cdot \mathbf{r}_d\right) d^2\kappa_\perp. \tag{7.22}$$

The Markov approximation is widely used in WPRM, e.g., references [178, 339]. It significantly simplifies derivations of closed-form equations for the statistical moments of a propagating field, results in equations which are almost identical to those derived without its use, and can be used for most problems within the framework of the parabolic equation method. The ranges of applicability of the Markov approximation, which are considered in detail elsewhere [339], are:

$$k^2 l^2 \langle \varepsilon_{\text{mov}}^2 \rangle \ll 1, \quad l_\perp \gg \sqrt{l_x/k}, \quad kr_{\text{coh}} \gg 1, \quad x \gg l_{\text{lag}}. \tag{7.23}$$

Here, l_x and $l_\perp$ are the scales of random inhomogeneities in the directions of the x-axis and in the plane perpendicular to it, $l = \left(l_x^2 + l_\perp^2 \right)^{1/2}$, r_{coh} is the coherence radius of the sound field determined with equation (7.176), and l_{lag} is the scale of largest inhomogeneities which affect a particular statistical moment of a sound field. For example, for the variance of phase fluctuations and the mean sound field $l_{\text{lag}} \sim L$, while for the mutual coherence function $l_{\text{lag}} \sim r_d$, where r_d is the distance between two points of observation.

In the derivations to follow, the statistical moments of a sound field will be expressed in terms of the effective spectrum $\Phi_{\text{eff}}(x_c, \mathbf{r}_c; 0, \boldsymbol{\kappa}_\perp)$. For the general case of a statistically inhomogeneous and anisotropic random medium, this spectrum is obtained from equation (7.18)

$$\Phi_{\text{eff}}(x_c, \mathbf{r}_c; 0, \boldsymbol{\kappa}_\perp) = \frac{4}{c_0^2} \left[\Phi_{\tilde{c}}(x_c, \mathbf{r}_c; 0, \boldsymbol{\kappa}_\perp) + \Phi_{11}(x_c, \mathbf{r}_c; 0, \boldsymbol{\kappa}_\perp) \right]. \tag{7.24}$$

We will assume that the random medium is *quasi-homogeneous*, i.e., it is homogeneous in any relatively small volume; however, its parameters such as the variances and length scales gradually change with the coordinate $\mathbf{R}_c = (x_c, \mathbf{r}_c)$. The effective spectrum $\Phi_{\text{eff}}(x_c, \mathbf{r}_c; 0, \boldsymbol{\kappa}_\perp)$ of a quasi-homogeneous, anisotropic random medium is also given by equation (7.24).

This equation can be simplified for a quasi-homogenous and isotropic random medium. In this case, $\Phi_{\tilde{c}}(\mathbf{R}_c; \kappa_x, \boldsymbol{\kappa}_\perp)$ depends only on the magnitude of the wave vector $\boldsymbol{\kappa} = (\kappa_x, \boldsymbol{\kappa}_\perp)$ and

$$\Phi_{11}(x_c, \mathbf{r}_c; \kappa_x, \boldsymbol{\kappa}_\perp) = \left(1 - \frac{\kappa_x^2}{\kappa^2} \right) \frac{E(x_c, \mathbf{r}_c; \kappa)}{4\pi\kappa^2}. \tag{7.25}$$

This formula is a generalization of equation (6.17) with $i = j = 1$ for the considered case of a quasi-homogenous random medium, for which the energy spectrum E can slowly depend on the coordinate $\mathbf{R}_c$. Using these results, we obtain:

$$\Phi_{\text{eff}}(x_c, \mathbf{r}_c; 0, \boldsymbol{\kappa}_\perp) = \frac{4}{c_0^2} \left[\Phi_{\tilde{c}}(x_c, \mathbf{r}_c; \boldsymbol{\kappa}_\perp) + \frac{E(x_c, \mathbf{r}_c; \boldsymbol{\kappa}_\perp)}{4\pi\kappa_\perp^2} \right]. \tag{7.26}$$

For a fully homogeneous, anisotropic random medium, the sound speed and velocity spectra do not depend on $\mathbf{R}_c$ and equation (7.24) simplifies to:

$$\Phi_{\text{eff}}(0, \boldsymbol{\kappa}_\perp) = \frac{4}{c_0^2} \left[\Phi_{\tilde{c}}(0, \boldsymbol{\kappa}_\perp) + \Phi_{11}(0, \boldsymbol{\kappa}_\perp) \right]. \tag{7.27}$$

Finally, for a homogeneous, isotropic random medium, we have

$$\Phi_{\text{eff}}(0, \kappa_\perp) = \frac{4}{c_0^2} \left[\Phi_{\tilde{c}}(\kappa_\perp) + \frac{E(\kappa_\perp)}{4\pi \kappa_\perp^2} \right]. \tag{7.28}$$

Substituting the value of $\Phi_{\tilde{c}}$ from equations (7.19), we obtain that in this case

$$\Phi_{\text{eff}}(0, \kappa_\perp) = \frac{\beta_c^2 \Phi_T(\kappa_\perp)}{T_0^2} + \frac{2\eta_c \beta_c \Phi_{CT}(\kappa_\perp)}{T_0} + \eta_c^2 \Phi_C(\kappa_\perp) + \frac{E(\kappa_\perp)}{\pi c_0^2 \kappa_\perp^2}. \tag{7.29}$$

7.1.4 Effective turbulence spectra

Here, we present particular forms of the effective spectrum $\Phi_{\text{eff}}(0, \kappa_\perp)$ given by equation (7.29) for the von Kármán, Kolmogorov, and Gaussian models of homogenous, isotropic turbulence considered in section 6.2.

7.1.4.1 Von Kármán spectrum

For the von Kármán turbulence model, the spectra $\Phi_T^{vK}(\kappa)$, $\Phi_{CT}^{vK}(\kappa)$, $\Phi_{CT}^{vK}(\kappa)$, $E^{vK}(\kappa)$ are given with equations (6.26)–(6.29). Substituting the values of these spectra into equation (7.29), we obtain

$$\Phi_{\text{eff}}^{vK}(0, \kappa_\perp) = \frac{\Gamma(11/6)}{\pi^{3/2}\Gamma(1/3)} \left[\frac{\beta_c^2 \sigma_T^2 L_T^3}{T_0^2 \left(1 + \kappa_\perp^2 L_T^2\right)^{11/6}} + \frac{2\eta_c \beta_c \sigma_{CT} L_{CT}^3}{T_0 \left(1 + \kappa_\perp^2 L_{CT}^2\right)^{11/6}} \right.$$
$$\left. + \frac{\eta_c^2 \sigma_C^2 L_C^3}{\left(1 + \kappa_\perp^2 L_C^2\right)^{11/6}} + \frac{22}{3} \frac{\sigma_v^2 \kappa_\perp^2 L_v^5}{c_0^2 \left(1 + \kappa_\perp^2 L_v^2\right)^{17/6}} \right]. \tag{7.30}$$

Here, σ_T^2, σ_{CT}^2, σ_C^2, σ_v^2 are the variances and L_T, L_{CT}, L_C, L_v the length scales of the corresponding random fields (see section 6.2.1), and Γ is the gamma function. Since the scales L_T, L_{CT}, and L_C generally are different, it is not possible to combine the first three terms in square brackets into one term which would describe a von Kármán spectrum for the sound speed fluctuations.

It follows from equation (7.30) that the relative contributions from the scalar random fields ($\tilde{T}$, $\tilde{C}$, and their cross correlation) to $\Phi_{\text{eff}}^{vK}(0, \kappa_\perp)$ have the same functional dependence on $\kappa_\perp$, while the contribution from the component $\tilde{v}_x$ of velocity fluctuations has a different dependence on $\kappa_\perp$. As a result, the relative contributions from scalar and vector random fields to the statistical moments of a sound field, generally, have different dependencies on parameters of the problem. On the other hand, the contributions from the scalar random fields to these statistical moments have the same dependence on these parameters. The latter result is used in this chapter to simplify formulas for the statistical moments of a sound field. In these formulas, we account only for temperature fluctuations; the contributions from the random field $\tilde{C}$ and the cross correlation between $\tilde{C}$ and $\tilde{T}$ can be readily obtained by replacing the corresponding coefficients according to the first three terms in the square brackets in equation (7.30). Also note that if in equation (7.30) the variances and length scales depend on the spatial coordinate $\mathbf{R}_c = (x_c, \mathbf{r}_c)$ (as is the case for atmospheric turbulence), the effective spectrum is a particular case of equation (7.26), which applies to a quasi-homogeneous, isotropic random field. Most of these conclusions are also valid for the Kolmogorov and Gaussian spectra considered next.

7.1.4.2   Kolmogorov spectrum

Substituting the spectra $\Phi_T^K(\kappa)$, $\Phi_{CT}^K(\kappa)$, $\Phi_{CT}^K(\kappa)$, $E^K(\kappa)$ given by equation (6.34) into equation (7.29), we obtain the effective spectrum for the Kolmogorov model

$$\Phi_{\text{eff}}^K(0, \kappa_\perp) = \mathcal{Q} C_{\text{eff}}^2 \kappa_\perp^{-11/3}. \tag{7.31}$$

Here, $\mathcal{Q}$ is the numerical coefficient determined by equation (6.35) and C_{eff}^2 is the *effective* structure-function parameter,

$$C_{\text{eff}}^2 = \frac{\beta_c^2 C_T^2}{T_0^2} + \frac{2\eta_c \beta_c \varrho_{CT} C_C C_T}{T_0} + \eta_c^2 C_C^2 + \frac{22}{3} \frac{C_v^2}{c_0^2}. \tag{7.32}$$

In this formula, C_T^2, C_C^2, C_v^2 are the structure-function parameters of the corresponding random fields and ϱ_{CT} is the correlation coefficient of the random fields $\widetilde{C}$ and $\widetilde{T}$ (section 6.2.2). The first three terms in equation (7.32) can be combined into the structure-function parameter of the sound speed fluctuations $C_{\widetilde{c}}^2$. It can be shown from equation (6.65) that

$$C_{\widetilde{c}}^2 = \frac{1}{4} \left(\frac{\beta_c^2 C_T^2}{T_0^2} + \frac{2\eta_c \beta_c \varrho_{CT} C_C C_T}{T_0} + \eta_c^2 C_C^2 \right). \tag{7.33}$$

Using this result, the effective structure-function parameter can be written as

$$C_{\text{eff}}^2 = 4 \frac{C_{\widetilde{c}}^2}{c_0^2} + \frac{22}{3} \frac{C_v^2}{c_0^2}. \tag{7.34}$$

For the Kolmogorov model, the difference between relative contributions from scalar and vector random fields to the effective spectrum is manifested in the coefficient 22/3 in front of C_v^2 [290]; see equations (7.32) and (7.34). If $\widetilde{v}_x$ were a scalar random field, the value of this coefficient would be 4. In the older theories of sound propagation in random moving media (e.g., references [52, 188, 374]), the incorrect value 4 of this coefficient was used so that the effective structure-function parameter had the form

$$C_{\text{old}}^2 = 4 \frac{C_{\widetilde{c}}^2}{c_0^2} + 4 \frac{C_v^2}{c_0^2}. \tag{7.35}$$

Other than the coefficient 22/3, the relative contributions from scalar and vector random fields to $\Phi_{\text{eff}}^K(0, \kappa_\perp)$ have the same dependence on $\kappa_\perp$. As a result, the contributions of these random fields to the statistical moments of a sound field have the same dependence on the propagation range x, wavenumber k, and sensor separation r_d.

7.1.4.3 Gaussian spectrum

Section 6.2.3 provides the spectra $\Phi_T^G(\kappa)$, $\Phi_{CT}^G(\kappa)$, $\Phi_{CT}^G(\kappa)$, $E^G(\kappa)$ for the Gaussian model. Substituting the values of these spectra into equation (7.29), we obtain the effective spectrum

$$
\Phi_{\text{eff}}^G(0, \kappa_\perp) = \frac{1}{8\pi^{3/2}} \left[\frac{\beta_c^2 \sigma_T^2 \mathcal{L}_T^3}{T_0^2} \exp\left(-\frac{\kappa_\perp^2 \mathcal{L}_T^2}{4}\right) \right.
$$
$$
+ \frac{2\eta_c \beta_c \sigma_{CT} \mathcal{L}_{CT}^3}{T_0} \exp\left(-\frac{\kappa_\perp^2 \mathcal{L}_{CT}^2}{4}\right) + \eta_c^2 \sigma_C^2 \mathcal{L}_C^3 \exp\left(-\frac{\kappa_\perp^2 \mathcal{L}_C^2}{4}\right)
$$
$$
\left. + \frac{\sigma_v^2 \kappa_\perp^2 L_v^5}{c_0^2} \exp\left(-\frac{\kappa_\perp^2 \mathcal{L}_v^2}{4}\right) \right]. \quad (7.36)
$$

Here, $\mathcal{L}_T$, $\mathcal{L}_{CT}$, $\mathcal{L}_C$, $\mathcal{L}_v$ are the length scales associated with the Gaussian turbulence model.

The Gaussian spectra had been used widely for comparing experimental data with theoretical results on sound propagation through a turbulent atmosphere [87, 88, 182, 247, 248]. Following the older theories of sound propagation (see text after equation (7.4)), in these and some other papers, the last term in the square brackets was used in the form $(4\sigma_v^2 L_v^3/c_0^2) \exp\left(-\kappa_\perp^2 \mathcal{L}_v^2/4\right)$. As a result, the contributions from temperature and velocity fluctuations to the statistical moments had the same functional dependence on the parameters. We will see later in this chapter that this is not the case for the correct effective spectrum (7.36).

7.1.5 Spectral representation of the effective correlation function

Because the phase and amplitude fluctuations will be analyzed in the spectral domain, we consider now the spectral representation of the effective correlation function.

The random field $\varepsilon_{\text{mov}}(x, \mathbf{r})$ is expressed as a two-dimensional Fourier integral:

$$
\varepsilon_{\text{mov}}(x, \mathbf{r}) = \int \widehat{\varepsilon}_{\text{mov}}(x, \mathbf{q}) \exp\left(i\mathbf{q} \cdot \mathbf{r}\right) d^2q, \quad (7.37)
$$

where $\widehat{\varepsilon}_{\text{mov}}(x, \mathbf{q})$ is the two-dimensional Fourier transform of $\varepsilon_{\text{mov}}(x, \mathbf{r})$. The correlation function of $\widehat{\varepsilon}_{\text{mov}}(x, \mathbf{q})$ is defined as:

$$
\widehat{B}_{\text{eff}}(x_1, \mathbf{q}_1; x_2, \mathbf{q}_2) = \langle \widehat{\varepsilon}_{\text{mov}}(x_1, \mathbf{q}_1) \widehat{\varepsilon}_{\text{mov}}(x_2, \mathbf{q}_2) \rangle. \quad (7.38)
$$

Using equation (7.37), $\widehat{\varepsilon}_{\text{mov}}(x, \mathbf{q})$ can be expressed in terms of $\varepsilon_{\text{mov}}(x, \mathbf{r})$. Substituting the result into equation (7.38), we have

$$
\widehat{B}_{\text{eff}}(x_1, \mathbf{q}_1; x_2, \mathbf{q}_2) = \frac{1}{16\pi^4} \int \int B_{\text{eff}}(x_1, \mathbf{r}_1; x_2, \mathbf{r}_2)
$$
$$
\times \exp\left(-i\mathbf{q}_1 \cdot \mathbf{r}_1 - i\mathbf{q}_2 \cdot \mathbf{r}_2\right) d^2r_1 d^2r_2. \quad (7.39)
$$

On the right-hand side of this formula, we introduce the center $\mathbf{R}_c = (x_c, \mathbf{r}_c)$ and difference $\mathbf{R}_d = (x_d, \mathbf{r}_d)$ coordinates (see equation (7.13)) and replace $\mathcal{B}_{\text{eff}}(x_1, \mathbf{r}_1; x_2, \mathbf{r}_2)$ with $\mathcal{B}_{\text{eff}}(x_c, \mathbf{r}_c; x_d, \mathbf{r}_d)$. The latter function is replaced with the Markov approximation, equation (7.22). In the resulting formula, the integral over $\mathbf{r}_d$ is equal to $4\pi^2 \delta(\boldsymbol{\kappa}_\perp - (\mathbf{q}_1 - \mathbf{q}_2)/2)$. Calculating the integral over $\boldsymbol{\kappa}_\perp$, we obtain

$$\widehat{\mathcal{B}}_{\text{eff}}(x_1, \mathbf{q}_1; x_2, \mathbf{q}_2) = \frac{\delta(x_1 - x_2)}{2\pi} \int \Phi_{\text{eff}}\left(x_1, \mathbf{r}_c; 0, \frac{\mathbf{q}_1 - \mathbf{q}_2}{2}\right)$$
$$\times \exp\left[-i(\mathbf{q}_1 + \mathbf{q}_2) \cdot \mathbf{r}_c\right] d^2 r_c. \quad (7.40)$$

On the right-hand side of this expression, we substituted with $x_d = x_1 - x_2$ and took into account that $x_c = (x_1 + x_2)/2 = x_1$.

7.2 Phase and log-amplitude fluctuations for arbitrary spectra

The phase and amplitude of a sound wave propagating in a random medium fluctuate. In this section, using the Rytov method, the variances and correlation functions of these fluctuations are obtained for plane and spherical sound waves. The spectra of the sound speed and velocity fluctuations are arbitrary.

7.2.1 Rytov method

The Rytov method is widely used in WPRM for calculations of the statistical characteristics of the phase and log-amplitude fluctuations. It generalizes the geometrical acoustics (optics) approximation by accounting for diffraction. It was originally developed by S. M. Rytov for a deterministic problem, namely light diffraction by ultrasound [338].

In the Rytov method, a solution equation (7.6) is sought in the form

$$A(x, \mathbf{r}) = \exp\left(\Psi(x, \mathbf{r})\right), \quad (7.41)$$

where $\Psi(x, \mathbf{r})$ is the *complex phase*. Substituting this formula into equation (7.6), we obtain a nonlinear equation for $\Psi(x, \mathbf{r})$:

$$2ik\frac{\partial \Psi}{\partial x} + (\nabla_\perp \Psi)^2 + \nabla_\perp^2 \Psi + k^2 \varepsilon_{\text{mov}} \Psi = 0. \quad (7.42)$$

The initial condition to this equation is formulated in the plane $x = 0$ and is obtained with the use of equations (7.41) and (7.7):

$$\Psi(x = 0, \mathbf{r}) = \ln p_0(\mathbf{r}). \quad (7.43)$$

Equation (7.42) is solved by an iterative technique. We write

$$\Psi(x, \mathbf{r}) = \Psi_0(x, \mathbf{r}) + \Psi_1(x, \mathbf{r}) + \Psi_2(x, \mathbf{r}) + \dots \, . \tag{7.44}$$

Here, the functions $\Psi_n(x, \mathbf{r})$ are of order $\varepsilon_{\text{mov}}^n$; in other words: $\Psi_0 \sim \varepsilon_{\text{mov}}^0 = 1$, $\Psi_1 \sim \varepsilon_{\text{mov}}^1 \equiv \varepsilon_{\text{mov}}$, $\Psi_2 \sim \varepsilon_{\text{mov}}^2$, etc. The initial conditions for Ψ_n can be determined with equations (7.42)–(7.44):

$$\Psi_0(x = 0, \mathbf{r}) = \ln p_0(\mathbf{r}), \tag{7.45}$$

$$\Psi_i(x = 0, \mathbf{r}) = 0, \qquad i = 1, 2, \dots \, . \tag{7.46}$$

Closed-form equations for Ψ_n are obtained by substituting series (7.44) into equation (7.42) and equating coefficients of the same power in ε_{mov}. For the first term of this series, we obtain the following equation:

$$2ik\frac{\partial \Psi_0}{\partial x} + (\nabla_\perp \Psi_0)^2 + \nabla_\perp^2 \Psi_0 = 0. \tag{7.47}$$

The $\Psi_0(x, \mathbf{r})$ is the complex phase of a sound wave propagating in a medium without random inhomogeneities. It is related to the complex amplitude $A^{(0)}(x, \mathbf{r})$ of this wave by a formula similar to equation (7.41)

$$A^{(0)}(x, \mathbf{r}) = \exp\left(\Psi_0(x, \mathbf{r})\right). \tag{7.48}$$

The complex amplitude $A^{(0)}(x, \mathbf{r})$ satisfies equation (7.6) with $\varepsilon_{\text{mov}}(x, \mathbf{r}) = 0$. The equation for $\Psi_1(x, \mathbf{r})$ is given by:

$$2ik\frac{\partial \Psi_1}{\partial x} + 2\nabla_\perp \Psi_0 \cdot \nabla_\perp \Psi_1 + \nabla_\perp^2 \Psi_1 + k^2 \varepsilon_{\text{mov}} = 0. \tag{7.49}$$

In this equation, we write

$$\Psi_1(x, \mathbf{r}) = F(x, \mathbf{r}) \exp\left(-\Psi_0(x, \mathbf{r})\right), \tag{7.50}$$

where $F(x, \mathbf{r})$ is a new unknown function. Substituting this equation into equation (7.49), we obtain an equation for F:

$$2ik\frac{\partial F}{\partial x} + \nabla_\perp^2 F + k^2 \varepsilon_{\text{mov}} A^{(0)} = 0, \tag{7.51}$$

with the initial condition $F(0, \mathbf{r}) = 0$. This equation can be readily solved, with result

$$F(x, \mathbf{r}) = \frac{k^2}{4\pi} \int_0^x \frac{dx'}{x - x'} \int \varepsilon_{\text{mov}}(x', \mathbf{r}') A^{(0)}(x', \mathbf{r}') \exp\left(\frac{ik(\mathbf{r} - \mathbf{r}')^2}{2(x - x')}\right) d^2 r'. \tag{7.52}$$

We eliminate F between equations (7.50) and (7.52) and take into account the fact that $\exp(\Psi_0) = A^{(0)}$. As a result, we obtain an explicit formula for the second term of the series (7.44):

$$\Psi_1(x, \mathbf{r}) = \frac{k^2}{4\pi} \int_0^x \frac{dx'}{x - x'} \int \varepsilon_{\text{mov}}(x', \mathbf{r}') \frac{A^{(0)}(x', \mathbf{r}')}{A^{(0)}(x, \mathbf{r})} \exp\left(\frac{ik(\mathbf{r} - \mathbf{r}')^2}{2(x - x')}\right) d^2r'.$$

(7.53)

When calculating the correlation functions of phase and log-amplitude fluctuations using the Rytov method, the fluctuations in the complex phase are often approximated by the term Ψ_1. This is called the *first* Rytov approximation. The term Ψ_2 in the series (7.44) corresponds to the *second* Rytov approximation. It is needed, for example, to calculate the mean values of the phase and log-amplitude fluctuations.

The term Ψ_1 can be recast as

$$\Psi_1(x, \mathbf{r}) = \chi(x, \mathbf{r}) + i\phi(x, \mathbf{r}),$$

(7.54)

where the real functions $\phi(x, \mathbf{r})$ and $\chi(x, \mathbf{r})$ denote the *phase* and *log-amplitude* fluctuations. These functions can be expressed in terms of the complex phase fluctuations:

$$\phi(x, \mathbf{r}) = \frac{1}{2i} \left[\Psi_1(x, \mathbf{r}) - \Psi_1^*(x, \mathbf{r})\right],$$

(7.55)

$$\chi(x, \mathbf{r}) = \frac{1}{2} \left[\Psi_1(x, \mathbf{r}) + \Psi_1^*(x, \mathbf{r})\right].$$

(7.56)

The *transverse correlation* functions of phase and log-amplitude fluctuations at two points of observation $(x, \mathbf{r}_1)$ and $(x, \mathbf{r}_2)$ (figure 7.1) are defined as

$$B_\phi(x; \mathbf{r}_1, \mathbf{r}_2) = \langle \phi(x, \mathbf{r}_1)\phi(x, \mathbf{r}_2)\rangle,$$

(7.57)

$$B_\chi(x; \mathbf{r}_1, \mathbf{r}_2) = \langle \chi(x, \mathbf{r}_1)\chi(x, \mathbf{r}_2)\rangle.$$

(7.58)

The decorrelation of the phase and log-amplitude fluctuations along the x-axis is usually much less than that in the transverse direction and is not considered here.

Substituting with equations (7.55) and (7.56) and replacing Ψ_1 with the right-hand side of equation (7.53), it is straightforward to express B_ϕ and B_χ in terms of the effective correlation function $\langle\varepsilon_{\text{mov}}(x_1, \mathbf{r}_1)\varepsilon_{\text{mov}}(x_2, \mathbf{r}_2)\rangle$. The resulting expressions, however, require further simplifications. The corresponding manipulations are very involved [72]. These manipulations are significantly simplified, however, when performed in the spectral domain, as will next be demonstrated.

7.2.2   Spectral domain

The complex phase fluctuation is expressed as a two-dimensional Fourier integral:

$$\Psi_1(x, \mathbf{r}) = \int \widehat{\Psi}_1(x, \mathbf{q}) \exp(i\mathbf{q} \cdot \mathbf{r}) \, d^2q,$$

(7.59)

where $\widehat{\Psi}_1(x, \mathbf{q})$ is the two-dimensional spatial spectrum of $\Psi_1(x, \mathbf{r})$. The phase and log-amplitude fluctuations are expressed similarly:

$$\phi(x, \mathbf{r}) = \int \widehat{\phi}(x, \mathbf{q}) \exp\left(i\mathbf{q} \cdot \mathbf{r}\right) d^2q, \tag{7.60}$$

$$\chi(x, \mathbf{r}) = \int \widehat{\chi}(x, \mathbf{q}) \exp\left(i\mathbf{q} \cdot \mathbf{r}\right) d^2q. \tag{7.61}$$

Using these formulas and equations (7.55) and (7.56), the two-dimensional spatial spectra of the phase and log-amplitudes fluctuations can be expressed in terms of $\widehat{\Psi}_1$:

$$\widehat{\phi}(x, \mathbf{q}) = \frac{1}{2i}\left[\widehat{\Psi}_1(x, \mathbf{q}) - \widehat{\Psi}_1^*(x, -\mathbf{q})\right], \tag{7.62}$$

$$\widehat{\chi}(x, \mathbf{q}) = \frac{1}{2}\left[\widehat{\Psi}_1(x, \mathbf{q}) + \widehat{\Psi}_1^*(x, -\mathbf{q})\right]. \tag{7.63}$$

In equations (7.57) and (7.58) for the correlation functions of the phase and log-amplitude fluctuations, we express $\phi(x, \mathbf{r})$ and $\chi(x, \mathbf{r})$ as Fourier integrals (7.60) and (7.61). As a result, we have

$$B_\phi(x; \mathbf{r}_1, \mathbf{r}_2) = \int\int \langle\widehat{\phi}(x, \mathbf{q}_1)\widehat{\phi}(x, \mathbf{q}_2)\rangle \exp\left(i\mathbf{q}_1 \cdot \mathbf{r}_1 + i\mathbf{q}_2 \cdot \mathbf{r}_2\right) d^2q_1 \, d^2q_2, \tag{7.64}$$

$$B_\chi(x; \mathbf{r}_1, \mathbf{r}_2) = \int\int \langle\widehat{\chi}(x, \mathbf{q}_1)\widehat{\chi}(x, \mathbf{q}_2)\rangle \exp\left(i\mathbf{q}_1 \cdot \mathbf{r}_1 + i\mathbf{q}_2 \cdot \mathbf{r}_2\right) d^2q_1 \, d^2q_2. \tag{7.65}$$

To proceed further, one needs to specify the waveform emitted by the source. In what follows, we will consider propagation of initially plane and spherical sound waves.

7.2.3 Plane wave propagation

The case of plane wave propagation results in relatively simple formulas for the statistical moments of a sound field which still capture the underlying physics of wave propagation in a random medium. Plane wave propagation is pertinent to sound sources (such as aircrafts, rockets, and meteors) located well above the atmospheric boundary layer (ABL), in which sound impinging on the ABL can be approximately considered as planar.

For an initially plane wave, $A^{(0)}(x, \mathbf{r}) = $ constant. In this case, equation (7.53) for the complex phase fluctuation simplifies to

$$\Psi_1(x, \mathbf{r}) = \frac{k^2}{4\pi} \int_0^x \frac{dx'}{x - x'} \int \varepsilon_{\text{mov}}(x', \mathbf{r}') \exp\left(\frac{ik(\mathbf{r} - \mathbf{r}')^2}{2(x - x')}\right) d^2r'. \tag{7.66}$$

Using equation (7.59), $\widehat{\Psi}_1(x, \mathbf{q})$ is expressed in terms of $\Psi_1(x, \mathbf{r})$. Substituting with equation (7.66) and replacing $\varepsilon_{\text{mov}}(x, \mathbf{r})$ by the right-hand side of equation (7.37), we obtain

$$\widehat{\Psi}_1(x, \mathbf{q}) = \frac{ik}{2} \int_0^x \widehat{\varepsilon}_{\text{mov}}(x', \mathbf{q}) \exp\left(-\frac{i(x - x')q^2}{2k}\right) dx'. \tag{7.67}$$

The random field $\widehat{\varepsilon}_{\text{mov}}(x, \mathbf{q})$ appearing in this formula has the following property

$$\widehat{\varepsilon}^*_{\text{mov}}(x, \mathbf{q}) = \widehat{\varepsilon}_{\text{mov}}(x, -\mathbf{q}). \tag{7.68}$$

This equality can be proved by complex conjugation of both sides of equation (7.37) and taking into account that $\widehat{\varepsilon}_{\text{mov}}(x, \mathbf{r})$ is a real function. Conjugating both sides of equation (7.67), using equation (7.68), and replacing $\mathbf{q}$ with $-\mathbf{q}$, we obtain

$$\widehat{\Psi}^*_1(x, -\mathbf{q}) = -\frac{ik}{2} \int_0^x \widehat{\varepsilon}_{\text{mov}}(x', \mathbf{q}) \exp\left(\frac{i(x - x')q^2}{2k}\right) dx'. \tag{7.69}$$

Substituting this equation and (7.67) into equations (7.62) and (7.63), we obtain the spatial spectra of the phase and log-amplitude fluctuations

$$\widehat{\phi}(x, \mathbf{q}) = \frac{k}{2} \int_0^x \widehat{\varepsilon}_{\text{mov}}(x', \mathbf{q}) \cos\left(\frac{(x - x')q^2}{2k}\right) dx', \tag{7.70}$$

$$\widehat{\chi}(x, \mathbf{q}) = \frac{k}{2} \int_0^x \widehat{\varepsilon}_{\text{mov}}(x', \mathbf{q}) \sin\left(\frac{(x - x')q^2}{2k}\right) dx'. \tag{7.71}$$

Substituting the values of $\widehat{\phi}(x, \mathbf{q})$ and $\widehat{\chi}(x, \mathbf{q})$ into equations (7.64) and (7.65), respectively, we express the correlation functions of the phase and log-amplitude fluctuations in terms of the correlation function of the random field $\widehat{\varepsilon}_{\text{mov}}(x', \mathbf{q})$:

$$B_\phi(x; \mathbf{r}_1, \mathbf{r}_2) = \frac{k^2}{4} \int_0^x dx_1 \int_0^x dx_2 \int d^2q_1 \int \langle \widehat{\varepsilon}_{\text{mov}}(x_1, \mathbf{q}_1) \widehat{\varepsilon}_{\text{mov}}(x_2, \mathbf{q}_2) \rangle$$
$$\times \cos\left(\frac{(x - x_1)q_1^2}{2k}\right) \cos\left(\frac{(x - x_2)q_2^2}{2k}\right) \exp\left(i\mathbf{q}_1 \cdot \mathbf{r}_1 + i\mathbf{q}_2 \cdot \mathbf{r}_2\right) d^2q_2, \tag{7.72}$$

$$B_\chi(x; \mathbf{r}_1, \mathbf{r}_2) = \frac{k^2}{4} \int_0^x dx_1 \int_0^x dx_2 \int d^2q_1 \int \langle \widehat{\varepsilon}_{\text{mov}}(x_1, \mathbf{q}_1) \widehat{\varepsilon}_{\text{mov}}(x_2, \mathbf{q}_2) \rangle$$
$$\times \sin\left(\frac{(x - x_1)q_1^2}{2k}\right) \sin\left(\frac{(x - x_2)q_2^2}{2k}\right) \exp\left(i\mathbf{q}_1 \cdot \mathbf{r}_1 + i\mathbf{q}_2 \cdot \mathbf{r}_2\right) d^2q_2. \tag{7.73}$$

The only difference between these correlation functions is that the cosine functions in the formula for B_ϕ are replaced with the sine functions of the same arguments in the formula for B_χ. In these formulas, the correlation function $\langle \widehat{\varepsilon}_{\mathrm{mov}}(x_1, \mathbf{q}_1) \widehat{\varepsilon}_{\mathrm{mov}}(x_2, \mathbf{q}_2) \rangle \equiv \widehat{B}_{\mathrm{eff}}(x_1, \mathbf{q}_1; x_2, \mathbf{q}_2)$ is given by the right-hand side of equation (7.40). Substituting with this equation, calculating the integral over x_2, and introducing the center $\mathbf{r}_c$ and difference $\mathbf{r}_d$ coordinates (equations (7.14) and (7.15)), we have for the correlation function of phase fluctuations

$$
B_\phi(x; \mathbf{r}_c, \mathbf{r}_d) = \frac{k^2}{8\pi} \int_0^x dx_1 \int d^2 q_1 \int d^2 q_2 \int \Phi_{\mathrm{eff}} \left(x_1, \mathbf{r}_c'; 0, \frac{\mathbf{q}_1 - \mathbf{q}_2}{2} \right)
$$
$$
\times \cos \left(\frac{(x - x_1) q_1^2}{2k} \right) \cos \left(\frac{(x - x_1) q_2^2}{2k} \right)
$$
$$
\times \exp \left[i \left(\mathbf{q}_1 + \mathbf{q}_2 \right) \cdot \left(\mathbf{r}_c - \mathbf{r}_c' \right) + i \left(\mathbf{q}_1 - \mathbf{q}_2 \right) \cdot \mathbf{r}_d / 2 \right] d^2 r_c'. \quad (7.74)
$$

Here, $B_\phi(x; \mathbf{r}_c, \mathbf{r}_d) \equiv B_\phi(x; \mathbf{r}_1, \mathbf{r}_2)$. In this formula, the domains essential for integration over $\mathbf{q}_1$ and $\mathbf{q}_2$ are given by: $q_1^2 < k/x$ and $q_2^2 < k/x$. Indeed, outside these domains the cosine functions oscillate rapidly, thus canceling the contributions to B_ϕ. Similarly, the domain essential for integration over $\mathbf{r}_c'$ is: $r_c' < 1/|\mathbf{q}_1 + \mathbf{q}_2| < \sqrt{x/k}$. If the scale L_c of the effective spectrum $\Phi_{\mathrm{eff}}(x_c, \mathbf{r}_c'; 0, (\mathbf{q}_1 - \mathbf{q}_2)/2)$ as a function of $\mathbf{r}_c'$ is greater than $\sqrt{x/k}$, in the domain essential for integration over $\mathbf{r}_c'$ the effective spectrum remains almost constant and the argument $\mathbf{r}_c'$ can be replaced with 0. In this case, the integral over $\mathbf{r}_c'$ can be approximated as

$$
\int \Phi_{\mathrm{eff}} \left(x_1, \mathbf{r}_c'; 0, \frac{\mathbf{q}_1 - \mathbf{q}_2}{2} \right) \exp \left[-i \left(\mathbf{q}_1 + \mathbf{q}_2 \right) \cdot \mathbf{r}_c' \right] d^2 r_c'
$$
$$
= 4\pi^2 \delta(\mathbf{q}_1 + \mathbf{q}_2) \Phi_{\mathrm{eff}} \left(x_1, 0; 0, \frac{\mathbf{q}_1 - \mathbf{q}_2}{2} \right). \quad (7.75)
$$

We substitute this result into equation (7.74) and calculate the integral over $\mathbf{q}_2$:

$$
B_\phi(x; \mathbf{r}_d) = \frac{\pi k^2}{2} \int_0^x dx' \int \Phi_{\mathrm{eff}}(x', 0; 0, \boldsymbol{\kappa}_\perp) \cos^2 \left(\frac{(1 - x'/x) \kappa_\perp^2}{2\kappa_F^2} \right) e^{i \boldsymbol{\kappa}_\perp \cdot \mathbf{r}_d} d^2 \kappa_\perp.
$$
$$
(7.76)
$$

Here, $\kappa_F = \sqrt{k/x}$ is a characteristic value of the wavenumber, which is inversely proportional to the scale of the first Fresnel zone $\sqrt{\lambda x}$. Introducing a new integration variable $\eta = x_1/x$, which corresponds to the normalized propagation path, we obtain the correlation function of phase fluctuations of a plane sound wave propagating through a quasi-homogeneous, anisotropic random medium with the sound speed and velocity fluctuations:

$$\mathcal{B}_\phi(x; \mathbf{r}_d) = \frac{\pi k^2 x}{2} \int_0^1 d\eta \int \Phi_{\text{eff}}(\eta x, 0; 0, \boldsymbol{\kappa}_\perp) \cos^2\left(\frac{(1-\eta)\kappa_\perp^2}{2\kappa_F^2}\right) e^{i\boldsymbol{\kappa}_\perp \cdot \mathbf{r}_d} d^2\kappa_\perp.$$

$$\text{(7.77)}$$

Similarly, starting with equation (7.73), we obtain the correlation function of log-amplitude fluctuations

$$\mathcal{B}_\chi(x; \mathbf{r}_d) = \frac{\pi k^2 x}{2} \int_0^1 d\eta \int \Phi_{\text{eff}}(\eta x, 0; 0, \boldsymbol{\kappa}_\perp) \sin^2\left(\frac{(1-\eta)\kappa_\perp^2}{2\kappa_F^2}\right) e^{i\boldsymbol{\kappa}_\perp \cdot \mathbf{r}_d} d^2\kappa_\perp.$$

$$\text{(7.78)}$$

It follows from these formulas that both correlation functions do not depend on the center coordinates $\mathbf{r}_c$. In a quasi-homogeneous random medium, the effective spectrum Φ_{eff} in equations (7.76)–(7.78) gradually changes along the sound propagation path. The difference between the correlation functions of phase and log-amplitude fluctuations is that $\mathcal{B}_\phi$ contains a squared-cosine function in the integrand, while $\mathcal{B}_\chi$ contains a squared-sine function of the same argument. Due to the squared-cosine function, the correlation function $\mathcal{B}_\phi$ is affected by the largest inhomogeneities in the medium. On the other hand, due to the squared-sine function, the scale of largest inhomogeneities affecting the correlation function $\mathcal{B}_\chi$ is of order $1/\kappa_F$. For a motionless random medium, equations (7.77)–(7.78) have been presented elsewhere [153].

The variances of the phase and log-amplitude fluctuations, $\langle \phi^2 \rangle$ and $\langle \chi^2 \rangle$, are obtained by setting $\mathbf{r}_d = 0$ in equations (7.77) and (7.78):

$$\langle \phi^2 \rangle = \frac{\pi k^2 x}{2} \int_0^1 d\eta \int \Phi_{\text{eff}}(\eta x, 0; 0, \boldsymbol{\kappa}_\perp) \cos^2\left(\frac{(1-\eta)\kappa_\perp^2}{2\kappa_F^2}\right) d^2\kappa_\perp, \quad \text{(7.79)}$$

$$\langle \chi^2 \rangle = \frac{\pi k^2 x}{2} \int_0^1 d\eta \int \Phi_{\text{eff}}(\eta x, 0; 0, \boldsymbol{\kappa}_\perp) \sin^2\left(\frac{(1-\eta)\kappa_\perp^2}{2\kappa_F^2}\right) d^2\kappa_\perp. \quad \text{(7.80)}$$

In the Rytov method, there is no restriction on the value of the variance $\langle \phi^2 \rangle$; the variance $\langle \chi^2 \rangle$ must remain less than 1. (For more details, see reference [339].)

Equations (7.77)–(7.80) are valid for a quasi-homogeneous, anisotropic random medium. For a quasi-homogeneous, isotropic random medium, Φ_{eff} in these formulas is given by equation (7.26), and for a homogeneous, anisotropic random medium, it is given by equation (7.27). For the latter two cases, equations (7.77) and (7.78) can be simplified; the corresponding results can be readily obtained and are not presented here. For the case of a homogeneous and isotropic random medium, when Φ_{eff} is given by equation (7.29), the correlation functions become:

$$\mathcal{B}_\phi(x; r_d) = \frac{\pi^2 k^2 x}{2} \int_0^\infty \Phi_{\text{eff}}(0, \kappa_\perp) J_0(\kappa_\perp r_d) \left(1 + \frac{\kappa_F^2}{\kappa_\perp^2} \sin \frac{\kappa_\perp^2}{\kappa_F^2}\right) \kappa_\perp \, d\kappa_\perp,$$

(7.81)

$$\mathcal{B}_\chi(x; r_d) = \frac{\pi^2 k^2 x}{2} \int_0^\infty \Phi_{\text{eff}}(0, \kappa_\perp) J_0(\kappa_\perp r_d) \left(1 - \frac{\kappa_F^2}{\kappa_\perp^2} \sin \frac{\kappa_\perp^2}{\kappa_F^2}\right) \kappa_\perp \, d\kappa_\perp,$$

(7.82)

and the variances of the phase and log-amplitude fluctuations are given by

$$\langle \phi^2 \rangle = \frac{\pi^2 k^2 x}{2} \int_0^\infty \Phi_{\text{eff}}(0, \kappa_\perp) \left(1 + \frac{\kappa_F^2}{\kappa_\perp^2} \sin \frac{\kappa_\perp^2}{\kappa_F^2}\right) \kappa_\perp \, d\kappa_\perp,$$

(7.83)

$$\langle \chi^2 \rangle = \frac{\pi^2 k^2 x}{2} \int_0^\infty \Phi_{\text{eff}}(0, \kappa_\perp) \left(1 - \frac{\kappa_F^2}{\kappa_\perp^2} \sin \frac{\kappa_\perp^2}{\kappa_F^2}\right) \kappa_\perp \, d\kappa_\perp.$$

(7.84)

For the Kolmogorov spectrum, the variance and correlation function of the phase fluctuations cannot be determined, since the corresponding integrals diverge. For this spectrum, phase fluctuations can be characterized by the structure function of phase fluctuations $D_\phi(x; \mathbf{r}_1, \mathbf{r}_2) = \langle [\phi(x, \mathbf{r}_1) - \phi(x, \mathbf{r}_2)]^2 \rangle$. A simple way to derive a formula for D_ϕ is to assume that B_ϕ exists and use equation (6.12) relating the structure and correlation functions:

$$\mathcal{D}_\phi(x; \mathbf{r}_d) = 2[\mathcal{B}_\phi(x; 0) - \mathcal{B}_\phi(x; \mathbf{r}_d)],$$

(7.85)

where $\mathbf{r}_d = \mathbf{r}_1 - \mathbf{r}_2$. This formula is valid for a homogeneous and isotropic random medium. Substituting $\mathcal{B}_\phi$ from equation (7.81) into this formula, one obtains

$$\mathcal{D}_\phi(x; r_d) = \pi^2 k^2 x \int_0^\infty \Phi_{\text{eff}}(0, \kappa_\perp) \left[1 - J_0(\kappa_\perp r_d)\right] \left(1 + \frac{\kappa_F^2}{\kappa_\perp^2} \sin \frac{\kappa_\perp^2}{\kappa_F^2}\right) \kappa_\perp \, d\kappa_\perp.$$

(7.86)

It can be shown that this formula is valid even if B_ϕ cannot be determined.

In equations (7.81)–(7.84) and (7.86), the terms proportional to $\sin(\kappa_\perp^2/\kappa_F^2)$ describe diffraction of a sound wave by random inhomogeneities. For a relatively short propagation range when geometrical acoustics is valid, the argument $\kappa_\perp^2/\kappa_F^2$ is small and the sine function can be decomposed into the Taylor series. For the phase fluctuations, $\sin(\kappa_\perp^2/\kappa_F^2)$ can be approximated as $\kappa_\perp^2/\kappa_F^2$, and for the log-amplitude fluctuations $\sin(\kappa_\perp^2/\kappa_F^2) = \kappa_\perp^2/\kappa_F^2 - \kappa_\perp^6/(6\kappa_F^6)$. As a result of these approximations, equations (7.81)–(7.84) and (7.86) provide formulas for the variances and correlation functions of phase and log-amplitude fluctuations in the geometrical acoustics limit.

7.2.4 Spherical wave propagation

In a medium without random inhomogeneities (i.e., $\varepsilon_{\text{mov}} = 0$), the sound pressure due to a unit-amplitude point source located at the origin of the coordinate system is given by equation (6.93) with $\mathbf{R}_0 = 0$. In the parabolic equation approximation, this equation takes the form

$$p^{(0)}(x, \mathbf{r}) = \frac{1}{x} \exp\left(ikx + \frac{ikr^2}{2x}\right). \tag{7.87}$$

Using this formula and equation (7.1), we have for the complex amplitude appearing in equation (7.53): $A^{(0)}(x, \mathbf{r}) = (1/x) \exp\left[ikr^2/(2x)\right]$. As a result, the complex phase fluctuation Ψ_1 for an initially spherical wave is

$$\Psi_1(x, \mathbf{r}) = \frac{xk^2}{4\pi} \exp\left(-\frac{ikr^2}{2x}\right) \int_0^x \frac{dx'}{x'(x - x')}$$

$$\times \int \varepsilon_{\text{mov}}(x', \mathbf{r}') \exp\left[\frac{ik(\mathbf{r} - \mathbf{r}')^2}{2(x - x')} + \frac{ik(\mathbf{r}')^2}{2x'}\right] d^2r'. \tag{7.88}$$

In this formula, Ψ_1 and ε_{mov} are replaced by their values from equations (7.59) and (7.37), respectively. Solving the resulting equation for $\widehat{\Psi}_1$, we have

$$\widehat{\Psi}_1(x, \mathbf{q}) = \frac{ik}{2} \int_0^x \widehat{\varepsilon}_{\text{mov}}\left(x', \frac{x\mathbf{q}}{x'}\right) \frac{x^2}{(x')^2} \exp\left[-\frac{ix(x - x')q^2}{2kx'}\right] dx'. \tag{7.89}$$

Conjugating both sides of this formula and using equation (7.68), we obtain

$$\widehat{\Psi}_1^*(x, -\mathbf{q}) = -\frac{ik}{2} \int_0^x \widehat{\varepsilon}_{\text{mov}}\left(x', \frac{x\mathbf{q}}{x'}\right) \frac{x^2}{(x')^2} \exp\left[\frac{ix(x - x')q^2}{2kx'}\right] dx'. \tag{7.90}$$

Substituting the values of $\widehat{\Psi}_1(x, \mathbf{q})$ and $\widehat{\Psi}_1^*(x, -\mathbf{q})$ into equations (7.62) and (7.63), we obtain the spatial spectra of the phase and log-amplitude fluctuations:

$$\widehat{\phi}(x, \mathbf{q}) = \frac{k}{2} \int_0^x \widehat{\varepsilon}_{\text{mov}}\left(x', \frac{x\mathbf{q}}{x'}\right) \cos\left(\frac{x(x - x')q^2}{2kx'}\right) \left(\frac{x}{x'}\right)^2 dx', \tag{7.91}$$

$$\widehat{\chi}(x, \mathbf{q}) = \frac{k}{2} \int_0^x \widehat{\varepsilon}_{\text{mov}}\left(x', \frac{x\mathbf{q}}{x'}\right) \sin\left(\frac{x(x - x')q^2}{2kx'}\right) \left(\frac{x}{x'}\right)^2 dx'. \tag{7.92}$$

Substituting the value of $\widehat{\phi}$ into equation (7.64), we have, for the correlation function of phase fluctuations,

$$B_\phi(x; \mathbf{r}_1, \mathbf{r}_2) = \frac{k^2}{4} \int_0^x dx_1 \int_0^x dx_2 \int d^2q_1 \int \exp\left(i\mathbf{q}_1 \cdot \mathbf{r}_1 + i\mathbf{q}_2 \cdot \mathbf{r}_2\right)$$

$$\times \cos\left(\frac{x(x - x_1)q_1^2}{2kx_1}\right) \cos\left(\frac{x(x - x_2)q_2^2}{2kx_2}\right) \frac{x^4}{x_1^2 x_2^2}$$

$$\times \langle \widehat{\varepsilon}_{\text{mov}}(x_1, x\mathbf{q}_1/x_1) \widehat{\varepsilon}_{\text{mov}}(x_2, x\mathbf{q}_2/x_2) \rangle \, d^2q_2. \tag{7.93}$$

In this formula, the integration variables $\mathbf{q}_1$ and $\mathbf{q}_2$ are replaced with $\mathbf{a}_1 = (x/x_1)\mathbf{q}_1$ and $\mathbf{a}_2 = (x/x_2)\mathbf{q}_2$, respectively, and the correlation function $\langle \widehat{\varepsilon}_{\mathrm{mov}}(x_1, \mathbf{a}_1)\widehat{\varepsilon}_{\mathrm{mov}}(x_2, \mathbf{a}_2)\rangle$ is replaced with the right-hand side of equation (7.40). Calculating the integral over x_2 and introducing the center $\mathbf{r}_c$ and difference $\mathbf{r}_d$ coordinates (equations (7.14) and (7.15)), we obtain

$$
B_\phi(x; \mathbf{r}_c, \mathbf{r}_d) = \frac{k^2}{8\pi} \int_0^x dx_1 \int d^2 a_1 \int d^2 a_2 \int \Phi_{\mathrm{eff}}\left(x_1, \mathbf{r}_c'; 0, \frac{\mathbf{a}_1 - \mathbf{a}_2}{2}\right)
$$
$$
\times \cos\left(\frac{x_1(x - x_1)a_1^2}{2kx}\right) \cos\left(\frac{x_1(x - x_1)a_2^2}{2kx}\right)
$$
$$
\times \exp\left\{i\left[(\mathbf{a}_1 + \mathbf{a}_2) \cdot \left(\mathbf{r}_c\frac{x_1}{x} - \mathbf{r}_c'\right) + (\mathbf{a}_1 - \mathbf{a}_2) \cdot \mathbf{r}_d\frac{x_1}{2x}\right]\right\} d^2 r_c'. \quad (7.94)
$$

As for plane wave propagation, we assume that the scale L_c of $\Phi_{\mathrm{eff}}(x_c, \mathbf{r}_c; 0, \boldsymbol{\kappa}_\perp)$ as a function of $\mathbf{r}_c$ is greater than $\sqrt{x/k}$. In this case, the integral over $\mathbf{r}_c'$ can be approximated by the right-hand side of equation (7.75). Calculating the integral over $\mathbf{a}_2$ in the resulting formula, introducing a new integration variable $\eta = x_1/x$, and denoting $\mathbf{a}_1 = \boldsymbol{\kappa}_\perp$, we obtain a formula for the correlation function of the phase fluctuations of a spherical sound wave propagating through a quasi-homogeneous, anisotropic random medium:

$$
B_\phi(x; \mathbf{r}_d) = \frac{\pi k^2 x}{2} \int_0^1 d\eta \int \Phi_{\mathrm{eff}}(\eta x, 0; 0, \boldsymbol{\kappa}_\perp)
$$
$$
\times \cos^2\left(\frac{\eta(1 - \eta)\kappa_\perp^2}{2\kappa_F^2}\right) \exp\left(i\eta\boldsymbol{\kappa}_\perp \cdot \mathbf{r}_d\right) d^2\kappa_\perp. \quad (7.95)
$$

The following formula for the correlation function of the log-amplitude fluctuations of a spherical wave can be derived analogously:

$$
B_\chi(x; \mathbf{r}_d) = \frac{\pi k^2 x}{2} \int_0^1 d\eta \int \Phi_{\mathrm{eff}}(\eta x, 0; 0, \boldsymbol{\kappa}_\perp)
$$
$$
\times \sin^2\left(\frac{\eta(1 - \eta)\kappa_\perp^2}{2\kappa_F^2}\right) \exp\left(i\eta\boldsymbol{\kappa}_\perp \cdot \mathbf{r}_d\right) d^2\kappa_\perp. \quad (7.96)
$$

These formulas coincide with those for a plane wave propagation (equations (7.77) and (7.78)), if the arguments of the exponential and trigonometric functions are divided by η. Formulas for the variances of phase and log-amplitude fluctuations, $\langle \phi^2\rangle$ and $\langle \chi^2\rangle$, of a spherical sound wave are obtained by setting $\mathbf{r}_d = 0$ in equations (7.95) and (7.96). For a motionless random medium, equations (7.95) and (7.96) coincide with equation (2.75) in reference [207] or equation (17-57) in reference [178].

For the case of a homogeneous and isotropic random medium, when Φ_{eff} is given by equation (7.29), the correlation functions of phase and log-amplitude fluctuations simplify to

$$
\mathcal{B}_\phi(x, r_d) = \frac{\pi^2 k^2 x}{2} \int_0^1 d\eta \int_0^\infty \Phi_{\text{eff}}(0, \kappa_\perp) J_0(\eta \kappa_\perp r_d)
$$
$$
\times \left[1 + \cos\left(\frac{\eta(1-\eta)\kappa_\perp^2}{\kappa_F^2} \right) \right] \kappa_\perp \, d\kappa_\perp, \quad (7.97)
$$

$$
\mathcal{B}_\chi(x, r_d) = \frac{\pi^2 k^2 x}{2} \int_0^1 d\eta \int_0^\infty \Phi_{\text{eff}}(0, \kappa_\perp) J_0(\eta \kappa_\perp r_d)
$$
$$
\times \left[1 - \cos\left(\frac{\eta(1-\eta)\kappa_\perp^2}{\kappa_F^2} \right) \right] \kappa_\perp \, d\kappa_\perp. \quad (7.98)
$$

The variances of phase and log-amplitude fluctuations are given by

$$
\langle \phi^2 \rangle = \frac{\pi^2 k^2 x}{2} \int_0^1 d\eta \int_0^\infty \Phi_{\text{eff}}(0, \kappa_\perp) \left[1 + \cos\left(\frac{\eta(1-\eta)\kappa_\perp^2}{\kappa_F^2} \right) \right] \kappa_\perp \, d\kappa_\perp,
$$
$$
(7.99)
$$
$$
\langle \chi^2 \rangle = \frac{\pi^2 k^2 x}{2} \int_0^1 d\eta \int_0^\infty \Phi_{\text{eff}}(0, \kappa_\perp) \left[1 - \cos\left(\frac{\eta(1-\eta)\kappa_\perp^2}{\kappa_F^2} \right) \right] \kappa_\perp \, d\kappa_\perp.
$$
$$
(7.100)
$$

Using equations (7.85) and (7.97), one obtains the structure function of phase fluctuations of a spherical sound wave propagating through a homogeneous and isotropic random medium:

$$
\mathcal{D}_\phi(x, r_d) = \pi^2 k^2 x \int_0^1 d\eta \int_0^\infty \Phi_{\text{eff}}(0, \kappa_\perp) \left[1 - J_0(\eta \kappa_\perp r_d) \right]
$$
$$
\times \left[1 + \cos\left(\frac{\eta(1-\eta)\kappa_\perp^2}{\kappa_F^2} \right) \right] \kappa_\perp \, d\kappa_\perp. \quad (7.101)
$$

In the geometrical acoustics approximation in equations (7.97), (7.99), and (7.101) for the phase fluctuations, a cosine function is replaced by 1. In equations (7.98) and (7.100) for the log-amplitude fluctuations, it is replaced by $1 - \alpha^2/2$, where $\alpha = \eta(1-\eta)\kappa_\perp^2/\kappa_F^2$.

7.3 Phase and log-amplitude fluctuations for the turbulence spectra

Here, we use the theory developed in the previous section to derive the variances and correlation functions of phase and log-amplitude fluctuations of plane and spherical sound waves propagating through homogeneous and isotropic turbulence described by the von Kármán, Kolmogorov, and Gaussian spectra. To do so, we must calculate the integrals over η and $\kappa_\perp$ considered in section 7.2. For the Kolmogorov spectra, the calculations are similar to those for the case of sound propagation through a motionless random medium. The formulas presented below enable the analysis of the dependence of the variances and correlation functions on parameters of the problem. For some statistical moments, although analytical expressions cannot be obtained, the moments can be readily calculated numerically.

7.3.1 Statistical moments of plane waves

7.3.1.1 Von Kármán spectrum

Substituting the von Kármán effective spectrum $\Phi_{\text{eff}}(0, \kappa_\perp)$ given by equation (7.30) into equations (7.83) and (7.84), and calculating the integrals over $\kappa_\perp$, one obtains the variances of phase and log-amplitude fluctuations:

$$\langle \phi^2 \rangle = \frac{\pi^{1/2}\Gamma(5/6)k^2 x}{4\Gamma(1/3)} \left[[1 + E(D_T)] \frac{\beta_c^2 \sigma_T^2 L_T}{T_0^2} + [1 + H(D_v)] \frac{4\sigma_v^2 L_v}{c_0^2} \right],$$
(7.102)

$$\langle \chi^2 \rangle = \frac{\pi^{1/2}\Gamma(5/6)k^2 x}{4\Gamma(1/3)} \left[[1 - E(D_T)] \frac{\beta_c^2 \sigma_T^2 L_T}{T_0^2} + [1 - H(D_v)] \frac{4\sigma_v^2 L_v}{c_0^2} \right].$$
(7.103)

Here, $D_T = x/(kL_T^2)$ and $D_v = x/(kL_v^2)$ are the *wave parameters* corresponding to the temperature and velocity fluctuations. The limiting case $D_T, D_v \ll 1$ corresponds to the geometrical acoustics approximation. The opposite limiting case $D_T, D_v \gg 1$ is the range of applicability of Fraunhofer diffraction. In equations (7.102) and (7.103), the functions $E(D_T)$ and $H(D_v)$ are given by

$$E(D_T) = {}_2F_3\left(1, \frac{1}{2}; \frac{3}{2}, \frac{7}{12}, \frac{1}{12}; -\frac{D_T^2}{4}\right)$$
$$- \frac{12\pi \sin(\pi/12)}{11\Gamma(5/6)} D_T^{5/6} {}_1F_2\left(\frac{11}{12}; \frac{1}{2}, \frac{23}{12}; -\frac{D_T^2}{4}\right)$$
$$+ \frac{12\pi \cos(\pi/12)}{17\Gamma(5/6)} D_T^{11/6} {}_1F_2\left(\frac{17}{12}; \frac{3}{2}, \frac{29}{12}; -\frac{D_T^2}{4}\right), \quad (7.104)$$

$$H(D_v) = \frac{2\pi}{\Gamma(5/6)} D_v^{5/6} \sin\left(D_v - \frac{\pi}{12}\right) + {}_1F_2\left(1; \frac{1}{12}, \frac{7}{12}; -\frac{D_v^2}{4}\right), \quad (7.105)$$

where ${}_1F_2$ and ${}_2F_3$ are the generalized hypergeometric functions. As explained in section 7.1.4, to simplify equations (7.102) and (7.103), we set $\sigma_C = \sigma_{CT} =$

0. The results for finite variances σ_C and σ_{CT} can be readily obtained by replacing the subscript T in equations (7.102) and (7.103) with subscripts C and CT and using the coefficients in front of the variances consistent with equation (7.30). It follows from equations (7.102) and (7.103) that the relative contributions from temperature and velocity fluctuations to $\langle \phi^2 \rangle$ and $\langle \chi^2 \rangle$ have different dependences on the wave parameters D_T and D_v. This difference is due to the fact that temperature and velocity fluctuations are scalar and vector random fields, respectively, and scatter sound in the forward direction differently. (See also the text following equations (7.4) and (7.30).)

Figures 7.2 and 7.3 depict the variances of phase and log-amplitude fluctuations, $\langle \phi^2 \rangle$ and $\langle \chi^2 \rangle$, calculated with equations (7.102) and (7.103) for the case of sound propagation in the atmosphere when the coefficient $\beta_c = 1$ (section 6.3). The variances are potted versus the surface heat flux Q_H for three values of the friction velocity $u_* = 0.1$ m/s, 0.3 m/s, and 0.6 m/s, which are representative of light, moderate, and strong wind. The limiting values of Q_H (0 W/m^2 and 200 W/m^2) correspond to cloudy and sunny conditions. The propagation range and height above the ground are $x = 1$ km and $z = 20$ m, respectively, the acoustic frequency is $f = kc_0/(2\pi) = 150$ Hz, the height of the ABL $z_i = 1$ km, and the surface temperature is $T_s = 18$ °C. These parameters enable determination (section 6.2.4) of the values of σ_T^2, σ_v^2, L_T, and L_v appearing in equations (7.102) and (7.103). It follows from figures 7.2 and 7.3 that the variance of phase fluctuations $\langle \phi^2 \rangle$ is in the range 0.09–27.6, while the variance of log-amplitude fluctuations $\langle \chi^2 \rangle$ in the range 0.02–0.98. Both variances increase with increasing heat flux and friction velocity. The values of $\langle \chi^2 \rangle$ for strong wind are of the order of 1, indicating that the Rytov method might not be valid in this case.

Substitution of the von Kármán effective spectrum $\Phi_{\text{eff}}(0, \kappa_\perp)$ into equations (7.81) and (7.82) for the correlation functions of phase and log-amplitude fluctuations results in the integrals over $\kappa_\perp$, which cannot be evaluated analytically.

7.3.1.2   Kolmogorov spectrum

The variances of phase and log-amplitude fluctuations are obtained by substituting the Kolmogorov effective spectrum $\Phi_{\text{eff}}(0, \kappa_\perp)$, equation (7.31), into equations (7.83) and (7.84). For the phase variance, the integral diverges so that $\langle \phi^2 \rangle \to \infty$ (cannot be determined). For the log-amplitude variance, calculating the integral over $\kappa_\perp$ similarly to that in reference [339], where the case of a motionless random medium was considered, one obtains:

$$\langle \chi^2 \rangle = \frac{\sqrt{3\pi} \sin(\pi/12)\Gamma(1/3)}{2^{1/3} 22} k^{7/6} x^{11/6} \left(\frac{\beta_c^2 C_T^2}{T_0^2} + \frac{22}{3} \frac{C_v^2}{c_0^2} \right). \tag{7.106}$$

The value of the numerical coefficient in this formula is 0.077. In equation (7.106) and other equations for the statistical moments of a sound field for the Kolmogorov spectra to be presented, the relative contributions from temperature and velocity fluctuations to the statistical moments have the same

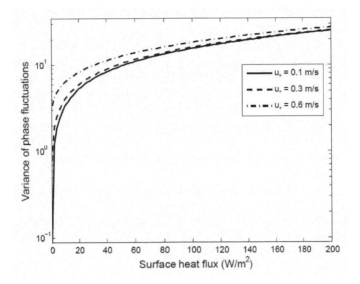

FIGURE 7.2
Variance of phase fluctuations of a plane sound wave versus the surface heat Q_H flux for three values of the friction velocity u_*. The propagation range is $x = 1$ km and the acoustic frequency $f = 150$ Hz.

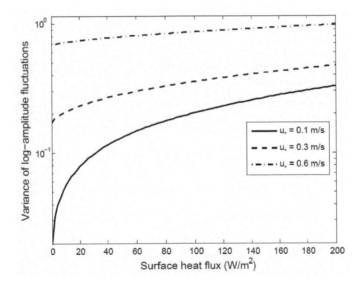

FIGURE 7.3
Variance of log-amplitude fluctuations of a plane sound wave versus the surface heat Q_H flux for three values of the friction velocity u_*. The propagation range is $x = 1$ km and the acoustic frequency $f = 150$ Hz.

dependencies on x, k, and, for the correlation and coherence functions, r_d. In all these equations, the coefficient in front of C_v^2 is $22/3$.

The results for $\langle \phi^2 \rangle$ and $\langle \chi^2 \rangle$ can also be obtained as limiting cases of equations (7.102) and (7.103) valid for the von Kármán spectra. Indeed, assuming that $L_T, L_v \to 0$ (so that $D_T, D_v \ll 1$), expanding the functions $E(D_T)$ and $H(D_v)$ into series in D_T and D_v, and expressing σ_T^2 and σ_v^2 in terms of C_T^2 and C_v^2 (equation (6.36)), it can be shown that $\langle \phi^2 \rangle \to \infty$ and $\langle \chi^2 \rangle$ is given by equation (7.106). In this limiting case, the inhomogeneities in the inertial subrange affect the phase and log-amplitude fluctuations and, as one would expect, the results for the von Kármán spectra coincide with those for the Kolmogorov spectra.

The formula for the correlation function of log-amplitude fluctuations is obtained by substituting the Kolmogorov effective spectrum $\Phi_{\text{eff}}(0, \kappa_\perp)$ into equation (7.82) and calculating the integral over $\kappa_\perp$ similarly to that in reference [374]

$$\mathcal{B}_\chi(x; r_d) = \langle \chi^2 \rangle b_\chi(x; r_d). \tag{7.107}$$

Here, $\langle \chi^2 \rangle$ is given by equation (7.106) and $b_\chi(x; r_d)$ is the normalized correlation function:

$$b_\chi(x; r_d) = \mathrm{Re}\, {}_1F_1\left(-\frac{11}{6}; 1; \frac{i\kappa_F^2 r_d^2}{4}\right) - \cot\frac{\pi}{12}\, \mathrm{Im}\, {}_1F_1\left(-\frac{11}{6}; 1; \frac{i\kappa_F^2 r_d^2}{4}\right)$$
$$-\frac{11}{6\Gamma\left(\frac{11}{6}\right)\sin\frac{\pi}{12}}\left(\frac{\kappa_F^2 r_d^2}{4}\right)^{5/6}, \tag{7.108}$$

where ${}_1F_1$ is the confluent hypergeometric function.

For the Kolmogorov spectrum, $\mathcal{B}_\phi(x; r_d)$ cannot be determined since $\langle \phi^2 \rangle \to \infty$. The structure function of phase fluctuations is obtained by substituting $\Phi_{\text{eff}}(0, \kappa_\perp)$ from equation (7.31) into equation (7.86) and calculating the integral over $\kappa_\perp$ [374]:

$$\mathcal{D}_\phi(x; r_d) = \left\{0.73k^2 x r_d^{5/3} - 0.154k^{7/6}x^{11/6}\left[1 - b_\chi(x, r_d)\right]\right\}\left(\frac{\beta_c^2 C_T^2}{T_0^2} + \frac{22}{3}\frac{C_v^2}{c_0^2}\right). \tag{7.109}$$

7.3.1.3   Gaussian spectrum

The effective spectrum $\Phi_{\text{eff}}(0, \kappa_\perp)$ for the Gaussian model is given by equation (7.36). The variances of phase and log-amplitude fluctuations are obtained by substituting this spectrum into equations (7.83) and (7.84) and calculating the resulting integrals:

$$\langle \phi^2 \rangle = \frac{\sqrt{\pi}k^2 x}{8}\left[\left(1 + \frac{\arctan \mathcal{D}_T}{\mathcal{D}_T}\right)\frac{\beta_c^2 \sigma_T^2 \mathcal{L}_T}{T_0^2} + \left(1 + \frac{1}{1 + \mathcal{D}_v^2}\right)\frac{4\sigma_v^2 \mathcal{L}_v}{c_0^2}\right], \tag{7.110}$$

$$\langle \chi^2 \rangle = \frac{\sqrt{\pi} k^2 x}{8} \left[\left(1 - \frac{\arctan \mathcal{D}_T}{\mathcal{D}_T} \right) \frac{\beta_c^2 \sigma_T^2 \mathcal{L}_T}{T_0^2} + \left(1 - \frac{1}{1 + \mathcal{D}_v^2} \right) \frac{4 \sigma_v^2 \mathcal{L}_v}{c_0^2} \right].$$
(7.111)

Here, $\mathcal{D}_T = 4x / \left(k \mathcal{L}_T^2 \right)$ and $\mathcal{D}_v = 4x / \left(k \mathcal{L}_v^2 \right)$ are the wave parameters pertinent to the Gaussian spectra. It follows from these equations that the relative contributions from temperature and velocity fluctuations to $\langle \phi^2 \rangle$ and $\langle \chi^2 \rangle$ have different dependencies on the wave parameters $\mathcal{D}_T$ and $\mathcal{D}_v$. Equations (7.110) and (7.111) have been given elsewhere for $\sigma_v^2 = 0$ [339].

In the limiting case $\mathcal{D}_T, \mathcal{D}_v \gg 1$, equations (7.110) and (7.111) simplify to

$$\langle \phi^2 \rangle = \langle \chi^2 \rangle = \frac{\sqrt{\pi} k^2 x}{8} \left(\frac{\beta_c^2 \sigma_T^2 \mathcal{L}_T}{T_0^2} + \frac{4 \sigma_v^2 \mathcal{L}_v}{c_0^2} \right).$$
(7.112)

These formulas can also be obtained as a limiting case of equations (7.102) and (7.103) for the von Kármán spectra. Assuming that $\mathcal{D}_T, \mathcal{D}_v \gg 1$, taking the asymptotic values of functions E and H for large values of the arguments, and replacing the scales L_T, L_v with $\mathcal{L}_T$, $\mathcal{L}_v$ (equations (6.48) and (6.49)), it can be shown that $\langle \phi^2 \rangle$ and $\langle \chi^2 \rangle$ are given by equation (7.112). In this limiting case, the large-scale inhomogeneities in the energy subrange affect the variances of phase and log-amplitude fluctuations and, as one would expect, the results for the von Kármán spectra coincide with those for the Gaussian spectra. Due to the condition $\langle \chi^2 \rangle < 1$ of the validity of the Rytov method, equation (7.112) might not be applicable to the variance of log-amplitude fluctuations in this case.

Substituting the Gaussian effective spectrum $\Phi_{\text{eff}}(0, \kappa_\perp)$ into equations (7.81) and (7.82), and calculating the integrals over $\kappa_\perp$, one obtains the correlation functions of the phase and log-amplitude fluctuations:

$$B_\phi(x; r_d) = \frac{\sqrt{\pi} k^2 x}{8} \left\{ \left[e^{-r_d^2 / \mathcal{L}_T^2} - \frac{1}{\mathcal{D}_T} \operatorname{Im} E_1 \left(\frac{r_d^2 e^{i \arctan \mathcal{D}_T}}{\mathcal{L}_T^2 \sqrt{1 + \mathcal{D}_T^2}} \right) \right] \frac{\beta_c^2 \sigma_T^2 \mathcal{L}_T}{T_0^2} \right.$$

$$+ \left[\left(1 - \frac{r_d^2}{\mathcal{L}_v^2} \right) e^{-r_d^2 / \mathcal{L}_v^2} + \frac{\exp \left(-\frac{r_d^2}{\mathcal{L}_v^2 (1 + \mathcal{D}_v^2)} \right)}{1 + \mathcal{D}_v^2} \right.$$

$$\left. \left. \times \left[\cos \left(\frac{r_d^2 \mathcal{D}_v}{\mathcal{L}_v^2 (1 + \mathcal{D}_v^2)} \right) - \frac{1}{\mathcal{D}_v} \sin \left(\frac{r_d^2 \mathcal{D}_v}{\mathcal{L}_v^2 (1 + \mathcal{D}_v^2)} \right) \right] \right] \frac{4 \sigma_v^2 \mathcal{L}_v}{c_0^2} \right\}, \quad (7.113)$$

$$B_\chi(x; r_d) = \frac{\sqrt{\pi} k^2 x}{8} \left\{ \left[e^{-r_d^2 / \mathcal{L}_T^2} + \frac{1}{\mathcal{D}_T} \operatorname{Im} E_1 \left(\frac{r_d^2 e^{i \arctan \mathcal{D}_T}}{\mathcal{L}_T^2 \sqrt{1 + \mathcal{D}_T^2}} \right) \right] \frac{\beta_c^2 \sigma_T^2 \mathcal{L}_T}{T_0^2} \right.$$

$$+ \left[\left(1 - \frac{r_d^2}{\mathcal{L}_v^2} \right) e^{-r_d^2 / \mathcal{L}_v^2} - \frac{\exp \left(-\frac{r_d^2}{\mathcal{L}_v^2 (1 + \mathcal{D}_v^2)} \right)}{1 + \mathcal{D}_v^2} \right.$$

$$\left. \left. \times \left[\cos \left(\frac{r_d^2 \mathcal{D}_v}{\mathcal{L}_v^2 (1 + \mathcal{D}_v^2)} \right) - \frac{1}{\mathcal{D}_v} \sin \left(\frac{r_d^2 \mathcal{D}_v}{\mathcal{L}_v^2 (1 + \mathcal{D}_v^2)} \right) \right] \right] \frac{4 \sigma_v^2 \mathcal{L}_v}{c_0^2} \right\}. \quad (7.114)$$

In these formulas, E_1 is the exponential-integral function, the imaginary part of which can be expressed in the form

$$\operatorname{Im} E_1 \left(\frac{r^2 e^{i \arctan \mathcal{D}}}{\mathcal{L}^2 \sqrt{1 + \mathcal{D}^2}} \right) = - \int_{r^2/(\mathcal{L}^2(1+\mathcal{D}^2))}^{\infty} y^{-1} e^{-y} \sin(y\mathcal{D}) \, dy.$$

Equations (7.113) and (7.114) are given for $\sigma_v^2 = 0$ elsewhere [72].

7.3.2 Statistical moments of spherical waves

7.3.2.1 Von Kármán spectrum

Substitution of the von Kármán effective spectrum (7.30) into equations (7.97)–(7.100) for the correlation functions and variances of phase and log-amplitude fluctuations of a spherical sound wave results in the integrals which cannot be evaluated analytically. The integrals can be evaluated numerically, however.

7.3.2.2 Kolmogorov spectrum

The variance of log-amplitude fluctuations is obtained by substituting the Kolmogorov effective spectrum $\Phi_{\text{eff}}(0, \kappa_\perp)$, equation (7.31), into equation (7.100). Evaluating the integrals over η and $\kappa_\perp$ similarly to reference [374], where the case of a motionless random medium is considered, we have

$$\langle \chi^2 \rangle = 0.031 k^{7/6} x^{11/6} \left(\frac{\beta_c^2 C_T^2}{T_0^2} + \frac{22}{3} \frac{C_v^2}{c_0^2} \right). \tag{7.115}$$

This result is identical to equation (7.106) for a plane wave, except for a different value of the numerical coefficient; the variance for a spherical wave is 2.48 times smaller than that for a plane wave.

The correlation function of log-amplitude fluctuations is obtained by substituting the Kolmogorov effective spectrum $\Phi_{\text{eff}}(0, \kappa_\perp)$ into equation (7.98) and calculating the integral over $\kappa_\perp$ similarly to that in reference [178]:

$$\mathcal{B}_\chi(x; r_d) = 0.544 k^{7/6} x^{11/6} \left(\frac{\beta_c^2 C_T^2}{T_0^2} + \frac{22}{3} \frac{C_v^2}{c_0^2} \right)$$
$$\times \operatorname{Re} \int_0^1 {}_1F_1 \left(-\frac{5}{6}; 1; \frac{i\eta \kappa_F^2 r_d^2}{4(1-\eta)} \right) [i\eta(1-\eta)]^{5/6} \, d\eta. \tag{7.116}$$

Substitution of $\Phi_{\text{eff}}(0, \kappa_\perp)$ for the Kolmogorov turbulence model into formulas for $\langle \phi^2 \rangle$ and $\mathcal{B}_\phi(x, r)$ results in divergent integrals as for plane wave propagation. The structure function of phase fluctuations for spherical wave propagation is given by

$$\mathcal{D}_\phi(x; r_d) = 0.544 k^{7/6} x^{11/6} \left(\frac{\beta_c^2 C_T^2}{T_0^2} + \frac{22}{3} \frac{C_v^2}{c_0^2} \right) \int_0^1 \left\{ 0.670 (\eta \kappa_F r_d)^{5/3} \right.$$

$$\left. - \left[1 - {}_1F_1 \left(-\frac{5}{6}; 1; \frac{i \eta \kappa_F^2 r_d^2}{4(1 - \eta)} \right) \right] 2[i\eta(1 - \eta)]^{5/6} \right\} d\eta. \quad (7.117)$$

This formula can be obtained by substituting $\Phi_{\text{eff}}(0, \kappa_\perp)$ from equation (7.31) into equation (7.101) and calculating the integral over $\kappa_\perp$ similarly to the case of a motionless random medium [178].

7.3.2.3 Gaussian spectrum

The variances of phase and log-amplitude fluctuations can be calculated by substituting equation (7.36) for the Gaussian effective spectrum $\Phi_{\text{eff}}(0, \kappa_\perp)$ into equations (7.99) and (7.100). After evaluating the resulting integrals, we obtain

$$\langle \phi^2 \rangle = \frac{\sqrt{\pi} k^2 x}{8} \left\{ [1 + M(\mathcal{D}_T)] \frac{\beta_c^2 \sigma_T^2 \mathcal{L}_T}{T_0^2} + [1 + N(\mathcal{D}_v)] \frac{4 \sigma_v^2 \mathcal{L}_v}{c_0^2} \right\}, \quad (7.118)$$

$$\langle \chi^2 \rangle = \frac{\sqrt{\pi} k^2 x}{8} \left\{ [1 - M(\mathcal{D}_T)] \frac{\beta_c^2 \sigma_T^2 \mathcal{L}_T}{T_0^2} + [1 - N(\mathcal{D}_v)] \frac{4 \sigma_v^2 \mathcal{L}_v}{c_0^2} \right\}. \quad (7.119)$$

Here, the functions $M(\mathcal{D}_T)$ and $N(\mathcal{D}_v)$ are given by

$$M(\mathcal{D}_T) = \frac{\arctan \sqrt{\frac{2}{\Omega_T}} + \frac{\Omega_T \Delta_T}{2} \ln \frac{1 + \Delta_T \sqrt{2\Omega_T}}{1 - \Delta_T \sqrt{2\Omega_T}}}{\Delta_T^2 (\Omega_T + 1) \sqrt{8\Omega_T}},$$

$$N(\mathcal{D}_v) = \frac{\Omega_v (\Omega_v + 2)}{2(\Omega_v + 1)^2} \left[1 + \frac{\sqrt{2\Omega_v}(\Omega_v + 3) \arctan \sqrt{\frac{2}{\Omega_v}}}{4(\Omega_v + 1)} \right.$$

$$\left. + \frac{\Delta_v \sqrt{2\Omega_v}(\Omega_v - 1)(\Omega_v + 2) \ln \frac{1 + \Delta_v \sqrt{2\Omega_v}}{1 - \Delta_v \sqrt{2\Omega_v}}}{8(\Omega_v + 1)} \right],$$

where $\Delta_T = \mathcal{D}_T/4$, $\Omega_T = \sqrt{1 + \Delta_T^{-2}} - 1$ and $\Delta_v = \mathcal{D}_v/4$, $\Omega_v = \sqrt{1 + \Delta_v^{-2}} - 1$. Equations (7.118) and (7.119) with $\sigma_v^2 = 0$ have been derived elsewhere [190]. In the limiting case $\mathcal{D}_T, \mathcal{D}_v \gg 1$, equations (7.118) and (7.119) simplify and coincide with equation (7.112). Thus, in this limiting case, the variances of phase and log-amplitude fluctuations of a spherical wave coincide with those for a plane wave.

Substituting the Gaussian effective spectrum into equations (7.97) and

(7.98) and calculating the integral over $\kappa_\perp$, we obtain the correlation functions of phase and log-amplitude fluctuations

$$
\begin{aligned}
\mathcal{B}_\phi(x; r_d) = \frac{\sqrt{\pi} k^2 x}{8} \int_0^1 & \left\{ \left[e^{-\rho_T^2} + \frac{e^{-\rho_T^2/d_T}}{d_T} \right. \right. \\
& \times \left[\cos\left(\frac{\rho_T^2 \mathcal{D}_T \tau}{d_T} \right) + \mathcal{D}_T \tau \sin\left(\frac{\rho_T^2 \mathcal{D}_T \tau}{d_T} \right) \right] \bigg] \frac{\beta_c^2 \sigma_T^2 \mathcal{L}_T}{T_0^2} \\
+ & \left[(1 - \rho_v^2) e^{-\rho_v^2} + \frac{e^{-\rho_v^2/d_v}}{d_v^3} \left[(1 - \mathcal{D}_v^4 \tau^4 + \rho_v^2(3d_v - 4)) \cos\left(\frac{\rho_v^2 \mathcal{D}_v \tau}{d_v} \right) \right. \right. \\
& \left. \left. + \mathcal{D}_v(2d_v + \rho_v^2(d_v - 4)) \tau \sin\left(\frac{\rho_v^2 \mathcal{D}_v \tau}{d_v} \right) \right] \right] \frac{4 \sigma_v^2 \mathcal{L}_v}{c_0^2} \bigg\} \, d\eta, \quad (7.120)
\end{aligned}
$$

$$
\begin{aligned}
\mathcal{B}_\chi(x; r_d) = \frac{\sqrt{\pi} k^2 x}{8} \int_0^1 & \left\{ \left[e^{-\rho_T^2} - \frac{e^{-\rho_T^2/d_T}}{d_T} \right. \right. \\
& \times \left[\cos\left(\frac{\rho_T^2 \mathcal{D}_T \tau}{d_T} \right) + \mathcal{D}_T \tau \sin\left(\frac{\rho_T^2 \mathcal{D}_T \tau}{d_T} \right) \right] \bigg] \frac{\beta_c^2 \sigma_T^2 \mathcal{L}_T}{T_0^2} \\
+ & \left[(1 - \rho_v^2) e^{-\rho_v^2} - \frac{e^{-\rho_v^2/d_v}}{d_v^3} \left[(1 - \mathcal{D}_v^4 \tau^4 + \rho_v^2(3d_v - 4)) \cos\left(\frac{\rho_v^2 \mathcal{D}_v \tau}{d_v} \right) \right. \right. \\
& \left. \left. + \mathcal{D}_v \tau (2d_v + \rho_v^2(d_v - 4)) \sin\left(\frac{\rho_v^2 \mathcal{D}_v \tau}{d_v} \right) \right] \right] \frac{4 \sigma_v^2 \mathcal{L}_v}{c_0^2} \bigg\} \, d\eta. \quad (7.121)
\end{aligned}
$$

Here, $\tau = \eta(1 - \eta)$, $\rho_T = r_d \eta / \mathcal{L}_T$, $d_T = 1 + \mathcal{D}_T^2 \tau^2$, and $\rho_v = r_d \eta / \mathcal{L}_v$, $d_v = 1 + \mathcal{D}_v^2 \tau^2$. In these formulas, the remaining integrals over η cannot be evaluated analytically.

7.4   Statistical moments of the sound field

In many applications, it is important to know the effect of a random medium on the sound field $p(x, \mathbf{r})$ rather than on its phase and amplitude. In this section, we present a derivation of closed-form equations for the statistical moments of the sound field $p(x, \mathbf{r})$. The moments may be of any order. Using these equations, analytical formulas for the mean sound field and the mutual coherence functions for arbitrary spectra of random inhomogeneities are obtained.

7.4.1   Statistical moments of arbitrary order

A general moment of order $n + m$ of the complex sound pressure is defined as

$$\Gamma_{n,m}(x; \mathbf{r}_1, ..., \mathbf{r}_n; \mathbf{r}'_1, ..., \mathbf{r}'_m) = \langle p(x, \mathbf{r}_1)...p(x, \mathbf{r}_n)p^*(x, \mathbf{r}'_1)...p^*(x, \mathbf{r}'_m)\rangle.$$
(7.122)

Here, n can be $1, 2, \ldots$, and m can be $0, 1, 2, \ldots$. If $m = 0$, p^* does not appear in this formula. In equation (7.122), the sound-pressure fields are taken at the same range x since their decorrelation is usually much less along the x-axis than in the transverse direction. The transverse-longitudinal mutual coherence function is investigated elsewhere [293].

Some particular cases of $\Gamma_{n,m}$ are of special interest. The moment $n = 1$, $m = 0$ is the *mean sound field*. Keeping in mind that the field is complex-valued, the mean field will become small when the phases are randomized, i.e., the field is incoherent. Hence this moment represents the coherently propagating part of the field. The second-order moment $n = 1$, $m = 1$ is often called the *mutual coherence function*; it is one of the most important statistical characteristics of the field, as it indicates the cross correlation between different locations in space. The fourth-order moment $n = 2$ and $m = 2$ enables analysis of the intensity fluctuations of the sound field. Other moments $\Gamma_{n,m}$ such that $n, m > 2$ can be measured with phased microphone arrays.

7.4.2   Derivation of closed-form equations

The derivation of the closed-form equations for $\Gamma_{n,m}$ is similar to that in reference [339], where the case of electromagnetic wave propagation in a motionless random medium was considered. In this section, the analysis is generalized to sound propagation in a statistically inhomogeneous, anisotropic random medium with the sound speed and velocity fluctuations.

We start by introducing the function $\widetilde{\Gamma}_{n,m}$:

$$\widetilde{\Gamma}_{n,m}(x; \mathbf{r}_1, ..., \mathbf{r}_n; \mathbf{r}'_1, ..., \mathbf{r}'_m) = p(x, \mathbf{r}_1)...p(x, \mathbf{r}_n)p^*(x, \mathbf{r}'_1)...p^*(x, \mathbf{r}'_m). \quad (7.123)$$

The statistical moment $\Gamma_{n,m}$ is the mean value of this function: $\Gamma_{n,m} = \langle \widetilde{\Gamma}_{n,m} \rangle$. The factor $p(x, \mathbf{r}_1)$ on the right-hand side of equation (7.123) satisfies equation (7.8), which is recast here as

$$2ik\frac{\partial p(x, \mathbf{r}_1)}{\partial x} + L_{\mathbf{r}_1}\, p(x, \mathbf{r}_1) + k^2\varepsilon_{\mathrm{mov}}(x, \mathbf{r}_1)p(x, \mathbf{r}_1) = 0.$$
(7.124)

Here, the operator $L_{\mathbf{r}_1}$ acts on the transverse coordinates:

$$L_{\mathbf{r}} = 2k^2 + \nabla_\perp^2 \equiv 2k^2 + \frac{\partial^2}{\partial \mathbf{r}^2}.$$
(7.125)

Equation (7.124) is multiplied by $p(x, \mathbf{r}_2)...p(x, \mathbf{r}_n)p^*(x, \mathbf{r}'_1)...p^*(x, \mathbf{r}'_m)$. Since this function does not depend on $\mathbf{r}_1$, the resulting equation can be written as

$$2ik\frac{\partial p(x,\mathbf{r}_1)}{\partial x}p(x,\mathbf{r}_2)...p(x,\mathbf{r}_n)p^*(x,\mathbf{r}_1')...p^*(x,\mathbf{r}_m')$$

$$+\left[L_{\mathbf{r}_1}+k^2\varepsilon_{\mathrm{mov}}(x,\mathbf{r}_1)\right]\widetilde{\Gamma}_{n,m}(x)=0.\quad(7.126)$$

For simplicity, in this equation and some others to follow, the functional arguments $\mathbf{r}_i$ and $\mathbf{r}_j'$ are omitted. Equations for $\partial p(x,\mathbf{r}_i)/\partial x$, where $i=2,...,n$, can be derived analogously.

Taking the complex conjugate of both sides of equation (7.124), we obtain the equation for $p^*(x,\mathbf{r}_1')$:

$$2ik\frac{\partial p^*(x,\mathbf{r}_1')}{\partial x}-L_{\mathbf{r}_1'}^*\,p^*(x,\mathbf{r}_1')-k^2\varepsilon_{\mathrm{mov}}(x,\mathbf{r}_1')p^*(x,\mathbf{r}_1')=0.\quad(7.127)$$

In this equation, $L_{\mathbf{r}_1'}^*=L_{\mathbf{r}_1'}$. Nevertheless, for the purposes of Chapter 8, in derivations below we will retain the conjugation in the operator $L_{\mathbf{r}_1'}^*$. Both sides of equation (7.127) are multiplied by the function $p(x,\mathbf{r}_1)...p(x,\mathbf{r}_n)p^*(x,\mathbf{r}_2')...p^*(x,\mathbf{r}_m')$. Taking into account that this function does not depend on $\mathbf{r}_1'$, we can write the resulting equation as

$$2ikp(x,\mathbf{r}_1)...p(x,\mathbf{r}_n)\frac{\partial p^*(x,\mathbf{r}_1')}{\partial x}p^*(x,\mathbf{r}_2')...p^*(x,\mathbf{r}_m')$$

$$-\left[L_{\mathbf{r}_1'}^*+k^2\varepsilon_{\mathrm{mov}}(x,\mathbf{r}_1')\right]\widetilde{\Gamma}_{n,m}(x)=0.\quad(7.128)$$

Analogously, we can derive equations for $\partial p^*(x,\mathbf{r}_j')/\partial x$, where $j=2,...,n$.

The closed-form equation for $\widetilde{\Gamma}_{n,m}$ is obtained by adding together all equations for $\partial p(x,\mathbf{r}_i)/\partial x$, where $i=1,...,n$, and all equations for $\partial p^*(x,\mathbf{r}_j')/\partial x$, where $j=1,...,m$:

$$2ik\frac{\partial\widetilde{\Gamma}_{n,m}(x)}{\partial x}+M_{n,m}(x)\widetilde{\Gamma}_{n,m}(x)+k^2\zeta_{n,m}(x)\widetilde{\Gamma}_{n,m}(x)=0.\quad(7.129)$$

Here the random variable $\zeta_{n,m}$ is given by

$$\zeta_{n,m}(x;\mathbf{r}_1,...,\mathbf{r}_n;\mathbf{r}_1',...,\mathbf{r}_m')=\varepsilon_{\mathrm{mov}}(x,\mathbf{r}_1)+...+\varepsilon_{\mathrm{mov}}(x,\mathbf{r}_n)$$

$$-\varepsilon_{\mathrm{mov}}(x,\mathbf{r}_1')-...-\varepsilon_{\mathrm{mov}}(x,\mathbf{r}_m'),\quad(7.130)$$

and the operator $M_{n,m}$ is

$$M_{n,m}(x;\mathbf{r}_1,...,\mathbf{r}_n;\mathbf{r}_1',...,\mathbf{r}_m')=L_{\mathbf{r}_1}+...+L_{\mathbf{r}_n}-L_{\mathbf{r}_1'}^*-...-L_{\mathbf{r}_m'}^*.\quad(7.131)$$

We next integrate equation (7.129) with respect to x and write the result in the following form

$$2ik\widetilde{\Gamma}_{n,m}(x)-2ik\exp\left(\frac{ik}{2}\int_0^x\zeta_{n,m}(x_1)\,dx_1\right)\widetilde{\Gamma}_{n,m}(x=0)$$

$$=-\int_0^x\exp\left(\frac{ik}{2}\int_{x'}^x\zeta_{n,m}(x_1)\,dx_1\right)M_{n,m}(x')\widetilde{\Gamma}_{n,m}(x')\,dx'.\quad(7.132)$$

On the right-hand side of this equation, the functions $\tilde{\Gamma}_{n,m}(x')$ and $\exp\left[(ik/2)\int_{x'}^{x}\zeta_{n,m}(x_1)\,dx_1\right]$ do not correlate in the Markov approximation equation (7.22). We average both sides of equation (7.132), denote $\Pi(x,x') = \left\langle\exp\left[(ik/2)\int_{x'}^{x}\zeta_{n,m}(x_1)\,dx_1\right]\right\rangle$, and recall that $\left\langle\tilde{\Gamma}_{n,m}\right\rangle = \Gamma_{n,m}$. As a result, we obtain the closed-form equation for the statistical moment of order $n+m$:

$$2ik\Gamma_{n,m}(x) - 2ik\Pi(x,0)\Gamma_{n,m}(x=0) = -\int_{0}^{x}\Pi(x,x')M_{n,m}(x')\Gamma_{n,m}(x')\,dx'.$$

$$(7.133)$$

Let ε_{mov} be a random Gaussian field. Then, $\epsilon = (ik/2)\int_{x'}^{x}\zeta_{n,m}(x_1)\,dx_1$ appearing in the formula for $\Pi(x,x')$ is also a random Gaussian field, for which the following equality holds: $\langle\exp(\epsilon)\rangle = \exp\left(\langle\epsilon\rangle^2/2\right)$. Using this equality, we have

$$\Pi(x,x') = \exp\left[-\frac{k^2}{8}\int_{x'}^{x}\int_{x'}^{x}\langle\zeta_{n,m}(x_1)\zeta_{n,m}(x_2)\rangle\,dx_1\,dx_2\right].$$

$$(7.134)$$

Consider $\langle\zeta_{n,m}(x_1)\zeta_{n,m}(x_2)\rangle$ appearing in this expression. Substituting $\zeta_{n,m}$ from equation (7.130) and using the Markov approximation (7.22), it can be shown that

$$\langle\zeta_{n,m}(x_1)\zeta_{n,m}(x_2)\rangle = \delta(x_1 - x_2)Q_{n,m}(x_1),$$

$$(7.135)$$

where

$$Q_{n,m}(x;\mathbf{r}_1,...,\mathbf{r}_n;\mathbf{r}'_1,...,\mathbf{r}'_m) = \sum_{i=1}^{n}\sum_{j=1}^{n}b'_{\text{eff}}(x;\mathbf{r}_i,\mathbf{r}_j)$$
$$+ \sum_{i=1}^{m}\sum_{j=1}^{m}b'_{\text{eff}}(x;\mathbf{r}'_i,\mathbf{r}'_j) - 2\sum_{i=1}^{n}\sum_{j=1}^{m}b'_{\text{eff}}(x;\mathbf{r}_i,\mathbf{r}'_j). \quad (7.136)$$

In this formula, $b'_{\text{eff}}(x;\mathbf{r}_1,\mathbf{r}_2)$ is the transverse correlation function calculated at the coordinates $\mathbf{r}_1$ and $\mathbf{r}_2$:

$$b'_{\text{eff}}(x;\mathbf{r}_1,\mathbf{r}_2) = b'_{\text{eff}}(x;\mathbf{r}_c + \mathbf{r}_d/2,\mathbf{r}_c - \mathbf{r}_d/2) \equiv b_{\text{eff}}(x,\mathbf{r}_c;\mathbf{r}_d), \quad (7.137)$$

where $\mathbf{r}_c = (\mathbf{r}_1 + \mathbf{r}_2)/2$, $\mathbf{r}_d = \mathbf{r}_1 - \mathbf{r}_2$, and $b_{\text{eff}}(x,\mathbf{r}_c;\mathbf{r}_d)$ is determined with equation (7.20). Substituting the value of $\Pi(x,x')$ into equation (7.133), we obtain

$$2ik\Gamma_{n,m}(x) - 2ik\exp\left[-\frac{k^2}{8}\int_{0}^{x}Q_{n,m}(x_1)\,dx_1\right]\Gamma_{n,m}(x=0)$$
$$= -\int_{0}^{x}\exp\left[-\frac{k^2}{8}\int_{x'}^{x}Q_{n,m}(x_1)\,dx_1\right]M_{n,m}(x')\Gamma_{n,m}(x')\,dx'. \quad (7.138)$$

We multiply both sides of this equation by $\exp\left[(k^2/8)\int_0^x Q_{n,m}(x_1)\,dx_1\right]$ and, then, differentiate with respect to x. As a result, we obtain the closed-form equation for the statistical moment of order $n + m$ in a statistically inhomogeneous, anisotropic random medium:

$$\left[\frac{\partial}{\partial x} - \frac{i}{2k}M_{n,m}(x) + \frac{k^2}{8}Q_{n,m}(x)\right]\Gamma_{n,m}(x;\mathbf{r}_1,...,\mathbf{r}_n;\mathbf{r}_1',...,\mathbf{r}_m') = 0. \quad (7.139)$$

The initial condition to this equation is formulated in the plane $x = 0$ and can obtained by making use of equations (7.7) and (7.122):

$$\Gamma_{n,m}(x = 0;\mathbf{r}_1,...,\mathbf{r}_n;\mathbf{r}_1',...,\mathbf{r}_m') = p_0(\mathbf{r}_1)...p_0(\mathbf{r}_n)p_0^*(\mathbf{r}_1')...p_0^*(\mathbf{r}_m'). \quad (7.140)$$

7.4.3   Mean sound field

The sound field p can be expressed as a sum

$$p(x,\mathbf{r}) = \langle p(x,\mathbf{r})\rangle + \widetilde{p}(x,\mathbf{r}).$$

Here, $\langle p(x,\mathbf{r})\rangle = \Gamma_{1,0}(x,\mathbf{r})$ is the mean field and $\widetilde{p}(x,\mathbf{r})$ is the fluctuating part of the field, the mean value of which is zero. In the initial plane $x = 0$, $\langle p(0,\mathbf{r})\rangle = p(0,\mathbf{r})$ and $\widetilde{p}(0,\mathbf{r}) = 0$. As a sound wave propagates through a random medium, $\langle p(x,\mathbf{r})\rangle$ decreases, while $\left\langle|\widetilde{p}(x,\mathbf{r})|^2\right\rangle$ increases. At a sufficiently large range x from the source, $\langle p\rangle$ becomes negligibly small and the acoustic energy is concentrated in the fluctuating part of the field $\widetilde{p}$.

The mean sound field is expressed in the following form:

$$\langle p(x,\mathbf{r})\rangle = \langle A(x,\mathbf{r})\rangle\exp(ikx), \quad (7.141)$$

where $\langle A\rangle$ is the mean complex amplitude. An equation for $\langle A\rangle$ is obtained by setting $n = 1$ and $m = 0$ in equation (7.139):

$$\left[\frac{\partial}{\partial x} - \frac{i}{2k}\nabla_\perp^2 + \frac{k^2}{8}b_{\text{eff}}'(x;\mathbf{r},\mathbf{r})\right]\langle A(x,\mathbf{r})\rangle = 0. \quad (7.142)$$

The initial condition is $\langle A(0,\mathbf{r})\rangle = p_0(\mathbf{r})$. In this equation, $b_{\text{eff}}'(x;\mathbf{r},\mathbf{r}) \equiv b_{\text{eff}}(x,\mathbf{r};0)$, see equation (7.137).

Equation (7.142) can be solved for $\langle A(x,\mathbf{r})\rangle$. To this end, the mean complex amplitude is expressed as a Fourier integral

$$\langle A(x,\mathbf{r})\rangle = \int\widehat{A}(x,\mathbf{q})\exp(i\mathbf{q}\cdot\mathbf{r})\,d^2q. \quad (7.143)$$

Here, $\widehat{A}(x,\mathbf{q})$ is the two-dimensional Fourier transform of $\langle A(x,\mathbf{r})\rangle$. Similarly,

$$b_{\text{eff}}(x,\mathbf{r};0) = \int\widehat{b}_{\text{eff}}(x,\mathbf{q})\exp(-i\mathbf{q}\cdot\mathbf{r})\,d^2q, \quad (7.144)$$

where $\widehat{b}_{\text{eff}}(x, \mathbf{q})$ is the spectral density of $b_{\text{eff}}(x, \mathbf{r}; 0)$. In the integrand of this equation, the minus sign in the exponent is chosen for convenience of subsequent calculations. The values of $\langle A(x, \mathbf{r}) \rangle$ and $b_{\text{eff}}(x, \mathbf{r}; 0)$ are substituted into equation (7.142). As a result, we obtain an integro-differential equation for $\widehat{A}$:

$$\frac{\partial \widehat{A}(x, \mathbf{q})}{\partial x} + \frac{iq^2}{2k} \widehat{A}(x, \mathbf{q}) + \frac{k^2}{8} \int \widehat{A}(x, \mathbf{q} + \mathbf{q}_1) \widehat{b}_{\text{eff}}(x, \mathbf{q}_1) \, d^2 q_1 = 0. \quad (7.145)$$

The function $\widehat{b}_{\text{eff}}(x, \mathbf{q})$ appearing in this equation can be expressed in terms of $b_{\text{eff}}(x, \mathbf{r}; 0)$ using equation (7.144). Then, $b_{\text{eff}}(x, \mathbf{r}; 0)$ is substituted with the right-hand side of equation (7.21) with $\mathbf{r}_d = 0$. As a result, we obtain

$$\frac{\partial \widehat{A}(x, \mathbf{q})}{\partial x} + \frac{iq^2}{2k} \widehat{A}(x, \mathbf{q}) + \frac{k^2}{16\pi} \int d^2 q_1 \int d^2 \kappa_\perp$$
$$\times \int \widehat{A}(x, \mathbf{q} + \mathbf{q}_1) \Phi_{\text{eff}}(x, \mathbf{r}; 0, \kappa_\perp) e^{i\mathbf{q}_1 \cdot \mathbf{r}} \, d^2 r = 0. \quad (7.146)$$

Using equation (7.75) to calculate the integral over $\mathbf{r}$, we have

$$\left[\frac{\partial}{\partial x} + \frac{i\kappa q^2}{2k} + \frac{\pi k^2}{4} \int \Phi_{\text{eff}}(x, 0; 0, \kappa_\perp) \, d^2 \kappa_\perp \right] \widehat{A}(x, \mathbf{q}) = 0. \quad (7.147)$$

This equation can be readily solved for $\widehat{A}$:

$$\widehat{A}(x, \mathbf{q}) = \exp \left[-\frac{iq^2 x}{2k} - \frac{\pi k^2}{4} \int_0^x dx' \int \Phi_{\text{eff}}(x, 0; 0, \kappa_\perp) \, d^2 \kappa_\perp \right] \widehat{A}(0, \mathbf{q}). \quad (7.148)$$

Here, $\widehat{A}(0, \mathbf{q})$ is the value of $\widehat{A}(x, \mathbf{q})$ at $x = 0$. Substituting the value of the quantity $\widehat{A}(x, \mathbf{q})$ into equation (7.143), we obtain

$$\langle A(x, \mathbf{r}) \rangle = \exp \left[-\frac{\pi k^2}{4} \int_0^x dx' \int \Phi_{\text{eff}}(x', 0; 0, \kappa_\perp) \, d^2 \kappa_\perp \right]$$
$$\times \int \widehat{A}(0, \mathbf{q}) \exp \left(i\mathbf{q} \cdot \mathbf{r} - \frac{iq^2 x}{2k} \right) d^2 q. \quad (7.149)$$

The integral over $\mathbf{q}$ in this formula is equal to the complex amplitude $A^{(0)}(x, \mathbf{r})$ of a sound field propagating in a medium without random inhomogeneities in the parabolic equation approximation. Introducing the new integration variable $\eta = x'/x$ along the propagation path and substituting the value of $\langle A(x, \mathbf{r}) \rangle$ into equation (7.141), we obtain the mean sound field in a quasi-homogeneous, anisotropic random medium:

$$\langle p(x, \mathbf{r}) \rangle = \exp (-\gamma x) \, p^{(0)}(x, \mathbf{r}). \quad (7.150)$$

Here, $p^{(0)}(x, \mathbf{r}) = A^{(0)}(x, \mathbf{r}) \exp(ikx)$ is the sound field in a medium without random inhomogeneities and γ is the extinction coefficient of the mean field:

$$\gamma = \frac{\pi k^2}{4} \int_0^1 d\eta \int \Phi_{\text{eff}}(\eta x, 0; 0, \boldsymbol{\kappa}_\perp) \, d^2\kappa_\perp. \qquad (7.151)$$

According to equation (7.150), the mean sound field attenuates exponentially with range x and the extinction coefficient does not depend on the sound field $p^{(0)}(x, \mathbf{r})$.

Comparing the extinction coefficient with the variances of phase and log-amplitude fluctuations, we obtain a useful relationship:

$$\gamma = \frac{\langle \phi^2 \rangle + \langle \chi^2 \rangle}{2x}. \qquad (7.152)$$

This relationship is valid for both plane waves, when $\langle \phi^2 \rangle$ and $\langle \chi^2 \rangle$ are given by equations (7.79) and (7.80), and for spherical waves, when they are given by equations (7.95) and (7.96) with $\mathbf{r}_d = 0$.

For a quasi-homogeneous, isotropic random medium and for a homogeneous, anisotropic random medium, Φ_{eff} in equation (7.151) is given by equations (7.26) and (7.27), respectively. For a homogeneous and isotropic random medium, when Φ_{eff} is given by equation (7.29), the extinction coefficient simplifies to

$$\gamma = \frac{\pi^2 k^2}{2} \int_0^\infty \Phi_{\text{eff}}(0, \kappa_\perp) \kappa_\perp \, d\kappa_\perp. \qquad (7.153)$$

It follows from this formula and equation (7.151) that the extinction coefficient γ is proportional to the integral of the effective spectrum over the wave vector $\boldsymbol{\kappa}_\perp$, i.e., to the integral length scale defined by equations (6.46) and (6.47). Since most of the turbulence energy is in the large-scale eddies, the extinction coefficient is affected mainly by these eddies.

7.4.4   Mutual coherence function for an arbitrary waveform

The transverse mutual coherence function is defined as

$$\Gamma_{1,1}(x; \mathbf{r}_1, \mathbf{r}_2) = \langle p(x; \mathbf{r}_1) p^*(x; \mathbf{r}_2) \rangle. \qquad (7.154)$$

It describes the coherence of the sound field $p(x; \mathbf{r})$ at two points of observation located at the same range x from the source. Setting $n = 1$ and $m = 1$ in equation (7.139), we obtain an equation for the mutual coherence function:

$$\left[\frac{\partial}{\partial x} - \frac{i}{2k} \left(\frac{\partial^2}{\partial \mathbf{r}_1^2} - \frac{\partial^2}{\partial \mathbf{r}_2^2} \right) + \frac{k^2}{8} Q_{1,1}(x; \mathbf{r}_1, \mathbf{r}_2) \right] \Gamma_{1,1}(x; \mathbf{r}_1, \mathbf{r}_2) = 0. \qquad (7.155)$$

Here, the function $Q_{1,1}$ is given by

$$Q_{1,1}(x; \mathbf{r}_1, \mathbf{r}_2) = b'_{\text{eff}}(x; \mathbf{r}_1, \mathbf{r}_1) + b'_{\text{eff}}(x; \mathbf{r}_2, \mathbf{r}_2) - 2b'_{\text{eff}}(x; \mathbf{r}_1, \mathbf{r}_2). \qquad (7.156)$$

The initial condition for equation (7.155) is

$$\Gamma_{1,1}(0; \mathbf{r}_1, \mathbf{r}_2) = p_0(\mathbf{r}_1)p_0^*(\mathbf{r}_2). \tag{7.157}$$

Below we present an approximate solution of equation (7.155) using an approach similar to that in reference [339], where propagation in a homogeneous motionless random medium was considered.

As before, we introduce the center $\mathbf{r}_c = (\mathbf{r}_1 + \mathbf{r}_2)/2$ and difference $\mathbf{r}_d = \mathbf{r}_1 - \mathbf{r}_2$ coordinates. Then $b'_{\text{eff}}(x; \mathbf{r}_1, \mathbf{r}_2) \equiv b_{\text{eff}}(x, \mathbf{r}_c; \mathbf{r}_d)$ (see equation (7.137)), $\Gamma_{1,1}(x; \mathbf{r}_1, \mathbf{r}_2) = \Gamma_{1,1}(x; \mathbf{r}_c + \mathbf{r}_d/2, \mathbf{r}_c - \mathbf{r}_d/2) \equiv \Gamma(x; \mathbf{r}_c, \mathbf{r}_d)$, and $Q_{1,1}(x; \mathbf{r}_1, \mathbf{r}_2) = Q_{1,1}(x; \mathbf{r}_c + \mathbf{r}_d/2, \mathbf{r}_c - \mathbf{r}_d/2) \equiv Q(x; \mathbf{r}_c, \mathbf{r}_d)$. As a result, equation (7.155) takes the form:

$$\left[\frac{\partial}{\partial x} - \frac{i}{k} \frac{\partial}{\partial \mathbf{r}_d} \cdot \frac{\partial}{\partial \mathbf{r}_c} + \frac{k^2}{8} Q(x; \mathbf{r}_c, \mathbf{r}_d) \right] \Gamma(x; \mathbf{r}_c, \mathbf{r}_d) = 0. \tag{7.158}$$

Here

$$Q(x; \mathbf{r}_c, \mathbf{r}_d) = b_{\text{eff}}(x; \mathbf{r}_c + \mathbf{r}_d/2, 0) + b_{\text{eff}}(x; \mathbf{r}_c - \mathbf{r}_d/2, 0) - 2b_{\text{eff}}(x; \mathbf{r}_c, \mathbf{r}_d). \tag{7.159}$$

Substituting $b_{\text{eff}}(x; \mathbf{r}_c, \mathbf{r}_d)$ from equation (7.21), we have

$$Q(x; \mathbf{r}_c, \mathbf{r}_d) = 2\pi \int \left[\Phi_{\text{eff}}(x, \mathbf{r}_c + \mathbf{r}_d/2; 0, \boldsymbol{\kappa}_\perp) \right.$$
$$\left. + \Phi_{\text{eff}}(x, \mathbf{r}_c - \mathbf{r}_d/2; 0, \boldsymbol{\kappa}_\perp) - 2\Phi_{\text{eff}}(x, \mathbf{r}_c; 0, \boldsymbol{\kappa}_\perp) e^{i\boldsymbol{\kappa}_\perp \cdot \mathbf{r}_d} \right] d^2\kappa_\perp. \tag{7.160}$$

The initial condition to equation (7.158) reads:

$$\Gamma(0; \mathbf{r}_c, \mathbf{r}_d) = p_0(\mathbf{r}_c + \mathbf{r}_d/2)p_0^*(\mathbf{r}_c - \mathbf{r}_d/2). \tag{7.161}$$

Let us express the mutual coherence function as a Fourier integral

$$\Gamma(x; \mathbf{r}_c, \mathbf{r}_d) = \int \widehat{\Gamma}(x; \mathbf{q}, \mathbf{r}_d) \exp(i\mathbf{q} \cdot \mathbf{r}_c) \, d^2q, \tag{7.162}$$

where $\widehat{\Gamma}(x; \mathbf{q}, \mathbf{r}_d)$ is the Fourier transform of $\Gamma(x; \mathbf{r}_c, \mathbf{r}_d)$. Similarly,

$$Q(x; \mathbf{r}_c, \mathbf{r}_d) = \int \widehat{Q}(x; \mathbf{q}, \mathbf{r}_d) \exp(-i\mathbf{q} \cdot \mathbf{r}_c) \, d^2q. \tag{7.163}$$

Substituting the values of $\Gamma(x; \mathbf{r}_c, \mathbf{r}_d)$ and $Q(x; \mathbf{r}_c, \mathbf{r}_d)$ into equation (7.158), we obtain the following integro-differential equation for $\widehat{\Gamma}$:

$$\frac{\partial \widehat{\Gamma}(x; \mathbf{q}, \mathbf{r}_d)}{\partial x} + \frac{\mathbf{q}}{k} \cdot \frac{\partial \widehat{\Gamma}(x; \mathbf{q}, \mathbf{r}_d)}{\partial \mathbf{r}_d} + \frac{k^2}{8} \int \widehat{\Gamma}(x; \mathbf{q}_1 + \mathbf{q}, \mathbf{r}_d) \widehat{Q}(x; \mathbf{q}_1, \mathbf{r}_d) \, d^2q_1 = 0. \tag{7.164}$$

This equation can be recast in the equivalent form:

$$\frac{\partial}{\partial x}\left[\exp\left(\frac{x\mathbf{q}}{k}\cdot\frac{\partial}{\partial\mathbf{r}_d}\right)\widehat{\Gamma}(x;\mathbf{q},\mathbf{r}_d)\right]$$
$$=-\frac{k^2}{8}\exp\left(\frac{x\mathbf{q}}{k}\cdot\frac{\partial}{\partial\mathbf{r}_d}\right)\int\widehat{\Gamma}(x;\mathbf{q}_1+\mathbf{q},\mathbf{r}_d)\widehat{Q}(x;\mathbf{q}_1,\mathbf{r}_d)\,d^2q_1. \quad (7.165)$$

(This transformation can be proved by operating with $\partial/\partial x$ on the terms in square brackets of the latter equation.) The pseudo-differential operator appearing in equation (7.165) has the following property:

$$\exp\left(\frac{x\mathbf{q}}{k}\cdot\frac{\partial}{\partial\mathbf{r}_d}\right)\Phi(\mathbf{r}_d)=\Phi\left(\mathbf{r}_d+\frac{x\mathbf{q}}{k}\right). \quad (7.166)$$

Here, $\Phi(\mathbf{r}_d)$ is an arbitrary function. Using this equality, equation (7.165) can be recast as

$$\frac{\partial\widehat{\Gamma}(x;\mathbf{q},\mathbf{r}_d+x\mathbf{q}/k)}{\partial x}=-\frac{k^2}{8}\int\widehat{\Gamma}(x;\mathbf{q}_1+\mathbf{q},\mathbf{r}_d+x\mathbf{q}/k)$$
$$\times\widehat{Q}(x;\mathbf{q}_1,\mathbf{r}_d+x\mathbf{q}/k)\,d^2q_1. \quad (7.167)$$

The quantity $\widehat{Q}(x;\mathbf{q},\mathbf{r}_d)$ in this equation can be expressed in the form

$$\widehat{Q}(x;\mathbf{q},\mathbf{r}_d)=\frac{1}{2\pi}\int\int e^{i\mathbf{q}\cdot\mathbf{r}'_c}[\Phi_{\text{eff}}(x,\mathbf{r}'_c+\mathbf{r}_d/2;0,\boldsymbol{\kappa}_\perp)$$
$$+\Phi_{\text{eff}}(x,\mathbf{r}'_c-\mathbf{r}_d/2;0,\boldsymbol{\kappa}_\perp)-2e^{i\boldsymbol{\kappa}_\perp\cdot\mathbf{r}_d}\Phi_{\text{eff}}(x,\mathbf{r}'_c;0,\boldsymbol{\kappa}_\perp)]\,d^2\kappa_\perp\,d^2r'_c. \quad (7.168)$$

(To obtain this formula, $\widehat{Q}(x;\mathbf{q},\mathbf{r}_d)$ is expressed in terms of the function $Q(x;\mathbf{r}_c,\mathbf{r}_d)$ using equation (7.163); then, $Q(x;\mathbf{r}_c,\mathbf{r}_d)$ is replaced by the right-hand side of equation (7.160).) Substitution of $\widehat{Q}$ into equation (7.167) yields

$$\frac{\partial\widehat{\Gamma}(x;\mathbf{q},\mathbf{r}_d+x\mathbf{q}/k)}{\partial x}=-\frac{k^2}{16\pi}\int d^2q_1\int d^2\kappa_\perp\int\widehat{\Gamma}(x;\mathbf{q}_1+\mathbf{q},\mathbf{r}_d+x\mathbf{q}/k)$$
$$\times[\Phi_{\text{eff}}(x,\mathbf{r}'_c+(\mathbf{r}_d+x\mathbf{q}/k)/2;0,\boldsymbol{\kappa}_\perp)+\Phi_{\text{eff}}(x,\mathbf{r}'_c-(\mathbf{r}_d+x\mathbf{q}/k)/2;0,\boldsymbol{\kappa}_\perp)$$
$$-2e^{i\boldsymbol{\kappa}_\perp\cdot(\mathbf{r}_d+x\mathbf{q}/k)}\Phi_{\text{eff}}(x,\mathbf{r}'_c;0,\boldsymbol{\kappa}_\perp)]e^{i\mathbf{q}_1\cdot\mathbf{r}'_c}\,d^2r'_c. \quad (7.169)$$

It can be shown that the scale of $\widehat{\Gamma}(x;\mathbf{q}_1+\mathbf{q},\mathbf{r}_d+x\mathbf{q}/k)$ as a function of $\mathbf{q}_1$ is $\sqrt{k/x}$. Since we are assuming that the scale L_c of $\Phi_{\text{eff}}(x,\mathbf{r}'_c;0,\boldsymbol{\kappa}_\perp)$ as a function of $\mathbf{r}'_c$ is much greater than $\sqrt{x/k}$, the factor $\exp(i\mathbf{q}_1\cdot\mathbf{r}'_c)$ in equation (7.169) oscillates rapidly in the region $r'_c>L_c$ so that the contribution from this region to the integral over $\mathbf{r}'_c$ can be ignored. Therefore, this integral can be calculated with equation (7.75). As a result, equation (7.169) takes the form:

$$\frac{\partial \widehat{\Gamma}(x; \mathbf{q}, \mathbf{r}_d + x\mathbf{q}/k)}{\partial x} = -\frac{k^2}{8} Q'(x; 0, \mathbf{r}_d + x\mathbf{q}/k) \widehat{\Gamma}(x; \mathbf{q}, \mathbf{r}_d + x\mathbf{q}/k), \quad (7.170)$$

where

$$Q'(x; 0, \mathbf{r}_d) = 4\pi \int \Phi_{\text{eff}}(x, 0; 0, \boldsymbol{\kappa}_\perp) \left[1 - e^{i\boldsymbol{\kappa}_\perp \cdot \mathbf{r}_d} \right] d^2\kappa_\perp. \quad (7.171)$$

Equation (7.170) can be readily solved:

$$\widehat{\Gamma}(x; \mathbf{q}, \mathbf{r}_d + x\mathbf{q}/k) = \exp\left[-\frac{k^2}{8} \int_0^x Q'(x'; 0, \mathbf{r}_d + x'\mathbf{q}/k) \, dx' \right] \widehat{\Gamma}(0, \mathbf{q}, \mathbf{r}_d). \quad (7.172)$$

Here, $\widehat{\Gamma}(0; \mathbf{q}, \mathbf{r}_d)$ is the value of $\widehat{\Gamma}(x; \mathbf{q}, \mathbf{r}_d)$ at $x = 0$. Introducing a new vector $\mathbf{r}'_d = \mathbf{r}_d + \mathbf{q}x/k$ in equation (7.172), we have

$$\widehat{\Gamma}(x; \mathbf{q}, \mathbf{r}'_d) = \exp\left[-\frac{k^2}{8} \int_0^x Q'(x'; 0, \mathbf{r}'_d + (x' - x)\mathbf{q}/k) \, dx' \right] \widehat{\Gamma}(0, \mathbf{q}, \mathbf{r}'_d - x\mathbf{q}/k). \quad (7.173)$$

In this equation, $\mathbf{r}'_d$ is replaced with $\mathbf{r}_d$ and $\widehat{\Gamma}(0; \mathbf{q}, \mathbf{r}_d)$ is expressed in terms of $\Gamma(0; \mathbf{r}_c, \mathbf{r}_d)$ using equation (7.162) with $x = 0$. As a result, we have

$$\Gamma(x; \mathbf{r}_c, \mathbf{r}_d) = \frac{1}{4\pi^2} \int \int \Gamma(0; \mathbf{r}'_c, \mathbf{r}_d - x\mathbf{q}/k)$$

$$\times \exp\left[i\mathbf{q} \cdot (\mathbf{r}_c - \mathbf{r}'_c) - \frac{k^2}{8} \int_0^x Q'(x'; 0, \mathbf{r}_d - (x - x')\mathbf{q}/k) \, dx' \right] d^2q \, d^2r'_c. \quad (7.174)$$

In this formula, $\Gamma(0; \mathbf{r}'_c, \mathbf{r}_d - x\mathbf{q}/k)$ is replaced by equation (7.161) and Q' with equation (7.171). Replacing $\mathbf{q}$ with a new integration variable $\mathbf{r}'_d = \mathbf{r}_d - x\mathbf{q}/k$, we obtain the mutual coherence function in a quasi-homogeneous, anisotropic random medium:

$$\Gamma(x; \mathbf{r}_c, \mathbf{r}_d) = \frac{k^2}{4\pi^2 x^2} \int \int p_0 \left(\mathbf{r}'_c + \mathbf{r}'_d/2 \right) p_0^* \left(\mathbf{r}'_c - \mathbf{r}'_d/2 \right)$$

$$\times \exp\left\{ \frac{ik (\mathbf{r}_d - \mathbf{r}'_d) \cdot (\mathbf{r}_c - \mathbf{r}'_c)}{x} - \frac{k^2 \pi x}{2} \int_0^1 d\eta \int \Phi_{\text{eff}}(\eta x, 0; 0, \boldsymbol{\kappa}_\perp) \right.$$

$$\left. \times \left[1 - \exp\left[i\boldsymbol{\kappa}_\perp \cdot (\eta \mathbf{r}_d + (1 - \eta) \mathbf{r}'_d) \right] \right] d^2\kappa_\perp \right\} d^2r'_d \, d^2r'_c. \quad (7.175)$$

This equation is valid for an arbitrary initial waveform in the plane $x = 0$. The mutual coherence functions of plane and spherical sound waves are considered below. It follows from equation (7.175) that the coherence between two points of observation decreases with increasing k, x, and r_d.

The coherence radius r_{coh} is a very useful parameter for describing the spatial decay of the coherence of the sound field. It is defined as the separation r_d between two points of observation at which the magnitude of the mutual coherence function decreases by a factor $1/e$:

$$|\Gamma(x; \mathbf{r}_c, \mathbf{r}_{\text{coh}})| = e^{-1} |\Gamma(x; \mathbf{r}_c, 0)|. \tag{7.176}$$

For an anisotropic random medium, the coherence radius is generally a vector. For microphone separations beyond $\mathbf{r}_{\text{coh}}$, coherent signal processing is normally infeasible.

If $|\mathbf{r}_1 - \mathbf{r}_2|$ is sufficiently large, the sound fields $p(x, \mathbf{r}_1)$ and $p^*(x, \mathbf{r}_2)$ at two points of observation are uncorrelated:

$$\Gamma(x; \mathbf{r}_1, \mathbf{r}_2) \cong \langle p(x, \mathbf{r}_1) \rangle \langle p^*(x, \mathbf{r}_2) \rangle = \exp\left(-2\gamma x\right) p^{(0)}(x, \mathbf{r}_1) \left(p^{(0)}(x, \mathbf{r}_2)\right)^*. \tag{7.177}$$

This equation provides a useful relationship between Γ and $\langle p \rangle$ in this limiting case.

7.4.5   Plane wave coherence

For the initially plane wave at $x = 0$, in equation (7.175) $p_0 = $ constant. In this equation, the integral over $\mathbf{r}'_c$ is equal to $4\pi^2 \delta \left[(k/x)(\mathbf{r}_d - \mathbf{r}'_d)\right]$. The remaining integral over $\mathbf{r}'_d$ can be readily calculated, with result

$$\Gamma(x; \mathbf{r}_d) = |p_0|^2 \exp\left[-\frac{\pi k^2 x}{2} \int_0^1 d\eta \int \Phi_{\text{eff}}(\eta x, 0; 0, \boldsymbol{\kappa}_\perp)\left(1 - e^{i\boldsymbol{\kappa}_\perp \cdot \mathbf{r}_d}\right) d^2\kappa_\perp\right]. \tag{7.178}$$

For plane wave propagation, the mutual coherence function does not depend on $\mathbf{r}_c$. For a motionless random medium, this equation is essentially the same as equation (3.21) from reference [207].

The mutual coherence function, equation (7.178), obtained with the parabolic equation method can be expressed in terms of the correlation functions of phase and log-amplitude fluctuations of a plane sound wave calculated with the Rytov method:

$$\Gamma(x; r_d) = |p_0|^2 \exp\left\{-[\mathcal{B}_\phi(x; 0) - \mathcal{B}_\phi(x; r_d) + \mathcal{B}_\chi(x; 0) - \mathcal{B}_\chi(x; r_d)]\right\}. \tag{7.179}$$

This relationship can be verified by substituting the values of $\mathcal{B}_\phi$ and $\mathcal{B}_\chi$ from equations (7.77) and (7.78), respectively.

For a quasi-homogeneous, isotropic random medium, Φ_{eff} in equation (7.178) is given by equation (7.26), and for a homogeneous, anisotropic random medium, it is given by equation (7.27). For the case of a homogeneous and isotropic random medium, when Φ_{eff} is determined with equation (7.29), the formula for the mutual coherence function simplifies

$$\Gamma(x; r_d) = |p_0|^2 \exp\left[-\pi^2 k^2 x \int_0^\infty \Phi_{\text{eff}}(0, \kappa_\perp)\left[1 - J_0(\kappa_\perp r_d)\right] \kappa_\perp \, d\kappa_\perp\right]. \quad (7.180)$$

Here, J_0 is the Bessel function.

7.4.6 Spherical wave coherence

In free space, the sound pressure due to the unit-amplitude point source located at the origin of the coordinate system is given by equation (7.87) in the parabolic equation approximation. It can be shown from that equation that the initial sound field due to this source is $p_0(\mathbf{r}) = (i2\pi/k)\delta(\mathbf{r})$. Substituting the value of $p_0(\mathbf{r})$ into equation (7.175) and calculating the integrals over $\mathbf{r}'_c$ and $\mathbf{r}'_d$, we obtain

$$\Gamma(x; \mathbf{r}_c, \mathbf{r}_d) = \frac{1}{x^2} \exp\left[\frac{ik\mathbf{r}_c \cdot \mathbf{r}_d}{x}\right.$$
$$\left. -\frac{\pi k^2 x}{2} \int_0^1 d\eta \int \Phi_{\text{eff}}(\eta x, 0; 0, \boldsymbol{\kappa}_\perp)\left(1 - e^{i\eta\boldsymbol{\kappa}_\perp \cdot \mathbf{r}_d}\right) d^2\kappa_\perp\right]. \quad (7.181)$$

For a motionless random medium, this formula coincides with equation (3.28) in reference [207]. Similarly to the plane wave case, the spherical wave mutual coherence can be expressed in terms of the correlation functions of phase and log-amplitude fluctuations:

$$\Gamma(x; \mathbf{r}_c, \mathbf{r}_d) = \frac{1}{x^2} \exp\left\{\frac{ik\mathbf{r}_c \cdot \mathbf{r}_d}{x}\right.$$
$$\left. - \left[\mathcal{B}_\phi(x; 0) - \mathcal{B}_\phi(x; r_d) + \mathcal{B}_\chi(x; 0) - \mathcal{B}_\chi(x; r_d)\right]\right\}. \quad (7.182)$$

Here, $\mathcal{B}_\phi(x; r_d)$ and $\mathcal{B}_\chi(x; r_d)$ are given with equations (7.95) and (7.96), respectively.

For the case of a homogeneous, isotropic random medium, equation (7.181) simplifies

$$\Gamma(x; \mathbf{r}_c, \mathbf{r}_d) = \frac{1}{x^2} \exp\left[\frac{ik\mathbf{r}_c \cdot \mathbf{r}_d}{x}\right.$$
$$\left. - \pi^2 k^2 x \int_0^1 d\eta \int_0^\infty \Phi_{\text{eff}}(0, \kappa_\perp)[1 - J_0(\eta\kappa_\perp r_d)]\kappa_\perp \, d\kappa_\perp\right]. \quad (7.183)$$

7.5 Mean sound field and mutual coherence for the turbulence spectra

In this section, we present formulas for the extinction coefficient of the mean sound field for the von Kármán, Kolmogorov, and Gaussian spectra of homogeneous and isotropic turbulence. Formulas for the mutual coherence functions of plane and spherical sound waves for these turbulence models are also presented.

7.5.1 Extinction coefficient of the mean field

7.5.1.1 Von Kármán spectrum

The extinction coefficient γ of the mean sound field is given by equation (7.153). Substituting $\Phi_{\text{eff}}(0, \kappa_\perp)$ for the von Kármán spectra, equation (7.30), into this equation, and calculating the integral over $\kappa_\perp$, we obtain

$$\gamma = \gamma_T + \gamma_v = \frac{\pi^{1/2}\Gamma(5/6)k^2}{4\Gamma(1/3)} \left(\frac{\beta_c^2 \sigma_T^2 L_T}{T_0^2} + \frac{4\sigma_v^2 L_v}{c_0^2} \right). \tag{7.184}$$

Here, γ_T and γ_v are the relative contributions from temperature and velocity fluctuations to the extinction coefficient. These contributions have the same dependence on the sound wavenumber (k), the variances $(\sigma_T^2$ and $\sigma_v^2)$, and the von Kármán length scales $(L_T$ and $L_v)$. As indicated after equation (7.153), γ is affected mainly by the energy-containing subrange of turbulence.

The extinction length of the mean sound field is defined as $1/\gamma$. Figure 7.4 depicts the extinction length calculated with equation (7.184) for the case of sound propagation in the atmosphere. This quantity is plotted versus the surface heat flux Q_H for three values of the friction velocity, $u_* = 0.1$ m/s, 0.3 m/s, and 0.6 m/s, which correspond to light, moderate, and strong wind. The acoustic frequency for these calculations is $f = 500$ Hz, the propagation height $z = 20$ m, the height of the ABL $z_i = 1$ km, and the surface temperature $T_g = 18\,°C$. From these values, we can determine σ_T^2, σ_v^2, L_T, and L_v appearing in equation (7.184) (section 6.2.4). It follows from figure 7.4 that the extinction length varies between 6.3 m and 1.7 km. It decreases with increasing heat flux and friction velocity. Since γ is proportional to f^2, extinction lengths for acoustic frequencies f other than $f = 500$ Hz can be readily obtained from the figure.

7.5.1.2 Kolmogorov spectrum

Substitution of $\Phi_{\text{eff}}(0, \kappa_\perp)$ from equation (7.31) into equation (7.153) for γ results in a divergent integral; that is $\gamma \to \infty$. The outcome is the same with equation (7.184), if $L_T, L_v \to \infty$. Because it is inapplicable to the energy-

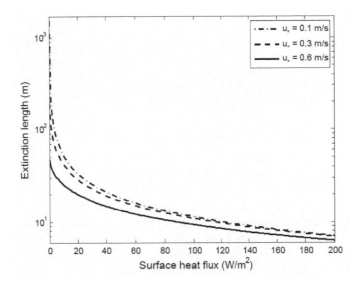

FIGURE 7.4
Extinction length, $1/\gamma$, of the mean sound field versus the surface heat flux Q_H for three values of the friction velocity u_*. The acoustic frequency is $f = 500$ Hz.

containing subrange, the Kolmogorov spectrum cannot be used to predict the extinction coefficients.

7.5.1.3 Gaussian spectrum

For the Gaussian spectra, the extinction coefficient is calculated with equations (7.36) and (7.153):

$$\gamma = \gamma_T + \gamma_v = \frac{\sqrt{\pi}k^2}{8}\left(\frac{\beta_c^2\sigma_T^2\mathcal{L}_T}{T_0^2} + \frac{4\sigma_v^2\mathcal{L}_v}{c_0^2}\right). \tag{7.185}$$

Here, γ_T and γ_v are the relative contributions from temperature and velocity fluctuations to γ. Using equations (6.48) and (6.49), $\mathcal{L}_T$ and $\mathcal{L}_v$ can be expressed in terms of L_T and L_v, respectively. In this case, the extinction coefficient for the Gaussian spectra coincides with that for the von Kármán spectra.

7.5.2 Mutual coherence function for plane waves

7.5.2.1 Von Kármán spectrum

The mutual coherence function for plane sound waves is given by equation (7.180). Substituting equation (7.30) for von Kármán effective spectrum and calculating the integral over $\kappa_\perp$, one obtains

$$
\Gamma(x; r_d) = |p_0|^2 \exp \left\{ -2\gamma_T x \left[1 - \frac{\Gamma(1/6)}{\pi} \left(\frac{r_d}{2L_T} \right)^{5/6} K_{5/6} \left(\frac{r_d}{L_T} \right) \right] \right.
$$
$$
\left. -2\gamma_v x \left[1 - \frac{\Gamma(1/6)}{\pi} \left(\frac{r_d}{2L_v} \right)^{5/6} \left(K_{5/6} \left(\frac{r_d}{L v} \right) - \frac{r_d}{2L_v} K_{1/6} \left(\frac{r_d}{L v} \right) \right) \right] \right\}.
$$
$$(7.186)$$

Here, γ_T and γ_v are determined with equation (7.184), and $K_{1/6}$ and $K_{5/6}$ are the modified Bessel functions of the second kind. Equation (7.186) has been derived elsewhere for $\gamma_v = 0$ [112]. The relative contributions from temperature and velocity fluctuations to $\Gamma(x; r_d)$ have different dependences on r_d/L_T and r_d/L_v. The coherence radius can obtained by equating the exponential in equation (7.186) to one, replacing r_d with $r_{\rm coh}$, and solving for $r_{\rm coh}$. For the von Kármán spectra, there is no analytical formula for $r_{\rm coh}$; however, the coherence radius can be readily calculated numerically.

Solid and dash-dotted lines in figure 7.5 show the magnitude of the normalized mutual coherence function versus the sensor separation r_d. The coherence is normalized by its value at $r_d = 0$. It is plotted for moderate wind ($u_* = 0.3$ m/s) and two values of the surface heat flux, corresponding to cloudy ($Q_H = 0$ W/m^2) and sunny ($Q_H = 200$ W/m^2) conditions. The propagation range is $x = 1$ km; other parameters of the calculations are the same as for figure 7.4. It follows from figure 7.5 that the coherence gradually decreases with increasing sensor separation r_d and it is greater for cloudy than for sunny conditions. The coherence radii $r_{\rm coh}$ for these two cases are 6.3 m and 3.1 m, respectively.

In the limiting case $r_d \gg L_T, L_v$, equation (7.186) becomes

$$
\Gamma(x; r_d) = |p_0|^2 \exp \left[-2(\gamma_T + \gamma_v)x \right]. \tag{7.187}
$$

Comparing with equation (7.177), we conclude that the sound field at two widely separated observation points is uncorrelated.

7.5.2.2 Kolmogorov spectrum

Substituting $\Phi_{\rm eff}(0, \kappa_\perp)$ given by equation (7.31) into equation (7.180), and calculating the integral over $\kappa_\perp$ similarly to that for the case of a motionless random medium [178], we have

$$
\Gamma(x; r_d) = |p_0|^2 \exp \left[-Bk^2 r_d^{5/3} x \left(\frac{\beta_c^2 C_T^2}{T_0^2} + \frac{22}{3} \frac{C_v^2}{c_0^2} \right) \right]. \tag{7.188}
$$

Here, $B = [\sqrt{3}\Gamma^2(1/3)]/[5 \cdot 2^{7/3}\Gamma(2/3)] \approx 0.364$ is a numerical coefficient. For the Kolmogorov spectra, the coherence radius can be calculated analytically, with result:

$$
r_{\rm coh} = \left[Bk^2 x \left(\frac{\beta_c^2 C_T^2}{T_0^2} + \frac{22}{3} \frac{C_v^2}{c_0^2} \right) \right]^{-3/5}. \tag{7.189}
$$

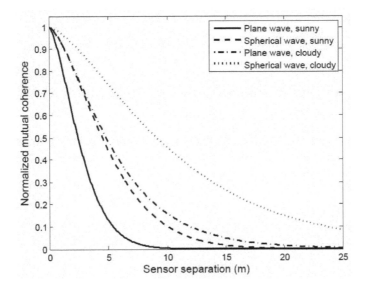

FIGURE 7.5

Magnitude of the normalized mutual coherence functions of plane and spherical sound waves versus the sensor separation r_d for moderate wind ($u_* = 0.3$ m/s). The propagation range is $x = 1$ km and the acoustic frequency $f = 500$ Hz. For cloudy conditions, $Q_H = 0$ W/m², and for sunny conditions, $Q_H = 200$ W/m².

This formula provides an explicit dependence of r_{coh} on parameters of the problem.

In the limiting case $r_d \ll L_T, L_v$, the modified Bessel functions in equation (7.186) can be expanded into series in r_d/L_T and r_d/L_v. Retaining only the first nonvanishing terms in the exponent in this equation, replacing γ_T and γ_v by their values from equation (7.184), and replacing the variances σ_T^2, σ_v^2 with the structure-function parameters C_T^2, C_v^2 (see equation (6.36)), one obtains a formula for the coherence coinciding with equation (7.188). This result is to be expected, since the Kolmogorov spectra are the limiting cases of the von Kármán spectra for large values of L_T and L_v.

7.5.2.3 Gaussian spectrum

Using equations (7.180) and (7.36), one obtains a formula for the mutual coherence function in the form

$$\Gamma(x; r_d) = |p_0|^2 \exp\left\{-2\gamma_T x \left(1 - e^{-r_d^2/\mathcal{L}_T^2}\right)\right.$$

$$\left. -2\gamma_v x \left[1 - \left(1 - \frac{r_d^2}{\mathcal{L}_v^2}\right) e^{-r_d^2/\mathcal{L}_v^2}\right]\right\}. \quad (7.190)$$

Here, γ_T and γ_v are given by equation (7.185). This formula is given for $\sigma_v^2 = 0$ elsewhere [72]. In the limiting case $r_d \gg L_T, L_v$, equation (7.190) coincides with equation (7.187), as it should.

7.5.3 Mutual coherence function for spherical waves

7.5.3.1 Von Kármán spectrum

The mutual coherence function of a spherical wave is given by equation (7.183). Substituting the von Kármán effective spectrum, equation (7.30), and calculating the integral over $\kappa_\perp$, we have

$$
\Gamma(x; \mathbf{r}_c, \mathbf{r}_d) = \frac{1}{x^2} \exp \left\{ \frac{ik\mathbf{r}_c \cdot \mathbf{r}_d}{x} \right.
$$
$$
- \frac{2\gamma_T x L_T}{r_d} \int_0^{r_d/L_T} \left[1 - \frac{2^{1/6}\eta^{5/6}}{\Gamma(5/6)} K_{5/6}(\eta) \right] d\eta
$$
$$
\left. - \frac{2\gamma_v x L_v}{r_d} \int_0^{r_d/L_v} \left[1 - \frac{2^{1/6}\eta^{5/6}}{\Gamma(5/6)} \left(K_{5/6}(\eta) - \frac{\eta}{2} K_{1/6}(\eta) \right) \right] d\eta \right\}. \quad (7.191)
$$

The remaining integrals over η cannot be evaluated analytically.

Dashed and dotted lines in figure 7.5 correspond to the magnitude of the normalized mutual coherence function of a spherical wave calculated with equation (7.191). Parameters of the calculations are the same as for a plane wave. It follows from the figure that the spherical wave coherence is greater than the plane wave coherence. The coherence radii r_{coh} of a spherical wave are 5.7 m and 12.0 m for sunny and cloudy conditions, respectively.

Figure 7.6 depicts the coherence radii r_{coh} calculated with equation (7.191) for the case of sound propagation in the atmosphere. The coherence radii are plotted versus the surface heat flux Q_H for three values of the friction velocity $u_* = 0.1$ m/s, 0.3 m/s, and 0.6 m/s. The propagation range is $x = 1$ km; other parameters of the problem are the same as for figure 7.4. The coherence radii vary between 3.5 m and 42.3 m. For $u_* = 0.1$ m/s and $Q_H = 0$ W/m^2, the magnitude of the normalized mutual coherence function remains greater than $1/e$ so that the coherence radius cannot be determined. It follows from figure 7.6 that the coherence radii decrease with increasing heat flux and friction velocity. Most of the values of the coherence radii are within the inertial subrange of turbulence and, hence, can (at least by order of magnitude) be calculated with the Kolmogorov spectra; see equation (7.193) below. This equation enables estimation of r_{coh} for propagation ranges x and acoustic frequencies f different from those in figure 7.6.

7.5.3.2 Kolmogorov spectrum

Substituting $\Phi_{\text{eff}}(0, \kappa_\perp)$ for the Kolmogorov spectra into equation (7.183), and calculating the integrals over η and $\kappa_\perp$ similarly to that in reference [178], one

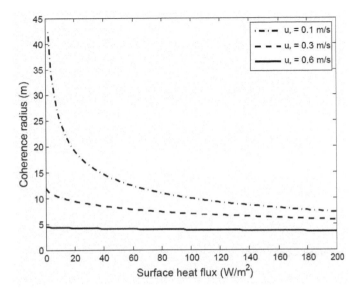

FIGURE 7.6
Coherence radius of a spherical sound wave versus the surface heat flux Q_H for three values of the friction velocity u_*. The propagation range is $x = 1$ km and the acoustic frequency $f = 500$ Hz.

obtains

$$\Gamma(x; \mathbf{r}_c, \mathbf{r}_d,) = \frac{1}{x^2} \exp\left[\frac{i k \mathbf{r}_c \cdot \mathbf{r}_d}{x} - \frac{3}{8} B k^2 r_d^{5/3} x \left(\frac{\beta_c^2 C_T^2}{T_0^2} + \frac{22}{3}\frac{C_v^2}{c_0^2}\right)\right]. \quad (7.192)$$

Here, B is the same coefficient as in equation (7.188) for the coherence function of a plane wave. The coherence radius for a spherical wave is given by

$$r_{\text{coh}} = \left[\frac{3}{8} B k^2 x \left(\frac{\beta_c^2 C_T^2}{T_0^2} + \frac{22}{3}\frac{C_v^2}{c_0^2}\right)\right]^{-3/5}. \quad (7.193)$$

It is larger than that for a plane wave by a factor of $(8/3)^{3/5} \cong 1.8$.

Analogously to the plane wave coherence, it can be shown that in the limiting case $r_d \ll L_T, L_v$, equation (7.191) for the mutual coherence function for the von Kármán spectra coincides with the Kolmogorov spectra, equation (7.192).

7.5.3.3   Gaussian spectrum

For this spectrum, the mutual coherence function for spherical waves is obtained from equations (7.183) and (7.36) in the form

$$
\Gamma(x; \mathbf{r}_c, \mathbf{r}_d) = \frac{1}{x^2} \exp \left\{ \frac{ik\mathbf{r}_c \cdot \mathbf{r}_d}{x} - 2\gamma_T x \left[1 - \frac{\sqrt{\pi} \mathcal{L}_T}{2r_d} \mathrm{erf} \left(\frac{r_d}{\mathcal{L}_T} \right) \right] \right.
$$
$$
\left. - 2\gamma_v x \left[1 - \frac{\sqrt{\pi} \mathcal{L}_v}{4r_d} \mathrm{erf} \left(\frac{r_d}{\mathcal{L}_v} \right) - \frac{1}{2} \exp \left(-\frac{r_d^2}{\mathcal{L}_v^2} \right) \right] \right\}, \quad (7.194)
$$

where γ_T and γ_v are given by equation (7.185). Equation (7.194) for $\gamma_v = 0$ is given elsewhere [88].

7.6   Experimental data on sound propagation in random moving media

In this section, measurements of the mutual coherence function of a spherical sound wave after passing through a turbulent jet are presented and discussed. Then, experiments about sound propagation in a turbulent atmosphere and ocean are considered.

7.6.1   Sound propagation through a turbulent jet

The mutual coherence function of a sound wave propagating through a turbulent jet was measured in a well-controlled laboratory experiment [34]. Laboratory experiments enable relatively accurate measurements of the statistical moments of a sound field. Experiments in the atmosphere and ocean are more challenging due to the variability of the turbulence characteristics in time and along the propagation path.

Figure 7.7 depicts a schematic of the experimental setup [34]. A turbulent jet was emitted from a nozzle with horizontal and vertical dimensions of 1 m and 8 cm, respectively. The mean speed v of the jet varied between 7.9 m/s and 12.2 m/s. The variance of the velocity fluctuations was proportional to the mean speed squared: $\sigma_v^2 = \beta^2 v^2$, where $\beta = 0.217$. The spectrum of velocity fluctuations was well approximated by a von Kármán spectrum with the outer scale $L_v = 8.97$ cm. An acoustic source was located at the distance $x_1 = 0.25$ m from the left boundary of the jet (if viewed as in figure 7.7). The source emitted a spherical wave with frequency $f = 25$ kHz or 40 kHz. Two microphones were placed near the right boundary of the jet at a distance $x = 1.25$ m from the source. The microphones were located along the vertical z-axis; the separation r_d between them varied from 2 cm to 14 cm. The jet emitted from the nozzle extended in the vertical direction; at the

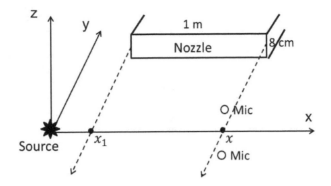

FIGURE 7.7
Schematic of the experimental setup of sound propagation through a turbulent jet [34]. Dashed lines indicate horizontal boundaries of the jet emitted from the nozzle and its direction of propagation. The source is situated in the center of the coordinate system (x, y, z). Two microphones (circles) are located along the z-axis at the right boundary of the jet.

sound propagation path it was confined within ± 8 cm from the x-axis. (The vertical extension was defined as the distance from the center of the jet at which v decreased by a factor of 2.) In the experiment, the mutual coherence function $\Gamma(x; r_d)$ was measured.

Older theories of sound propagation in random moving media (see the text following equation (7.4)) were unable to explain the measured mutual coherence function. Comparison between the experimental data and theoretical predictions was done 14 years after the experiment [291], when the theory presented in this chapter had become available. For the considered setup, the mutual coherence function is given by equation (7.191) with $r_c = 0$ and $\gamma_T = 0$:

$$
\Gamma(x; r_d) = \frac{1}{x^2} \exp\left\{ -\frac{2\gamma_v x L_v}{r_d} \right.
$$
$$
\times \int_{x_1 r_d/(x L_v)}^{r_d/L_v} \left[1 - \frac{2^{1/6} \eta^{5/6}}{\Gamma(5/6)} \left(K_{5/6}(\eta) - \frac{\eta}{2} K_{1/6}(\eta) \right) \right] d\eta \right\}. \quad (7.195)
$$

Here, the lower limit of integration reflects the fact that there was no turbulence for sound propagation from the source to the left boundary of the jet and the extinction coefficient γ_v is determined with equation (7.184). Due to different values of the variance $\sigma_v^2 = \beta^2 v^2$ and acoustic wavenumber $k = 2\pi f/c_0$ in the experimental trials, γ_v varied between 0.35 m^{-1} and 2.13 m^{-1}. The mean value of the extinction coefficient was $\overline{\gamma}_v = 1.2$ m^{-1}. To present results for the mutual coherence function of different trials as one experimental curve, we consider the normalized mutual coherence function, $\Gamma(x; r_d)/\Gamma(x; 0)$, raised to

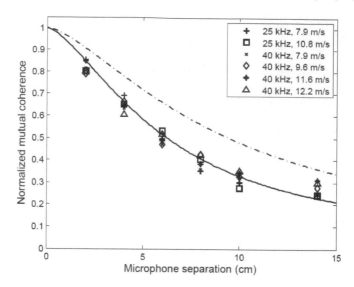

FIGURE 7.8
Normalized mutual coherence function raised to the power $\overline{\gamma}_v/\gamma_v$, $[\Gamma(x;r_d)/\Gamma(x;0)]^{\overline{\gamma}_v/\gamma_v}$, versus the microphone separation r_d [291]. Symbols are experimental data for different values of the acoustic frequency f and jet speed v. Solid and dash-dotted lines are the theoretical predictions based on equations (7.196) and (7.197).

the power $\overline{\gamma}_v/\gamma_v$. Using equation (7.195), we have

$$[\Gamma(x;r_d)/\Gamma(x;0)]^{\overline{\gamma}_v/\gamma_v} = \exp\left\{-\frac{2\overline{\gamma}_v x L_v}{r_d}\right.$$
$$\left.\times \int_{x_1 r_d/(xL_v)}^{r_d/L_v} \left[1 - \frac{2^{1/6}\eta^{5/6}}{\Gamma(5/6)}\left(K_{5/6}(\eta) - \frac{\eta}{2}K_{1/6}(\eta)\right)\right] d\eta\right\}. \quad (7.196)$$

The solid line in figure 7.8 is the dependence of $[\Gamma(x;r_d)/\Gamma(x;0)]^{\overline{\gamma}_v/\gamma_v}$ on the microphone separation r_d obtained with this formula. Symbols correspond to $[\Gamma(x;r_d)/\Gamma(x;0)]^{\overline{\gamma}_v/\gamma_v}$ obtained from experimental data for different runs. The theoretical prediction is in good agreement with the experimental data. Note that for the microphone separation $r_d = 14$ cm, the experimental data are located above the theoretical prediction. For this separation, the microphones were situated at the distance $r_d/2 = 7$ cm from the center of the jet, which was comparable to the jet's vertical extension of 8 cm. In this case, sound propagated through less intense turbulence than that near the jet's center, resulting in nearly the same values of the coherence as those for $r_d = 10$ cm.

The dash-dotted line in figure 7.8 is the theoretical prediction for $[\Gamma(x;r_d)/\Gamma(x;0)]^{\overline{\gamma}_v/\gamma_v}$ based on the older theories of sound propagation in random moving media. The dash-dotted line deviates noticeably from the prediction based on equation (7.196) and cannot explain experimental data. This line was obtained in accordance with the older theories: the normalized variance of temperature fluctuations σ_T^2/T_0^2 in the mutual coherence function of a sound wave propagating through a scalar random field, which is given by equation (7.191) with $r_c = 0$ and $\gamma_v = 0$, was replaced with $4\sigma_v^2/c_0^2$. As a result, one obtains the following formula

$$[\Gamma(x; r_d)/\Gamma(x; 0)]^{\overline{\gamma}_v/\gamma_v}$$

$$= \exp\left\{ -\frac{2\overline{\gamma}_v x L_v}{r_d} \int_{x_1 r_d/(x L_v)}^{r_d/L_v} \left[1 - \frac{2^{1/6}\eta^{5/6}}{\Gamma(5/6)} K_{5/6}(\eta)\right] d\eta \right\}. \quad (7.197)$$

The dash-dotted line in figure 7.8 is the theoretical prediction based on this formula.

7.6.2 Sound propagation in a turbulent atmosphere

Brown and Hall [52] and Tatarskii [374] overview experimental studies of line-of-sight sound propagation through a turbulent atmosphere, which were performed from the late 1940s to the mid-1970s. At that time, the main interest was in the analysis of phase and log-amplitude fluctuations of sound waves at frequencies and propagation ranges for which these fluctuations were affected by turbulence in the inertial subrange. The experiments tested the theories of wave propagation through random media which were then being developed. The variance and correlation function of log-amplitude fluctuations, $\langle \chi^2 \rangle$ and $B_\chi(x; r_d)$, and the structure function of phase fluctuations, $D_\phi(x; r_d)$, were usually measured. (In some experiments, the temporal correlation function of log-amplitude fluctuations and the temporal structure function of phase fluctuations were measured; then they were converted to $B_\chi(x; r_d)$ and $D_\phi(x; r_d)$ using Taylor's frozen-turbulence hypothesis.) The experiments confirmed the theoretical dependencies of $\langle \chi^2 \rangle$, $B_\chi(x; r_d)$, and $D_\phi(x; r_d)$ on the propagation range x, microphone separation r_d, and sound wavenumber k, which are described with equations for the Kolmogorov spectra in section 7.3. These dependencies are not affected by the value of the numerical coefficient in front of C_v^2/c_0^2. As far as we know, no attempt was made during that time to check experimentally the value of this coefficient.

In the period after the mid-1970s, many additional experimental studies were conducted. However, in most of these experiments the statistical characteristics of sound signals were affected not only by atmospheric turbulence, but also by other factors such as multipath propagation, reflection from an impedance ground, sound scattering into a refractive shadow zone, e.g., references [24, 87, 156, 182, 274]. Line-of-sight propagation has been studied in just a few experiments. In reference [432], the phase and amplitude fluctuations of sound waves were measured with an acoustic tomographic array. The short pulses employed in tomography (section 3.7) usually enable separation of direct signals from speakers to microphones from those reflected from the ground. Therefore, though intended for different purposes, such experiments can also be used to study the statistical characteristics of acoustic signals for line-of-sight sound propagation.

Reference [88] reports on an experiment in which a speaker was placed on the asphalt airport runway and several microphones were located above the runway at various heights to achieve line-of-sight propagation. The speaker transmitted five pure tones between 250 Hz and 4 kHz; the propagation range was up to 300 m. The variances of phase and log-amplitude fluctuations, $\langle \phi^2 \rangle$ and $\langle \chi^2 \rangle$, the mutual coherence function $\Gamma(x; r_d)$, and the *longitudinal* coherence function were measured. The latter was calculated by cross-correlating signals recorded by a reference microphone located 2 m from the speaker and a microphone at range x; for this geometry, the longitudinal coherence is proportional to the mean sound field $\langle p \rangle$. The Gaussian spectra of turbulence were used to explain the measured statistical moments. These spectra might be appropriate for the theoretical predictions of the mean sound field and the variance of phase fluctuations at large x (see the text after equation (7.112)), which are affected by large-scale inhomogeneities. For the statistical moments of the sound field which are affected by turbulence in the inertial subrange (such as $\langle \chi^2 \rangle$ and $\Gamma(x; r_d)$), the Gaussian spectra are inappropriate.

For theoretical predictions of sound propagation through a turbulent atmosphere, it seems reasonable to use the von Kármán spectrum, since it includes both the inertial and energy subranges of turbulence. This was done in reference [249], where the mutual coherence function $\Gamma(x; r_d)$ of a spherical sound wave was measured. The propagation range was $x = 100$ m at the height above the ground which decreased linearly from $z = 16$ m to 10.5 m. Five narrowband signals in the range 650 Hz–5 kHz were transmitted. The maximum transverse separation between microphones was $r_d = 2$ m. The measured coherence was compared to equation (7.191), in which σ_v^2 was replaced with C_v^2 using equation (6.36). The best fit between theoretical predictions and experimental data enabled determination of the structure-function parameter $C_v^2 = 0.09$ m$^{4/3}$/s^2 and the outer scale $L_v \sim 60$–70 m. (The temperature fluctuations were relative small and ignored in the analysis.) The values of these parameters measured with a sonic anemometer were close to those obtained from the acoustic measurements: $C_v^2 = 0.088$ m$^{4/3}$/s^2 and $L_v \sim 100$ m. These results indicate the feasibility of acoustic remote sensing of turbulence parameters. Note that for the maximum separation between the microphones $r_d = 2$ m, the mutual coherence function was affected mainly by turbulence in the inertial subrange. As a result, C_v^2 was reconstructed from acoustic measurements more accurately than L_v.

7.6.3   Sound propagation in a turbulent ocean

Short-range, high-frequency, line-of-sight sound propagation is used for remote sensing of the oceanic environment, and for testing theories of wave propagation in random media. In the acoustic scintillation method [113, 126], signals from a distant source are correlated at two hydrophones with a relatively small spatial separation. The time-lag of the maximum in the cross correlation function is used to determine the transverse current. In reciprocal acoustic transmission (section 1.3.6), path-averaged values of the sound speed and current velocity fluctuations can be measured [250]. By measuring the statistical moments of forward propagating signals (e.g., the variance of log-amplitude fluctuations or the angle of arrival fluctuations) one can retrieve statistical characteristics of turbulent oceanic flows [98].

In the following, we describe an experimental study of sound propagation through turbulent flow produced by strong tidal currents in Cordova Channel near British Columbia [99]. The experiment was specifically designed to discern the contributions from scalar and vector random fields to the statistical moments of sound signals. Transmitters at $f = 67.567$ kHz and receivers were located on both sides of the Cordova Channel; the propagation range was $x = 695$ m at the depth of 19.5 m. Reciprocal transmissions were used to determine the path-averaged values of the structure-function parameters of the current velocity fluctuations C_v^2 and sound speed fluctuations $C_{\tilde{c}}^2$ in the inertial subrange. (In the experiment, the salinity fluctuations might have had a noticeable contribution to $C_{\tilde{c}}^2$.)

Using C_v^2 and $C_{\tilde{c}}^2$ from reciprocal transmissions, the path-averaged values of the effective structure-function parameter, C_{eff}^2, determined with equation (7.34) were calculated. Also the values of the structure-function parameter used in the older theories, C_{old}^2, given by equation (7.35) were obtained. This approach for determining C_{eff}^2 and C_{old}^2 does not rely on predictions of the theories of sound propagation in random moving media.

In the considered experiment, the variance of log-amplitude fluctuations of a spherical wave, $\langle \chi^2 \rangle$, given by equation (7.115) and the *wave structure function* of a plane wave, $D_w(x, r_d)$, were also measured for the case of forward propagation. The latter function is related to the mutual coherence function of a plane wave by the formula: $\Gamma(x; r_d) = \exp\left[-D_w(x, r_d)/2\right]$. Using equation (7.188), we obtain

$$D_w(x, r_d) = 2Bk^2 r_d^{5/3} x C_{\text{eff}}^2. \tag{7.198}$$

Here, $\beta_c^2 C_T^2 / T_0^2 + 22 C_v^2 / (3c_0^2)$ was replaced with C_{eff}^2. A similar replacement can also be made in equation (7.115) for $\langle \chi^2 \rangle$. From the measurements of $\langle \chi^2 \rangle$ and $D_w(x, r_d)$, the path-averaged values of the effective structure-function parameter were determined and denoted as C_{meas}^2. For the theory presented in this chapter, $C_{\text{meas}}^2 = C_{\text{eff}}^2$; for the older theories, $C_{\text{meas}}^2 = C_{\text{old}}^2$.

The experimental data obtained in reference [99] are plotted in figures 7.9(a) and (b). In these figures, the vertical axes correspond to C_{old}^2 and C_{eff}^2 obtained from reciprocal acoustic transmission and the horizontal axis corresponds to C_{meas}^2. The upper and lower plots are the experimental tests of the older theories and the theory presented in this chapter. For a theory to be

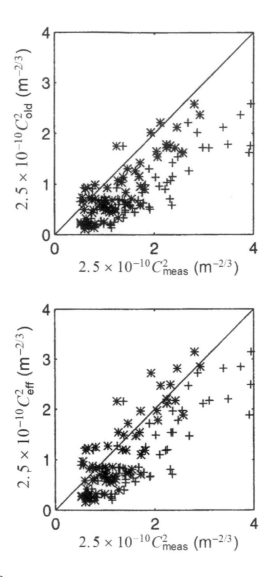

FIGURE 7.9
Comparison between different values (C^2_{old} and C^2_{eff}) of the effective structure-function parameter and its measurements (C^2_{meas}) for sound propagation through a turbulent flow in Cordova Channel. The values of C^2_{old} and C^2_{eff} are determined from reciprocal acoustic transmission. C^2_{meas} are retrieved from the measurements of the variance of log-amplitude fluctuations ($+$) and the wave structure function ($*$). (Reproduced with permission from reference [99]. Copyright 1998, Acoustical Society of America.)

valid, the experimental points should be close to the solid lines which correspond to the dependencies $C_{meas}^2 = C_{old}^2$ (upper plot) and $C_{meas}^2 = C_{eff}^2$ (lower plot). In figure 7.9(a), most of the experimental data are located below the solid line and do not support the older theories. In figure 7.9(b), the experimental data are close to the solid line, especially for the case of the wave structure function. The small deviation of the data for the log-amplitude fluctuations from the solid line can be explained by the fact that the measurements based on reciprocal transmissions and $D_w(x, r_d)$ had uniform weighting along the propagation path, while those for $\langle \chi^2 \rangle$ were weighted toward the center of the channel where the current might have been stronger.

8

Multipath sound propagation in a random moving medium

In Chapter 7, we considered line-of-sight propagation, for which (in the absence of random inhomogeneities) sound propagates to the observation point along a single, straight line path. Scattering by random inhomogeneities in the direction of sound propagation was analyzed. In this chapter, we will consider multipath sound propagation in a random moving medium, meaning that waves propagate to the observation point along two or more paths. Multipath propagation can occur due to reflection of a sound wave from a surface or by refraction from mean gradients in the ambient fields. It can also occur because of sound scattering by random inhomogeneities in all directions, as is the case when the sound wavelength is of order or greater than the outer scale of turbulence.

If a source and receiver are located above the ground at a relatively short distance, the sound field is the sum of the direct and ground reflected waves. Atmospheric turbulence significantly affects the interference of these two arrivals. This phenomenon is considered in section 8.1. In section 8.2, closed-form equations for the statistical moments of arbitrary order of a sound field propagating above an impedance ground in a refractive, turbulent atmosphere are derived. The equations obtained can be solved numerically with techniques presented in Chapter 11. In sections 8.3 and 8.4, we consider the theory of multiple scattering, which enables one to describe sound scattering at large angles, including backscattering. Starting with a Helmholtz-type equation, formulas for the mean Green's function and mean sound field are obtained. A closed-form equation for the mutual coherence function of the sound field fluctuations is derived. With some approximations, this equation reduces to the equation of radiative transfer.

In this chapter, we consider propagation and scattering of a monochromatic sound wave with the frequency ω. The dependence of the sound pressure on time t, given by the factor $\exp(-i\omega t)$, is omitted.

8.1 Interference of the direct and surface-reflected waves in a random medium

In the atmosphere, acoustic sources (such as cars, trains, and airplanes) and receivers (human listeners or microphones) are usually elevated above the ground. At relatively short propagation ranges, for which refraction can be ignored, the sound field at a receiver is the sum of the wave propagating directly to the receiver and the wave reflected from the ground. Due to the interference of these two waves, the resulting sound field as a function of range or acoustic frequency may exhibit local maxima and minima. In a turbulent atmosphere, the phase and amplitude of the direct and ground-reflected waves fluctuate. This results in partial destruction of the interference between the two waves and may increase the mean intensity at the interference minima by tens of dB. In earlier works [175, 87, 86], this problem was studied using heuristic approaches. A rigorous, energy-conserving approach based on the Rytov method was suggested by Clifford and Lataitis [82]. However, they utilized older theories of sound propagation in random moving media and did not distinguish between sound scattering by scalar and vector random fields, which, as discussed in Chapter 7, scatter the field differently. Furthermore, as indicated in reference [347], they ignored the ground reflection of the scattered waves. In this section, following references [297, 347], we will study the interference of the direct and ground-reflected waves in a turbulent atmosphere. The solution applies more generally in aeroacoustics to a source and receiver located above a flat surface in a turbulent flow.

8.1.1 Geometry and starting equations

The geometry of the problem is shown in figure 8.1. A source and receiver are located above an impedance ground. The plane $z = 0$ of the Cartesian coordinate system $\mathbf{R} = (x, y, z)$ coincides with the surface of the ground and the z-axis is directed upward. The source and receiver coordinates are $\mathbf{R}_s = (0, 0, h_s)$ and $\mathbf{R}_r = (x_r, 0, h_r)$, respectively. Here, h_s and h_r are the source and receiver heights above the ground and x_r is the distance between them along the x-axis, called the range. The subscripts s and r correspond to source and receiver. In formulations below, we assume that the mean value T_0 of the temperature does not depend on the height z and the mean wind velocity is zero. The sound speed, temperature, and velocity fluctuations are denoted as $\widetilde{c}(\mathbf{R})$, $\widetilde{T}(\mathbf{R})$, and $\widetilde{\mathbf{v}}(\mathbf{R})$, respectively. The functions $\widetilde{c}(\mathbf{R})$, and $\widetilde{T}(\mathbf{R})$ are related by equation (6.65).

In the absence of atmospheric turbulence, the sound pressure at the receiver due to a unit-amplitude source (section 4.2.1) is given by [16, 345]:

$$p = p_{1,0} + p_{2,0} = \frac{1}{R_1} \exp(ikR_1) + \frac{Q}{R_2} \exp(ikR_2). \tag{8.1}$$

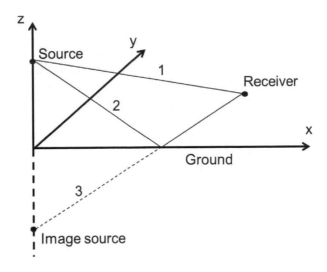

FIGURE 8.1
Geometry of the problem. A source and receiver are located above the ground.
Lines 1, 2, and 3 correspond to the direct wave from the source to receiver,
the wave reflected from the ground, and the wave from the image source to
receiver.

Here, $p_{1,0}$ and $p_{2,0}$ correspond to the sound fields of the direct and ground-
reflected waves, $R_1 = \sqrt{x_r^2 + (h_s - h_r)^2}$ and $R_2 = \sqrt{x_r^2 + (h_s + h_r)^2}$ are
the path lengths of these waves, k is the sound wavenumber, and Q is the
spherical-wave reflection coefficient, which can also be interpreted as the apparent strength of the image source. The indices 1 and 2 refer to the direct
and ground-reflected waves. In the approximation of the near-grazing sound
propagation when $x_r \gg h_s, h_r$ and for the locally reacting surface, Q is given
by [15, 273]

$$Q = \frac{Z \sin \alpha - 1 + 2F(d)}{Z \sin \alpha + 1}, \tag{8.2}$$

where Z is the normalized specific impedance of the ground, α is the grazing
angle of the wave incident on the ground, and

$$F(d) = \left[1 + i\sqrt{\pi}d \exp(-d^2)\mathrm{erfc}(-id)\right] \tag{8.3}$$

is the boundary loss factor. Here, $d = \sqrt{ikR_2/2}\,(1/Z + \sin \alpha)$ is the *numerical
distance* and erfc is the complementary error function. The quantity pp^* can be
readily calculated with equations (8.1)–(8.3). This quantity may have multiple
maxima and minima due to the interference of the direct and ground-reflected
waves. This interference pattern depends on k and x_r.

8.1.2   Mean-squared sound pressure

In a turbulent atmosphere, p becomes a random field. We will analyze the mean-squared sound pressure $\langle pp^* \rangle$, where the brackets $\langle \ \rangle$ denote averaging over an ensemble of realizations of $\widetilde{T}$ and $\widetilde{\mathbf{v}}$. Theoretical formulations are developed for an equivalent geometry of the problem, when the impedance ground is removed, a turbulent atmosphere fills the whole space, and the field from the actual source is summed to that produced by the image source of strength Q located at the point $\mathbf{R}_{im} = (0, 0, -h_s)$ (figure 8.1), where the subscript im stands for the image source. For this new geometry, the sound speed, temperature, and velocity fluctuations are symmetrical with respect to the plane $z = 0$, e.g., $\widetilde{c}(x, y, -z) = \widetilde{c}(x, y, z)$.

For the new geometry, the sound field at the receiver is again a sum of two terms:

$$p(\mathbf{R}) = p_1(\mathbf{R}) + p_2(\mathbf{R}), \tag{8.4}$$

where p_1 and p_2 are the sound fields due to the source and the image source, respectively. We will use the Rytov method (section 7.2.1) to calculate the mean-squared sound pressure. Following this method, we write $p_1(\mathbf{R}) = p_{1,0}(\mathbf{R}) \exp[\Psi_1(\mathbf{R})]$ and $p_2(\mathbf{R}) = p_{2,0}(\mathbf{R}) \exp[\Psi_2(\mathbf{R})]$. Here, $p_{1,0}$ and $p_{2,0}$ are the sound fields in a non-turbulent atmosphere determined with equation (8.1), and Ψ_1 and Ψ_2 are the complex phases due to the actual and image sources in a turbulent atmosphere. With these notations, equation (8.4) reads

$$p(\mathbf{R}) = p_{1,0}(\mathbf{R}) \exp[\Psi_1(\mathbf{R})] + p_{2,0}(\mathbf{R}) \exp[\Psi_2(\mathbf{R})]. \tag{8.5}$$

Multiplying the sound pressure by its conjugate, averaging the resulting expression over the ensemble of realizations, and substituting the values of $p_{1,0}$ and $p_{2,0}$, one obtains

$$\langle p(\mathbf{R})p^*(\mathbf{R}) \rangle = \frac{1}{R_1^2} \langle e^{\Psi_1 + \Psi_1^*} \rangle + \frac{|Q|^2}{R_2^2} \langle e^{\Psi_2 + \Psi_2^*} \rangle$$

$$+ \frac{|Q|}{R_1 R_2} \left[e^{ik(R_1 - R_2) + i\varphi} \langle e^{\Psi_1^* + \Psi_2} \rangle + e^{-ik(R_1 - R_2) - i\varphi} \langle e^{\Psi_1 + \Psi_2^*} \rangle \right], \tag{8.6}$$

where $Q = |Q|e^{i\varphi}$. (Strictly speaking, equation (8.6) is valid for a specular reflection.) For sound propagation in a turbulent atmosphere, the acoustic energy is conserved; therefore $\langle p_1 p_1^* \rangle = p_{1,0} p_{1,0}^*$. Substituting $p_1 = p_{1,0} \exp(\Psi_1)$ into this formula, we obtain $\langle \exp(\Psi_1 + \Psi_1^*) \rangle = 1$. Similarly, $\langle \exp(\Psi_2 + \Psi_2^*) \rangle = 1$. The last two equalities allow us to simplify the first and second terms on the right-hand side of equation (8.6).

Consider the quantity $\langle \exp(\Psi_1^* + \Psi_2) \rangle$ appearing in the third term in equation (8.6). Similarly to equation (7.54), we write $\Psi_1 = \chi_1 + i\phi_1$ and $\Psi_2 = \chi_2 + i\phi_2$, where χ_1 and ϕ_1 are the log-amplitude and phase fluctuations in the sound field due to the source, and χ_2 and ϕ_2 are those for the image source. We assume that the temperature and velocity fluctuations have Gaussian (normal) distributions. In this case, the random fields χ_1, ϕ_1, χ_2, and ϕ_2 also have Gaussian statistics [339]. For a random field μ with a Gaussian distribution, the following equality holds:

$$\langle \exp(\mu) \rangle = \exp\left[\frac{1}{2}\langle(\mu - \langle\mu\rangle)^2\rangle + \langle\mu\rangle\right] = \exp\left[\frac{\langle\mu^2\rangle}{2} - \frac{\langle\mu\rangle^2}{2} + \langle\mu\rangle\right]. \quad (8.7)$$

Using this equality, we have

$$\langle \exp(\Psi_1^* + \Psi_2) \rangle = \langle \exp(\chi_1 - i\phi_1 + \chi_2 + i\phi_2) \rangle$$
$$= \exp\left[\frac{1}{2}\left\langle(\chi_1 - i\phi_1 + \chi_2 + i\phi_2)^2\right\rangle - \frac{1}{2}\left(\langle\chi_1\rangle - i\langle\phi_1\rangle + \langle\chi_2\rangle + i\langle\phi_2\rangle\right)^2\right.$$
$$\left. + \langle\chi_1\rangle - i\langle\phi_1\rangle + \langle\chi_2\rangle + i\langle\phi_2\rangle\right]. \quad (8.8)$$

In this formula, $\left\langle(\chi_1 - i\phi_1 + \chi_2 + i\phi_2)^2\right\rangle$ is recast as

$$\left\langle(\chi_1 - i\phi_1 + \chi_2 + i\phi_2)^2\right\rangle = \langle\chi_1^2\rangle - \langle\phi_1^2\rangle + \langle\chi_2^2\rangle - \langle\phi_2^2\rangle$$
$$- 2i\langle\chi_1\phi_1\rangle + 2\langle\chi_1\chi_2\rangle + 2i\langle\chi_1\phi_2\rangle - 2i\langle\phi_1\chi_2\rangle + 2\langle\phi_1\phi_2\rangle + 2i\langle\chi_2\phi_2\rangle. \quad (8.9)$$

Each term on the right-hand side of this formula can be calculated using the *first* Rytov approximation considered in section 7.2. All these terms are of order ε_{mov}^2, where $\varepsilon_{mov} = -2(\widetilde{c} + \widetilde{v}_x)/c_0$ is twice the fluctuation of the acoustic refractive index in the atmosphere (equation (7.3)). Here, $\widetilde{v}_x$ is the x-component of the velocity fluctuations and c_0 is the reference value of the sound speed.

In the first Rytov approximation, the quantities $\langle\chi_1\rangle$, $\langle\phi_1\rangle$, $\langle\chi_2\rangle$, $\langle\phi_2\rangle$ appearing in equation (8.8) are zero. These quantities can be calculated in the *second* Rytov approximation which corresponds to the third term on the right-hand side of the series expansion of the complex phase, equation (7.44). (Note that Ψ_1 and Ψ_2 have different meanings in section 7.4 than in this section.) A simpler way to calculate $\langle\chi_1\rangle$ is to consider the equality $\langle\exp(\Psi_1 + \Psi_1^*)\rangle = 1$. Substituting $\Psi_1 = \chi_1 + i\phi_1$ into this equality, we have $\langle\exp(2\chi_1)\rangle = 1$. Using equation (8.7), this result can be recast as

$$\exp\left[2\langle\chi_1^2\rangle - 2\langle\chi_1\rangle^2 + 2\langle\chi_1\rangle\right] = 1, \quad (8.10)$$

or $\langle\chi_1^2\rangle - \langle\chi_1\rangle^2 + \langle\chi_1\rangle = 0$. It follows from this equation that

$$\langle\chi_1\rangle = -\langle\chi_1^2\rangle + O(\varepsilon_{mov}^4), \quad (8.11)$$

and $\langle \chi_1 \rangle^2 = O(\varepsilon_{mov}^4)$. In the derivations to follow, we retain terms of order ε_{mov}^2 and omit terms of order ε_{mov}^4.

To obtain a formula for $\langle \phi_1 \rangle$, consider the mean sound field due to the source:

$$\langle p_1 \rangle = \frac{1}{R_1} \exp\left(ikR_1\right) \langle \exp\left(\chi_1 + i\phi_1\right)\rangle. \tag{8.12}$$

Using equation (8.7) and taking into account that $\langle \chi_1 \rangle^2$ and $\langle \phi_1 \rangle^2$ are of order ε_{mov}^4, we have

$$\langle p_1 \rangle = \frac{1}{R_1} \exp\left(ikR_1\right) \exp\left(-\frac{\langle \chi_1^2 \rangle}{2} - \frac{\langle \phi_1^2 \rangle}{2} + i\langle \chi_1 \phi_1 \rangle + i\langle \phi_1 \rangle\right). \tag{8.13}$$

In section 7.4, the mean sound field was calculated using the parabolic equation method. The resulting expression for the mean field, given by equations (7.150) and (7.152), coincides with equation (8.13) if in the latter equation the sum of the last two terms in the exponential is zero. Thus,

$$\langle \phi_1 \rangle = -\langle \chi_1 \phi_1 \rangle. \tag{8.14}$$

Similarly we obtain formulas for $\langle \chi_2 \rangle$ and $\langle \phi_2 \rangle$:

$$\langle \chi_2 \rangle = -\langle \chi_2^2 \rangle, \quad \langle \phi_2 \rangle = -\langle \chi_2 S_2 \rangle. \tag{8.15}$$

Equations (8.9), (8.11), (8.14), and (8.15) are substituted into equation (8.8). Taking into account that the terms $(\langle \chi_1 \rangle - i\langle \phi_1 \rangle + \langle \chi_2 \rangle + i\langle \phi_2 \rangle)^2$ are of order ε_{mov}^4, we obtain

$$\langle \exp\left(\Psi_1^* + \Psi_2\right)\rangle = \exp\left[\langle \chi_1 \chi_2 \rangle + \langle \phi_1 \phi_2 \rangle\right.$$
$$\left. - \frac{\langle \chi_1^2 \rangle + \langle \phi_1^2 \rangle + \langle \chi_2^2 \rangle + \langle \phi_2^2 \rangle}{2} + i(\langle \chi_1 \phi_2 \rangle - \langle \chi_2 \phi_1 \rangle)\right]. \tag{8.16}$$

The term $\langle \exp\left(\Psi_1 + \Psi_2^*\right)\rangle$ also appearing in equation (8.6) is the complex conjugate of $\langle \exp\left(\Psi_1^* + \Psi_2\right)\rangle$. Substituting both terms into this equation, we obtain the mean-squared sound pressure [82, 297]:

$$\langle pp^* \rangle = \frac{1}{R_1^2} + \frac{|Q|^2}{R_2^2} + \frac{2|Q|C_{coh}}{R_1 R_2} \cos\left[k(R_1 - R_2) + \varphi + \langle \chi_1 S_2 \rangle - \langle \chi_2 S_1 \rangle\right]. \tag{8.17}$$

Here, the *coherence factor* C_{coh} describes the coherence between the wave emitted by the source and that emitted by the image source:

$$C_{coh} = \exp\left(\langle \chi_1 \chi_2 \rangle + \langle \phi_1 \phi_2 \rangle - \frac{\langle \chi_1^2 \rangle + \langle \chi_2^2 \rangle + \langle \phi_1^2 \rangle + \langle \phi_2^2 \rangle}{2}\right). \tag{8.18}$$

For $Q = 1$, the same result was obtained in reference [347]. Theoretical estimates [82] and numerical calculations [347] show that the term $\langle \chi_1 S_2 \rangle - \langle \chi_2 S_1 \rangle$ in the cosine function in equation (8.17) is small and can be ignored.

8.1.3 Coherence factor

In this subsection, we derive an expression for the coherence factor based on the narrow-angle parabolic equation. For near-grazing sound propagation when $x_r \gg h_s, h_r$, the complex phases $\Psi_1 = \chi_1 + i\phi_1$ and $\Psi_2 = \chi_2 + i\phi_2$ can be determined with equation (7.53) with properly chosen complex amplitudes $A^{(0)}(x, y, z)$ of the emitted waves in a non-turbulent atmosphere. For the source located at $\mathbf{R}_s = (0, 0, h_s)$, the complex amplitude is given by

$$A^{(0)}(x, y, z) = \frac{1}{x} \exp\left[\frac{iky^2}{2x} + \frac{ik(z - h_s)^2}{2x}\right]. \tag{8.19}$$

This result can be obtained from equation (7.87) by considering a coordinate system in which the source coordinates are $(0, 0, h_s)$. Substituting this formula into equation (7.53), we obtain

$$\Psi_1 = \frac{k^2 x_r}{4\pi} \int_0^{x_r} \frac{dx}{x(x_r - x)} \int_{-\infty}^{\infty} dy \int_{-\infty}^{\infty} \varepsilon_{\text{mov}}(x, y, z)$$
$$\times \exp\left\{\frac{ikx_r\left[y^2 + (z - h_s + (h_s - h_r)x/x_r)^2\right]}{2x(x_r - x)}\right\} dz. \tag{8.20}$$

For the image source located at $\mathbf{R}_{im} = (0, 0, -h_s)$, the complex amplitude $A^{(0)}(x, y, z)$ is given by equation (8.19) if $z - h_s$ is replaced by $z + h_s$. Substituting the resulting formula for $A^{(0)}$ into equation (7.53), we have

$$\Psi_2 = \frac{k^2 x_r}{4\pi} \int_0^{x_r} \frac{dx}{x(x_r - x)} \int_{-\infty}^{\infty} dy \int_{-\infty}^{\infty} \varepsilon_{\text{mov}}(x, y, z)$$
$$\times \exp\left\{\frac{ikx_r\left[y^2 + (z + h_s - (h_s + h_r)x/x_r)^2\right]}{2x(x_r - x)}\right\} dz. \tag{8.21}$$

In the formulas for Ψ_1 and Ψ_2, the random field $\varepsilon_{\text{mov}}(x, y, z)$ is symmetrical with respect to the plane $z = 0$.

Using equations (8.20) and (8.21), we can determine the log-amplitude fluctuations (χ_1 and χ_2) and the phase fluctuations (ϕ_1 and ϕ_2) in the sound waves emitted by the actual and image sources. Then, the statistical moments of these fluctuations appearing in equation (8.18) can be calculated [297, 347]. As a result, for the coherence factor we have

$$C_{\text{coh}} = \exp\left\{-\frac{k^2 x_r}{4h} \int_0^h [b'_{\text{eff}}(0, 0) - b'_{\text{eff}}(0, z)] \, dz\right\}. \tag{8.22}$$

Here, $h = 2h_s h_r / (h_s + h_r)$ is the maximum separation between the direct path from the source to receiver and the path reflected from the ground, in the approximation of near-grazing propagation. The transverse correlation function $b'_{\text{eff}}(y, z) \equiv b'_{\text{eff}}(\mathbf{r})$ is determined with equations (7.21) and (7.137). Here, $\mathbf{r} = (y, z)$. For a statistically homogeneous random medium, $b'_{\text{eff}}(\mathbf{r})$ is given by

$$b'_{\text{eff}}(\mathbf{r}) = 2\pi \int \Phi_{\text{eff}}(0, \boldsymbol{\kappa}_\perp) \exp\left(i\boldsymbol{\kappa}_\perp \cdot \mathbf{r}\right) d^2\kappa_\perp, \tag{8.23}$$

where $\Phi_{\text{eff}}(\kappa_x, \boldsymbol{\kappa}_\perp)$ is the three-dimensional spectrum of the correlation function of the random field $\varepsilon_{\text{mov}}(x, \mathbf{r})$. In this chapter, if the limits of integration are not indicated, they are assumed to be from $-\infty$ to $+\infty$.

Substituting b'_{eff} into equation (8.22), we express the coherence factor in terms of the effective spectrum Φ_{eff}:

$$C_{\text{coh}} = \exp\left\{-\frac{\pi k^2 x_r}{2} \int_0^1 d\eta \int \Phi_{\text{eff}}(0, \boldsymbol{\kappa}_\perp)\left[1 - e^{i\eta\kappa_z h}\right] d^2\kappa_\perp\right\}, \tag{8.24}$$

where κ_z is the component of the vector $\boldsymbol{\kappa}_\perp = (\kappa_y, \kappa_z)$. For a statistically isotropic random medium, $\Phi_{\text{eff}}(0, \boldsymbol{\kappa}_\perp)$ depends only on the magnitude of the vector $\boldsymbol{\kappa}_\perp$. In this case, equation (8.24) simplifies to

$$C_{\text{coh}} = \exp\left\{-\pi^2 k^2 x_r \int_0^1 d\eta \int_0^\infty \Phi_{\text{eff}}(0, \kappa_\perp)\left[1 - J_0\left(\eta\kappa_\perp h\right)\right] \kappa_\perp \, d\kappa_\perp\right\}. \tag{8.25}$$

The coherence factor is closely related to the mutual coherence function of a spherical sound wave $\Gamma(x; \mathbf{r}_c; \mathbf{r}_d)$ given by equation (7.181), where x is the propagation range, and $\mathbf{r}_c$ and $\mathbf{r}_d$ are the transverse center and difference coordinates of two points of observation (equations (7.14) and (7.15)). We denote the mutual coherence function in a non-turbulent atmosphere as $\Gamma_0(x; \mathbf{r}_c; \mathbf{r}_d)$; it is given by equation (7.181) with $\Phi_{\text{eff}} = 0$. Comparing equations (8.24) and (7.181), we obtain

$$C_{\text{coh}} = \frac{\Gamma(x_r; \mathbf{r}_c; 0, h)}{\Gamma_0(x_r; \mathbf{r}_c; 0, h)}. \tag{8.26}$$

Thus, the coherence factor C_{coh} is equal to the normalized mutual coherence function of a spherical sound wave for line-of-sight propagation when two points of observation are separated by the distance h. This result is valid for both isotropic and anisotropic random media.

In section 7.5.3, the mutual coherence function of spherical waves was calculated for the von Kármán, Kolmogorov, and Gaussian spectra of temperature and velocity fluctuations. The corresponding results and equation (8.26) readily enable determination of the coherence factor. For example, for the von

Kármán spectra, using equation (7.191), we have

$$C_{\text{coh}} = \exp \left\{ -\frac{2\gamma_T x_r L_T}{h} \int_0^{h/L_T} \left[1 - \frac{2^{1/6}\eta^{5/6}}{\Gamma(5/6)} K_{5/6}(\eta) \right] d\eta \right.$$
$$\left. -\frac{2\gamma_v x_r L_v}{h} \int_0^{h/L_v} \left[1 - \frac{2^{1/6}\eta^{5/6}}{\Gamma(5/6)} \left(K_{5/6}(\eta) - \frac{\eta}{2}K_{1/6}(\eta) \right) \right] d\eta \right\}. \quad (8.27)$$

Here, L_T and L_v are the outer scales of temperature and velocity fluctuations, γ_T and γ_v are the extinction coefficients corresponding to these fluctuations determined by equation (7.184), and $K_{1/6}(\eta)$ and $K_{5/6}(\eta)$ are the modified Bessel functions of the second kind. It follows from equation (8.27) that C_{coh} decreases with increasing propagation range and acoustic frequency. As C_{coh} decreases, the effect of turbulence on the interference of the direct and reflected waves increases.

8.1.4 Numerical results and experimental data

In this subsection, the effect of atmospheric turbulence on the interference of the direct and ground-reflected waves is studied numerically, and experimental data relevant to this phenomenon are discussed.

Consider the *transmission loss* TL of a point source above the ground

$$\text{TL} = 10 \log \frac{\langle pp^* \rangle}{p_{\text{free}} p_{\text{free}}^*}, \quad (8.28)$$

where p_{free} is the sound pressure at 1 m from the source in free space. (Definitions of the transmission loss and other related quantities are given in section 10.1.)

Figure 8.2 depicts TL versus the acoustic frequency $f = kc_0/(2\pi)$ for sound propagation above an acoustically hard surface with $Q = 1$. The solid line corresponds to sound propagation in a non-turbulent atmosphere and the dashed and dotted lines to a turbulent atmosphere. These results are obtained with equations (8.17) and (8.27) for the source and receiver heights $h_s = 6$ m and $h_r = 2$ m, respectively, and the distance between them $x_r = 100$ m. The height of the atmospheric boundary layer is $z_i = 1$ km, the surface temperature is $T_s = 18\,^\circ\text{C}$, the friction velocity $u_* = 0.3$ m/s is representative of moderate wind, and the surface heat flux is $Q_H = 0$ W/m^2 or $Q_H = 200$ W/m^2, which corresponds to cloudy or sunny conditions. These meteorological parameters enable determination (section 6.2.4) of the variances and length scales σ_T^2, σ_v^2, L_T, and L_v appearing in equation (8.27) and in equation (7.184) for γ_T and γ_v. (When calculating the variances and length scales, the height above the ground is chosen as $h_s/2$.)

It follows from figure 8.2 that in a non-turbulent atmosphere above an acoustically hard ground, the transmission loss TL is very small at the interference minima. Atmospheric turbulence results in a significant increase in

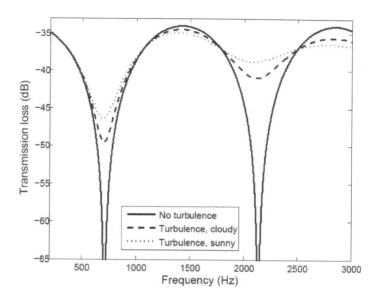

FIGURE 8.2
Transmission loss versus the acoustic frequency for sound propagation in an atmosphere near an acoustically hard ground. The source and receiver heights are 6 m and 2 m, respectively; the distance between them is 100 m. A solid line corresponds to the case of a non-turbulent atmosphere. Dashed and dotted lines correspond to a turbulent atmosphere with cloudy ($Q_H = 0$ W/m^2) and sunny conditions ($Q_H = 200$ W/m^2) with moderate wind ($u_* = 0.3$ m/s).

TL at these minima, especially for higher frequencies. The interference maxima are also affected by turbulence, but to a lesser extent. These effects are more pronounced for sunny conditions (when turbulence is stronger) than for cloudy conditions.

Ingard and Maling [175] were probably the first to study experimentally the interference of the direct and ground-reflected waves in a turbulent atmosphere. In their experiments, a source and receiver were located at 1.2 m above a perfectly reflecting surface, the propagation range varied from about 2 m to 70 m, and the acoustic frequencies were 500 Hz, 1 kHz, and 2 kHz. The experiments showed that the interference minima are affected by atmospheric turbulence even for relatively weak turbulence.

Reference [87] reports on sound-pressure level measurements above an asphalt runway at a small airport near Ottawa. A source was located at 1.2 m above the ground, and a microphone at 0.6 m or 1.2 m. The measurements were done at three ranges: 15 m, 30 m, and 45 m. The acoustic frequency varied between 1 kHz and 6 kHz. Experimental data were compared with theoretical predictions based on the Gaussian spectra of temperature and wind velocity fluctuations and the older theories of sound scattering by scalar and

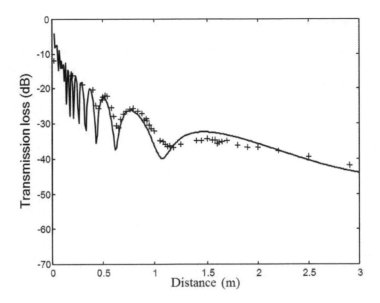

FIGURE 8.3

Transmission loss versus the distance between a source and microphone in a laboratory experiment of sound propagation through thermal turbulence near an acoustically hard wall. Symbols correspond to experimental data. A solid line is the theoretical prediction based on the von Kármán spectrum of temperature fluctuations. (Reproduced with permission from reference [297]. Copyright 2001, Acoustical Society of America.)

vector random fields (section 7.1.1). As pointed out in Chapters 6 and 7, the Gaussian spectra generally are inappropriate for atmospheric acoustics.

In reference [297], theoretical predictions based on the von Kármán spectrum were compared with experimental data obtained in a large anechoic chamber [36]. In the experiment, thermal turbulence was created by a large, horizontal grid. The spectrum of temperature fluctuations was well approximated by a von Kármán spectrum with $\sigma_T^2/T_0^2 = 4.9 \times 10^{-5}$ and $L_T = 0.147$ m. An acoustically hard wall was placed perpendicular to the grid. A sound source and microphone were located, respectively, at $h_s = 10$ cm and $h_r = 7$ cm from the wall, a reference microphone was 2 cm from the source, and the acoustic frequency was $f = 40$ kHz. The measured transmission loss TL is plotted versus the horizontal distance x_r in figure 8.3. Several local maxima and minima are clearly seen in the figure. The solid line corresponds to the theoretical predictions based on equations (8.17), (8.27), and (8.28), where p_{free} is the sound pressure measured by the reference microphone. The theoretical predictions generally agree with experimental data.

8.2 Statistical moments of the sound field above an impedance boundary in a refractive, turbulent medium

Sound propagation in the atmosphere is affected simultaneously by several factors, such as refraction due to the mean vertical profiles of temperature and wind velocity, scattering by turbulence, and interaction with the impedance ground (see section 1.2.3 and figure 1.4). Refraction by the mean profiles and reflection from the ground can result in multipath propagation. All these factors can be incorporated into Monte Carlo simulations of sound propagating through random realizations of temperature and velocity fluctuations, by the parabolic equation and other methods (Chapters 9 and 11).

In this section, an alternative approach for this problem is presented [422]. We derive closed-form equations for the statistical moments of the sound field propagating near an impedance ground in a refractive, turbulent atmosphere. The equations derived can then be solved numerically. This approach enables better understanding of the physics of the problem, and potentially offers faster numerical calculations. The interaction of sound waves with the ground is accounted for via boundary conditions. Without these boundary conditions, the derived equations describe sound propagation in a refractive, random medium. Though sound propagation in an atmosphere is explicitly considered, the results obtained can be used in other fields of acoustics.

8.2.1 Geometry and starting equations

Let a sound source be located in the plane $x = 0$ of the Cartesian coordinate system (x, y, z). The surface of the ground coincides with the plane $z = 0$. Receivers are located close to the x-axis. The sound speed $c(\mathbf{R}) = \bar{c}(\mathbf{R}) + \tilde{c}(\mathbf{R})$ and velocity $\mathbf{v}(\mathbf{R}) = \bar{\mathbf{v}}(\mathbf{R}) + \tilde{\mathbf{v}}(\mathbf{R})$ depend on the spatial coordinates $\mathbf{R} = (x, y, z)$. Here, the overline and tilde denote the mean and fluctuating components of the corresponding fields.

Sound propagation is described with the parabolic equation (2.110) for the complex amplitude $A(\mathbf{R})$ of the sound field $p(\mathbf{R})$. In this equation, we substitute with $\mathbf{v} = \bar{\mathbf{v}} + \tilde{\mathbf{v}}$ and $\varepsilon = \bar{\varepsilon} + \tilde{\varepsilon}$, where $\bar{\varepsilon}$ and $\tilde{\varepsilon}$ are the mean and fluctuating components of the function $\varepsilon = c_0^2/c^2 - 1$. As a result, one obtains

$$2ik\frac{\partial A}{\partial x} + \nabla_\perp^2 A + k^2 \left[\bar{\varepsilon} - \frac{2\bar{v}_x}{c_0} + \frac{2i}{\omega}\bar{\mathbf{v}}_\perp \cdot \nabla_\perp\right] A$$
$$+ k^2 \varepsilon_{\mathrm{mov}} A + \frac{2ik}{c_0}\tilde{\mathbf{v}}_\perp \cdot \nabla_\perp A = 0. \quad (8.29)$$

Here, $\nabla_\perp = (\partial/\partial y, \partial/\partial z)$, $\varepsilon_{\mathrm{mov}} = \tilde{\varepsilon} - 2\tilde{v}_x/c_0$, the vectors $\bar{\mathbf{v}}$ and $\tilde{\mathbf{v}}$ are expressed in the form $\bar{\mathbf{v}} = (\bar{v}_x, \bar{\mathbf{v}}_\perp)$ and $\tilde{\mathbf{v}} = (\tilde{v}_x, \tilde{\mathbf{v}}_\perp)$, where $\bar{v}_x, \tilde{v}_x$ and $\bar{\mathbf{v}}_\perp, \tilde{\mathbf{v}}_\perp$ are the components along the x-axis and perpendicular to it. To an accuracy of $\tilde{c}/c_0$, the function $\tilde{\varepsilon} = -2\tilde{c}/c_0$ so that the random field $\varepsilon_{\text{mov}} = -2\left(\tilde{c} + \tilde{v}_x\right)/c_0$ coincides with that considered in detail for line-of-sight propagation, equation (7.3). Except for the third term describing refraction by the mean profiles, equation (8.29) coincides with the parabolic equation (7.2). Similar to the latter equation, we omit the term proportional to $\tilde{\mathbf{v}}_\perp$. To an accuracy $\bar{v}_x/c_0$, in equation (8.29) $\bar{\varepsilon} - 2\bar{v}_x/c_0 = c_0^2/c_{\text{eff}}^2 - 1$, where $c_{\text{eff}} = \bar{c} + \bar{v}_x$ is the mean value of the effective sound speed. Substituting with $A(x, \mathbf{r}) = p(x, \mathbf{r})\exp(-ikx)$, we obtain the parabolic equation for the sound field in a refractive, random medium:

$$\left[2ik\frac{\partial}{\partial x} + \widetilde{L}_{\mathbf{r}}(x, \mathbf{r}) + k^2\varepsilon_{\text{mov}}(x, \mathbf{r})\right]p(x, \mathbf{r}) = 0. \tag{8.30}$$

Here, the vector $\mathbf{r} = (y, z)$ indicates the transverse coordinates (with respect to sound propagation). The operator $\widetilde{L}_{\mathbf{r}}$ acts on the transverse coordinates $\mathbf{r}$ and is given by

$$\widetilde{L}_{\mathbf{r}}(x, \mathbf{r}) = k^2\left[1 + \frac{c_0^2}{c_{\text{eff}}^2(x, \mathbf{r})}\right] + \frac{2ik}{c_0}\overline{\mathbf{v}}_\perp(x, \mathbf{r}) \cdot \nabla_\perp + \nabla_\perp^2. \tag{8.31}$$

The three terms on the right-hand side of this formula describe, respectively, refraction (through the effective sound speed), advection of sound with the mean crosswind, and diffraction of sound waves.

Equation (8.30) differs from the parabolic equation (7.124) only in the form of the operators $\widetilde{L}_{\mathbf{r}}$ and $L_{\mathbf{r}}$. The former operator accounts for sound refraction and advection, while the latter does not. The initial condition for equation (8.31) coincides with equation (7.7) and is given by $p(x = 0, \mathbf{r}) = p_0(\mathbf{r})$.

Another difference between the formulation here and that in Chapter 7 is the impedance boundary condition at the surface:

$$\left.\frac{\partial p(x, y, z)}{\partial z}\right|_{z=0} = -ik\beta(\omega)\,p(x, y, z)|_{z=0}. \tag{8.32}$$

Here, $\beta(\omega)$ is the normalized admittance of the ground, being equal to $\varrho_0 c_0/Z_s$, where ϱ_0 is the density of the air at $z = 0$ and Z_s is the specific impedance of the ground, which is assumed here to be locally reacting.

8.2.2 Statistical moments of arbitrary order

As in Chapter 7, we consider here the statistical moment of order $n + m$ of the sound pressure:

$$\Gamma_{n,m}(x; \mathbf{r}_1, ..., \mathbf{r}_n; \mathbf{r}_1', ..., \mathbf{r}_m') = \langle p(x, \mathbf{r}_1)...p(x, \mathbf{r}_n)p^*(x, \mathbf{r}_1')...p^*(x, \mathbf{r}_m')\rangle. \tag{8.33}$$

Starting with the parabolic equation (8.30) and employing the same approach as used to derive equation (7.139), we obtain a closed-form equation for $\Gamma_{n,m}$

in a refractive, turbulent atmosphere above an impedance ground:

$$\left[\frac{\partial}{\partial x} - \frac{i}{2k}\widetilde{M}_{n,m}(x) + \frac{k^2}{8}Q_{n,m}(x)\right]\Gamma_{n,m}(x; \mathbf{r}_1, ..., \mathbf{r}_n; \mathbf{r}'_1, ..., \mathbf{r}'_m) = 0. \quad (8.34)$$

In this equation, the function $Q_{n,m}$ is a linear combination of the transverse correlation functions $b'_{\text{eff}}(x; \mathbf{r}, \mathbf{r}')$ of the random field $\varepsilon_{\text{mov}}(x, \mathbf{r})$ and is given by equation (7.136). The operator

$$\widetilde{M}_{n,m}(x; \mathbf{r}_1, ..., \mathbf{r}_n; \mathbf{r}'_1, ..., \mathbf{r}'_m) = \widetilde{L}_{\mathbf{r}_1} + ... + \widetilde{L}_{\mathbf{r}_n} - \widetilde{L}^*_{\mathbf{r}'_1} - ... - \widetilde{L}^*_{\mathbf{r}'_m} \quad (8.35)$$

is similar to the operator $M_{n,m}$ considered in Chapter 7. The initial condition for $\Gamma_{n,m}$ is given by equation (7.140).

To derive the boundary conditions for $\Gamma_{n,m}$ at $z = 0$, equation (8.32) is recast as

$$\frac{\partial p(x, y_i, z_i)}{\partial z_i}\bigg|_{z_i=0} = -ik\beta p(x, y_i, z_i)|_{z_i=0}, \quad (8.36)$$

where $i = 1, ..., n$. Multiplying both sides of this equation by $p(x, \mathbf{r}_1)...p(x, \mathbf{r}_{i-1})p(x, \mathbf{r}_{i+1})...p(x, \mathbf{r}_n)p^*(x, \mathbf{r}'_1)...p^*(x, \mathbf{r}'_m)$ and averaging the resulting equation, we have

$$\frac{\partial\Gamma_{n,m}(x; \mathbf{r}_1, ..., \mathbf{r}_n; \mathbf{r}'_1, ..., \mathbf{r}'_m)}{\partial z_i}\bigg|_{z_i=0} = -ik\beta\Gamma_{n,m}\bigg|_{z_i=0}, \quad i = 1, ..., n. \quad (8.37)$$

Taking the complex conjugate of both sides of equation (8.36), and changing to the prime coordinates, one obtains

$$\frac{\partial p^*(x, y'_j, z'_j)}{\partial z'_j}\bigg|_{z'_j=0} = ik\beta^* p^*(x, y'_j, z'_j)|_{z'_j=0}, \quad (8.38)$$

where $j = 1, ..., m$. Both sides of this equation are multiplied by $p(x, \mathbf{r}_1)...p(x, \mathbf{r}_n)p^*(x, \mathbf{r}'_1)...p^*(x, \mathbf{r}'_{j-1})p(x, \mathbf{r}'_{j+1})...p^*(x, \mathbf{r}'_m)$. Averaging the result, one obtains

$$\frac{\partial\Gamma_{n,m}(x; \mathbf{r}_1, ..., \mathbf{r}_n; \mathbf{r}'_1, ..., \mathbf{r}'_m)}{\partial z'_j}\bigg|_{z'_j=0} = ik\beta^*\Gamma_{n,m}|_{z'_j=0}, \quad j = 1, ..., m. \quad (8.39)$$

Equations (8.37) and (8.39) comprise the boundary conditions on $\Gamma_{n,m}$.

8.2.3   Mean sound field

The mean sound field $\langle p(x, \mathbf{r})\rangle$ corresponds to $n = 1$ and $m = 0$. Substituting the values of $M_{1,0}$ and $Q_{1,0}$ into equation (8.34), we have

$$\left[\frac{\partial}{\partial x} - \frac{i}{2k}\widetilde{L}_{\mathbf{r}} + \frac{k^2}{8}b'_{\text{eff}}(x; \mathbf{r}, \mathbf{r})\right]\langle p(x, \mathbf{r})\rangle = 0. \quad (8.40)$$

The boundary condition for the mean sound field is obtained from equation (8.37):

$$\left.\frac{\partial \langle p(x,y,z)\rangle}{\partial z}\right|_{z=0} = -ik\beta \langle p(x,y,z)\rangle|_{z=0}. \tag{8.41}$$

Equation (8.40) is similar to equations (7.141) and (7.142) obtained in Chapter 7 and can be solved by the approach considered in that chapter. As a result, we obtain the mean sound field near an impedance ground in a refractive, turbulent atmosphere

$$\langle p(x,\mathbf{r})\rangle = \exp(-\gamma x)p^{(0)}(x,\mathbf{r}). \tag{8.42}$$

Here, γ is the extinction coefficient of the mean sound field given by equation (7.151) for a quasi-homogeneous, anisotropic turbulence and by equation (7.153) for the homogeneous and isotropic case. The detailed analysis of γ is presented in Chapter 7. The $p^{(0)}(x,\mathbf{r})$ is the sound field propagating above an impedance ground in a refractive atmosphere without turbulence, which satisfies the parabolic equation (8.30) with $\varepsilon_{\text{mov}} = 0$ and the boundary condition, equation (8.32). This field can be calculated numerically with techniques presented in Chapter 11.

8.2.4 Mutual coherence function

The mutual coherence function $\Gamma_{1,1}(x;\mathbf{r}_1,\mathbf{r}_2) = \langle p(x,\mathbf{r}_1)p^*(x,\mathbf{r}_2)\rangle$ satisfies equation (8.34) with $n = 1$, $m = 1$:

$$\left[\frac{\partial}{\partial x} - \frac{i}{2k}\widetilde{M}_{1,1} + \frac{k^2}{8}Q_{1,1}(x;\mathbf{r}_1,\mathbf{r}_2)\right]\Gamma_{1,1}(x;\mathbf{r}_1,\mathbf{r}_2) = 0. \tag{8.43}$$

Here, the function $Q_{1,1}$ is determined with equation (7.156) and the operator $\widetilde{M}_{1,1}$ is given by:

$$\widetilde{M}_{1,1} = k^2 \left[\frac{c_0^2}{c_{\text{eff}}^2(x,\mathbf{r}_1)} - \frac{c_0^2}{c_{\text{eff}}^2(x,\mathbf{r}_2)}\right]$$
$$+ \frac{2ik}{c_0}\left[\overline{\mathbf{v}}_\perp(x,\mathbf{r}_1)\cdot\nabla_{\mathbf{r}_1} + \overline{\mathbf{v}}_\perp(x,\mathbf{r}_2)\cdot\nabla_{\mathbf{r}_2}\right] + \nabla_{\mathbf{r}_1}^2 - \nabla_{\mathbf{r}_2}^2, \tag{8.44}$$

where $\nabla_{\mathbf{r}_1} = (\partial/\partial y_1, \partial/\partial z_1)$ and similarly for $\nabla_{\mathbf{r}_2}$. The initial condition for equation (8.43) is $\Gamma_{1,1}(0;\mathbf{r}_1,\mathbf{r}_2) = p_0(\mathbf{r}_1)p_0^*(\mathbf{r}_2)$. The boundary conditions are obtained with equations (8.37) and (8.39):

$$\left.\frac{\partial\Gamma_{1,1}(x;y_1,z_1,y_2,z_2)}{\partial z_1}\right|_{z_1=0} = -ik\beta\,\Gamma_{1,1}(x;y_1,z_1,y_2,z_2)|_{z_1=0}, \tag{8.45}$$

$$\left.\frac{\partial\Gamma_{1,1}(x;y_1,z_1,y_2,z_2)}{\partial z_2}\right|_{z_2=0} = ik\beta^*\,\Gamma_{1,1}(x;y_1,z_1,y_2,z_2)|_{z_2=0}. \tag{8.46}$$

Equation (8.43) and the corresponding boundary conditions describe the mutual coherence function of a sound wave propagating above an impedance ground in a refractive, turbulent atmosphere. Generally, this equation cannot be solved analytically. Techniques for numerical solution have been developed [422, 65] and applied to the analysis of sound scattering into refractive shadow zones [65] and assessment of acoustic "climates" [66]. These techniques are considered in Chapter 11.

8.2.5 Propagation without boundaries

In a random moving medium without boundaries, equation (8.43) has analytical solutions for two cases. In the first case, $\bar{c}$ is constant, $\overline{\mathbf{V}}_\perp = 0$, and $\bar{v}_x(z) = \bar{v}_0 + \alpha z^2$, where α is a parameter. For this situation the x-axis is the axis of a waveguide. A solution of equation (8.43) is given by equation (7.128) in reference [290].

In the second case, the mean sound speed $\bar{c}$ and mean wind velocity $\overline{\mathbf{V}} = (\bar{v}_x, \overline{\mathbf{V}}_\perp)$ do not depend on the spatial coordinates so that the effective sound speed $c_{\text{eff}} = \bar{c} + \bar{v}_x$ is constant. This case enables one to study the effect of sound advection on the mutual coherence function. In the equations above, we set $c_{\text{eff}} = c_0$. Then, the operator $\widetilde{M}_{1,1}$ simplifies to

$$\widetilde{M}_{1,1} = \frac{2ik}{c_0} \left[\overline{\mathbf{V}}_\perp(x, \mathbf{r}_1) \cdot \nabla_{\mathbf{r}_1} + \overline{\mathbf{V}}_\perp(x, \mathbf{r}_2) \cdot \nabla_{\mathbf{r}_2} \right] + \nabla_{\mathbf{r}_1}^2 - \nabla_{\mathbf{r}_2}^2. \tag{8.47}$$

We introduce the new transverse coordinates

$$\boldsymbol{\varrho}_1 = \mathbf{r}_1 - x\overline{\mathbf{V}}_\perp/c_0, \qquad \boldsymbol{\varrho}_2 = \mathbf{r}_2 - x\overline{\mathbf{V}}_\perp/c_0 \tag{8.48}$$

and the new unknown function $\widetilde{\Gamma}_{1,1}$, which is related to the mutual coherence function by

$$\Gamma_{1,1}(x; \mathbf{r}_1, \mathbf{r}_2) = \Gamma_{1,1}(x; \boldsymbol{\varrho}_1 + x\overline{\mathbf{V}}_\perp/c_0, \boldsymbol{\varrho}_2 + x\overline{\mathbf{V}}_\perp/c_0) \equiv \widetilde{\Gamma}_{1,1}(x; \boldsymbol{\varrho}_1, \boldsymbol{\varrho}_2). \tag{8.49}$$

Using this relationship in equation (8.43) and substituting the value of the operator $\widetilde{M}_{1,1}$, we obtain an equation for $\widetilde{\Gamma}_{1,1}$:

$$\left[\frac{\partial}{\partial x} + \frac{i}{2k} \left(\nabla_{\boldsymbol{\varrho}_2}^2 - \nabla_{\boldsymbol{\varrho}_1}^2 \right) \right.$$
$$\left. + \frac{k^2}{8} Q_{1,1} \left(\boldsymbol{\varrho}_1 + \frac{x\overline{\mathbf{V}}_\perp}{c_0}, \boldsymbol{\varrho}_2 + \frac{x\overline{\mathbf{V}}_\perp}{c_0} \right) \right] \widetilde{\Gamma}_{1,1}(x; \boldsymbol{\varrho}_1, \boldsymbol{\varrho}_2) = 0. \tag{8.50}$$

The initial condition is $\widetilde{\Gamma}_{1,1}(0; \boldsymbol{\varrho}_1, \boldsymbol{\varrho}_2) = p_0(\boldsymbol{\varrho}_1)p_0^*(\boldsymbol{\varrho}_2)$.

Equation (8.50) has a similar form as equation (7.155), the solution of which is given by equation (7.175). Transforming back to the transverse coordinates $\mathbf{r}_1$ and $\mathbf{r}_2$, and introducing the center $\mathbf{r}_c = (\mathbf{r}_1 + \mathbf{r}_2)/2$ and difference

$\mathbf{r}_d = \mathbf{r}_1 - \mathbf{r}_2$ coordinates, we have

$$\Gamma(x; \mathbf{r}_c, \mathbf{r}_d) = \frac{k^2}{4\pi^2 x^2} \int \int p_0 \left(\mathbf{r}'_c + \mathbf{r}'_d/2\right) p_0^* \left(\mathbf{r}'_c - \mathbf{r}'_d/2\right)$$

$$\times \exp \left\{ \frac{ik \left(\mathbf{r}_d - \mathbf{r}'_d\right) \cdot \left(\mathbf{r}_c - \mathbf{r}'_c - \frac{x\overline{\mathbf{v}}_\perp}{c_0}\right)}{x} - \frac{k^2 \pi x}{2} \int_0^1 d\eta \int \Phi_{\text{eff}}(\eta x, 0; 0, \boldsymbol{\kappa}_\perp) \right.$$

$$\left. \times \left[1 - \exp\left[i\boldsymbol{\kappa}_\perp \cdot \left(\eta \mathbf{r}_d + (1-\eta)\,\mathbf{r}'_d\right)\right]\right] d^2\kappa_\perp \right\} d^2 r'_d \, d^2 r'_c. \quad (8.51)$$

This formula for the mutual coherence function accounts for advection of sound by the mean crosswind (through the term in the exponential, proportional to $x\overline{\mathbf{v}}_\perp/c_0$). Equation (8.51) has been obtained for the case in which $\overline{c}$ and $\overline{\mathbf{v}}$ are constant, i.e., there is no refraction of sound. It may also be applied in a refractive atmosphere, if there is a single, dominant propagation path. Along this path, the mutual coherence function can be calculated in many cases by the same approach as that used to derive equation (8.51). As a result, one obtains that the coherence function is given approximately by this equation if the range x is replaced with the length of the actual path.

For an initially plane sound wave, $p_0 = $ constant. Calculating the integrals over $\mathbf{r}'_d$ and $\mathbf{r}'_c$ in equation (8.51), it can be shown that the mutual coherence function $\Gamma(x; \mathbf{r}_d)$ is still given by equation (7.178) obtained in Chapter 7. Thus, as expected, the crosswind does not affect the plane wave coherence.

For an initially spherical wave, $p_0(\mathbf{r}) = (i2\pi/k)\delta(\mathbf{r})$. Using this formula in equation (8.51) and calculating the integrals over $\mathbf{r}'_d$ and $\mathbf{r}'_c$, one obtains

$$\Gamma(x; \mathbf{r}_c, \mathbf{r}_d) = \frac{1}{x^2} \exp \left[\frac{ik\mathbf{r}_d \cdot (\mathbf{r}_c - x\overline{\mathbf{v}}_\perp/c_0)}{x} \right.$$

$$\left. - \frac{\pi k^2 x}{2} \int_0^1 d\eta \int \Phi_{\text{eff}}(\eta x, 0; 0, \boldsymbol{\kappa}_\perp) \left(1 - e^{i\eta \boldsymbol{\kappa}_\perp \cdot \mathbf{r}_d}\right) d^2\kappa_\perp \right]. \quad (8.52)$$

It follows from this formula that the crosswind $\overline{\mathbf{v}}_\perp$ affects the phase of the mutual coherence function.

8.3 Theory of multiple scattering: mean field

In Chapter 7 and sections 8.1 and 8.2, we considered scattering in the direction of sound propagation. In a random medium, such scattering occurs if the sound wavelength λ is smaller than the scale L of largest inhomogeneities (the outer scale of turbulence). We used the parabolic equations (7.8) and (8.30) for the analysis of the statistical moments of a sound field.

The parabolic equation method is unable to describe scattering at large angles. If $\lambda \ll L$, such scattering is relatively weak. However, it might be important in some problems such as sound scattering into refractive shadow zones or behind barriers. If $\lambda \gtrsim L$, sound is scattered significantly by random inhomogeneities in all directions.

In this section, we will account for sound scattering at large angles, including backscattering. Such scattering results in sound waves arriving at the observation point from different direction, i.e., multipath sound propagation in a random medium. The starting equation of the analysis will be a Helmholtz-type equation (6.91). Using the theory of multiple scattering, we will obtain the mean Green's function and mean sound field in a random moving medium.

8.3.1   Formulation of the problem

Consider sound propagation in a random moving medium with fluctuations in the sound speed $\widetilde{c}(\mathbf{R})$, density $\widetilde{\varrho}(\mathbf{R})$, and velocity $\widetilde{\mathbf{v}}(\mathbf{R})$. The mean sound speed $\overline{c} = c_0$ and mean density $\overline{\varrho} = \varrho_0$ do not depend on the spatial coordinates $\mathbf{R}$. The mean medium velocity is zero. Propagation and scattering of the sound field $p(\mathbf{R})$ can then be described by equation (6.91), where we omit a "hat " above the functions p and Q.

We will assume that a point mass source is located at $\mathbf{R}_0$ and the function $Q(\mathbf{R})$ in equation (6.91) is given by $Q(\mathbf{R}) = -i\delta(\mathbf{R} - \mathbf{R}_0)/\varrho\omega$, where δ is the delta function. In this case, equation (6.91) takes the form

$$[\nabla^2 + k^2(1 + \varepsilon) - (\nabla \ln(\varrho/\varrho_0)) \cdot \nabla - \frac{2i}{\omega} \frac{\partial \widetilde{v}_i}{\partial x_j} \frac{\partial^2}{\partial x_i \partial x_j} + \frac{2ik}{c_0} \widetilde{\mathbf{v}} \cdot \nabla]p(\mathbf{R})$$

$$= \left(1 + \frac{i\widetilde{\mathbf{v}}(\mathbf{R}_0)}{\omega} \cdot \nabla\right) \delta(\mathbf{R} - \mathbf{R}_0). \quad (8.53)$$

Here, $\varrho = \varrho_0 + \widetilde{\varrho}$ is the total density and $\varepsilon = c_0^2/(c_0 + \widetilde{c})^2 - 1$. For simplicity, we will ignore the effect of a random medium on the sound emission and assume that $\widetilde{\mathbf{v}}(\mathbf{R}_0) = 0$. In this case, in a homogeneous motionless medium, a solution of equation (8.53) is the Green's function:

$$G_0(\mathbf{R} - \mathbf{R}_0) = -\frac{\exp(ik|\mathbf{R} - \mathbf{R}_0|)}{4\pi|\mathbf{R} - \mathbf{R}_0|}. \quad (8.54)$$

This function and the sound field $p^{(0)}$ due to the unit-amplitude point source, given by equation (6.93), are related by $p^{(0)}(\mathbf{R} - \mathbf{R}_0) = -4\pi p^{(0)}(\mathbf{R} - \mathbf{R}_0)$.

In a random moving medium, a solution of equation (8.53) is also a Green's function, which is denoted as $G(\mathbf{R}, \mathbf{R}_0)$. This equation can be recast as an integral equation,

$$G(\mathbf{R}, \mathbf{R}_0) = G_0(\mathbf{R} - \mathbf{R}_0)$$

$$+ \int G_0(\mathbf{R} - \mathbf{R}') \left[2k^2 \frac{\widetilde{c}(\mathbf{R}')}{c_0} + \frac{\nabla'\widetilde{\varrho}(\mathbf{R}')}{\varrho_0} \cdot \nabla'\right.$$

$$\left. + \frac{2i}{\omega} \frac{\partial \widetilde{v}_i(\mathbf{R}')}{\partial x_j'} \frac{\partial^2}{\partial x_i'\partial x_j'} - \frac{2ik}{c_0}\widetilde{\mathbf{v}}(\mathbf{R}') \cdot \nabla'\right] G(\mathbf{R}', \mathbf{R}_0)\, d^3R'. \quad (8.55)$$

Here, $\nabla' = \partial/\partial\mathbf{R}' = (\partial/\partial x', \partial/\partial y', \partial/\partial z')$, and the functions ε and $\ln(\varrho/\varrho_0)$ were replaced with their values from equation (6.92). Equation (8.55) differs from the integral equation (6.95) only by the amplitude of the point source.

In equation (8.55), we express the sound speed and density fluctuations using equations (6.65) and (6.66)

$$\frac{\tilde{c}}{c_0} = \frac{\beta_c \tilde{T}}{2T_0} + \frac{\eta_c \tilde{C}}{2}, \quad \frac{\tilde{\varrho}}{\varrho_0} = \frac{\beta_\varrho \tilde{T}}{T_0} + \eta_\varrho \tilde{C}. \tag{8.56}$$

Here, $\tilde{T}$ and $\tilde{C}$ are the temperature fluctuations and the fluctuations in the concentration of the component dissolved in the medium. The coefficients β_c, η_c, β_ϱ, and η_ϱ were studied in Chapter 6.

Formulations for the statistical moments of the sound field are complicated by the differential operators appearing in equation (8.55). To simplify the analysis, we use spectral representations of the functions in this equation:

$$G_0(\mathbf{R} - \mathbf{R}_0) = \int\int \delta(\boldsymbol{\kappa} - \boldsymbol{\kappa}_0) g_0(\kappa) e^{i(\boldsymbol{\kappa}\cdot\mathbf{R} - \boldsymbol{\kappa}_0\cdot\mathbf{R}_0)} \, d^3\kappa \, d^3\kappa_0, \tag{8.57}$$

$$G(\mathbf{R}, \mathbf{R}_0) = 8\pi^3 \int\int g_0(\kappa) g_0(\kappa_0) g(\boldsymbol{\kappa}, \boldsymbol{\kappa}_0) e^{i(\boldsymbol{\kappa}\cdot\mathbf{R} - \boldsymbol{\kappa}_0\cdot\mathbf{R}_0)} \, d^3\kappa \, d^3\kappa_0, \tag{8.58}$$

$$\left\{\tilde{T}(\mathbf{R}), \tilde{C}(\mathbf{R}), \tilde{v}_i(\mathbf{R})\right\} = \int \left\{\widehat{T}(\boldsymbol{\kappa}), \widehat{C}(\boldsymbol{\kappa}), \widehat{v}_i(\boldsymbol{\kappa})\right\} e^{i\boldsymbol{\kappa}\cdot\mathbf{R}} \, d^3\kappa. \tag{8.59}$$

Here, $g(\boldsymbol{\kappa}, \boldsymbol{\kappa}_0)$, $\widehat{T}(\boldsymbol{\kappa})$, $\widehat{C}(\boldsymbol{\kappa})$, and $\widehat{v}_i(\boldsymbol{\kappa})$ are the quantities appearing in the corresponding integral transforms, and $g_0(\kappa) = 1/(8\pi^3(k^2 - \kappa^2))$. Substitution of equations (8.57)–(8.59) into equation (8.55) results in an integral equation for the quantity $g(\boldsymbol{\kappa}, \boldsymbol{\kappa}_0)$:

$$g(\boldsymbol{\kappa}, \boldsymbol{\kappa}') = (k^2 - \kappa^2)\delta(\boldsymbol{\kappa} - \boldsymbol{\kappa}') + \int \frac{\widehat{\mu}(\boldsymbol{\kappa}, \boldsymbol{\kappa}_1)}{k^2 - \kappa_1^2} g(\boldsymbol{\kappa}_1, \boldsymbol{\kappa}') \, d^3\kappa_1. \tag{8.60}$$

In this equation, $\widehat{\mu}$ is a linear combination of the random fields $\widehat{T}$, $\widehat{C}$, and $\widehat{v}_i$:

$$\begin{aligned}
\widehat{\mu}(\boldsymbol{\kappa}, \boldsymbol{\kappa}_1) = &[k^2 \beta_c - \beta_\varrho \boldsymbol{\kappa}_1 \cdot (\boldsymbol{\kappa} - \boldsymbol{\kappa}_1)] \frac{\widehat{T}(\boldsymbol{\kappa} - \boldsymbol{\kappa}_1)}{T_0} \\
&+ [k^2 \eta_c - \eta_\varrho \boldsymbol{\kappa}_1 \cdot (\boldsymbol{\kappa} - \boldsymbol{\kappa}_1))] \widehat{C}(\boldsymbol{\kappa} - \boldsymbol{\kappa}_1) \\
&+ 2[k^2 + \boldsymbol{\kappa}_1 \cdot (\boldsymbol{\kappa} - \boldsymbol{\kappa}_1)] \boldsymbol{\kappa}_1 \cdot \frac{\widehat{\mathbf{v}}(\boldsymbol{\kappa} - \boldsymbol{\kappa}_1)}{\omega}.
\end{aligned} \tag{8.61}$$

Scalar products of the vectors $\boldsymbol{\kappa}$, $\boldsymbol{\kappa}_1$, and $\widehat{\mathbf{v}}$ in this formula replace the differential operators appearing in equation (8.55).

Replacing the function $g(\boldsymbol{\kappa}_1, \boldsymbol{\kappa}')$ on the right-hand side of equation (8.60) by its value given by the left-hand side of this equation and repeating this procedure, we express g in the form of a perturbation series,

$$
g(\kappa, \kappa') = (k^2 - \kappa^2)\delta(\kappa - \kappa') + \widehat{\mu}(\kappa, \kappa') + \int \frac{\widehat{\mu}(\kappa, \kappa_1)\widehat{\mu}(\kappa_1, \kappa')}{k^2 - \kappa_1^2} d^3\kappa_1
$$

$$
+ \int d^3\kappa_1 \frac{\widehat{\mu}(\kappa, \kappa_1)}{k^2 - \kappa_1^2} \int \frac{\widehat{\mu}(\kappa_1, \kappa_2)\widehat{\mu}(\kappa_2, \kappa')}{k^2 - \kappa_2^2} d^3\kappa_2 + \int d^3\kappa_1 \frac{\widehat{\mu}(\kappa, \kappa_1)}{k^2 - \kappa_1^2}
$$

$$
\times \int d^3\kappa_2 \frac{\widehat{\mu}(\kappa_1, \kappa_2)}{k^2 - \kappa_2^2} \int \frac{\widehat{\mu}(\kappa_2, \kappa_3)\widehat{\mu}(\kappa_3, \kappa')}{k^2 - \kappa_3^2} d^3\kappa_3 + \ldots . \quad (8.62)
$$

This series describes the multiple scattering process in a random medium. The first term on the right-hand side corresponds to propagation in a medium without random inhomogeneities, the second term describes the single-scattered sound field, the third term describes the sound field scattered twice by the random field $\widehat{\mu}$, and so forth.

8.3.2   Mean Green's function

Averaging both sides of equation (8.58) over an ensemble of realizations of the random field $\widehat{\mu}$, we relate the mean Green's function $\overline{G}(\mathbf{R}, \mathbf{R}_0)$ to the mean value of $g(\kappa, \kappa_0)$:

$$
\overline{G}(\mathbf{R}, \mathbf{R}_0) = 8\pi^3 \int \int g_0(\kappa) g_0(\kappa_0) \overline{g}(\kappa, \kappa_0) e^{i(\kappa \cdot \mathbf{R} - \kappa_0 \cdot \mathbf{R}_0)} d^3\kappa \, d^3\kappa_0. \quad (8.63)
$$

(To simplify the derivations below, the quantities related to the mean Green's function are denoted with overlines over the corresponding notations, e.g., $\overline{g}(\kappa, \kappa_0) \equiv \langle g(\kappa, \kappa_0) \rangle$.) The quantity $\overline{g}$ can be obtained by averaging both sides of equation (8.62). The random field $\widehat{\mu}(\kappa, \kappa_1)$ appearing on the right-hand side of this equation is proportional to the random fields $\widehat{T}(\kappa)$, $\widehat{C}(\kappa)$, and $\widehat{v}_i(\kappa)$ (equation (8.61)). In what follows we assume that $\widetilde{T}(\mathbf{R})$, $\widetilde{C}(\mathbf{R})$, and $\widetilde{v}_i(\mathbf{R})$ are random fields with normal (Gaussian) distributions. Then, $\widehat{\mu}(\kappa, \kappa_1)$ is also a Gaussian random field. All odd moments of $\widehat{\mu}$ are equal to zero, and all even moments can be expressed in terms of the second moment $\langle \widehat{\mu}(\kappa, \kappa_1)\widehat{\mu}(\kappa', \kappa_2) \rangle$. This second moment can be calculated by substituting with equation (8.61). Similarly to Chapter 6, we assume that $\langle \widetilde{v}_i \widetilde{T} \rangle = \langle \widetilde{v}_i \widetilde{C} \rangle = 0$. Using the equality

$$
\langle \widehat{v}_i(\kappa_1)\widehat{v}_j(\kappa_2) \rangle = \delta(\kappa_1 + \kappa_2)\Phi_{ij}(\kappa_1) \quad (8.64)
$$

which is valid for the statistically homogeneous velocity fluctuations $\widetilde{\mathbf{v}} = (\widetilde{v}_1, \widetilde{v}_2, \widetilde{v}_3)$, and using analogous equalities for the statistically homogeneous fields $\widetilde{T}$ and $\widetilde{C}$, one obtains

$$
\langle \widehat{\mu}(\kappa, \kappa_1)\widehat{\mu}(\kappa', \kappa_2) \rangle = \delta(\kappa - \kappa_1 + \kappa' - \kappa_2)U(\kappa, \kappa_1, \kappa'). \quad (8.65)
$$

Here, the function U is given by

$$
U(\boldsymbol{\kappa}, \boldsymbol{\kappa}_1, \boldsymbol{\kappa}')
$$
$$
= [k^2\beta_c - \beta_\varrho \boldsymbol{\kappa}_1 \cdot (\boldsymbol{\kappa} - \boldsymbol{\kappa}_1)][k^2\beta_c - \beta_\varrho(\boldsymbol{\kappa}_1 - \boldsymbol{\kappa}) \cdot (\boldsymbol{\kappa} - \boldsymbol{\kappa}_1 + \boldsymbol{\kappa}')]\frac{\Phi_T(\boldsymbol{\kappa} - \boldsymbol{\kappa}_1)}{T_0^2}
$$
$$
+ \{[k^2\beta_c - \beta_\varrho \boldsymbol{\kappa}_1 \cdot (\boldsymbol{\kappa} - \boldsymbol{\kappa}_1)][k^2\eta_c - \eta_\varrho(\boldsymbol{\kappa}_1 - \boldsymbol{\kappa}) \cdot (\boldsymbol{\kappa} - \boldsymbol{\kappa}_1 + \boldsymbol{\kappa}')]
$$
$$
+ [k^2\beta_c - \beta_\varrho(\boldsymbol{\kappa}_1 - \boldsymbol{\kappa}) \cdot (\boldsymbol{\kappa} - \boldsymbol{\kappa}_1 + \boldsymbol{\kappa}')][k^2\eta_c - \eta_\varrho \boldsymbol{\kappa}_1 \cdot (\boldsymbol{\kappa} - \boldsymbol{\kappa}_1)]\}\frac{\Phi_{CT}(\boldsymbol{\kappa} - \boldsymbol{\kappa}_1)}{T_0}
$$
$$
+ [k^2\eta_c - \eta_\varrho \boldsymbol{\kappa}_1 \cdot (\boldsymbol{\kappa} - \boldsymbol{\kappa}_1)][k^2\eta_c - \eta_\varrho(\boldsymbol{\kappa}_1 - \boldsymbol{\kappa}) \cdot (\boldsymbol{\kappa} - \boldsymbol{\kappa}_1 + \boldsymbol{\kappa}')]\Phi_C(\boldsymbol{\kappa} - \boldsymbol{\kappa}_1)
$$
$$
+ [k^2 + (\boldsymbol{\kappa} - \boldsymbol{\kappa}_1) \cdot \boldsymbol{\kappa}_1][k^2 + (\boldsymbol{\kappa}_1 - \boldsymbol{\kappa}) \cdot (\boldsymbol{\kappa} - \boldsymbol{\kappa}_1 + \boldsymbol{\kappa}')]\kappa_{1,i}\kappa_{2,j}\frac{4\Phi_{ij}(\boldsymbol{\kappa} - \boldsymbol{\kappa}_1)}{\omega^2},
$$
$$
\tag{8.66}
$$

where $\kappa_{1,i}$ and $\kappa_{2,j}$ are components of the vectors $\boldsymbol{\kappa}_1$ and $\boldsymbol{\kappa}_2 = \boldsymbol{\kappa} - \boldsymbol{\kappa}_1 + \boldsymbol{\kappa}'$.

Now we average both sides of equation (8.62). Using the properties of the random field $\widehat{\mu}$, we conclude that the second and fourth terms on the right-hand side of this equation are zero. The fifth term contains the fourth moment $\langle \widehat{\mu}_1 \widehat{\mu}_2 \widehat{\mu}_3 \widehat{\mu}_4 \rangle$ (where the subscripts are shortcuts for different arguments of $\widehat{\mu}$). Since $\widehat{\mu}$ is a Gaussian random field, its fourth moment can be expressed in terms of the second moments:

$$
\langle \widehat{\mu}_1 \widehat{\mu}_2 \widehat{\mu}_3 \widehat{\mu}_4 \rangle = \langle \widehat{\mu}_1 \widehat{\mu}_2 \rangle \langle \widehat{\mu}_3 \widehat{\mu}_4 \rangle + \langle \widehat{\mu}_1 \widehat{\mu}_3 \rangle \langle \widehat{\mu}_2 \widehat{\mu}_4 \rangle + \langle \widehat{\mu}_1 \widehat{\mu}_4 \rangle \langle \widehat{\mu}_2 \widehat{\mu}_3 \rangle. \tag{8.67}
$$

The right-hand side of this relationship contains all possible combinations of the products of the second moments of $\widehat{\mu}$. Other terms in equation (8.62) are handled similarly. As a result, we have

$$
\bar{g}(\boldsymbol{\kappa}, \boldsymbol{\kappa}') = (k^2 - \kappa^2)\delta(\boldsymbol{\kappa} - \boldsymbol{\kappa}') + \int \frac{\langle \widehat{\mu}(\boldsymbol{\kappa}, \boldsymbol{\kappa}_1)\widehat{\mu}(\boldsymbol{\kappa}_1, \boldsymbol{\kappa}')\rangle}{k^2 - \kappa_1^2} d^3\kappa_1 + \int \frac{d^3\kappa_1}{k^2 - \kappa_1^2}
$$
$$
\times \int d^3\kappa_2 \frac{\langle \widehat{\mu}(\boldsymbol{\kappa}, \boldsymbol{\kappa}_1)\widehat{\mu}(\boldsymbol{\kappa}_1, \boldsymbol{\kappa}_2)\rangle}{k^2 - \kappa_2^2} \int \frac{\langle \widehat{\mu}(\boldsymbol{\kappa}_2, \boldsymbol{\kappa}_3)\widehat{\mu}(\boldsymbol{\kappa}_3, \boldsymbol{\kappa}')\rangle}{k^2 - \kappa_3^2} d^3\kappa_3 + \dots.
$$
$$
\tag{8.68}
$$

The second term on the right-hand side of this formula corresponds to the field twice scattered by $\widehat{\mu}$. The field scattered four times is a sum of three terms corresponding to the terms on the right-hand side of equation (8.67). Only the first of these terms is presented in equation (8.68).

Using the *diagram technique*, the series in equation (8.68) can be written in the following form [290]:

$$
\bar{g}(\boldsymbol{\kappa}, \boldsymbol{\kappa}') = \frac{(k^2 - \kappa^2)^2}{k^2 - \kappa^2 - D(\boldsymbol{\kappa})}\delta(\boldsymbol{\kappa} - \boldsymbol{\kappa}'). \tag{8.69}
$$

Here, the function $D(\boldsymbol{\kappa})$ is related to the *mass-operator kernel* $Q(\boldsymbol{\kappa}, \boldsymbol{\kappa}') = \delta(\boldsymbol{\kappa} - \boldsymbol{\kappa}')D(\boldsymbol{\kappa})$. The mass-operator kernel $Q(\boldsymbol{\kappa}, \boldsymbol{\kappa}')$ is a subseries of terms in the right-hand side of equation (8.68). In the Bourret approximation, only the first term in this subseries is retained. In this approximation,

$$D(\boldsymbol{\kappa}) = \int \frac{U(\boldsymbol{\kappa}, \boldsymbol{\kappa}_1, \boldsymbol{\kappa}_1)}{k^2 - \kappa_1^2} d^3\kappa_1. \tag{8.70}$$

Though the function $D(\boldsymbol{\kappa})$ cannot be calculated exactly, equation (8.69) is an exact result.

Substituting the value of $\overline{g}$ into equation (8.63), one obtains the mean Green's function in a statistically homogeneous random moving medium:

$$\overline{G}(\mathbf{R} - \mathbf{R}_0) = \int \frac{\exp(i\boldsymbol{\kappa} \cdot (\mathbf{R} - \mathbf{R}_0))}{8\pi^3[k^2 - \kappa^2 - D(\boldsymbol{\kappa})]} d^3\kappa. \tag{8.71}$$

8.3.3   Isotropic random medium

Here we consider the case of a statistically isotropic random medium and assume that $\nabla \cdot \widetilde{\mathbf{v}} = 0$. In this case, in equation (8.66) for $U(\boldsymbol{\kappa}, \boldsymbol{\kappa}_1, \boldsymbol{\kappa}')$, the functions $\Phi_T(\boldsymbol{\kappa})$, $\Phi_{CT}(\boldsymbol{\kappa})$, and $\Phi_C(\boldsymbol{\kappa})$ depend only on the magnitude of the vector $\boldsymbol{\kappa}$, and $\Phi_{ij}(\boldsymbol{\kappa})$ is given by equation (6.17). Using this equation, the last term in equation (8.66) can be expressed in the form

$$4[k^2 + \boldsymbol{\kappa}_1 \cdot (\boldsymbol{\kappa} - \boldsymbol{\kappa}_1)][k^2 + (\boldsymbol{\kappa}_1 - \boldsymbol{\kappa}) \cdot (\boldsymbol{\kappa} - \boldsymbol{\kappa}_1 + \boldsymbol{\kappa}')][(\boldsymbol{\kappa}_1 \cdot \boldsymbol{\kappa}')(\boldsymbol{\kappa} - \boldsymbol{\kappa}_1)^2$$

$$- (\boldsymbol{\kappa}_1 \cdot (\boldsymbol{\kappa} - \boldsymbol{\kappa}_1))(\boldsymbol{\kappa}' \cdot (\boldsymbol{\kappa} - \boldsymbol{\kappa}_1))] \frac{E(|\boldsymbol{\kappa} - \boldsymbol{\kappa}_1|)}{4\pi\omega^2(\boldsymbol{\kappa} - \boldsymbol{\kappa}_1)^4}. \tag{8.72}$$

Here, $E(\kappa)$ is the energy spectrum of velocity fluctuations.

It can be shown that in a homogeneous and isotropic random medium, the function $D(\boldsymbol{\kappa})$ depends only on a magnitude of the vector $\boldsymbol{\kappa}$. In this case, equation (8.71) for the mean Green's function can be simplified. To this end, we introduce the spherical coordinates κ, ν, and φ of the vector $\boldsymbol{\kappa}$. Here, ν is the angle between the vectors $\boldsymbol{\kappa}$ and $\mathbf{R} - \mathbf{R}_0$, and φ is the angle in the direction of projection of the vector $\mathbf{R}$ on the plane perpendicular to the vector $\mathbf{R} - \mathbf{R}_0$. Calculating the integrals over ν and φ in equation (8.71), one obtains

$$\overline{G}(|\mathbf{R} - \mathbf{R}_0|) = -\frac{i}{4\pi^2|\mathbf{R} - \mathbf{R}_0|} \int_{-\infty}^{\infty} \frac{\exp(i\kappa|\mathbf{R} - \mathbf{R}_0|)}{k^2 - \kappa^2 - D(\kappa)} \kappa \, d\kappa. \tag{8.73}$$

To calculate the integral over κ, the path of integration is extended to include an infinitely large semicircle in the upper half-plane of the complex values of κ. The integral along this semicircle is equal to zero. Therefore, the integral in equation (8.73) is given by a sum of residues of the integrand at its poles located in the upper half-plane. These poles κ_n are determined by the equation

$$k^2 - \kappa_n^2 - D(\kappa_n) = 0. \tag{8.74}$$

It can be shown [339] that in the Bourret approximation, $|D| \ll k^2$. Then,

$\kappa_n \approx \pm k$, so that $D(\kappa_n)$ in equation (8.74) can be replaced approximately by $D(k)$. In this case, the solution of this equation is given by $\kappa_n = \pm k N_{\text{eff}}$, where

$$N_{\text{eff}} = 1 - D(k)/2k^2. \tag{8.75}$$

It follows from the subsequent analysis that $\text{Im}\, D(k) < 0$. Therefore, there is only one pole $\kappa_n = k N_{\text{eff}}$ in the upper half-space. Calculating the residue at this pole, one obtains the mean Green's function

$$\overline{G}(|\mathbf{R} - \mathbf{R}_0|) = -\frac{\exp(ik N_{\text{eff}}|\mathbf{R} - \mathbf{R}_0|)}{4\pi|\mathbf{R} - \mathbf{R}_0|}. \tag{8.76}$$

This formula can be obtained from equation (8.54) for the Green's function $G_0(\mathbf{R} - \mathbf{R}_0)$ in a homogeneous motionless medium by replacing k with $k N_{\text{eff}}$. Therefore, N_{eff} can be termed the *effective refractive index* of a random moving medium.

8.3.4 Effective refractive index

We now calculate N_{eff} in the Bourret approximation. Substituting the value of $U(\boldsymbol{\kappa}, \boldsymbol{\kappa}_1, \boldsymbol{\kappa}_1)$ from equation (8.66) into equation (8.70), taking into account equation (8.72), and introducing a new integration variable $\boldsymbol{\kappa}_2 = \boldsymbol{\kappa}_1 - \boldsymbol{\kappa}$, one obtains:

$$
\begin{aligned}
D(\boldsymbol{\kappa}) = \int \frac{1}{k^2 + i0 - (\boldsymbol{\kappa} + \boldsymbol{\kappa}_2)^2} &\Big\{ [k^2 \beta_c + \beta_\varrho \boldsymbol{\kappa}_2 \cdot (\boldsymbol{\kappa} + \boldsymbol{\kappa}_2)] \\
\times [k^2 \beta_c - \beta_\varrho \boldsymbol{\kappa} \cdot \boldsymbol{\kappa}_2] \frac{\Phi_T(\boldsymbol{\kappa}_2)}{T_0^2} &+ 4[k^2 - \boldsymbol{\kappa}_2 \cdot (\boldsymbol{\kappa} + \boldsymbol{\kappa}_2)] \\
\times [k^2 + \boldsymbol{\kappa} \cdot \boldsymbol{\kappa}_2][k^2 \kappa_2^2 - (\boldsymbol{\kappa} \cdot \boldsymbol{\kappa}_2)^2] &\frac{E(\boldsymbol{\kappa}_2)}{4\pi \omega^2 \kappa_2^4} \Big\} d^3 \kappa_2.
\end{aligned} \tag{8.77}
$$

In the denominator of the integrand in this equation, k^2 has been replaced by $k^2 + i0$ in accordance with the principle of infinitesimal absorption, and it has been assumed that $\widetilde{C} = 0$ for simplicity. We introduce the spherical coordinates κ_2, ν, and φ of the vector $\boldsymbol{\kappa}_2$ in equation (8.77). Here, ν is the angle between $\boldsymbol{\kappa}$ and $\boldsymbol{\kappa}_2$, and the angle φ is in the direction of projection of the vector $\boldsymbol{\kappa}_2$ on the plane perpendicular to the vector $\boldsymbol{\kappa}$. Calculating the integral over φ, setting $\kappa = k$, and introducing a new integration variable $\zeta = \cos \nu$ yields

$$
\begin{aligned}
D(k) = -2\pi \int_0^\infty d\kappa_2 \int_{-1}^1 \frac{\kappa_2}{\kappa_2 + 2k\zeta - i0} &\Big\{ [k^2 \beta_c + \beta_\varrho(\kappa_2^2 + k\kappa_2\zeta)][k^2 \beta_c \\
-k\beta_\varrho \kappa_2 \zeta] \frac{\Phi_T(\kappa_2)}{T_0^2} + 4(1 - \zeta^2)(k^2 - \kappa_2^2 - k\kappa_2\zeta)(k^2 + k\kappa_2\zeta) &\frac{E(\kappa_2)}{4\pi c_0^2 \kappa_2^2} \Big\} d\zeta.
\end{aligned} \tag{8.78}
$$

In this formula, the integrals over ζ have the form

$$I_n = \int_{-1}^{1} \frac{\zeta^n}{\zeta + \kappa_2/2k - i0} \, d\zeta, \tag{8.79}$$

where $n = 0, 1, 2, 3,$ or 4. The integrand has no singularity due to the term $i0$. Therefore, all integrals I_n can be readily calculated. In the resulting equations for I_n, the term $i0$ is set to zero. For example, I_0 is given by:

$$I_0 = \ln \frac{\kappa_2 + 2k}{\kappa_2 - 2k} = \ln \left| \frac{\kappa_2 + 2k}{\kappa_2 - 2k} \right| + i\pi H(2k - \kappa_2), \tag{8.80}$$

where H is a Heaviside step function: $H(x) = 1$ for $x \geq 0$, and $H(x) = 0$ for $x < 0$. Substituting the values of the integrals I_n into equation (8.78) yields a formula for $D(k)$. Using this formula and equation (8.75), one obtains the effective refractive index of a random moving medium

$$N_{\text{eff}} = N_1 + iN_2. \tag{8.81}$$

Here, the real part of N_{eff} is given by

$$\begin{aligned}
N_1 = 1 + \frac{\pi k}{2} \int_0^\infty &\left\{ \left[\left(\beta_c + \frac{\beta_\varrho \kappa^2}{2k^2} \right)^2 \ln \left| \frac{\kappa + 2k}{\kappa - 2k} \right| - \frac{\beta_\varrho^2 \kappa^3}{k^3} \right] \frac{\Phi_T(\kappa)}{T_0^2} \right. \\
&+ 2 \left[\left(\beta_c + \frac{\beta_\varrho \kappa^2}{2k^2} \right) \left(\eta_c + \frac{\eta_\varrho \kappa^2}{2k^2} \right) \ln \left| \frac{\kappa + 2k}{\kappa - 2k} \right| - \frac{\beta_\varrho \eta_\varrho \kappa^3}{k^3} \right] \frac{\Phi_{CT}(\kappa)}{T_0} \\
&+ \left[\left(\eta_c + \frac{\eta_\varrho \kappa^2}{2k^2} \right)^2 \ln \left| \frac{\kappa + 2k}{\kappa - 2k} \right| - \frac{\eta_\varrho^2 \kappa^3}{k^3} \right] \Phi_C(\kappa) + 4 \left[\left(1 - \frac{\kappa^2}{2k^2} \right)^2 \right. \\
&\left. \left. \times \left(1 - \frac{\kappa^2}{4k^2} \right) \ln \left| \frac{\kappa + 2k}{\kappa - 2k} \right| + \frac{\kappa}{k} - \frac{5\kappa^3}{3k^3} + \frac{\kappa^5}{4k^5} \right] \frac{E(\kappa)}{4\pi c_0^2 \kappa^2} \right\} \kappa \, d\kappa.
\end{aligned} \tag{8.82}$$

The imaginary part of the effective refractive index is

$$\begin{aligned}
N_2 = \frac{\pi^2 k}{2} \int_0^{2k} &\left[\left(\beta_c + \frac{\beta_\varrho \kappa^2}{2k^2} \right)^2 \frac{\Phi_T(\kappa)}{T_0^2} \right. \\
&+ 2 \left(\beta_c + \frac{\beta_\varrho \kappa^2}{2k^2} \right) \left(\eta_c + \frac{\eta_\varrho \kappa^2}{2k^2} \right) \frac{\Phi_{CT}(\kappa)}{T_0} + \left(\eta_c + \frac{\eta_\varrho \kappa^2}{2k^2} \right)^2 \Phi_C(\kappa) \\
&\left. + 4 \left(1 - \frac{\kappa^2}{2k^2} \right)^2 \left(1 - \frac{\kappa^2}{4k^2} \right) \frac{E(\kappa)}{4\pi c_0^2 \kappa^2} \right] \kappa \, d\kappa.
\end{aligned} \tag{8.83}$$

In these formulas, we have restored the dependence of N_{eff} on $\Phi_C(\kappa)$ and $\Phi_{CT}(\kappa)$.

8.3.5 Extinction coefficient

Due to the imaginary part of the effective refractive index, the mean Green's function decreases exponentially with increasing distance from the source. The extinction coefficient of the mean Green's function is

$$\gamma = k \operatorname{Im} N_{\mathrm{eff}} = k N_2. \tag{8.84}$$

Multiplying both sides of equation (8.83) by k and setting $\kappa = 2k \sin(\theta/2)$, where θ is a new integration variable, the extinction coefficient can be expressed in the form

$$\gamma = \sigma_0/2 = \pi \int_0^\pi \sin(\theta)\sigma(\theta)\, d\theta. \tag{8.85}$$

Here, $\sigma(\theta)$ is the sound scattering cross section, given by equation (6.115), and σ_0 is the total scattering cross section, equation (6.112). It follows from equation (8.85) that the relationship $\sigma_0 = 2\gamma$, known for a motionless random medium [339], is also valid in the case of a random moving medium.

Let us evaluate γ in the limiting case $\lambda \ll L$. We also assume that the spectra $\Phi_T(\kappa)$, $\Phi_{CT}(\kappa)$, $\Phi_C(\kappa)$, and $E(\kappa)$ decrease rapidly enough if κ is increased. Such spectra are, for example, those corresponding to the von Kármán, Kolmogorov, and Gaussian models (section 6.2), and those for which the dependence on κ is given by $(1 + \kappa^2 L^2)^{-\nu}$ with $\nu > 1$. For such spectra, the main contribution to the integral in equation (8.83) comes from integration over small values of κ, corresponding to large-scale inhomogeneities with scales $l \gg \lambda$. The small-scale inhomogeneities ($l \lesssim \lambda$) do not contribute significantly to this integral. As a result, in equation (8.83) the terms proportional to β_ϱ and η_ϱ can be equated to zero, the two expressions in parentheses before $E(\kappa)$ are approximately equal to 1, and the upper-limit of integration can be extended to ∞. With these approximations, the extinction coefficient is given by

$$\gamma = \frac{\pi^2 k^2}{2} \int_0^\infty \Phi_{\mathrm{eff}}(0, \kappa)\kappa\, d\kappa. \tag{8.86}$$

This formula coincides with equation (7.153) for γ, obtained in the parabolic equation approximation. This result justifies the use of the parabolic equation in the case $\lambda \ll L$.

Now consider the opposite limiting case $\lambda \gtrsim L$, when the parabolic equation cannot be employed. In this case, in equation (8.83) the arguments of the spectra can be replaced by 0. After calculating the integral over κ, one obtains

$$
\gamma = \pi^2 k^4 \Bigg[\left(\beta_c^2 + \frac{4}{3}\beta_\varrho^2 + 2\beta_c\beta_\varrho \right) \frac{\Phi_T(0)}{T_0^2}
$$
$$
+ 2 \left(\eta_c\beta_c + \frac{4}{3}\eta_\varrho\beta_\varrho + \eta_\varrho\beta_c + \eta_c\beta_\varrho \right) \frac{\Phi_{CT}(0)}{T_0}
$$
$$
+ \left(\eta_c^2 + \frac{4}{3}\eta_\varrho^2 + 2\eta_c\eta_\varrho \right) \Phi_C(0) + \frac{\widetilde{E}}{6\pi c_0^2} \Bigg], \quad (8.87)
$$

where $\widetilde{E} = \lim_{\kappa \to 0} \left[E(\kappa)/\kappa^2 \right]$. It follows from this equation that γ is not proportional to the effective spectrum Φ_{eff}. Note that in the parabolic equation approximation, all statistical moments of the sound field are proportional to Φ_{eff}.

For an arbitrary ratio between λ and L, the extinction coefficient γ for the von Kármán spectra can be calculated numerically using equations (8.83) and (8.84). For the Kolmogorov spectra, $\gamma \to \infty$ since the integral on the right-hand side of equation (8.83) diverges.

Substituting $\Phi_T(\kappa)$ and $E(\kappa)$ for the Gaussian spectra into equation (8.83), assuming for simplicity that $\Phi_{CT}(\kappa) = \Phi_C(\kappa) = 0$ and $\beta_c = -\beta_\varrho = 1$ (this case corresponds to sound propagation in a dry atmosphere) and calculating the integral over κ, one obtains

$$
\gamma = \left[1 - \frac{4}{k^2\mathcal{L}_T^2} + \frac{8}{k^4\mathcal{L}_T^4} - \left(1 + \frac{4}{k^2\mathcal{L}_T^2} + \frac{8}{k^4\mathcal{L}_T^4} \right) e^{-k^2\mathcal{L}_T^2} \right] \frac{\sqrt{\pi}k^2\sigma_T^2\mathcal{L}_T}{8T_0^2}
$$
$$
+ \left[1 - \frac{10}{k^2\mathcal{L}_v^2} + \frac{48}{k^4\mathcal{L}_v^4} - \frac{96}{k^6\mathcal{L}_v^6} + \left(1 + \frac{10}{k^2\mathcal{L}_v^2} + \frac{48}{k^4\mathcal{L}_v^4} + \frac{96}{k^6\mathcal{L}_v^6} \right) e^{-k^2\mathcal{L}_v^2} \right]
$$
$$
\times \frac{\sqrt{\pi}k^2\sigma_v^2\mathcal{L}_v}{2c_0^2}. \quad (8.88)
$$

Here, $\mathcal{L}_T$ and $\mathcal{L}_v$ are the temperature and velocity scales pertinent to the Gaussian spectra, and σ_T^2 and σ_v^2 are the corresponding variances. If $k\mathcal{L}_T \gg 1$ and $k\mathcal{L}_v \gg 1$, the extinction coefficient γ becomes

$$
\gamma = \frac{\sqrt{\pi}k^2}{8} \left(\frac{\sigma_T^2\mathcal{L}_T}{T_0^2} + \frac{4\sigma_v^2\mathcal{L}_v}{c_0^2} \right). \quad (8.89)
$$

As expected, this formula agrees with equation (7.185) for γ, obtained in the parabolic equation approximation. In the opposite limiting case $k\mathcal{L}_T \ll 1$ and $k\mathcal{L}_v \ll 1$, the extinction coefficient is given by

$$
\gamma = \frac{\sqrt{\pi}k^4}{24} \left(\frac{\sigma_T^2\mathcal{L}_T^3}{T_0^2} + \frac{2k^2\sigma_v^2\mathcal{L}_v^5}{5c_0^2} \right). \quad (8.90)
$$

It follows from this formula that if $\sigma_T^2/T_0^2 \sim \sigma_v^2/c_0^2$, the contribution from the temperature fluctuations to the extinction coefficient is much greater than that from velocity fluctuations.

8.3.6   Mean sound field

We begin by expressing the sound field $p(\mathbf{R})$ as a convolution of the Green's function as follows:

$$p(\mathbf{R}) = \int G(\mathbf{R}, \mathbf{R}')Q(\mathbf{R}')d^3R' = \int \widehat{p}(\boldsymbol{\kappa}) \exp\left(i\boldsymbol{\kappa} \cdot \mathbf{R}\right) d^3\kappa. \qquad (8.91)$$

Here, the function Q characterizes the distribution of sources ($Q = \delta(\mathbf{R} - \mathbf{R}_0)$ for a point source located at $\mathbf{R}_0$), and $\widehat{p}(\boldsymbol{\kappa})$ is the Fourier transform of the sound field. Averaging both sides of equation (8.91), one obtains the mean sound field:

$$\langle p(\mathbf{R}) \rangle = \int \overline{G}(\mathbf{R} - \mathbf{R}')Q(\mathbf{R}') \, d^3R' = \int \langle \widehat{p}(\boldsymbol{\kappa}) \rangle \exp\left(i\boldsymbol{\kappa} \cdot \mathbf{R}\right) d^3\kappa, \qquad (8.92)$$

where $\langle \widehat{p}(\boldsymbol{\kappa}) \rangle$ is the mean Fourier transform. This quantity can be obtained using equations (8.92) and (8.71):

$$\langle \widehat{p}(\boldsymbol{\kappa}) \rangle = \frac{1}{8\pi^3(k^2 - \kappa^2 - D(\boldsymbol{\kappa}))} \int Q(\mathbf{R}) \exp\left(-i\boldsymbol{\kappa} \cdot \mathbf{R}\right) d^3R. \qquad (8.93)$$

8.4   Theory of multiple scattering: mutual coherence function

In this section, using the theory of multiple scattering, we derive a closed-form equation for the mutual coherence function of the sound field fluctuations. With some approximations, this equation reduces to the equation of radiative transfer, which is used in different fields of physics.

8.4.1   Closed-form equation for the mutual coherence function

In order to derive an equation for the mutual coherence function of the sound field fluctuations, we first consider two point sources located at $\mathbf{R}_0$ and $\mathbf{R}_0'$. The mutual coherence function of the sound field due to these sources is given by

$$\Gamma_G(\mathbf{R}, \mathbf{R}'; \mathbf{R}_0, \mathbf{R}_0') = \langle G(\mathbf{R}, \mathbf{R}_0)G^*(\mathbf{R}', \mathbf{R}_0') \rangle. \qquad (8.94)$$

Substituting with equation (8.58) yields

$$\Gamma_G(\mathbf{R}, \mathbf{R}'; \mathbf{R}_0, \mathbf{R}_0') = (2\pi)^6 \int \int \int \int g_0(\boldsymbol{\kappa})g_0(\boldsymbol{\kappa}_0)g_0^*(\boldsymbol{\kappa}')g_0^*(\boldsymbol{\kappa}_0')$$

$$\times \exp[i(\boldsymbol{\kappa} \cdot \mathbf{R} - \boldsymbol{\kappa}_0 \cdot \mathbf{R}_0 - \boldsymbol{\kappa}' \cdot \mathbf{R}' + \boldsymbol{\kappa}_0' \cdot \mathbf{R}_0')]$$

$$\times \Upsilon(\boldsymbol{\kappa}, \boldsymbol{\kappa}'; \boldsymbol{\kappa}_0, \boldsymbol{\kappa}_0') \, d^3\kappa \, d^3\kappa_0 \, d^3\kappa' \, d^3\kappa_0'. \qquad (8.95)$$

Here, Υ is the mutual coherence function of the quantity $g(\boldsymbol{\kappa}, \boldsymbol{\kappa}_0)$:

$$\Upsilon(\boldsymbol{\kappa}, \boldsymbol{\kappa}'; \boldsymbol{\kappa}_0, \boldsymbol{\kappa}_0') = \langle g(\boldsymbol{\kappa}, \boldsymbol{\kappa}_0) g^*(\boldsymbol{\kappa}', \boldsymbol{\kappa}_0') \rangle. \tag{8.96}$$

The function $g(\boldsymbol{\kappa}, \boldsymbol{\kappa}_0)$ is given by the series in the right-hand side of equation (8.62):

$$g(\boldsymbol{\kappa}, \boldsymbol{\kappa}_0) = (k^2 - \kappa^2)\delta(\boldsymbol{\kappa} - \boldsymbol{\kappa}_0) + \widehat{\mu}(\boldsymbol{\kappa}, \boldsymbol{\kappa}_0) + \int \frac{\widehat{\mu}(\boldsymbol{\kappa}, \boldsymbol{\kappa}_1)\widehat{\mu}(\boldsymbol{\kappa}_1, \boldsymbol{\kappa}_0)}{k^2 - \kappa_1^2} d^3\kappa_1$$

$$+ \int d^3\kappa_1 \frac{\widehat{\mu}(\boldsymbol{\kappa}, \boldsymbol{\kappa}_1)}{k^2 - \kappa_1^2} \int \frac{\widehat{\mu}(\boldsymbol{\kappa}_1, \boldsymbol{\kappa}_2)\widehat{\mu}(\boldsymbol{\kappa}_2, \boldsymbol{\kappa}_0)}{k^2 - \kappa_2^2} d^3\kappa_2 + \dots . \tag{8.97}$$

Here, the first four terms of the series are presented. The quantity $g^*(\boldsymbol{\kappa}', \boldsymbol{\kappa}_0')$ is obtained by complex conjugation of both sides of this formula and changing arguments:

$$g^*(\boldsymbol{\kappa}', \boldsymbol{\kappa}_0') = \left[k^2 - (\kappa')^2\right] \left(\delta(\boldsymbol{\kappa}' - \boldsymbol{\kappa}_0') + \widehat{\mu}^*(\boldsymbol{\kappa}', \boldsymbol{\kappa}_0') + \int \frac{\widehat{\mu}^*(\boldsymbol{\kappa}', \boldsymbol{\kappa}_1)\widehat{\mu}^*(\boldsymbol{\kappa}_1, \boldsymbol{\kappa}_0')}{k^2 - \kappa_1^2}\right.$$

$$\times d^3\kappa_1 + \int d^3\kappa_1 \frac{\widehat{\mu}^*(\boldsymbol{\kappa}', \boldsymbol{\kappa}_1)}{k^2 - \kappa_1^2} \int \frac{\widehat{\mu}^*(\boldsymbol{\kappa}_1, \boldsymbol{\kappa}_2)\widehat{\mu}^*(\boldsymbol{\kappa}_2, \boldsymbol{\kappa}_0')}{k^2 - \kappa_2^2} d^3\kappa_2 + \dots .$$

$$\tag{8.98}$$

Substituting these two series into equation (8.96), the mutual coherence function Υ is expressed as a series whose terms contain the statistical moments of the random fields of $\widehat{\mu}$ and $\widehat{\mu}^*$ of different order. Since $\widehat{\mu}$ has a normal distribution, all odd statistical moments are zero. Similarly to equation (8.67), all even statistical moments can be expressed in terms of the second moments $\langle \widehat{\mu}(\boldsymbol{\kappa}, \boldsymbol{\kappa}_1)\widehat{\mu}(\boldsymbol{\kappa}', \boldsymbol{\kappa}_2) \rangle$, $\langle \widehat{\mu}^*(\boldsymbol{\kappa}, \boldsymbol{\kappa}_1)\widehat{\mu}^*(\boldsymbol{\kappa}', \boldsymbol{\kappa}_2) \rangle$, or $\langle \widehat{\mu}(\boldsymbol{\kappa}, \boldsymbol{\kappa}_1)\widehat{\mu}^*(\boldsymbol{\kappa}', \boldsymbol{\kappa}_2) \rangle$. The first of these moments is given by equations (8.65) and (8.66). Since the right-hand sides of these equations contain only real functions, $\langle \widehat{\mu}^*(\boldsymbol{\kappa}, \boldsymbol{\kappa}_1)\widehat{\mu}^*(\boldsymbol{\kappa}', \boldsymbol{\kappa}_2) \rangle = \langle \widehat{\mu}(\boldsymbol{\kappa}, \boldsymbol{\kappa}_1)\widehat{\mu}(\boldsymbol{\kappa}', \boldsymbol{\kappa}_2) \rangle$. Finally, it can be shown that

$$\langle \widehat{\mu}(\boldsymbol{\kappa}, \boldsymbol{\kappa}_1)\widehat{\mu}^*(\boldsymbol{\kappa}', \boldsymbol{\kappa}_2) \rangle = \delta(\boldsymbol{\kappa} - \boldsymbol{\kappa}_1 - \boldsymbol{\kappa}' + \boldsymbol{\kappa}_2)W(\boldsymbol{\kappa}, \boldsymbol{\kappa}_1, \boldsymbol{\kappa}'). \tag{8.99}$$

Here, the function $W(\boldsymbol{\kappa}, \boldsymbol{\kappa}_1, \boldsymbol{\kappa}')$ is given by the right-hand side of equation (8.66) for $U(\boldsymbol{\kappa}, \boldsymbol{\kappa}_1, \boldsymbol{\kappa}')$ if $\boldsymbol{\kappa}'$ and Φ_{ij} are replaced by $-\boldsymbol{\kappa}'$ and $-\Phi_{ij}$, respectively.

Using the diagram technique, the closed-form equation for the mutual coherence function $\Upsilon(\boldsymbol{\kappa}, \boldsymbol{\kappa}'; \boldsymbol{\kappa}_0, \boldsymbol{\kappa}_0')$ can be derived [290]:

$$\Upsilon(\boldsymbol{\kappa}, \boldsymbol{\kappa}'; \boldsymbol{\kappa}_0, \boldsymbol{\kappa}_0') = \overline{g}(\boldsymbol{\kappa}, \boldsymbol{\kappa}_0)\overline{g}^*(\boldsymbol{\kappa}', \boldsymbol{\kappa}_0') + \int\int\int\int \Upsilon(\boldsymbol{\kappa}_2, \boldsymbol{\kappa}_2'; \boldsymbol{\kappa}_0, \boldsymbol{\kappa}_0')$$

$$\times \frac{\overline{g}(\boldsymbol{\kappa}, \boldsymbol{\kappa}_1)\overline{g}^*(\boldsymbol{\kappa}', \boldsymbol{\kappa}_1')\Lambda(\boldsymbol{\kappa}_1, \boldsymbol{\kappa}_1'; \boldsymbol{\kappa}_2, \boldsymbol{\kappa}_2')}{(k^2 - \kappa_1^2)[k - (\kappa_1')^2](k^2 - \kappa_2^2)[k^2 - (\kappa_2')^2]} d^3\kappa_1 \, d^3\kappa_1' \, d^3\kappa_2 \, d^3\kappa_2'.$$

$$\tag{8.100}$$

Here, the function Λ is called the *intensity-operator kernel*. As the mass-operator kernel, Λ is a subseries of the series which represents the mutual coherence function Υ in terms of multiple scattering. This subseries cannot be summed exactly. In the *ladder* approximation, only the first term in the subseries is retained:

$$\Lambda(\kappa, \kappa'; \kappa_0, \kappa_0') = \langle \hat{\mu}(\kappa, \kappa_0) \hat{\mu}^*(\kappa', \kappa_0') \rangle, \tag{8.101}$$

where the right-hand side is determined by equation (8.99).

Equation (8.100) can be simplified by writing $\overline{g}(\kappa, \kappa_1)$ in the form: $\overline{g}(\kappa, \kappa_1) = \delta(\kappa - \kappa_1)(k^2 - \kappa^2)\overline{u}(\kappa)$, where the function

$$\overline{u}(\kappa) = \frac{k^2 - \kappa^2}{k^2 - \kappa^2 - D(\kappa)}. \tag{8.102}$$

Substituting the value of $\overline{g}$ into the integrand of equation (8.100), one obtains

$$\Upsilon(\kappa, \kappa'; \kappa_0, \kappa_0') = \overline{g}(\kappa, \kappa_0)\overline{g}^*(\kappa', \kappa_0') + \overline{u}(\kappa)\overline{u}^*(\kappa')$$
$$\times \int \int \frac{\Lambda(\kappa, \kappa'; \kappa_1, \kappa_2)}{(k^2 - \kappa_1^2)(k^2 - \kappa_2^2)} \Upsilon(\kappa_1, \kappa_2; \kappa_0, \kappa_0') \, d^3\kappa_1 \, d^3\kappa_2. \tag{8.103}$$

This equation is a closed-form equation for the function Υ. If Υ is calculated, the mutual coherence function Γ_G can be obtained from equation (8.95). In general, equation (8.103) cannot be solved exactly.

The formulations above can also be applied for the distribution of sources, characterized by the function $Q(\mathbf{R})$. In this case, the sound field $p(\mathbf{R})$ is given by equation (8.91). Consider the fluctuating component of this sound field, $\widetilde{p}(\mathbf{R}) = p(\mathbf{R}) - \langle p(\mathbf{R}) \rangle$. The mutual coherence function of $\widetilde{p}(\mathbf{R})$ is defined as

$$\widetilde{\Gamma}(\mathbf{R}, \mathbf{R}') = \langle \widetilde{p}(\mathbf{R})\widetilde{p}^*(\mathbf{R}') \rangle = \int \int \widehat{\Gamma}(\kappa, \kappa') e^{i(\kappa \cdot \mathbf{R} - \kappa' \cdot \mathbf{R}')} \, d^3\kappa \, d^3\kappa', \tag{8.104}$$

where $\widehat{\Gamma}$ is the spectral density of the mutual coherence function. Using equations (8.91)–(8.93) and (8.103), it can be shown that $\widehat{\Gamma}(\kappa, \kappa')$ satisfies the integral equation,

$$\left[k^2 - \kappa^2 - D(\kappa) \right] \left[k^2 - (\kappa')^2 - D^*(\kappa') \right] \widehat{\Gamma}(\kappa, \kappa')$$
$$= \int \int \Lambda(\kappa, \kappa'; \kappa_1, \kappa_2) [\widehat{\Gamma}(\kappa_1, \kappa_2) + \overline{q}(\kappa_1)\overline{q}^*(\kappa_2)] \, d^3\kappa_1 \, d^3\kappa_2. \tag{8.105}$$

This equation also cannot be solved exactly. Making certain assumptions, it can be reduced to the equation of radiative transfer, as described in the next subsection.

8.4.2 Equation of radiative transfer

We introduce the center $\mathbf{R}_c = (\mathbf{R} + \mathbf{R}')/2$ and difference $\mathbf{R}_d = \mathbf{R} - \mathbf{R}'$ coordinates of the observation points and denote $\widetilde{\Gamma}(\mathbf{R}_c + \mathbf{R}_d/2, \mathbf{R}_c - \mathbf{R}_d/2) \equiv$

$\widetilde{\Gamma}_{cd}(\mathbf{R}_c, \mathbf{R}_d)$. The mutual coherence function $\widetilde{\Gamma}_{cd}$ is expressed as

$$\widetilde{\Gamma}_{cd}(\mathbf{R}_c, \mathbf{R}_d) = \oint J(\mathbf{R}_c, \mathbf{n}) \exp\left(i k \mathbf{n} \cdot \mathbf{R}_d\right) d\Omega(\mathbf{n}), \qquad (8.106)$$

where $\Omega(\mathbf{n})$ is the solid angle in the direction of the unit vector $\mathbf{n}$ and $J(\mathbf{R}_c, \mathbf{n})$ is the angular spectrum of the mutual coherence function. In the theory of radiative transfer, $J(\mathbf{R}_c, \mathbf{n})$ is often termed the *specific intensity*.

In equation (8.104), we replace the mutual coherence function with the right-hand side of equation (8.106). We also assume that the scale of variation of $\widetilde{\Gamma}_{cd}(\mathbf{R}_c, \mathbf{R}_d)$ with respect to $\mathbf{R}_c$ is greater than that with respect to $\mathbf{R}_d$. (This assumption is not restrictive and is considered in detail in reference [339].) In equation (8.105), we employ the ladder approximation (8.101) and the Bourret approximation (8.70). As a result, equation (8.105) can be transformed to the equation for the specific intensity:

$$\mathbf{n} \cdot \frac{\partial J(\mathbf{R}_c, \mathbf{n})}{\partial \mathbf{R}_c} + 2\gamma J(\mathbf{R}_c, \mathbf{n}) = \oint J(\mathbf{R}_c, \mathbf{n}_0) \sigma(\mathbf{n} - \mathbf{n}_0)\, d\Omega(\mathbf{n}_0)$$

$$+ \frac{\pi}{2} \int d^3 \kappa_0\, W(k\mathbf{n}, \boldsymbol{\kappa}_0, k\mathbf{n}) \int \overline{q}\left(\boldsymbol{\kappa}_0 + \frac{\boldsymbol{\kappa}}{2}\right) \overline{q}^*\left(\boldsymbol{\kappa}_0 - \frac{\boldsymbol{\kappa}}{2}\right) e^{i\boldsymbol{\kappa}\cdot\mathbf{R}}\, d^3\kappa. \qquad (8.107)$$

Here, $\sigma(\mathbf{n} - \mathbf{n}_0)$ is the sound scattering cross section given by equation (6.114). When deriving equation (8.107), we took into account that $W(k\mathbf{n}, k\mathbf{n}_0, k\mathbf{n}) = (2/\pi)\,\sigma(\mathbf{n} - \mathbf{n}_0)$.

Equation (8.107) has the form of the radiative transfer equation and describes redistribution of acoustic energy in different directions due to scattering. The specific intensity $J(\mathbf{R}_c, \mathbf{n})$ characterizes the flux of acoustic energy at the point $\mathbf{R}_c$ in the direction of the vector $\mathbf{n}$. The quantity 2γ plays a role of the extinction coefficient of the specific intensity. It follows from equations (6.112) and (8.85) that

$$2\gamma = \sigma_0 = \oint \sigma(\mathbf{n} - \mathbf{n}_0)\, d\Omega(\mathbf{n}_0). \qquad (8.108)$$

Therefore, the decrease of the specific intensity in a particular direction is caused only by scattering (i.e., there is no absorption of sound). The second term on the right-hand side of equation (8.107) results from the transformation of the coherent part of the sound field into the incoherent one.

In general, the equation of radiative transfer cannot be solved exactly. Numerical solution methods have been developed [239]. Equation (8.107) can be solved analytically in the small-angle approximation, when the scattering cross section $\sigma(\mathbf{n} - \mathbf{n}_0)$ decreases rapidly enough as a function of the angle between the vectors $\mathbf{n}$ and $\mathbf{n}_0$. Substituting the resulting value of $J(\mathbf{R}_c, \mathbf{n})$ into equation (8.106), one obtains the mutual coherence function $\widetilde{\Gamma}_{cd}(\mathbf{R}_c, \mathbf{R}_d)$ of the fluctuating component of the sound field. The mutual coherence function of the sound field $p(\mathbf{R})$ is then $\widetilde{\Gamma}_{cd}(\mathbf{R}, \mathbf{R}') + \langle p(\mathbf{R}) \rangle \langle p^*(\mathbf{R}') \rangle$. It can be shown that the latter coherence function coincides exactly with the coherence function given by equation (7.175), which was obtained using the parabolic equation and Markov approximations. This result, derived from the theory of multiple scattering, helps to justify these approximations.

Part III

Numerical methods for sound propagation in moving media

In Parts I and II, the theory of sound propagation and scattering in moving media was developed from first principles. Predictions were derived for certain idealized situations, such as refraction in a horizontally stratified medium, line-of-sight propagation through turbulence, and the sound scattering cross section for the inertial subrange of turbulence. While such analytical results are extremely useful for illuminating the physics underlying these phenomena, and may apply directly to some practical situations, often they are unable to account for the many complex and interacting factors impacting sound propagation in the atmosphere and ocean. Hence, in this part of the book, we consider numerical methods enabling application of the general theoretical developments from the first two parts. Although the example calculations deal specifically with outdoor sound propagation, largely because this is the area in which the authors have conducted most of their research, the techniques can be utilized in other environments.

While the focus here is on numerical treatments pertinent to inhomogeneous moving media, the complex and interdependent nature of the many environmental influences on sound propagation compels some discussion of related topics. Important, practical examples of complexities and feedbacks among outdoor sound propagation phenomenon, which are difficult to address without numerical methods, include:

- Refraction by nonlinear wind and temperature profiles. Analytical solutions for linear sound-speed profiles, as well as other idealized gradient functions, are known (section 3.5.2). However, in actuality the vertical profiles assume logarithmic and more complex shapes, for which analytical solutions are unavailable.

- Interactions between refraction and reflection of sound energy by the ground. For example, although it is typically the case that sound levels will be higher in conditions of downward refraction, if the ground surface is highly absorptive, as with coarse sand or snow, sound levels can actually be diminished.

- Interactions between stratification and turbulence in the propagation medium. In the atmosphere, wind shear causes refraction of sound (generally downward for downwind propagation, and upward for upwind propagation), while at the same time creating turbulence which scatters the sound. Furthermore, a super-adiabatic temperature gradient produces turbulence while also enhancing upward refraction. A temperature inversion suppresses turbulence while creating downward refraction.

- The direct and indirect influences of topographic variations, such as hills and buildings. Topography has a direct influence by reflecting the sound waves and creating acoustic shadows, and an indirect one by altering the atmospheric flow field and temperature stratification. For example, the wind accelerates as it passes over hills, and cold air tends to pool in valleys at night. Sound is also diffracted and scattered into non-line-of-sight regions behind hills and buildings.

- Height dependence (statistical inhomogeneity) of turbulence properties such as the outer length scales, structure-function parameters, and spectra. Although such height dependence is often neglected and very difficult to address analytically, it can have important implications for scattering into shadow regions and the coherence of signals.

Fortunately, advances in methods for computational acoustics occurring during the past several decades, including the parabolic equation (PE), wavenumber integration (specifically, the fast-field program, or FFP), finite-difference, time-domain (FDTD) method, pseudospectral time-domain (PSTD) method, and boundary-element method (BEM), allow many of the phenomena described above to be addressed realistically. Predictive methods for outdoor sound propagation may be broadly distinguished into four categories: (1) heuristic (or engineering) methods, (2) geometric acoustics (ray tracing and related high-frequency approximations), (3) wave-based, frequency-domain methods, and (4) wave-based, time-domain methods. The appropriate choice of a method depends greatly upon the application, goals of the analysis, and the computational resources. For example, if one wishes to numerically test the validity of an approximate theoretical solution, resources and time often permit highly intensive and accurate calculations. Many scenarios encountered in engineering practice, however, are too complex to address with rigorous calculation methods, and thus heuristic approaches are often adopted to speed up and standardize the predictions. Such a situation might involve evaluating the cost effectiveness of potential noise mitigation measures, such as construction of barriers and vegetative screens, prior to their purchase and installation.

Heuristic approaches are usually based on analysis of empirical and/or simulation data for the particular noise sources and mitigation measures of interest. Some heuristic models, such as Harmonoise [342, 385] (which has been widely used in Europe to map noise levels), utilize numerical propagation calculations for calibration. A general difficulty with heuristic approaches is that they may provide poor results when applied to situations for which they were not specifically designed or tested. Existing heuristic approaches often assume that adjustments associated with various propagation effects (such as refraction, shielding by barriers, and ground attenuation) are independent and sum linearly. While heuristic approaches are of interest in many applications, they are not considered further in this book, since our emphasis is on more rigorous approaches to calculating refraction and turbulent scattering. Attenborough et al. [16] provide more information and references on various heuristic models, along with a comparative discussion.

By *wave-based*, we mean methods based on the wave equation or related equations maintaining an explicit wavenumber dependence. Unlike ray tracing, the computational burden of the wave-based methods increases with frequency (decreasing wavelength), because the solution is usually calculated on a discrete spatial grid with resolution proportional to the wavelength. Wave-based solutions may be implemented in the frequency or time domain. The PE and FFP are generally implemented in the frequency domain. Since frequency-domain solutions can be calculated one frequency at a time, less memory is often required. For the time-domain methods, resolution of the temporal grid varies inversely with the maximum supported frequency.

Chapter 9 describes approaches to modeling random fields and turbulence that are particularly useful for generating inputs to sound propagation calculations. Ray tracing and its application to the near-ground atmosphere is described in Chapter 10. That chapter also provides some background material on definitions of sound levels and transmission loss, formulation of boundary conditions for the ground, and interactions of sound waves with porous media. Wave-based methods, in the frequency and time domains, are the subject of Chapters 11 and 12, respectively. Lastly, in Chapter 13, we explore techniques for incorporating randomness and uncertainty into propagation calculations, and for quantifying the accuracy of predictions when such uncertainties are present.

9

Numerical representation of random fields

Realistic, high-resolution representation of random variations in the propagation medium, such as atmospheric turbulence, is a particularly important and challenging aspect of simulating wave propagation in the atmosphere, ocean, and other environments [437]. One approach is to employ a physics-based simulation of the propagation medium, such as a computational fluid dynamics (CFD) simulation of a turbulent flow. Among the various classes of CFD, large-eddy simulation (LES) is generally most appropriate for capturing the dynamics of turbulence in the atmospheric boundary layer [129, 243, 255]. The main drawbacks of LES are its computational intensiveness (which may, in practice, far exceed the sound propagation calculation) and its limited resolution, which is typically no better than a few meters, and thus suitable only for sound propagation at low frequencies [417].

Turbulence can also be synthesized with *kinematic* approaches, meaning that the fields are synthesized from prescribed spatial statistics, rather than by attempting to solve equations describing the full fluid dynamics. In particular, we may endeavor to synthesize fields with prescribed second-order spatial statistics, such as spectra and correlations. Kinematic approaches have been commonly used in acoustical modeling since the 1990s, when propagation calculations advanced to the point where wave-based solutions became feasible in a medium having 2D or 3D spatial variability in the index-of-refraction [75, 136, 191]. Since the turbulence could not be simulated by CFD or directly measured to subwavelength resolution in the audible range, as needed for accurate acoustical calculations, kinematic approaches provided a viable alternative to achieving necessary spatial resolution. Other attractions of the kinematic approaches are their relative simplicity and computational efficiency. Although kinematic approaches do not capture the turbulent dynamics, this shortcoming may not be important for calculating second-order statistics of the scattered sound field, such as the mean-square sound pressure. On the other hand, if the goal is to simulate a realistic time series of an acoustic signal, kinematic approaches may be inappropriate. They cannot realistically capture phenomena involving interactions between the mean and turbulent components of the flow, such as wind gusts and thermal plumes.

Most kinematic approaches involve spectral synthesis from a superposition of spatial Fourier modes (harmonic functions) [75, 127, 136, 150, 238, 358]. The amplitudes of the modes are made proportional to the square root of the spectral density at that wavenumber, and the phases of the modes are

317

independently randomized. A similar method, based on orthogonal wavelets, has also been devised [200]. Application of an inverse transform (from the wavenumber to the spatial domain) then yields the random field consistent with the desired spectrum.

An alternative to the spectrally based methods, but nonetheless a kinematic method, involves representing the turbulence field as a collection of randomly positioned, eddy-like structures [92, 141, 247]. In principle, second-order statistics from the eddy and spectral methods can be made identical, by specifying the eddy characteristics appropriately [142]. The spatially localized nature of the eddies can aid physical intuition and provide computational benefits. When the ensemble of eddies is constructed in a self-similar manner using correct scalings for turbulence, the inertial subrange and other realistic spectral characteristics naturally emerge.

This chapter describes both the spectral (section 9.1) and eddy-based (section 9.2) kinematic methods. Interested readers are referred to other references [129, 243, 255] for descriptions of CFD of atmospheric turbulence, and for their application to sound propagation [68, 417].

9.1   Spectral methods

We will first consider spectral methods for homogeneous and isotropic turbulence. Even though actual turbulence rarely exhibits these properties, in some applications it may be reasonable to employ such a model. In particular, scattering may be dominated by eddies within the inertial subrange, which exhibit local isotropy (i.e., isotropy for small spatial separations). It may also be impractical to model anisotropy and inhomogeneity, due to the complexities and uncertainties involved. However, it is possible to extend the spectral approach to include anisotropy and inhomogeneity, as will be discussed in section 9.1.2.

9.1.1   Homogeneous and isotropic turbulence

As described in the introduction to this chapter, the basic spectral method involves synthesizing the field from a superposition of Fourier modes. The amplitudes are made proportional to the square root of the spectral density, and the phases independently randomized. The motivation for this approach follows directly from the definition of the (two-sided) spectral density of a random field (e.g., reference [29]):

$$\Phi\left(\boldsymbol{\kappa}\right) = \lim_{V \to \infty} \frac{(2\pi)^D}{V} \left\langle \left|\hat{s}\left(\boldsymbol{\kappa}\right)\right|^2 \right\rangle, \tag{9.1}$$

where D is the number of spatial dimensions (the cases $D = 2$ and $D = 3$ being of primary interest for calculations), V is the D-dimensional volume (or area) occupied by the field, κ is the wave vector, and $\hat{s}(\kappa)$ is the Fourier transform of a realization of the random field. The angle brackets indicate ensemble averaging. Simply setting

$$\hat{s}(\kappa) = \sqrt{\frac{V\Phi(\kappa)}{(2\pi)^D}} \exp\left[i\gamma(\kappa)\right], \tag{9.2}$$

where $\gamma(\kappa)$ is a random phase between 0 and 2π, ensures a random field possessing the desired spectral density. Then, application of an inverse Fourier transform yields the synthesized random field

$$s(\mathbf{R}) = \int_{-\infty}^{\infty} \hat{s}(\kappa)\, e^{i\kappa\cdot\mathbf{R}}\, d\kappa. \tag{9.3}$$

We call this method for synthesizing random fields the *random-phase method* (RPM).

In practice, the field must be synthesized from a finite number of Fourier modes, and thus the inverse transform, equation (9.3), is implemented in discretized form. The basic process will be described here for a realization in the two-dimensional vertical plane (x, z), as is often desired for sound propagation calculations; generalization to other orientations or numbers of dimensions is straightforward. In 2D, equation (9.3) becomes

$$\begin{aligned}
s(x, z) &= \int_{-\infty}^{\infty}\int_{-\infty}^{\infty} \hat{s}(\kappa_x, \kappa_z)\, e^{i(x\kappa_x + y\kappa_z)}\, d\kappa_x\, d\kappa_z \\
&\simeq \int_{-\Lambda_x/2}^{\Lambda_x/2}\int_{-\Lambda_z/2}^{\Lambda_z/2} \hat{s}(\kappa_x, \kappa_z)\, e^{i(x\kappa_x + y\kappa_z)}\, d\kappa_x\, d\kappa_z,
\end{aligned} \tag{9.4}$$

where $\pm\Lambda_x/2$ and $\pm\Lambda_z/2$ are finite limits at which the wavenumber integration is truncated. These limits should be at a wavenumber high enough to capture most of the energy in the spectrum. The wavenumber axes κ_x and κ_z can be discretized as follows:

$$\kappa_x = \kappa_{x,m} = -\Lambda_x/2 + m\Delta\kappa_x, \tag{9.5}$$

where $m = 0, 1, \ldots, M-1$ and $\Lambda_x = M\Delta\kappa_x$, and

$$\kappa_z = \kappa_{z,n} = -\Lambda_z/2 + n\Delta\kappa_z, \tag{9.6}$$

where $n = 0, 1, \ldots, N-1$ and $\Lambda_z = N\Delta\kappa_z$. The spatial axes are similarly discretized as $x = x_j = j\Delta x$, $j = 0, 1, \ldots, M-1$, and $z = z_k = k\Delta z$, $k = 0, 1, \ldots, N-1$. In textbooks on digital signal processing (e.g., reference [29]), it is shown that the length of the transform in one domain determines the resolution in the other. In our case, $L_x = M\Delta x = 2\pi/\Delta\kappa_x$ and $L_z = N\Delta z = 2\pi/\Delta\kappa_z$, respectively. Similarly, $\Lambda_x = 2\pi/\Delta x$ and $\Lambda_z = 2\pi/\Delta z$. We thus have

$$s(x_j, z_k) \simeq \frac{(2\pi)^2}{V} \sum_{m=0}^{M-1}\sum_{n=0}^{N-1} \hat{s}(\kappa_{x,m}, \kappa_{z,n})\, (-1)^{j+k}\, e^{i2\pi(mj/M + nk/N)}, \tag{9.7}$$

in which we have identified $V = L_x L_z$. Substituting $\hat{s}\,(m\Delta\kappa_x, n\Delta\kappa_z)$ as specified by equation (9.2), we have

$$s\,(x_j, z_k) \simeq \frac{2\pi}{\sqrt{V}}(-1)^{j+k} \sum_{m=0}^{M-1} \sum_{n=0}^{N-1} \sqrt{\Phi^{2D}\,(\kappa_{x,m}, \kappa_{z,n})}$$
$$\times\, e^{i\gamma_{mn}+i2\pi(mj/M+nk/N)}, \quad (9.8)$$

where the γ_{mn} are random phases. This result is in a form amenable to calculation with a 2D fast Fourier transform (FFT). One subtlety to be addressed is that, as formulated, the fields will be complex. If we wish to construct a real-valued field, the overall variance will be halved if we simply retain the real part of $s\,(x_j, z_k)$, since the variance of the imaginary part has been discarded. Hence, for the real-valued realization, we use $\sqrt{2}\mathrm{Re}[s\,(x_j, z_k)]$.

Three-dimensional spectra and correlation functions for turbulence were given in Chapter 6. However, sound propagation calculations are often performed in a vertical plane. The relationship between the 3D spectrum and the 2D spectrum for the vertical plane (x, z) can be derived by first setting the y-separation to zero in equation (6.3), which results in

$$B(x, 0, z) = \int_{-\infty}^{\infty} \int_{-\infty}^{\infty} \int_{-\infty}^{\infty} \Phi^{3D}\,(\kappa_x, \kappa_y, \kappa_z)\, e^{i(\kappa_x x + \kappa_z z)}\, d\kappa_x\, d\kappa_y\, d\kappa_z. \quad (9.9)$$

The 2D spectrum $\Phi^{2D}\,(\kappa_x, \kappa_z)$ is by definition the Fourier transform of $B(x, 0, z)$ in the (x, z)-plane. Hence, application of a 2D transformation to both sides of equation (9.9) yields

$$\Phi^{2D}\,(\kappa_x, \kappa_z) = \int_{-\infty}^{\infty} \Phi^{3D}\,(\kappa_x, \kappa_y, \kappa_z)\, d\kappa_y. \quad (9.10)$$

For a scalar such as temperature, substitution of the 3D von Kármán spectrum, equation (6.26), into the preceding equation results in

$$\Phi_T^{2D}\,(\kappa_x, \kappa_z) = \frac{\sigma_T^2 L_T^2}{3\pi\,(1 + \kappa^2 L_T^2)^{4/3}}, \quad (9.11)$$

in which $\kappa^2 = \kappa_x^2 + \kappa_z^2$. To obtain the 2D von Kármán spectrum for velocity fluctuations in the x-direction, we substitute equation (6.29) into (6.17), and then perform the integration in equation (9.10). The result is

$$\Phi_{11}^{2D}\,(\kappa_x, \kappa_z) = \frac{\sigma_v^2 L_v^2}{3\pi\,(1 + \kappa^2 L_v^2)^{7/3}} \left(\frac{4}{3}\kappa_z^2 L_v^2 + \frac{1 + \kappa^2 L_v^2}{2}\right). \quad (9.12)$$

Figure 9.1 shows example 2D realizations of turbulence fields by the RPM method, based on equations (9.11) and (9.12). In comparison to the scalar fluctuations, the velocity fluctuations have the appearance of being stretched along the orientation of the velocity component (the x-axis).

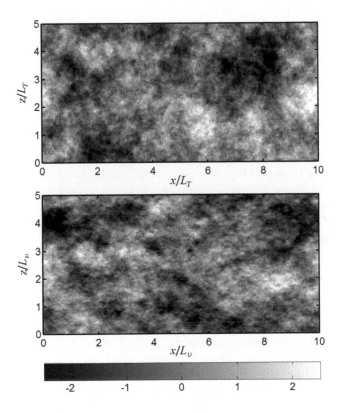

FIGURE 9.1
Two-dimensional realizations of fields based on the von Kármán turbulence spectrum for scalars (top) and velocity (bottom). The latter is for the x-component of the vector field. The spatial axes are normalized by the length scales L_T (top) or L_v (bottom). The gray scale represents the fluctuation normalized by σ_T (top) or σ_v (bottom).

9.1.2 Inhomogeneous and anisotropic turbulence

The methodology in the previous section assumed that the random field is statistically homogeneous. This is a significant limitation, for example, in surface-layer turbulence produced by wind shear, for which the outer length scale is proportional to height, thus making the statistics inherently inhomogeneous in the vertical direction. Therefore it is desirable to extend RPM to synthesize inhomogeneous turbulence structure. Such a method, called the *generalized random-phase method* (GRPM) [412], is described in this subsection.

The idea underlying GRPM is to synthesize the field using randomly phased empirical orthogonal functions (EOFs), which reduce to Fourier modes when the second-order statistics are homogeneous [231, 412], in which case GRPM and RPM are equivalent. The EOFs are defined as the eigenfunctions of the correlation function:

$$\int_V B\left(\mathbf{R}, \mathbf{R}'\right) \psi\left(\mathbf{R}'\right) d\mathbf{R}' = \lambda \psi\left(\mathbf{R}\right), \tag{9.13}$$

where $\mathbf{R}$ and $\mathbf{R}'$ are two points in space, ψ is an eigenfunction, and λ the corresponding eigenvalue. (Although the formulation here is for a single field, it can be readily extended to multiple, correlated fields, such as turbulent velocities, using tensors [412].) In general, the spatial coordinates $\mathbf{R}$ may consist of a mixture of homogeneous coordinates, for which the correlation depends only on the separation, and inhomogeneous coordinates. We indicate the former set of coordinates as $\mathbf{x}$ and the latter set as $\boldsymbol{\xi}$. Partitioning $\mathbf{R}$ in this manner, we have

$$\int_{V_\xi} \int_{V_x} B\left(\mathbf{x} - \mathbf{x}'; \boldsymbol{\xi}, \boldsymbol{\xi}'\right) \psi\left(\mathbf{x}'; \boldsymbol{\xi}'\right) d\mathbf{x}' \, d\boldsymbol{\xi}' = \lambda \psi\left(\mathbf{x}; \boldsymbol{\xi}\right). \tag{9.14}$$

Here, V_x is the H-dimensional volume for the homogeneous coordinates, and V_ξ is the $(D - H)$-dimensional volume for the inhomogeneous coordinates. We now apply a Fourier transform to the preceding equation with respect to the homogeneous coordinates only. Because the integral is a convolution with respect to the homogeneous coordinates, as a consequence of the well-known convolution theorem for Fourier transforms, it reduces to a multiplication after transformation. Remaining is an eigenvalue problem involving the inhomogeneous coordinates only:

$$\int_{V_\xi} \hat{B}\left(\boldsymbol{\kappa}; \boldsymbol{\xi}, \boldsymbol{\xi}'\right) \hat{\psi}\left(\boldsymbol{\kappa}; \boldsymbol{\xi}'\right) d\boldsymbol{\xi}' = \lambda\left(\boldsymbol{\kappa}\right) \hat{\psi}\left(\boldsymbol{\kappa}; \boldsymbol{\xi}\right). \tag{9.15}$$

In the preceding equation, functions that have been transformed with respect to the homogeneous coordinates are indicated by a hat, e.g., $\hat{\psi}$. The wavenumbers corresponding to the homogeneous coordinates are indicated by $\boldsymbol{\kappa}$. Note that the solution to the eigenvalue problem depends on $\boldsymbol{\kappa}$. The function $\hat{B}\left(\boldsymbol{\kappa}; \boldsymbol{\xi}, \boldsymbol{\xi}'\right)$ is a spectrum with regard to the homogeneous coordinates, and a correlation function with regard to the inhomogeneous ones. As a generalization of equation (9.1), it can be defined as

$$\hat{B}\left(\boldsymbol{\kappa}; \boldsymbol{\xi}, \boldsymbol{\xi}'\right) = \lim_{V_H \to \infty} \frac{(2\pi)^H}{V_H} \left\langle \hat{s}\left(\boldsymbol{\kappa}; \boldsymbol{\xi}\right) \hat{s}^*\left(\boldsymbol{\kappa}; \boldsymbol{\xi}'\right) \right\rangle, \tag{9.16}$$

where

$$\hat{s}\left(\boldsymbol{\kappa}; \boldsymbol{\xi}\right) = \frac{1}{(2\pi)^H} \int_{-\infty}^{\infty} s\left(\mathbf{x}; \boldsymbol{\xi}\right) e^{-i\boldsymbol{\kappa} \cdot \mathbf{x}} \, d\boldsymbol{\kappa} \tag{9.17}$$

is the Fourier transform of the random field with respect to the homogeneous coordinates only.

The next step in GRPM involves finding the eigenfunctions and eigenvalues of $(2\pi)^H \hat{B}(\boldsymbol{\kappa}; \boldsymbol{\xi}, \boldsymbol{\xi}')$. Let us designate the solutions of equation (9.15) as $\hat{\psi}_n(\boldsymbol{\kappa}; \boldsymbol{\xi})$, $\lambda_n(\boldsymbol{\kappa})$, where $n = 1, 2, \ldots$, and the set is arranged in order of decreasing eigenvalue. The eigenfunctions are orthonormal such that

$$\int_{V_\xi} \hat{\psi}_m(\boldsymbol{\kappa}; \boldsymbol{\xi}) \hat{\psi}_n^*(\boldsymbol{\kappa}; \boldsymbol{\xi}) \, d\boldsymbol{\xi} = \delta_{mn}, \tag{9.18}$$

where $\delta_{mn} = 1$ if $m = n$ and zero otherwise. We then expand the transformed random field $\hat{s}(\boldsymbol{\kappa}; \boldsymbol{\xi})$ as a series of the eigenfunctions:

$$\hat{s}(\boldsymbol{\kappa}; \boldsymbol{\xi}) = \sum_n a_n(\boldsymbol{\kappa}) \hat{\psi}_n(\boldsymbol{\kappa}; \boldsymbol{\xi}). \tag{9.19}$$

Substituting this expansion into equation (9.16), we have

$$\hat{B}(\boldsymbol{\kappa}; \boldsymbol{\xi}, \boldsymbol{\xi}') = \frac{(2\pi)^H}{V_H} \sum_m \sum_n \langle a_m(\boldsymbol{\kappa}) a_n^*(\boldsymbol{\kappa}) \rangle \hat{\psi}_m(\boldsymbol{\kappa}; \boldsymbol{\xi}) \hat{\psi}_n^*(\boldsymbol{\kappa}; \boldsymbol{\xi}'). \tag{9.20}$$

Multiplying both sides by $\hat{\psi}_j(\boldsymbol{\kappa}; \boldsymbol{\xi}')$, integrating over the volume V_ξ, and invoking orthonormality of the eigenfunctions, the previous result becomes

$$\int_{V_\xi} \hat{B}(\boldsymbol{\kappa}; \boldsymbol{\xi}, \boldsymbol{\xi}') \hat{\psi}_j(\boldsymbol{\kappa}; \boldsymbol{\xi}') \, d\boldsymbol{\xi}' = \frac{(2\pi)^H}{V_H} \sum_m \langle a_m(\boldsymbol{\kappa}) a_j^*(\boldsymbol{\kappa}) \rangle \hat{\psi}_m(\boldsymbol{\kappa}; \boldsymbol{\xi}).$$

Comparison with equation (9.15) now indicates

$$\frac{(2\pi)^H}{V_H} \sum_m \langle a_m(\boldsymbol{\kappa}) a_j^*(\boldsymbol{\kappa}) \rangle \hat{\psi}_m(\boldsymbol{\kappa}; \boldsymbol{\xi}) = \lambda_j(\boldsymbol{\kappa}) \hat{\psi}_j(\boldsymbol{\kappa}; \boldsymbol{\xi}).$$

Finally, multiplying by $\hat{\psi}_n^*(\boldsymbol{\kappa}; \boldsymbol{\xi})$, integrating over V_ξ, and invoking orthonormality one more time results in

$$\langle a_n(\boldsymbol{\kappa}) a_j^*(\boldsymbol{\kappa}) \rangle = \delta_{jn} \frac{V_H \lambda_j(\boldsymbol{\kappa})}{(2\pi)^H}. \tag{9.21}$$

This equation shows that the EOF expansion coefficients are mutually uncorrelated and that the expected value of the magnitude squared of each coefficient is proportional to the corresponding eigenvalue. The simplest way to choose the coefficients while satisfying this condition is to set

$$a_n(\boldsymbol{\kappa}) = \sqrt{\frac{V_H \lambda_n(\boldsymbol{\kappa})}{(2\pi)^H}} \exp\left[i\gamma_n(\boldsymbol{\kappa})\right], \tag{9.22}$$

in which $\gamma_n(\boldsymbol{\kappa})$ is a random phase between 0 and 2π. By selecting the expansion coefficients as indicated by equation (9.22), we ensure that the synthesized field is consistent with the model correlation function. In fact, substituting equation (9.22) into (9.21), we have

$$\hat{B}(\boldsymbol{\kappa};\boldsymbol{\xi},\boldsymbol{\xi}') = \sum_n \lambda_n(\boldsymbol{\kappa})\hat{\psi}_n(\boldsymbol{\kappa};\boldsymbol{\xi})\,\hat{\psi}_n^*(\boldsymbol{\kappa};\boldsymbol{\xi}'). \tag{9.23}$$

Setting $\boldsymbol{\xi} = \boldsymbol{\xi}'$, integrating over V_ξ, and applying equation (9.18), we find the simple result

$$\int_{V_\xi} \hat{B}(\boldsymbol{\kappa};\boldsymbol{\xi},\boldsymbol{\xi})\,d\xi = \sum_n \lambda_n(\boldsymbol{\kappa}). \tag{9.24}$$

That is, the variance, when integrated over V_ξ, equals the sum of the eigenvalues.

Having determined the coefficients in equation (9.19), the final step in GRPM is to apply an inverse Fourier transform to the synthesized field, thus converting it entirely to the spatial domain:

$$s(\mathbf{R}) = s(\mathbf{x};\boldsymbol{\xi}) = \int_{-\infty}^{\infty} \hat{s}(\boldsymbol{\kappa};\boldsymbol{\xi})\,e^{i\boldsymbol{\kappa}\cdot\mathbf{x}}\,d\boldsymbol{\kappa}. \tag{9.25}$$

GRPM of homogeneous fields is equivalent to RPM as described in section 9.1.1. In this case, the Fourier transform is applied to all coordinates (that is, there are no inhomogeneous coordinates $\boldsymbol{\xi}$), and the transformed eigenvalue problem corresponding to equation (9.15) is $\Phi(\boldsymbol{\kappa})\hat{\psi}(\boldsymbol{\kappa}) = \lambda(\boldsymbol{\kappa})\hat{\psi}(\boldsymbol{\kappa})$. Hence we simply set $\hat{\psi}(\boldsymbol{\kappa}) = 1$ and $\Phi(\boldsymbol{\kappa}) = \lambda(\boldsymbol{\kappa})$, and then equation (9.22) reduces to (9.2).

As discussed earlier in this subsection, sound propagation calculations are often performed in a vertical plane, and turbulence statistics in the atmosphere depend on height. Hence, a case of particular interest involves analyzing correlation functions that are inhomogeneous in z, homogeneous along one of the horizontal coordinates (x), and with zero separation along the other horizontal coordinate (y), i.e., $B(\mathbf{R},\mathbf{R}') = B(\Delta x, 0, z, z')$, where $\Delta x = x' - x$. For equation (9.15), we then have

$$\int_0^{L_z} \hat{B}(\kappa_x;0,z,z')\,\widehat{\psi}(\kappa_x,z')\,dz' = \lambda(\kappa_x)\,\widehat{\psi}(\kappa_x;z), \tag{9.26}$$

where

$$\hat{B}(\kappa_x;0,z,z') = \frac{1}{2\pi}\int_{-\infty}^{\infty} B(x;0,z,z')\exp(-i\kappa_x x)\,dx. \tag{9.27}$$

The random expansion coefficients, from equation (9.22), are

$$a_n(\kappa_x) = \sqrt{\frac{L_x\lambda_n(\kappa_x)}{2\pi}}\exp[i\gamma_n(\kappa_x)]. \tag{9.28}$$

For computational implementation, the eigenvalue problem is solved by discretizing the vertical integration domain in equation (9.26) into N levels and then numerically performing an eigenanalysis on the resulting matrix equation [258, 412]. Setting $z = z_n = n\Delta z$ (where $n = 1, 2, \ldots, N$ and $L_z = N\Delta z$), we have

$$\widehat{\mathbf{B}}(\kappa_x)\,\widehat{\boldsymbol{\psi}}(\kappa_x) = \bar{\lambda}(\kappa_x)\,\widehat{\boldsymbol{\psi}}(\kappa_x), \tag{9.29}$$

where $\widehat{\mathbf{B}}$ is an $N \times N$ matrix, $\widehat{\boldsymbol{\psi}}$ is an $N \times 1$ vector, and $\bar{\lambda} = \lambda/\Delta z$. The element of $\widehat{\mathbf{B}}$ at position (n, n') is $\hat{B}(\kappa_x; 0, z_n, z_{n'})$, whereas the element of $\widehat{\boldsymbol{\psi}}$ at position n is $\hat{\psi}(\kappa_x, z_n)$. A set of N eigenfunctions $\bar{\psi}_n(\kappa_x) = \hat{\psi}_n(\kappa_x)\sqrt{\Delta z}$ and N eigenvalues $\bar{\lambda}_n(\kappa_x)$ results from solving equation (9.29) with numerical methods producing orthonormal eigenfunctions. The expansion for the random field, from equation (9.19), is then

$$\hat{s}(\kappa; \xi) = \sum_n \sqrt{\frac{L_x \bar{\lambda}_n(\kappa_x)}{2\pi}}\, \bar{\psi}_n(\kappa_x) \exp\left[i\gamma_n(\kappa_x)\right]. \tag{9.30}$$

For a scalar von Kármán model, $\hat{B}(\kappa_x; 0, z, z')$ can be found by taking the inverse Fourier transform of equation (9.11) with respect to κ_z. The result is

$$\hat{B}_T(\kappa_x; 0, z, z') = \frac{2\sigma_T^2 L_T}{\sqrt{\pi}\Gamma(1/3)} \left(\frac{\zeta_T/2}{1 + \kappa_x^2 L_T^2}\right)^{5/6} K_{5/6}(\zeta_T), \tag{9.31}$$

in which $\zeta_T = (|z' - z|/L_T)\sqrt{1 + \kappa_x^2 L_T^2}$. For the velocity von Kármán model, we find from equation (9.12)

$$\hat{B}_{11}(\kappa_x; 0, z, z') = \frac{2\sigma_v^2 L_v}{\sqrt{\pi}\Gamma(1/3)} \left(\frac{\zeta_v/2}{1 + \kappa_x^2 L_v^2}\right)^{5/6} \left[K_{5/6}(\zeta_v) - \frac{\zeta_v}{2} K_{1/6}(\zeta_v)\right], \tag{9.32}$$

in which $\zeta_v = (|z' - z|/L_v)\sqrt{1 + \kappa_x^2 L_v^2}$.

Parameters for von Kármán spectra of the wind and temperature in the atmospheric surface layer are given by equations (6.50)–(6.53). Since several of the variance and length-scale parameters in these spectra are height dependent, realizations must be performed with GRPM. The situation considered here is for sunny daytime conditions (unstable stratification), $Q_H = 200$ W/m^2 and $z_i = 1000$ m, and with a moderate wind, $u_* = 0.3$. A few example EOFs for turbulent velocities are shown in figure 9.2. The appearance of the EOFs bears some similarity to harmonic functions, except that the amplitude and period vary with height.

Random realizations of temperature and wind fluctuations are shown in figure 9.3. For the temperature field, the increasing size of the length scale with height is evident, as is the decreasing variance. The length scale of the shear-induced velocity fluctuations similarly increases with height, although the

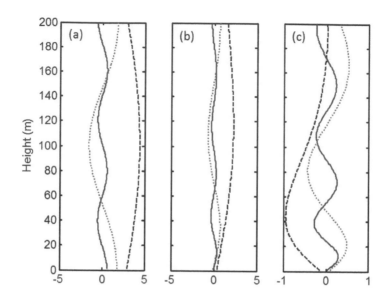

FIGURE 9.2
Example empirical orthogonal functions (EOFs) for turbulence velocity spectra, as scaled by the square roots of the corresponding eigenvalues. (Dimen-
sions after scaling are $m^{3/2}\,s^{-1}$.) The curves show $\sqrt{\lambda_1}\hat{\psi}_1$ (dashed line), $\sqrt{\lambda_3}\hat{\psi}_3$
(dotted line), and $\sqrt{\lambda_6}\hat{\psi}_6$ (solid line), all for $\hat{B}_{11}$ as calculated with the von
Kármán spectrum. (a) Buoyancy-induced turbulence at $\kappa_x = 2\pi/(1000\,\text{m})$. (b)
Shear-induced turbulence at $\kappa_x = 2\pi/(1000\,\text{m})$. (c) Shear-induced turbulence
at $\kappa_x = 2\pi/(250\,\text{m})$.

variance does not. The buoyancy-induced velocity fluctuations have height-
independent length scales and variance, and thus could have been synthesized
from a homogeneous method. Also shown in figure 9.3 is the quantity ε_{mov},
as calculated from equation (7.3), which represents the total turbulent fluctuation in the effective sound-speed approximation. In this case, ε_{mov} is most
strongly affected by the buoyancy-induced turbulence, although some impact
from the shear-induced turbulence is evident.

Let us consider an example application of the GRPM method to sound
propagation. In this example, the transmission loss (TL, see section 10.1)
is calculated using a narrow-angle parabolic equation (PE), which will be
discussed in more detail in section 11.2.1. The source height is 1.5 m and
the frequency is 250 Hz. Atmospheric parameters are characteristic of mostly
cloudy, high wind (neutral) conditions; specifically, $u_* = 0.6$ m/s, $Q_H = Q_E = 0$ W/m^2, and $z_i = 1000$ m. The atmospheric profiles are modeled using
MOST, as described in section 2.2.3. Ground properties are characteristic of

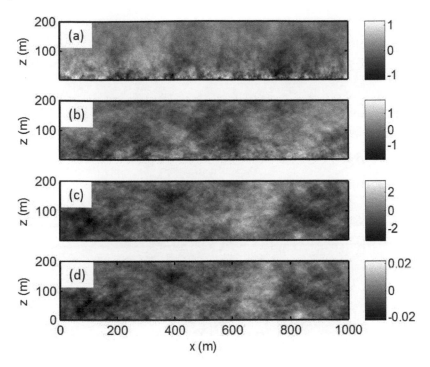

FIGURE 9.3
Two-dimensional realizations of turbulent fluctuations in the atmospheric surface layer based on von Kármán spectra with height-dependent model parameters. Considered are sunny daytime conditions, $Q_H = 200$ W/m^2 and
$z_i = 1000$ m, with a moderate wind, $u_* = 0.3$. (a) Temperature field (gray
scale in °K). (b) Wind velocity field, x-component, for shear-generated turbulence (gray scale in m/s). (c) Wind velocity field, x-component, for buoyancy-
generated turbulence (gray scale in m/s). (d) The field $\varepsilon_{\mathrm{mov}}$ (gray scale in
m/s).

loose soil with high grass, namely the static flow resistivity is $\sigma = 20$ kPa
s/m^2, the porosity $\Omega = 0.515$, and the roughness height $z_0 = 0.05$ m. With
the relaxation model to be described in section 10.2, these values yield a
normalized ground impedance $Z/\varrho_0 c_0 = 8.77 + i8.39$.

Shown in figure 9.4(a) is a random realization of the turbulent velocity
fluctuations for this atmospheric condition. The vertical axis is discretized
at a resolution of 0.5 m between the ground and an altitude of 100 m; thus
the eigenanalysis is performed on a 200×200 matrix. 300 horizontal modes
are retained. Next, figure 9.4(b) shows the PE calculation for propagation
through the velocity field, as determined by adding the turbulence field in
figure 9.4(a) to the logarithmic mean wind profile. Upwind of the source, a
strong shadow zone forms, into which sound waves are randomly scattered

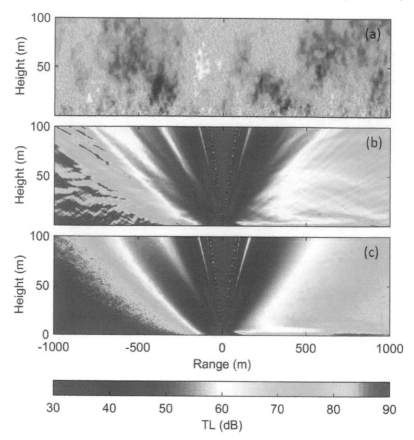

FIGURE 9.4
Turbulent velocity fluctuations and sound propagation in mostly cloudy, high-wind (neutral) atmospheric conditions. The source height is 1.5 m and the frequency 250 Hz. Upwind is to the left, and downwind to the right, with the source in the middle. (a) Random realization of the turbulent velocity field. (b) Narrow-angle PE calculation corresponding to the same realization shown in (a). (c) Average of PE calculations for 1024 random turbulence realizations.

by turbulence. Downwind, the sound levels are considerably higher, although substantial scattering is still evident. (The TL for this case, without turbulent scattering, is shown in figure 11.9(c).) Lastly, in figure 9.4(c) is shown the result of averaging (based on the squared sound pressure) 1024 random realizations created in the same manner as figure 9.4(b).

9.2 Eddy (quasi-wavelet) methods

Turbulence is often conceived as an assemblage of coherent eddy structures. Townsend [381], for example, conceptualized shear-layer turbulence with inclined cylindrical vortices, and linked this qualitative description to the observed properties of the correlation function of the velocity field. Several subsequent authors have explored eddy-based models in the context of wave scattering by turbulence [92, 141, 247]. An advantage of the eddy-based models, in comparison to the spectral approach described in section 9.1, is that the scatterers are spatially localized, which facilitates theoretical and numerical calculations [426], and can aid intuition of the wave scattering processes. Spatial localization also facilitates parallelized computational procedures for synthesizing random fields.

A further promising application for eddy-based models lies in addressing the spatial unevenness, or *intermittency*, of activity in random fields. Turbulent intermittency has been an active topic of research for several decades (e.g., references [204, 235, 279, 362]), as has been its impacts on wave propagation (e.g., references [122, 147, 165, 376, 434]). This phenomenon involves the concentration of turbulent activity into progressively smaller regions of space as large eddies decay into smaller ones. Spectral methods, in which the Fourier modes are modeled with independent phases, are incapable of replicating intermittency. Because eddy-based models are constructed in the spatial domain, however, clustering can be readily incorporated.

In this section, we formulate an eddy-based modeling approach which associates the eddies with objects called *quasi-wavelets* (QWs). Wavelets (e.g., reference [380]) can be convenient analogs for physical objects because they are spatially localized and their mathematical properties are well understood. The prefix *quasi* is added here to indicate that the functions with which we represent the objects are not, strictly speaking, always true wavelets. In particular, wavelet admissibility criteria such as possessing a zero mean may be relaxed, and the QWs may be spherically symmetric in multiple dimensions, which is not normal practice in wavelet analysis. Like ordinary wavelets, QWs are always based on translations and dilations of a parent function; however, their positions and orientations are randomized. No quasi-wavelet transform exists in the same sense of a wavelet transform.

The QW model, as presented in the following sections, had its origins in the *turbule* scattering model of Goedecke and Auvermann [141]. Over time, more rigorous connections were drawn between the turbule model and realistic properties of atmospheric turbulence, and the quasi-wavelet point-of-view was adopted. This led to formulas for appropriately selecting the QW parent function, the size distributions, number densities, and amplitudes [142, 143]. Approaches were developed for synthesizing turbulence fields with realistic correlations, anisotropy, and inhomogeneity [425]. Self-similarity and fractal concepts, as introduced by Mandelbrot [236], Frisch et al. [123], and other researchers, were integrated with the QW model [424, 431]. The following sections provide an overview and synthesis of these many previous articles on QW-based modeling of turbulence.

9.2.1   Self-similar scalar model

QWs, like ordinary wavelets, are derived from translations and dilations of a parent function that is localized in space. We assume for the time being that the QW parent function f is spherically symmetric. That is, f depends only on the magnitude χ of the vector $\boldsymbol{\chi} \equiv (\mathbf{R} - \mathbf{b}^{mn})/a_n$, where $\mathbf{R}$ is the spatial coordinate, $\mathbf{b}^{mn}$ is the location of the center of the QW, and a_n is a length scale, which can be taken as the radius of the QW or other convenient measure. The index n ($n = 1, \ldots, N$) indicates the *size class* of the QW, with $n = 1$ being the largest size (the *outer* scale) and $n = N$ the smallest (the *inner* scale). The index m indicates a particular QW belonging to the size class n ($m = 1, \ldots, N_n$, where N_n is the number of QWs in the size class n within the observation volume V). A superscript mn indexes a particular QW, whereas a subscript n indicates a property associated with all QWs of a particular size class.

The parent function is normalized such that

$$\int f^2(\chi)\, d^3\chi = 1. \tag{9.33}$$

(The integration limits in the preceding equation, and elsewhere in this section, are implicitly infinite unless otherwise indicated.) The following Gaussian parent function meets this normalization condition, and is convenient for analysis and rapid calculation on computers:

$$f(\chi) = \exp\left(-\frac{\pi\chi^2}{2}\right). \tag{9.34}$$

This parent function is illustrated in figure 9.5(a). Other choices for spherically symmetric parent functions are of course possible. The Gaussian parent function has the potential drawback of not possessing a zero mean, and thus not being localized in a spectral sense. An alternative that overcomes this drawback can be derived by applying the Laplacian operator to the Gaussian function, thus yielding the so-called "Mexican-hat" QW [425]:

$$f(\chi) = \frac{4}{15}(3 - \chi^2)\exp\left(-\frac{\pi\chi^2}{2}\right). \tag{9.35}$$

This parent function is illustrated in figure 9.5(b). Several other possibilities for parent functions have been considered, including an exponential, and one involving Bessel functions, which leads exactly to a von Kármán spectral model [142].

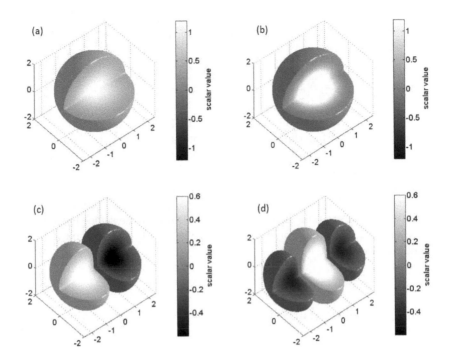

FIGURE 9.5
Scalar QWs derived from a Gaussian parent function. Sections of the QWs have been cut away to show the internal structure. Spatial coordinates are in a normalized system (χ in the text). (a) Monopole QW. (b) Twice-differentiated ("Mexican-hat") monopole QW. (c) Dipole QW ($\mathbf{d}^{\alpha n}$ horizontal). (d) Longitudinal quadrupole QW ($\mathbf{D}^{\alpha n}$ and $\mathbf{d}^{\alpha n}$ horizontal). (The QWs in (c) and (d) are discussed in section 9.2.5.)

Let us consider now the synthesis of a scalar field $Q(\mathbf{R})$. The field perturbation associated with an individual QW is written as

$$Q^{mn}(\mathbf{R}) = h^{mn} q_n f\left(\frac{|\mathbf{R} - \mathbf{b}^{mn}|}{a_n}\right), \qquad (9.36)$$

where h^{mn} is a random sign factor equal to $+1$ or -1 (with equal probability), and q_n is an amplitude factor for the size class. The total field is then constructed by summing the contributions of each individual QW in the ensemble:

$$Q(\mathbf{R}) = \sum_{n=1}^{N} \sum_{m=1}^{N_n} Q^{mn}(\mathbf{R}). \qquad (9.37)$$

The quantity $V_n = N_n a_n^3$ corresponds to the total effective volume occupied by QWs of size a_n; hence V_n/V may be regarded as the effective fraction of V occupied by QWs of size a_n. This motivates the definition of the *packing fraction* as

$$\phi_n = \frac{N_n a_n^3}{V} = \overline{N}_n a_n^3, \tag{9.38}$$

where $\overline{N}_n = N_n/V$ is the number density (number per unit volume) for the size class. Note that the packing fraction is unaffected by overlap among individual QWs; it simply reflects the number of QWs in a size class per unit volume, scaled by the effective size of QWs in that size class.

A self-similar ensemble of QWs is constructed according to rules that are independent of the size class n (excepting perhaps the largest and smallest size classes). Assuming self-similarity, the ratio of the scales between adjacent classes, a_{n+1}/a_n, must be independent of n. Thus an invariant ratio ℓ can be defined as follows:*

$$\ell = \frac{a_{n+1}}{a_n}. \tag{9.39}$$

The parameter ℓ can be thought of as the ratio of length scales between *generations* in the model.† From equation (9.39), it follows that

$$a_n = a_1 \ell^{n-1}. \tag{9.40}$$

We also define additional quantities associated with the QWs, such as amplitudes and energies, and relate these to the length scales and number densities. Ratios of such properties between adjacent size classes must be invariant in order to preserve self-similarity. The ratio of packing fractions between adjacent size classes,

$$\phi = \frac{\phi_{n+1}}{\phi_n}, \tag{9.41}$$

is of particular importance. A decreasing packing fraction ($\phi < 1$) indicates that activity is concentrated in less volume as the size of the QWs decreases; this relates to the fractal dimension of the process.‡ Presuming that ϕ_1 and ϕ are known, we can determine the packing fraction for any size class n as $\phi_n = \phi_1 \phi^{n-1}$.

The value of ϕ is dependent upon the choice of ℓ. To formulate this linkage,

---

*In previous formulations of QW models (e.g., reference [143]), the ratio a_{n+1}/a_n was set to $e^{-\mu}$, where μ is an adjustable parameter equal to $-\ln \ell$. Adoption of ℓ, as done here, helps to simplify the model presentation.

†In the construction of many fractals and the "beta" model for turbulence [123], for example, at each iteration of the construction a_{n+1} is set to $a_n/2$, which thus implies $\ell = 1/2$, and each generation is half the size of the preceding one.

‡When $\ell = 1/2$, the ratio ϕ_{n+1}/ϕ_n corresponds to the parameter β in the beta model for turbulence [123].

we first observe that self-similarity implies a power-law dependence on the length scale a. Defining $\phi(a)$ as the packing fraction at a (that is, $\phi(a_n) = \phi_n$), we have $\phi(a) \propto a^\beta$, where β is an invariant parameter. Then

$$\frac{\phi_{n+1}}{\phi_n} = \left(\frac{a_{n+1}}{a_n}\right)^\beta = \ell^\beta. \tag{9.42}$$

Comparing to equation (9.41), we deduce

$$\phi = \ell^\beta. \tag{9.43}$$

The ratio of number densities between adjacent classes, $\mathcal{N} = \overline{N}_{n+1}/\overline{N}_n$, follows from equations (9.38) and (9.39) as:

$$\mathcal{N} = \phi\ell^{-3}. \tag{9.44}$$

The parameters ϕ_1 and ϕ relate to the intermittency of the QW model construction. Mahrt [233] proposed classifying intermittency into *global* and *intrinsic* types; such a distinction is conceptually very useful in the context of the QW model. Global intermittency is associated with the production of objects at the outer (largest) scale. In regions of space where such large-scale production is absent, there is no cascade of energy, and thus activity at all scales is nearly absent. The density of large-scale production is controlled by the value of ϕ_1; a value much less than 1 implies strong global intermittency. Intrinsic intermittency, on the other hand, is associated with the cascade process itself. When the cascade process concentrates offspring into smaller regions of space relative to their parents, small objects tend to occur in clusters of activity, whereas relatively larger objects are relatively evenly distributed. The intrinsic intermittency is controlled by the value of ϕ; a small value implies that the QWs occupy less volume as the size scale decreases.

Each QW in the size class n has a characteristic amplitude q_n as indicated in equation (9.36). For self-similarity, the ratio $q = q_{n+1}/q_n$ must be invariant. Analogously to the packing fraction, we define a power-law exponent λ such that $q(a) \propto a^\lambda$. It follows that

$$q = \ell^\lambda. \tag{9.45}$$

Let us consider constraints between the scaling parameters pertinent to the modeling of turbulence. According to Kolmogorov's original second hypothesis [203], in steady-state turbulence the transfer rate of specific turbulent kinetic energy (TKE) from one scale to the next is invariant, and equal to the dissipation of TKE at the molecular scale, ε. Let us define ε_n as the net transfer rate of energy (the *flux*) from scale n to $n+1$. This quantity has dimensions velocity^2 time^{-1}, or length^2 time^{-3}. Defining τ_n as the time scale associated with the turbulent transfer from size class n to $n+1$, by dimensional analysis one has $\varepsilon_n = c_v\phi_n a_n^2/\tau_n^3$, where c_v is a dimensionless constant. The factor

ϕ_n adjusts for the volume occupied by the QWs, so that ε_n represents the transfer specific to the active volume. Setting $\varepsilon_{n+1} = \varepsilon_n = \varepsilon$, we find

$$\tau^3 = \phi\ell^2, \tag{9.46}$$

where $\tau = \tau_{n+1}/\tau_n$.

A similar argument can be made regarding the transfer rate of scalar variance from one scale to the next. Let us define θ as the rate of variance destruction at the molecular scale. The transfer rate of variance from scale n to $n + 1$, θ_n, must equal θ for all n. By dimensional analysis, $\theta_n = c_q\phi_n q_n^2/\tau_n$, where c_q is a dimensionless constant. Setting $\theta_{n+1} = \theta_n = \theta$, we find $\phi q^2 = \tau$. Equation (9.46) can then be used to eliminate τ, which results in

$$q^3 = \ell/\phi. \tag{9.47}$$

From equations (9.43) and (9.45), we thus have the following constraint for turbulence:[*]

$$\lambda = (1 - \beta)/3. \tag{9.48}$$

Up to this point, four distinct ratios between adjacent QW size classes have been introduced: ℓ, ϕ, $\mathcal{N}$, and q. Only three of these ratios may be regarded as independent, due to equation (9.44). (The ratio τ is not counted here, since it was introduced specifically for turbulence and was shown not to be independent.) Furthermore, one of the ratios must be regarded as a *reference* ratio for defining a generation of the QW process, which otherwise would be completely arbitrary. Hence only two of the ratios may be considered as truly adjustable model parameters. In the following, we will follow the precedent of reference [123] (and many others on construction of fractals) by setting ℓ to a fixed value, conventionally $1/2$. For the two adjustable parameters in the model, we select β and λ, since these power-law exponents are independent of the method for defining a generation. The remaining model parameters can now all be calculated from ℓ, β, and λ, using equations (9.43), (9.44), and (9.45). There are still a number of other free parameters in the model representing the largest size class, namely a_1, ϕ_1, and q_1, as well as a_N, representing the smallest size class. The number of size classes N can be determined from a_1 and a_N.

A consequence of constraining the size ratio between adjacent classes to a fixed value ℓ is that the resulting sequence of QW sizes may be undesirably coarse. To address this problem, *fractional* generations can be introduced to create a more gradual variation in the sizes. The basic idea is simply to allow the generation (class) index n in equation (9.40) and similar equations to assume non-integer values. Constant ratios are still maintained, so that the length scales still occur in a geometric series. Specifically, we allow $n = 1 + (j - $

---

[*]It will be shown, in section 9.2.3, that the case $\beta = 0$ and $\lambda = 1/3$ corresponds to Kolmogorov's well-known "$-11/3$ law" for the spectrum.

$1)/K$, where K is called the *scale densification factor* and $j = 1, 2, \ldots, NK$. A factor of K more size classes are present than would occur for $K = 1$. For example, a cascade process with three generations ($N = 3$) and no scale densification ($K = 1$) would have the three length scales a_1, $a_2 = \ell a_1$, and $a_3 = \ell^2 a_1$. When this process is densified to $K = 2$, additional size classes would be introduced at $a_{3/2} = \ell^{1/2} a_1$, $a_{5/2} = \ell^{3/2} a_1$, and $a_{7/2} = \ell^{5/2} a_1$. Note that these additional size classes form a new generational sequence, i.e., $a_{5/2} = \ell a_{3/2}$ and $a_{7/2} = \ell^2 a_{3/2}$.

The scale densification should be formulated so that it has an insignificant net impact on the number of QWs. This motivates the following generalization of the packing fraction, equation (9.38), for the densified model:[*]

$$\phi_n = \frac{K N_n a_n^3}{V} = K \overline{N}_n a_n^3. \tag{9.49}$$

When $K = 1$, this definition coincides with equation (9.38). When the number of size classes is increased by a factor K, however, N_n must be diminished by a factor of $1/K$ to preserve the packing fraction. Thus the scale densification does not alter the packing fractions.

9.2.2 Random field realizations

Having formulated the various scaling parameters of the QW model, and determined their interrelationships, we now turn to the problem of generating random realizations of fields from ensembles of QWs. Three basic types of model constructions may be considered [431]: *static*, *steady-state*, and *non-steady*. The static construction does not explicitly incorporate evolution in time; it provides a single, independent realization of the random medium. In the steady-state construction, the QWs are dynamically created and destroyed through a cascade process, but creation and destruction are in equilibrium for each size class. The non-steady construction generalizes the steady-state model to situations in which the production of eddies is unsteady, and hence the cascade process is not in equilibrium.

In this book we describe only the static model. The basic procedure for generating a static realization is as follows:

1. Select the parent function.

2. Select the desired values of the outer scale a_1, the ratio ℓ (normally $1/2$), the number of generations N, and the densification factor K. Set the scales of the size classes a_n using equation (9.40).

3. Select the desired values of the power-law exponents for the packing fraction and amplitude, β and λ. Calculate ϕ and q from equations (9.43) and (9.45). Set $\phi_n = \phi_1 \phi^{n-1}$ and $q_n = q_1 q^{n-1}$.

---

[*]The densification factor K was not incorporated into the packing fraction in earlier formulations of the QW model.

4. Use equation (9.49) to determine the number of QWs in each size class, N_n.

5. For each size class, generate N_n random QW positions (the $\mathbf{b}^{mn}$) and random sign factors (the h^{mn}).

6. Calculate the field associated with each QW (equation (9.36)) and sum these to obtain the total field (equation (9.37)).

For the random QW positions in step 3, many spatial random process yielding a mean value of N_n can conceivably be used. The simplest approach is to simulate the spatial placements of QWs using a spatially homogeneous Poisson process with spatial rate $\overline{N}_n = N_n/V$. Conceptually, the overall volume V is partitioned into a large number of subvolumes with size δV, such that $\delta V \overline{N}_n \ll 1$. (This condition is necessary so that the probability of more than a single QW occurring in the subvolume is negligible.) Realizations of the total number of QWs, of size class n occurring in V, then follow a Poisson distribution [218]:

$$P_{N_n}(\nu) = \frac{N_n^\nu e^{-N_n}}{\nu!}, \qquad (9.50)$$

where ν is the number of random QWs. Numerically, we could synthesize the positions of QWs for the Poisson process by a couple approaches. First, and more directly, we could partition V into subvolumes as previously described. Then, within each subvolume, a random number ϵ is drawn with uniform distribution between 0 and 1. If $\epsilon < \delta V \overline{N}_n$, a QW is placed within the subvolume. This process must be repeated for each subvolume. Second, but more efficiently, we could generate a random QW count ν from the Poisson distribution equation (9.50). The ν QWs are then placed at random locations in V using a uniform random number generator. This second approach circumvents partitioning and random number generation for a large number of small subvolumes. QW fields generated by either of these two approaches are fully disorganized, in the sense that the positions of the QWs are mutually independent and uniformly random over space.

Figure 9.6 illustrates the construction of a 6-generation static QW field. The sequence incorporates progressively more generations (1, 2, 3, and finally 6). These realizations were calculated with $\ell = 1/2$, $K = 4$, $\beta = 0$, $\lambda = 1/3$, $a_1 = 1$, $\phi_1 = 0.1$, and $q_1 = 1$. Individual QWs are clearly evident in the first few generations. After six generations, however, the individual contributions tend to be lost in the general random appearance of the field.

9.2.3   Second-order statistics

Second-order statistics (such as variances, correlation functions, and spectral densities) can be derived analytically for the QW model. The following presentation consolidates results from several previous articles [142, 143, 424, 425].

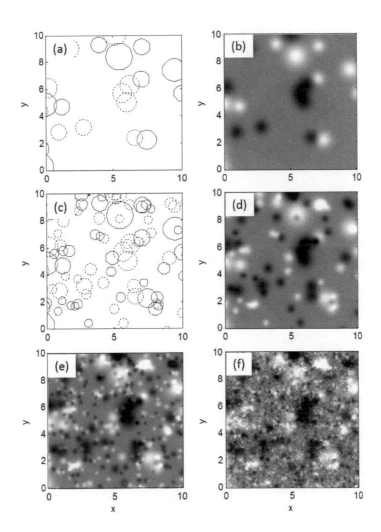

FIGURE 9.6

Construction of a static QW field with $N = 6$ generations and a densification factor $K = 4$. (a) Positions and radii of the QWs in the first generation. (b) Realization based only on the first generation. (c) Positions and radii of the QWs of the first two generations. (d) Realization based on the first two generations. (e) Realization based on the first three generations. (f) Final realization, based on all six generations. The initial class has size $a_1 = 1$ and amplitude $q_1 = 1$. In (a) and (c), dashed lines indicate negatively signed QWs; solid lines are positive. The gray scale ranges from -1 (black) to 1 (white).

We begin with the Fourier transform for a realization of the field, namely

$$\widetilde{Q}\left(\boldsymbol{\kappa}\right) = \frac{1}{\left(2\pi\right)^3} \int Q\left(\mathbf{R}\right) e^{-i\boldsymbol{\kappa}\cdot\mathbf{R}} d^3 R. \tag{9.51}$$

Since this transformation is a linear operation, equation (9.37) indicates that the spectra of the individual QWs add linearly to produce the total spectrum:

$$\widetilde{Q}\left(\boldsymbol{\kappa}\right) = \sum_{n=1}^{N}\sum_{m=1}^{N_n}\widetilde{Q}^{mn}\left(\boldsymbol{\kappa}\right). \tag{9.52}$$

Transformation of equation (9.36) yields the spectrum for an individual QW:

$$\widetilde{Q}^{mn}\left(\boldsymbol{\kappa}\right) = a_n^3 q_n \exp\left(-i\boldsymbol{\kappa}\cdot\mathbf{b}^{mn}\right) F\left(\kappa a_n\right), \tag{9.53}$$

where

$$F\left(y\right) = \frac{1}{\left(2\pi\right)^3}\int f\left(\chi\right) e^{-i\mathbf{y}\cdot\mathbf{\chi}} d^3\chi \tag{9.54}$$

is the spectrum of the parent function and $\mathbf{y} = \boldsymbol{\kappa} a$ is the normalized wave vector. For the Gaussian parent function, equation (9.34), we find

$$F\left(y\right) = \left(\sqrt{2\pi}\right)^{-3}\exp\left(-y^2/2\pi\right). \tag{9.55}$$

By Parseval's theorem (which equates variances in the spatial and wavenumber domains), the normalization condition, equation (9.33), requires

$$\left(2\pi\right)^3\int F^2\left(y\right)d^3 y = 1. \tag{9.56}$$

Substituting equation (9.52) into (9.1), and assuming statistical independence of the individual QWs, yields

$$\Phi\left(\boldsymbol{\kappa}\right) = \sum_{n=1}^{N}\Phi_n\left(\boldsymbol{\kappa}\right), \tag{9.57}$$

where

$$\Phi_n\left(\boldsymbol{\kappa}\right) = \left(2\pi\right)^3\frac{N_n}{V}q_n^2 a_n^6 F^2\left(\kappa a_n\right) = \frac{\left(2\pi\right)^3}{K}\phi_n q_n^2 a_n^3 F^2\left(\kappa a_n\right) \tag{9.58}$$

is the contribution to the spectral density from size class n. By integrating over wavenumber and applying equation (9.56), we find the contribution to the variance from size class n:

$$\sigma_n^2 = \int \Phi_n\left(\kappa\right)d^3 k = \frac{N_n}{V}q_n^2 a_n^3 = \frac{1}{K}\phi_n q_n^2. \tag{9.59}$$

The correlation function is defined by equation (6.3) as

$$B(\mathbf{R}) = \langle Q(\mathbf{R}_0) Q(\mathbf{R}_0 + \mathbf{R}) \rangle, \tag{9.60}$$

where $\mathbf{R}_0$ is a reference observation point. In a homogeneous, isotropic medium, the correlation function depends only on the magnitude of $\mathbf{R}$. By the Wiener–Khinchin theorem [29], the correlation function is the Fourier transform of the spectral density, namely

$$B(\mathbf{R}) = \int \Phi(\boldsymbol{\kappa}) e^{i\boldsymbol{\kappa}\cdot\mathbf{R}} d^3\kappa. \tag{9.61}$$

Since the Fourier transform is a linear operator, we can decompose $B(R)$ by size class as in equation (9.57), with result

$$B(R) = \sum_{n=1}^{N} B_n(R), \tag{9.62}$$

where, from equation (9.58),

$$B_n(R) = \frac{\phi_n q_n^2}{K} f_2\left(\frac{R}{a_n}\right), \tag{9.63}$$

in which we have defined

$$f_2(\chi) = (2\pi)^3 \int F^2(y) e^{i\mathbf{y}\cdot\boldsymbol{\chi}} d^3y. \tag{9.64}$$

The normalization condition, equation (9.56), implies that $f_2(0) = 1$. Hence $B_n(0) = \phi_n q_n^2/K$, which equals σ_n^2 as given by equation (9.59), as it must. For the Gaussian parent function, substitution with equation (9.55) (and the assistance of a standard table of Fourier transforms) yields

$$f_2(\chi) = \exp\left(-\frac{\pi\chi^2}{4}\right). \tag{9.65}$$

The overall spectral density, correlation function, and variance are found by summing over the size classes. In the limit of a highly densified representation ($K \to \infty$), the summation over classes transforms to an integration. Such an integral representation, with the QW size as the integration variable, can be derived as follows. Since $a_n = a_1 \ell^{n-1}$, we have $a_{n+\Delta n} = a_1 \ell^{n+\Delta n - 1}$, where $n + \Delta n = n + 1/K$ is the size class following n. We then find

$$\frac{a_{n+\Delta n}}{a_n} = \ell^{1/K}. \tag{9.66}$$

Defining $\Delta a_n = a_{n+\Delta n} - a_n$, it follows that $1 + \Delta a_n/a_n = \ell^{1/K}$. Taking the logarithm of both sides and assuming $\Delta a_n \ll a_n$, we have

$$\Delta a_n \simeq \frac{a_n}{K} \ln \ell. \tag{9.67}$$

Finally, through substitution of equation (9.38),

$$\Phi\left(\kappa\right) = \sum_{\text{classes}} \Phi_n\left(\kappa\right) = \frac{\left(2\pi\right)^3}{\ln\ell} \sum_{\text{classes}} \phi_n q_n^2 F^2\left(\kappa a_n\right) a_n^2 \Delta a_n, \tag{9.68}$$

$$B\left(R\right) = \sum_{\text{classes}} B_n\left(R\right) = \frac{1}{\ln\ell} \sum_{\text{classes}} \phi_n q_n^2 f_2\left(\frac{R}{a_n}\right) \frac{\Delta a_n}{a_n}, \tag{9.69}$$

and

$$\sigma^2 = \sum_{\text{classes}} \sigma_n^2 = \frac{1}{\ln\ell} \sum_{\text{classes}} \phi_n q_n^2 \frac{\Delta a_n}{a_n}. \tag{9.70}$$

The summations are to be performed over all size classes, including the non-integer classes. In the preceding equations, ϕ_n and q_n are implicitly functions of a_n, which follow from equations (9.43) and (9.45):

$$\phi_n = \phi_1\left(\frac{a_n}{a_1}\right)^\beta, \tag{9.71}$$

and

$$q_n = q_1\left(\frac{a_n}{a_1}\right)^\lambda. \tag{9.72}$$

Hence we have for the spectral density, correlation function, and variance, respectively,

$$\Phi\left(\kappa\right) = \frac{\left(2\pi\right)^3 \phi_1 q_1^2}{a_1^{\beta+2\lambda} \ln\ell} \sum_{\text{classes}} F^2\left(\kappa a_n\right) a_n^{\beta+2\lambda+2} \Delta a_n, \tag{9.73}$$

$$B\left(R\right) = \frac{\phi_1 q_1^2}{a_1^{\beta+2\lambda} \ln\ell} \sum_{\text{classes}} f_2\left(\frac{R}{a_n}\right) a_n^{\beta+2\lambda-1} \Delta a_n, \tag{9.74}$$

and

$$\sigma^2 = \frac{\phi_1 q_1^2}{a_1^{\beta+2\lambda} \ln\ell} \sum_{\text{classes}} a_n^{\beta+2\lambda-1} \Delta a_n. \tag{9.75}$$

Setting $y = \kappa a_n$ in equations (9.73) and (9.75), and taking the limit $\Delta a_n = \Delta y/\kappa \to 0$, results in

$$\Phi\left(\kappa\right) = \frac{\left(2\pi a_1\right)^3 \phi_1 q_1^2}{\left(\kappa a_1\right)^{3+\beta+2\lambda} \ln\ell} \int_{\kappa a_1}^{\kappa a_N} F^2\left(y\right) y^{\beta+2\lambda+2}\, dy \tag{9.76}$$

and

$$\sigma^2 = \frac{\phi_1 q_1^2}{\left(\kappa a_1\right)^{\beta+2\lambda} \ln\ell} \int_{\kappa a_1}^{\kappa a_N} y^{\beta+2\lambda-1}\, dy. \tag{9.77}$$

Similarly, setting $\chi = R/a_n$ in equation (9.74), and taking the limit $\Delta a_n = R/\Delta\chi \to 0$, results in

$$B\left(R\right) = -\frac{\phi_1 q_1^2 \left(R/a_1\right)^{\beta+2\lambda}}{\ln \ell} \int_{R/a_1}^{R/a_N} f_2\left(\chi\right) \chi^{-\beta-2\lambda-1} \, d\chi. \tag{9.78}$$

We observe that all second-order statistics are proportional to $\phi_1 q_1^2$. Hence decreasing the density of the QWs while increasing q_1^2 proportionately (such that $\phi_1 q_1^2$ is held constant) will not affect the second-order statistics of the field. However, such adjustments will affect higher-order statistics such as the kurtosis [143].

For the variance, the integration in equation (9.77) is easily performed, with result

$$\sigma^2 = -\frac{\phi_1 q_1^2}{(\beta + 2\lambda) \ln \ell} \left[1 - \left(\frac{a_N}{a_1}\right)^{\beta+2\lambda}\right]. \tag{9.79}$$

Since $\ln \ell$ is negative and $a_N < a_1$, the variance is always positive. The second term in square brackets is negligible when $a_N \ll a_1$, as is normally the situation of interest.*

The integrals for the spectral density and correlation function depend on the choice of parent function. For the Gaussian parent function, substitution of equation (9.55) into (9.76) yields

$$\Phi\left(\kappa\right) = -\frac{\phi_1 q_1^2 a_1^3}{2\pi^3 \ln \ell} \left(\frac{\kappa a_1}{\sqrt{\pi}}\right)^{-3-\beta-2\lambda}$$
$$\times \left\{\gamma\left[\frac{3+\beta+2\lambda}{2}, \frac{\kappa^2 a_1^2}{\pi}\right] - \gamma\left[\frac{3+\beta+2\lambda}{2}, \frac{\kappa^2 a_N^2}{\pi}\right]\right\}, \tag{9.80}$$

where

$$\gamma(s, x) = \int_0^x t^{s-1} e^{-t} \, dt \tag{9.81}$$

is the lower incomplete gamma function. Substitution of equation (9.65) into (9.78) yields

$$B\left(R\right) = -\frac{\phi_1 q_1^2}{2 \ln \ell} \left(\frac{\sqrt{\pi}R}{2a_1}\right)^{\beta+2\lambda}$$
$$\times \left\{\Gamma\left[-\frac{\beta+2\lambda}{2}, \frac{\pi R^2}{4a_1^2}\right] - \Gamma\left[-\frac{\beta+2\lambda}{2}, \frac{\pi R^2}{4a_N^2}\right]\right\}, \tag{9.82}$$

---

*This result is equivalent, for example, to equation (33) in reference [424], after making the replacement $-\ln \ell = K\mu$ and accounting for the incorporation of K into the packing fraction. Furthermore, $2/3 + \nu$ in that paper is equivalent to $\beta + 2\lambda$ here, and a change in the definition of the parent function leads to a constant factor of $8\pi^3$.

where $\Gamma(s,x) = \Gamma(s) - \gamma(s,x)$ is the upper incomplete gamma function.*

The correlation function and spectrum generally possess three regions with distinctive dependence on spatial separation or wavenumber: $R \gg a_1$ ($\kappa^{-1} \gg a_1$), called the *energy-containing* subrange, $a_1 \gg R \gg a_N$ ($a_1 \gg \kappa^{-1} \gg a_N$), called the *inertial* (self-similar) subrange, and $a_N \gg R$ ($a_N \gg \kappa^{-1}$), called the *dissipation* subrange. In order to determine the behavior of the spectrum in these three regions, we can apply the approximations $\gamma(s,x) \to x^s/s$ for $x \to 0$, and $\gamma(s,x) \to \Gamma(s)$ for $x \to \infty$, to the terms in involving a_1 and a_N in equation (9.80). For the energy-containing subrange, the small-argument approximation ($x \to 0$) applies to both terms; assuming $a_1 \gg a_N$, the term involving a_N becomes negligible in comparison to the one involving a_1, thus yielding $\Phi(\kappa) \simeq \sigma^2 a_1^3 \pi^{-3}(\beta + 2\lambda)/(3 + \beta + 2\lambda)$. For the inertial subrange, the large-argument approximation now applies to the term involving a_1, while the term involving a_N can still be neglected, yielding

$$\Phi(\kappa) \simeq \frac{\sigma^2 a_1^3 (\beta + 2\lambda)\Gamma[(3 + \beta + 2\lambda)/2]}{2\pi^3} \left(\frac{\kappa a_1}{\sqrt{\pi}}\right)^{-3-\beta-2\lambda}. \tag{9.83}$$

The spectrum is thus proportional to $\kappa^{-3-\beta-2\lambda}$ in the inertial subrange. For the dissipation subrange, the large-argument approximation applies to both terms, and they approximately cancel.

Considering next the correlation and structure functions (the latter having been defined by equation (6.12)), in the energy-containing subrange, the large-argument approximations apply, and we find that $B(R) \to 0$ and $D(R) \to 2\sigma^2$. In the inertial subrange, the small-argument approximation applies to the a_1 term and the large-argument approximation to the a_N term, with result

$$B(R) \simeq \sigma^2 \left[1 - \Gamma\left(1 - \frac{\beta + 2\lambda}{2}\right)\left(\frac{\sqrt{\pi}R}{2a_1}\right)^{\beta+2\lambda} \right]. \tag{9.84}$$

Thus, for the structure function,

$$D(R) \simeq 2\sigma^2 \Gamma\left(1 - \frac{\beta + 2\lambda}{2}\right)\left(\frac{\sqrt{\pi}R}{2a_1}\right)^{\beta+2\lambda}. \tag{9.85}$$

The structure function is thus proportional to $R^{\beta+2\lambda}$ in the inertial subrange. In the dissipation subrange, we find that $B(R) \to \sigma^2$ and $D(R) \to 0$.

Recalling that $\lambda = (1 - \beta)/3$ for 3D turbulence (equation (9.48)), the spectrum in the inertial subrange is proportional to $\kappa^{-11/3-\beta/3}$. When $\beta = 0$ (i.e., there is no intrinsic intermittency), the spectrum thus decays as $-11/3$, which is consistent with Kolmogorov's second hypothesis for turbulence [203]. The exponent $-\beta/3$ is an adjustment representing intrinsic intermittency. For

---

*The primary reason for switching from the lower to the upper incomplete gamma function, when writing an equation for the correlation function, is that the argument $s = -(\beta + 2\lambda)/2$ is negative. The upper function, unlike the lower, allows $s < 0$. Strictly speaking, the equation only holds for $R \neq 0$.

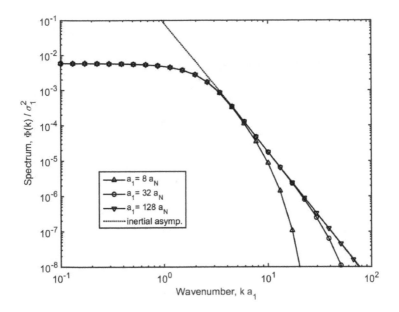

FIGURE 9.7
Spectral density of QW models for turbulence ($\lambda = 1/3$, $\beta = 0$, $\ell = 1/2$) as based upon a Gaussian parent function. Shown are curves for various values of the ratio a_1/a_N, along with the asymptotic result for the inertial subrange.

turbulence, the structure function is proportional to $R^{2/3+\beta/3}$. Comparing equation (9.85) to (6.38) with $\beta = 0$, we find for the structure-function parameter

$$C_T^2 = \frac{(2\pi)^{1/3}\Gamma(2/3)\sigma^2}{a_1^{2/3}}. \tag{9.86}$$

This result enables determination of a_1 in the scalar QW model, when given values for C_T^2 and σ^2.

Spectra for turbulence are shown in figure 9.7. The calculations are based on equation (9.80), with $\lambda = 1/3$ and $\beta = 0$. Curves for various values of a_1/a_N are shown. The spectra have been normalized by the variance for $a_1 \gg a_N$, namely $\sigma^2 = -(\phi_1 q_1^2)/[(\beta + 2\lambda)\ln\ell]$. An inertial subrange is evident when $a_1 \gtrsim 32a_N$. Figure 9.8 compares spectral calculations for discrete numbers of size classes (equations (9.57) and (9.58)) to the continuous calculation based on the integral (equation (9.80)), for the case $a_1 = 128a_N$. The densification factor K is stepped through the values 1, 4, and 16. For $K = 4$, the discrete calculations are quite close to the continuous calculation; for $K = 16$, the discrete calculation is nearly indistinguishable from the continuous one.

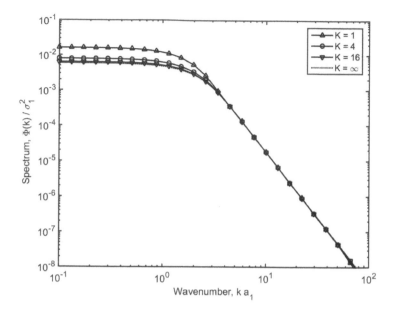

FIGURE 9.8
Spectral density of QW models for turbulence ($\lambda = 1/3$, $\beta = 0$, $\ell = 1/2$) as based upon a Gaussian parent function. Shown are curves for $a_1/a_N = 128$, along with calculations for discrete numbers of size classes with various densification factors ($K = 1$, 4, and 16). Also shown is the calculation based on the integral equation ($K \to \infty$).

9.2.4   Rotational quasi-wavelets for velocity fields

Discussion of the QW model so far has focused on simulation of scalar fields, such as turbulent fluctuations in temperature and humidity. Simulation of vector fields, such as turbulent velocity fluctuations, is also of interest. Since turbulence is a solenoidal (non-divergent) 3D vector field, it is natural to base the model on the curl of a vector parent function. Specifically, consider the vector potential $\mathbf{A}^{mn}(\mathbf{R})$ given by

$$\mathbf{A}^{mn}(\mathbf{R}) = \mathbf{\Omega}^{mn} a_n^2 \, f\left(\chi\right), \tag{9.87}$$

where $\mathbf{\Omega}^{mn}$ is the angular velocity vector of the QW. The factor a_n^2 in the definition provides dimensional consistency. In comparison to the scalar model, the vector QW has an orientation, as specified by $\mathbf{\Omega}^{mn}$. The resulting velocity field of the QW is

$$\mathbf{v}_{\mathrm{T}}^{mn}(\mathbf{R}) = \nabla \times \mathbf{A}^{mn}(\mathbf{R}). \tag{9.88}$$

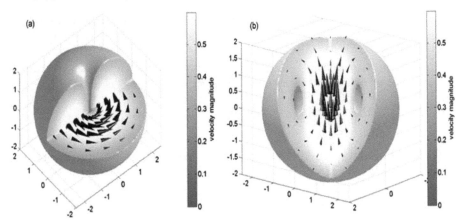

FIGURE 9.9
Velocity QWs derived from a Gaussian parent function. Sections of the QWs
have been cut away to show the internal structure. The angular velocity vector
is oriented vertically. Spatial coordinates are in a system normalized by the
length scale of the QW (χ in the text). Arrows indicate the magnitude and
direction of the flow. (a) Toroidal QW. (b) Poloidal QW.

The subscript T denotes here a *toroidal* QW, as will be explained shortly.
Writing out the curl of the potential leads to

$$\mathbf{v}_{\mathrm{T}}^{mn}(\mathbf{R}) = a_n^2 \nabla f(\chi) \times \mathbf{\Omega}^{mn} = a_n \theta(\chi)(\mathbf{\Omega}^{mn} \times \boldsymbol{\chi}), \tag{9.89}$$

in which $\theta(\chi) = -\chi^{-1}(df/d\chi)$. For the Gaussian parent function, equation (9.34), we find $\theta(\chi) = \pi \exp(-\pi\chi^2/2)$. A velocity QW constructed in
this manner is shown in figure 9.9(a). The flow circulates in planes perpendicular to the direction of $\mathbf{\Omega}^{mn}$ (along lines of constant latitude). The flow is
thus parallel to toroidal surfaces.

A complementary vector QW can be constructed from the curl of the
toroidal one [429]. Specifically, we set

$$\mathbf{v}_{\mathrm{P}}^{mn} = a_n \nabla \times [\nabla \times \mathbf{A}^{mn}(\mathbf{R})]. \tag{9.90}$$

By applying a number of vector identities, the velocity field is determined as

$$\begin{aligned}
\mathbf{v}_{\mathrm{P}}^{mn} &= 2a_n \theta \mathbf{\Omega}^{mn} + a_n \psi (\mathbf{\Omega}^{mn} \times \boldsymbol{\chi}) \times \boldsymbol{\chi} \\
&= a_n (2\theta + \chi^2 \psi) \mathbf{\Omega}^{mn} - (\boldsymbol{\chi} \cdot \mathbf{\Omega}^{mn}) \boldsymbol{\chi} \psi,
\end{aligned} \tag{9.91}$$

where $\psi(\chi) = -\chi^{-1}(d\theta/d\chi)$. For the Gaussian parent function, $\psi(\chi) = \pi^2 \exp(-\pi\chi^2/2)$. This QW is shown in figure 9.9(b). In the vertical orientation shown, there is a strong updraft in the middle, with weak, compensating
downdrafts along the sides. We call this QW *poloidal* since the flow is to and
from the poles. It resembles a vortex ring.

Taking the curl of the poloidal QW again yields a toroidal QW, and so forth. (The parent function of the new toroidal QW would be $2\theta(\chi)+\chi^2\psi(\chi)$, instead of $f(\chi)$.) Hence there are just two distinct varieties of solenoidal vector QWs.

The spectral density tensor $\Phi_{ij}(\boldsymbol{\kappa})$, representing the cross spectrum between the velocity fluctuations in the directions i and j, is defined by

$$\Phi_{ij}(\boldsymbol{\kappa}) = \lim_{V\to\infty} \frac{(2\pi)^3}{V} \langle \tilde{v}_i(\boldsymbol{\kappa})\tilde{v}_j^*(\boldsymbol{\kappa})\rangle. \tag{9.92}$$

For the toroidal velocity QWs, application of a Fourier transform to equation (9.89) leads to

$$\tilde{\mathbf{v}}_{\mathrm{T}}^{mn}(\boldsymbol{\kappa}) = i(\boldsymbol{\kappa}\times\boldsymbol{\Omega}^{mn})\exp(-i\boldsymbol{\kappa}\cdot\mathbf{b}^{mn})\,a_n^5 F(\kappa a_n). \tag{9.93}$$

The contribution of size class n to the spectral density tensor is then

$$\Phi_{ij,n}^{\mathrm{T}}(\boldsymbol{\kappa}) = \frac{(2\pi)^3}{K}\phi_n a_n^7 F^2(\kappa a_n)\langle[(\boldsymbol{\kappa}\times\boldsymbol{\Omega}^{mn})\cdot\mathbf{e}_i][(\boldsymbol{\kappa}\times\boldsymbol{\Omega}^{mn})\cdot\mathbf{e}_j]\rangle, \tag{9.94}$$

in which $\mathbf{e}_i$ is the unit vector along the ith coordinate axis. For vector QWs, the variances and covariances of the components of $\boldsymbol{\Omega}^{mn}$ comprise a 3×3 matrix (tensor); in general, each element of this matrix must be specified to complete the statistical model. For an isotropic model, however, the variances $\langle\Omega_i^{mn}\Omega_i^{mn}\rangle$ must all be equal, whereas the covariances $\langle\Omega_i^{mn}\Omega_j^{mn}\rangle$, $i\neq j$, must vanish. We thus define

$$\Omega_n^2 = \langle(\Omega_1^{mn})^2\rangle = \langle(\Omega_2^{mn})^2\rangle = \langle(\Omega_3^{mn})^2\rangle. \tag{9.95}$$

Expanding terms in equation (9.94) we then have, for example,

$$\Phi_{11,n}^{\mathrm{T}}(\boldsymbol{\kappa}) = \frac{(2\pi)^3}{K}\phi_n\Omega_n^2 a_n^7\left(\kappa_2^2+\kappa_3^2\right)F^2(\kappa a_n). \tag{9.96}$$

The other autospectra, $\Phi_{22,n}^{\mathrm{T}}$ and $\Phi_{33,n}^{\mathrm{T}}$, are given by similar equations with obvious changes in the wavenumber components. For the cross spectrum $\Phi_{13,n}$, we have

$$\Phi_{13,n}^{\mathrm{T}}(\boldsymbol{\kappa}) = -\frac{(2\pi)^3}{K}\phi_n\Omega_n^2 a_n^7\kappa_1\kappa_3 F^2(\kappa a_n), \tag{9.97}$$

and likewise for the other cross spectra. Note that the cross spectra are nonzero, even for the isotropic model.

The transformed velocity associated with a poloidal QW is

$$\begin{aligned}\tilde{\mathbf{v}}_{\mathrm{P}}^{mn}(\boldsymbol{\kappa}) &= ia_n\boldsymbol{\kappa}\times\tilde{\mathbf{v}}_{\mathrm{T}}^{mn}(\boldsymbol{\kappa})\\ &= \left[\boldsymbol{\Omega}^{mn}\kappa^2-\boldsymbol{\kappa}(\boldsymbol{\kappa}\cdot\boldsymbol{\Omega}^{mn})\right]\exp(-i\boldsymbol{\kappa}\cdot\mathbf{b}^{mn})\,a_n^5 F(\kappa a_n).\end{aligned} \tag{9.98}$$

We then find, for the spectral density tensor of poloidal QWs,

$$\Phi_{ij,n}^{P}(\boldsymbol{\kappa}) = \frac{(2\pi)^3}{K}\phi_n a_n^9 F^2(\kappa a_n)\big\langle \big[\Omega_i^{mn}\kappa^2 - \kappa_i\left(\boldsymbol{\kappa}\cdot\boldsymbol{\Omega}^{mn}\right)\big]$$
$$\times \big[\Omega_j^{mn}\kappa^2 - \kappa_j\left(\boldsymbol{\kappa}\cdot\boldsymbol{\Omega}^{mn}\right)\big]\big\rangle. \quad (9.99)$$

This result leads to

$$\Phi_{11,n}^{P}(\boldsymbol{\kappa}) = \frac{(2\pi)^3}{K}\phi_n\Omega_n^2 a_n^9\kappa^2\left(\kappa_2^2 + \kappa_3^2\right)F^2(\kappa a_n) \quad (9.100)$$

and

$$\Phi_{13,n}^{P}(\boldsymbol{\kappa}) = -\frac{(2\pi)^3}{K}\phi_n\Omega_n^2 a_n^9\kappa^2\kappa_1\kappa_3 F^2(\kappa a_n). \quad (9.101)$$

The spectrum of specific turbulent kinetic energy (TKE) was introduced in Chapter 6. By definition [25], $E(\kappa) = [\Phi_{11}(\boldsymbol{\kappa}) + \Phi_{22}(\boldsymbol{\kappa}) + \Phi_{33}(\boldsymbol{\kappa})]\left(2\pi\kappa^2\right)$. Thus the contribution to the energy spectrum from size class n, for toroidal QWs, follows from equation (9.96) as

$$E_n^{T}(\kappa) = \frac{2(2\pi)^4}{K}\phi_n\Omega_n^2 a_n^7\kappa^4 F^2(\kappa a_n). \quad (9.102)$$

For the poloidal QWs, equation (9.100) leads to

$$E_n^{P}(\kappa) = \frac{2(2\pi)^4}{K}\phi_n\Omega_n^2 a_n^9\kappa^6 F^2(\kappa a_n). \quad (9.103)$$

Note that the poloidal spectral density is just $(\kappa a_n)^2$ times the toroidal one. The variance associated with size class n can be determined from equation (6.18), with result

$$\sigma_n^2 = \frac{4\pi^{(1+p)/2}}{3}\Gamma\left(\frac{5+p}{2}\right)\phi_n\Omega_n^2 a_n^2, \quad (9.104)$$

where $p = 0$ for the toroidal and 2 for the poloidal QWs.

By dimensional arguments, the angular velocity Ω_n scales as q_n/a_n, where q_n represents the turbulent velocity scale for the size class. Hence, from equation (9.72),

$$\Omega_n = \Omega_1\left(\frac{a_n}{a_1}\right)^{\lambda-1}. \quad (9.105)$$

Following the procedure from the previous section, and substituting with equations (9.43) and (9.105), the summation over size classes becomes

$$E(\kappa) = \frac{2(2\pi)^4\phi_1\Omega_1^2}{a_1^{\beta+2\lambda-2}\ln\ell}\sum_{\text{classes}}(\kappa a_n)^{4+p}F^2(\kappa a_n)a_n^{\beta+2\lambda}\Delta a_n. \quad (9.106)$$

Setting $y = \kappa a_n$ and taking the limit $\Delta y \to 0$ results in

$$E(\kappa) = \frac{2(2\pi)^4 \phi_1 \Omega_1^2 a_1^3}{(\kappa a_1)^{\beta + 2\lambda + 1} \ln \ell} \int_{\kappa a_1}^{\kappa a_N} F^2(y) y^{\beta + 2\lambda + 4 + p} \, dy. \qquad (9.107)$$

For the Gaussian parent function, substitution of equation (9.55) into (9.107) yields

$$E(\kappa) = -\frac{2\pi^{p/2} \phi_1 \Omega_1^2 a_1^3}{\ln \ell} \left(\frac{\kappa a_1}{\sqrt{\pi}} \right)^{-\beta - 2\lambda - 1}$$

$$\times \left\{ \gamma \left[\frac{5 + p + \beta + 2\lambda}{2}, \frac{\kappa^2 a_1^2}{\pi} \right] - \gamma \left[\frac{5 + p + \beta + 2\lambda}{2}, \frac{\kappa^2 a_N^2}{\pi} \right] \right\}. \qquad (9.108)$$

Similarly, integrating equation (9.104) over size classes yields, for the Gaussian parent function,

$$\sigma_v^2 = -\frac{4\pi^{(1+p)/2} \Gamma((5+p)/2) \phi_1 \Omega_1^2 a_1^2}{3(\beta + 2\lambda) \ln \ell} \left[1 - \left(\frac{a_N}{a_1} \right)^{\beta + 2\lambda} \right]. \qquad (9.109)$$

At low wavenumber, $\kappa a_1 \ll 1$, the toroidal QWs produce the same wavenumber dependence as the von Kármán spectrum, namely $E(\kappa) \sim \kappa^4$. The spectrum for poloidal QWs, however, goes as $E(\kappa) \sim \kappa^6$. Within the inertial subrange, we find

$$E(\kappa) \simeq \frac{\sigma^2 a_1 3(\beta + 2\lambda) \Gamma[(5 + p + \beta + 2\lambda)/2]}{\sqrt{\pi} \Gamma[(5+p)/2]} \left(\frac{\kappa a_1}{\sqrt{\pi}} \right)^{-1 - \beta - 2\lambda}. \qquad (9.110)$$

For both the toroidal and poloidal QWs (with $\beta = 0$ and $\lambda = 1/3$), $E(\kappa) \sim \kappa^{-5/3}$ in the inertial subrange, as with the Kolmogorov and von Kármán spectrum. Comparing with equation (6.34), we have

$$C_v^2 = \frac{2(2\pi)^{1/3} \Gamma(2/3) \sigma^2}{(1 + p/3)(1 + p) a_1^{2/3}} \left[\frac{\sqrt{\pi}}{\Gamma(1/2 + p/2)} \right] \left[\frac{\Gamma(17/6 + p/2)}{\Gamma(17/6)} \right], \qquad (9.111)$$

For the toroidal QWs ($p = 0$), this reduces to $C_v^2 = 2(2\pi)^{1/3} \Gamma(2/3) \sigma^2 / a_1^{2/3}$.

Random field realizations for the velocity QWs proceed in the same manner as with the scalar QWs, except that random orientations must also be generated along with the random positions. A convenient method for generating randomly isotropic orientations is to first generate three independent values from a standard normal distribution (mean $m = 0$ and standard deviation $\sigma = 1$). These values can be interpreted as a random point (x, y, z), which lies in any direction relative to the origin with equal likelihood. The spherical coordinates of this point are $R = \sqrt{x^2 + y^2 + z^2}$, $\varphi = \arctan(y/x)$, and $\theta = \arccos(z/R)$. The angles (φ, θ) can thus serve as the random orientation of the QW.

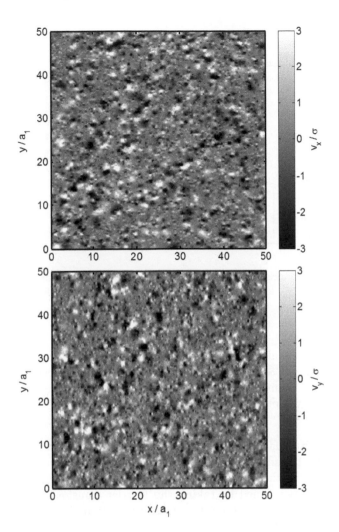

FIGURE 9.10
Two-dimensional realizations of fields for the toroidal QW model with $a_1 = 8a_N$. Top is for the x-component of the vector field; bottom is for the y-component. The spatial axes are normalized by a_1 and the velocity fluctuations are normalized by the standard deviation.

Figure 9.10 shows realizations of the x- and y-velocity components based on the toroidal QW model, with $a_1 = 8a_N$. The visualization is a horizontal plane of dimensions $50a_1 \times 50a_1$. (A buffer zone, with width of $10a_1$, is also included in the realization, but is not shown. The buffer eliminates edge effects.) A total of 4 million individual QWs are included in this realization.

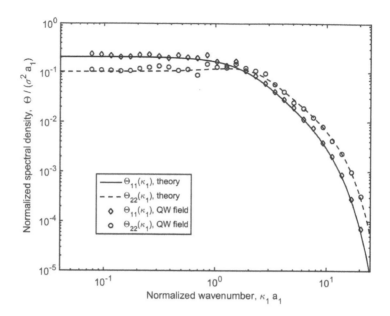

FIGURE 9.11
Comparison of simulated and predicted spectral densities for the toroidal QW model with $a_1 = 8a_N$. Shown are the longitudinal spectrum $\Theta_{11}(\kappa_1)$ and the lateral spectrum $\Theta_{22}(\kappa_1)$, both normalized by $\sigma^2 a_1$, as a function of the normalized wavenumber $\kappa_1 a_1$.

The 1D longitudinal and lateral (transverse) spectral densities (defined by equations (6.21) and (6.23), respectively), as calculated from the realization, are compared to theoretical predictions [142, 429] in figure 9.11. The calculations were performed by taking the Fourier transform of the field along the x-axis, and then averaging the magnitude squared of the transform (the power spectrum) along the y-axis. Except for some random variations at low wavenumber, the simulations match the theory.

9.2.5   Inhomogeneous and anisotropic turbulence

QWs can also be readily employed to create turbulence models with inhomogeneity and anisotropy. In particular, since the QW realizations are performed in the spatial domain, height-dependent variances and length scales (or structure-function parameters), as described in Chapter 6 of this book, can be incorporated without modification of the basic methodology described up to this point. In this regard, QWs provide a substantial benefit over the GRPM approach described in section 9.1.2.

The QW model also lends itself to modeling multiple, correlated fields.

Such correlations are a fundamental manifestation of momentum and heat transport by turbulence. Covariances between the vertical and horizontal fluctuations, $\langle \tilde{w}\tilde{u} \rangle$, are associated with a vertical flux of momentum between the air and ground; covariances between the vertical velocity and temperature, $\langle \tilde{w}\tilde{T} \rangle$, are associated with vertical flux of heat toward or away from the ground. In the QW model, such covariances can be created by suitable specification of parent functions and probability distributions for the orientations. In particular, reference [429] described a QW model for a shear layer with vertical momentum flux, whereas reference [425] described a QW model for an atmospheric surface layer with a vertical heat flux. The basic concepts of these models can be readily understood and programmed, even though derivation of the spectra associated with these models becomes quite involved.

The key to the shear-layer model is the anisotropic generalization of equation (9.95), the statistical model for the angular velocity of a QW, to favor certain orientations for the rotational axes. Such tilting of the vortices produces anisotropic variances and a momentum flux. For a vector QW, the variances and covariances of the components of $\boldsymbol{\Omega}^{mn}$ comprise a matrix, namely

$$
\begin{bmatrix}
\langle \Omega_1^{mn}\Omega_1^{mn} \rangle & \langle \Omega_1^{mn}\Omega_2^{mn} \rangle & \langle \Omega_1^{mn}\Omega_3^{mn} \rangle \\
\langle \Omega_1^{mn}\Omega_2^{mn} \rangle & \langle \Omega_2^{mn}\Omega_2^{mn} \rangle & \langle \Omega_2^{mn}\Omega_3^{mn} \rangle \\
\langle \Omega_1^{mn}\Omega_3^{mn} \rangle & \langle \Omega_2^{mn}\Omega_3^{mn} \rangle & \langle \Omega_3^{mn}\Omega_3^{mn} \rangle
\end{bmatrix}
$$
$$
= \begin{bmatrix}
\Omega_{n,1}^2 & \rho_{n,12}\Omega_{n,1}\Omega_{n,2} & \rho_{n,13}\Omega_{n,1}\Omega_{n,3} \\
\rho_{n,12}\Omega_{n,1}\Omega_{n,2} & \Omega_{n,2}^2 & \rho_{n,23}\Omega_{n,2}\Omega_{n,3} \\
\rho_{n,13}\Omega_{n,1}\Omega_{n,3} & \rho_{n,23}\Omega_{n,2}\Omega_{n,3} & \Omega_{n,3}^2
\end{bmatrix}, \quad (9.112)
$$

in which the $\Omega_{n,i}^2$ are variances of the angular-velocity components for the size class n and the $\rho_{n,ij}$ are correlation coefficients between the components. For the isotropic model, the variances match and the correlation coefficients all vanish.

When there is a tendency for Ω_1^{mn} and Ω_3^{mn} to have the same sign (either both positive or both negative, so that $\rho_{n,13} > 0$), there will be a negative momentum flux, as is characteristic of a shear layer. This is because a QW with such an orientation will have $\tilde{v}_1 > 0$ and $\tilde{v}_3 < 0$ on one half of the (x_1, x_3)-plane, and $\tilde{v}_1 < 0$ and $\tilde{v}_3 > 0$ on the other half. Hence the product $\tilde{v}_1\tilde{v}_3$ has the same sign throughout the (x_1, x_3)-plane.

In reference [429], the correlation coefficients were scaled according to the power law

$$
\rho_{n,ij} = \rho_{1,ij}\left(\frac{a_n}{a_1}\right)^{2/3}, \quad (9.113)
$$

which was shown to reproduce the observed $\kappa^{-7/3}$ dependence of turbulent cross spectra in the inertial subrange [186]. Predictions for variances and covariances in an atmospheric shear layer were also derived and found to compare well with empirical data. The comparisons provide some evidence favoring the poloidal QW model.

An important step in formulating models with correlations between scalar and velocity fields, as needed to produce a heat flux,, is the multipole generalization of the parent function for the scalar model, equation (9.36). Since such QWs are spherically symmetric, we may refer to equation (9.36) as a *monopole* model. A *dipole* model can be constructed by differentiating $f(\chi)$ along an axis passing through the origin of the QW. This creates a QW with a single axis of symmetry. On one side of the plane normal to this axis, and passing through the QW origin, the field is entirely positive. On the other side of the plane it is entirely negative. Designating the unit vector directed along this axis as $\mathbf{d}^{mn}$, we set

$$Q_{\mathrm{D}}^{mn}(\mathbf{R}) = h^{mn} q_n a_n \left(\mathbf{d}^{mn} \cdot \nabla \right) f(\chi). \tag{9.114}$$

One could continue applying derivatives to develop higher-order QWs. Quadrupoles are defined by a pair of orientation vectors $\mathbf{D}^{mn}$ and $\mathbf{d}^{mn}$ as

$$Q_{\mathrm{Q}}^{mn}(\mathbf{R}) = h^{mn} q_n a_n^2 \left(\mathbf{D}^{mn} \cdot \nabla \right) \left(\mathbf{d}^{mn} \cdot \nabla \right) f(\chi). \tag{9.115}$$

Figures 9.5(c) and (d) show dipole and quadrupole QWs. Specifically shown is a longitudinal quadrupole, for which $\mathbf{D}^{mn} = \mathbf{d}^{mn}$. Lateral quadrupoles involve orthogonal arrangements of the vectors $\mathbf{D}^{mn}$ and $\mathbf{d}^{mn}$.

A coupled field model involving scalar dipoles (or longitudinal quadrupoles) and rotational QWs can be modeled in essentially the same manner as the previously described shear-layer model. In this case, however, the cross correlations to be modeled are between components of $\mathbf{\Omega}^{mn}$ and $\mathbf{d}^{mn}$. Some couplings of scalar and velocity QWs are more useful than others. Coupling of a monopole scalar QW with a toroidal velocity QW, in particular, does not lead to a net scalar flux. This can be understood from figures 9.5(a) and 9.9(a): a flux in any direction on one side of the toroidal QW would be canceled by a flux of the opposite sign on the other side. A dipole scalar QW coupled with a toroidal velocity QW, however, can produce a net scalar flux. When the dipole in figure 9.5(c) is combined with the velocity field in figure 9.9(a), equal positive fluxes occur on both the left and right sides. A statistical preference for such orientations thus results in a flux. Based on this idea, an anisotropic model for free convection above a horizontal, heated surface was developed in reference [425]. Alternatively, scalar-flux models could be formulated by coupling longitudinal quadrupoles, figure 9.5(c), with poloidal QWs, figure 9.9(b). A statistical preference for the orientations shown in these figures would lead to relatively warm updrafts in the middle of the poloidal QWs, compensated by cool downdrafts away from the centers, and thus a net positive flux. Reversing the orientations of the QWs would produce structures with cool downdrafts in the middle.

10

Ray acoustics and ground interactions

The primary purpose of this chapter is to describe ray tracing methods for sound propagation in a moving medium, as based upon the equations developed in Part I of this book. This topic will be the focus of section 10.3. Prior to that section, however, some background will be provided on sound levels (section 10.1) and interactions of sound waves with the ground (section 10.2). These topics are pertinent to the discussion of ray tracing as well as numerical methods to follow in later chapters.

10.1 Sound levels and transmission loss

In this section, we introduce some basic notation and terminology that will be needed for subsequent discussions of numerical methods. In particular, we describe decibel (dB) units for sound levels, source levels, and transmission loss. The reader may refer to references [16, 198, 313, 345] and other textbooks on acoustics for more detailed introductions to these concepts.

As in previous chapters of this book, we indicate as $\hat{p}(\mathbf{R}, f)$ the complex pressure field at the observation point $\mathbf{R}$ for a harmonic source of frequency f. The pressure field in the time domain is calculated from

$$p(\mathbf{R}, t) = \mathrm{Re}\left[\hat{p}(\mathbf{R}, f)e^{-i\omega t}\right] = |\hat{p}| \cos(\omega t + \phi), \tag{10.1}$$

where $\hat{p}(\mathbf{R}, f) = |\hat{p}|e^{i\phi}$, $\mathrm{Re}[\,]$ indicates the real part, $\omega = 2\pi f$ is the angular frequency, and ϕ is the phase of $\hat{p}(\mathbf{R}, f)$. Let us indicate the time average of the squared pressure over a period of the sound wave, $T = 1/f$, as $\langle p^2(\mathbf{R}, t)\rangle_T$. This average can be shown to equal $|\hat{p}|^2/2$. The root-mean-square (rms) pressure is the square root of this average, namely $p_{\mathrm{rms}}(\mathbf{R}, f) = |\hat{p}|/\sqrt{2}$.

The sound-pressure level (SPL) is defined as the ratio of the received rms pressure to the reference pressure, p_{ref}, in dB:

$$\mathrm{SPL}(\mathbf{R}, f) = 20 \log \frac{p_{\mathrm{rms}}(\mathbf{R}, f)}{p_{\mathrm{ref}}} = 10 \log \frac{|\hat{p}(\mathbf{R}, f)|^2}{2p_{\mathrm{ref}}^2}. \tag{10.2}$$

For outdoor sound propagation, p_{ref} is normally set to $20\,\mu\mathrm{Pa}$.

To separate the SPL into contributions from the source amplitude and the propagation effects, we may in general write

$$\hat{p}(\mathbf{R}, f) = \hat{A}(f)\hat{g}(\mathbf{R}, f), \tag{10.3}$$

where $\hat{A}(f)$ is the complex source amplitude and $\hat{g}(\mathbf{R}, f)$ is the frequency-domain Green's function, i.e., the sound field produced by a unit-amplitude harmonic source. Hence equation (10.2) becomes

$$\text{SPL}(\mathbf{R}, f) = \text{SL}(f) + 20 \log |\hat{g}(\mathbf{R}, f)|, \tag{10.4}$$

where

$$\text{SL}(f) = 10 \log \frac{|\hat{A}(f)|^2}{2p_{\text{ref}}^2} \tag{10.5}$$

is the *source level*.

As an aid to interpretation, calculation results are often expressed relative to the field produced by a harmonic monopole point source radiating into free space. In a lossless, 3D medium with no boundaries, this field is given by

$$\hat{p}_{\text{free}}(\mathbf{R}, f) = \hat{A}(f)\frac{e^{ik(R-R_0)}}{R/R_0}, \tag{10.6}$$

where $k = 2\pi f/c$ is the wavenumber, c is the sound speed, and R_0 is a reference distance, which is usually set to 1 m. The sound-pressure level for propagation in free space follows as

$$\text{SPL}_{\text{free}}(\mathbf{R}, f) = \text{SL}(f) - 20 \log(R/R_0). \tag{10.7}$$

Next, we define the *relative level* at the receiver as

$$\begin{aligned}
\Delta\text{RL}(\mathbf{R}, f) &= \text{SPL}(\mathbf{R}, f) - \text{SPL}_{\text{free}}(\mathbf{R}, f) \\
&= 20 \log |\hat{g}(\mathbf{R}, f)| + 20 \log(R/R_0).
\end{aligned} \tag{10.8}$$

Note that the relative level is unaffected by the source level. In the literature on atmospheric acoustics, ΔRL is also called the *excess attenuation*. Sometimes, attenuation caused by air absorption over the distance R is included in the free-space reference solution.

The *transmission loss* is defined here as

$$\begin{aligned}
\text{TL}(\mathbf{R}, f) &= \text{SPL}(\mathbf{R}, f) - \text{SL}(f) = 20 \log |\hat{g}(\mathbf{R}, f)| \\
&= \Delta\text{RL}(\mathbf{R}, f) - 20 \log(R/R_0).
\end{aligned} \tag{10.9}$$

For a source radiating into free space, $\text{TL} = -20 \log(R/R_0)$, and the TL is zero at $R = R_0$. Thus the TL may also be interpreted as the level relative to that which would be observed at $R = R_0$ if the same source were radiating into free space. The TL is closely related to the relative level, although it does not include the compensation for spherical spreading. It is admittedly a misnomer to call the TL as defined by equation (10.9) a *loss*; positive values of the TL actually indicate a relative gain. Nonetheless, the terminology is conventional and we follow it here. (Some authors define the TL as the *negative* of the definition used here.)

The previous analysis may be generalized to non-harmonic signals by defining a frequency-dependent power spectrum $P(\mathbf{R}, f)$ (units Pa^2/Hz). First, $\langle p^2(\mathbf{R}, t)\rangle_T$ is filtered into a frequency band $[f - \Delta f/2, f + \Delta f/2]$. Then, taking the limit $\Delta f \to 0$ yields $P(\mathbf{R}, f)$.* For a harmonic signal, $P(\mathbf{R}, f) = |\hat{p}|^2/2$. The overall received level can then be found by integrating $P(\mathbf{R}, f)$ over frequency as follows:

$$\text{SPL}_W(\mathbf{R}) = 10 \log \left[\frac{1}{p_{\text{ref}}^2} \int_0^\infty W(f) P(\mathbf{R}, f)\, df \right]. \tag{10.10}$$

In the preceding equation, we have incorporated a frequency-dependent weighting function $W(f)$ for generality. Although calculations in this text are unweighted, A-, C-, and other weightings are commonly used to mimic human hearing response in noise-control applications [198].

Analogously to squared sound pressure, we can generalize the squared source amplitude $|\hat{A}|^2/2$ to a source spectral density $S(f)$. From equation 10.3, $P(\mathbf{R}, f) = S(f)|\hat{g}(\mathbf{R}, f)|^2$. The received level is thus

$$\text{SPL}_W(\mathbf{R}) = 10 \log \left[\frac{1}{p_{\text{ref}}^2} \int_0^\infty W(f) S(f) |\hat{g}(\mathbf{R}, f)|^2\, df \right]. \tag{10.11}$$

The source and propagation contributions become separate additive terms (as in equation (10.4)) only if either $S(f)$ or $\hat{g}(\mathbf{R}, f)$ is independent of frequency. The latter condition is satisfied for lossless propagation in free space, which from equation (10.6) yields

$$\text{SPL}_{W,\text{free}}(\mathbf{R}) = \text{SL}_W - 20 \log(R/R_0), \tag{10.12}$$

where

$$\text{SL}_W = 10 \log \left[\frac{1}{p_{\text{ref}}^2} \int_0^\infty W(f) S(f)\, df \right]. \tag{10.13}$$

Thus the overall (frequency-integrated) relative level at the receiver is

$$\Delta \text{RL}_W(\mathbf{R}) = \text{SPL}_W(\mathbf{R}) - \text{SL}_W + 20 \log(R/R_0). \tag{10.14}$$

For sound propagation calculations, equation (10.11) is often approximated by partitioning the integral into a finite number of frequency bands, and evaluating the integrand at the center frequency of each band:

$$\int_0^\infty W(f) S(f) |\hat{g}(\mathbf{R}, f)|^2\, df \approx \sum_{n=1}^N W(f_n) S(f_n) |\hat{g}(\mathbf{R}, f_n)|^2 \Delta f_n, \tag{10.15}$$

---

*Given here is the definition for the *one*-sided (as opposed to the *two*-sided) power spectral density.

in which the f_n are the center frequencies, the Δf_n are the bandwidths, and N is the number of bands. The bands may have constant width, or may have width proportional to their center frequencies, as with octave bands or one-third octave bands. For constant-width bands, we have $f_n = f_{\min} + (n-1/2)\Delta f$ and $\Delta f = (f_{\max} - f_{\min})/N$, where $f_{\min}$ and $f_{\max}$ are the lower and upper bounds on the frequency range used to approximate the integral. The bounds should encompass all of the significant energy in the source spectrum. For octave bands, we would normally specify a series center frequencies, such that $f_{n+1} = 2f_n$ (or, approximately so, when standardized center frequencies are used). The width of each band is $\Delta f_n = f_n(2^{1/2} - 2^{-1/2})$. One-third octave bands are handled similarly, except that $f_{n+1} = 2^{1/3}f_n$ and $\Delta f_n = f_n(2^{1/6} - 2^{-1/6})$.

10.2   Interactions with the ground

Although the emphasis of this book is on sound propagation within an inhomogeneous moving fluid, calculations of propagation in the atmosphere usually require accounting for interactions of the sound waves with the ground. Two broad approaches may be considered. The first involves explicit calculation of the field in the air only, and treatment of the ground interaction through an appropriate boundary condition (BC). The other involves explicit calculation of the waves within both the air and the ground, and enforcement of proper continuity conditions across the air/ground interface. The BC-based approach, by nature, usually involves approximations, although in many cases it provides very accurate results. In this section, we consider formulation of appropriate boundary conditions, particularly in the context of geometric acoustics. Later, in Chapters 11 and 12, we consider numerical approaches capable of explicit calculation of waves in the ground.

10.2.1   Reflections and image sources

Let us consider for now the idealized case in which the air and ground are homogeneous half spaces without any ambient fluid motion. The total sound field at a receiver may be conceptualized as the sum of sound traveling along a direct path from the source, plus sound traveling along a path that is reflected by the ground. This simple picture, which is illustrated in figure 8.1, is idealized in that it does not include waves propagating along the air/ground interface. However, it is often very reasonable to view the sound field as a sum of these two paths, so long as the angle θ_i made between the ray path and the ground (called the *grazing*, or *elevation*, angle) is not so small that the direct and ground-reflected waves nearly cancel [16, 313]. The complement of the grazing angle, $\psi_i = \pi/2 - \theta_i$, is normally called the *angle of incidence*;

this distinction should be kept in mind when comparing various equations in the literature.

The amplitude and phase of a harmonic disturbance will generally change upon reflection from the ground. Some important features of the ground reflection can be understood by first considering the case of plane-wave reflections at the interface between two homogeneous, motionless fluids; that is, we model the air as a fluid half-space with density ϱ_0 and sound speed c_0, and the ground as a fluid half-space with ϱ_g and sound speed c_g. Then, by enforcing continuity of the pressure field and particle displacement across the interface, it can be shown (e.g., reference [198], as well as equation (4.36), which was derived for the more general case of a moving medium) that the ratio of the reflected field to the incident field is

$$\mathcal{R}_p(\psi_i, \psi_t) = \frac{\varrho_g c_g \cos \psi_i - \varrho_0 c_0 \cos \psi_t}{\varrho_g c_g \cos \psi_i + \varrho_0 c_0 \cos \psi_t}, \tag{10.16}$$

where ψ_t is the transmission angle (the angle at which the plane waves travel in the ground), which satisfies Snell's law:

$$c_0 \sin \psi_t = c_g \sin \psi_i. \tag{10.17}$$

The quantity $\mathcal{R}_p(\psi_i, \psi_t)$ is called the *plane-wave reflection coefficient*. Equation (10.16) indicates that when $(\varrho_g c_g)/(\varrho_0 c_0) \gg 1$, $\mathcal{R}_p \simeq 1$ for nearly all incident angles. In this situation, the surface has the appearance of being a perfectly rigid reflector. When $(\varrho_g c_g)/(\varrho_0 c_0) \ll 1$, $\mathcal{R}_p \simeq -1$ for nearly all incident angles. This is called a *pressure-release* surface.

Generally speaking, the sound speed in the ground is substantially *lower* than in the air. This behavior, which may initially seem counterintuitive, is a consequence of viscous drag and thermal conduction as the sound waves interact with the small pores of the ground material.* Hence, in Snell's law, $c_g/c_0 \ll 1$, and as a consequence $\psi_t \ll 1$. This leads to the following approximation for equation (10.16):

$$\mathcal{R}_p(\psi_i, \psi_t) \simeq \mathcal{R}_p(\psi_i) = \frac{\varrho_g c_g \cos \psi_i - \varrho_0 c_0}{\varrho_g c_g \cos \psi_i + \varrho_0 c_0}. \tag{10.18}$$

Note that $\mathcal{R}_p$ is now independent of ψ_t. It depends only on ψ_i and the ratio of characteristic impedances of the two media, namely $(\varrho_g c_g)/(\varrho_0 c_0)$. Equation (10.18) is called the *normal reaction* approximation, because it amounts to assuming the transmitted waves propagate perpendicularly to the interface.

As we will see in section 10.2.2, when the ground is modeled as an effective fluid medium with dissipation, the characteristic impedance (product of the

---

*Modeling of the acoustic properties of porous media will be the topic of section 10.2.2. We are omitting from the discussion here the possible coupling of sound energy into the solid frame material, which can be significant, particularly at very low frequencies. In effect, the solid material is assumed to be perfectly rigid, which enables the ground to be treated as an effective fluid medium. See, for example, reference [16].

density and sound speed) becomes complex and frequency dependent. Hence, for a more general treatment, we should include frequency dependence in the reflection coefficient, which we indicate here as $\mathcal{R}_p = \mathcal{R}_p(\psi_i, \omega)$.

Since the pressure and particle displacement are continuous at the interface, and the waves are transmitted normally into the ground half space, the apparent BC must depend only on the component of the particle velocity in the air that is normal to the interface. The incident pressure field drives particle oscillations at the ground surface, which are impeded by the mismatch between $\varrho_0 c_0$ and $\varrho_g c_g$. We thus define the specific acoustic impedance as

$$Z(\omega) = \left. \frac{\hat{p}}{\hat{w}_{\text{in}}} \right|_{z=0}, \tag{10.19}$$

where $\hat{w}_{\text{in}}$ is the component of the particle velocity directed *into* the surface (the negative of $\hat{w}_z$). Here and in the remainder of this chapter, we use z for the vertical coordinate, and take the ground surface to be $z = 0$. For a plane wave, $\hat{p}$ and $\hat{w}_{\text{in}}$ can be readily calculated, resulting in [313]

$$\frac{Z(\omega) \cos \psi_i}{\varrho_0 c_0} = \frac{1 + \mathcal{R}_p(\psi_i, \omega)}{1 - \mathcal{R}_p(\psi_i, \omega)}. \tag{10.20}$$

Solving for $\mathcal{R}_p$, we find

$$\mathcal{R}_p(\psi_i, \omega) = \frac{\zeta(\omega) \cos \psi_i - 1}{\zeta(\omega) \cos \psi_i + 1} = \frac{\cos \psi_i - \beta(\omega)}{\cos \psi_i + \beta(\omega)}, \tag{10.21}$$

where $\zeta(\omega) = Z(\omega)/\varrho_0 c_0$ is the normalized surface impedance and $\beta(\omega) = \varrho_0 c_0 / Z(\omega)$ is the normalized surface admittance. If the real part of the impedance is positive, then $|\mathcal{R}_p| < 1$, thus indicating that the ground absorbs sound energy. A surface with impedance independent of ψ_i is referred to as a *locally reacting* surface. This behavior follows from the normal reaction approximation, but might be regarded as a more general assumption.

The concept of a reflection coefficient dependent only upon on the incident angle and surface impedance plays an important role in calculations of outdoor sound propagation. In practice, the concept is often adopted for spherical-wave interactions as well as for plane waves. Implicit to such a generalization is the notion that the reflection does not depend significantly upon the curvature of the wavefronts at the surface, which is generally reasonable when the heights of the source and receiver are small compared to their horizontal separation. Adopting the normal reaction and plane-wave approximations, we can write the total sound field in a homogeneous (non-turbulent and non-refractive) atmosphere as:

$$\hat{p}(\mathbf{R}, f) = \hat{A}(f) \left[\frac{e^{ikR_1}}{R_1} + \mathcal{R}_p(\psi_i, \omega) \frac{e^{ikR_2}}{R_2} \right], \tag{10.22}$$

where $\hat{A} = \hat{A}(f)$ is the complex source amplitude, $\mathbf{R} = (x, y, z)$ the receiver location (x and y being the horizontal coordinates, and z the height),

$R_1 = \sqrt{x^2 + y^2 + (z - h_s)^2}$ is the distance from the actual source to the observer, $R_2 = \sqrt{x^2 + y^2 + (z + h_s)^2}$ is the distance from the image source to the observer, and h_s is the height of the source above the ground. (Refer to figure 8.1.) The two terms on the right side of equation (10.22) may be interpreted as contributions from the actual source (direct path) and from an image source (reflected path).

We will see, in section 10.2.3, that $|\zeta(\omega)| \gg 1$ for many common ground materials, particularly at frequencies below several hundred Hz. In that case, for most incident angles, $\mathcal{R}_p \simeq 1$, and the surface appears rigid. Nonetheless, when the grazing angle θ_i becomes small, $\zeta(\omega) \cos \psi_i = \zeta(\omega) \sin \theta_i \simeq \zeta(\omega)\theta_i$ becomes small, and $\mathcal{R}_p \to -1$. Thus the surface transitions to a pressure-release behavior. The near cancellation between the direct and ground-reflected paths, as predicted by the plane-wave reflection coefficient, is questionable in such situations; in fact, the plane-wave assumption leads to the implausible prediction that the sound field is zero when the source and receiver are both on the ground. In such scenarios, the sound field must be dominated by other phenomena such as wavefront curvature effects and surface waves.

Formulas for the reflection coefficient of spherical wavefronts, Q, are also available, although the analysis is considerably more complicated than for plane waves; see, for example, equations (8.1)–(8.3) in this text. These formulas generally depend on a "numerical distance" d, as well as $\zeta(\omega)$ and θ_i. For the ray tracing calculations in section 10.3, we will nonetheless use the simpler plane-wave reflection coefficient. The primary justification is that, when refraction is present, the assumptions regarding the geometry of the wavefronts used to derive the spherical coefficient are no longer applicable. So, the more complicated formula for spherical wave reflections does not necessarily provide better results.

10.2.2 Propagation in porous materials

Ground surfaces such as soil, sand, and snow, and even asphalt and cement, are often modeled acoustically as porous materials, i.e., as a solid porous frame saturated by air. Propagation in such materials differs from propagation through the atmosphere in many key aspects. For one, porous materials are highly dissipative and dispersive, due to viscous and thermal diffusion processes occurring as the sound waves interact with the porous frame. Furthermore, the pores are generally too small to enable development of an ambient or turbulent flow. Although modeling of sound propagation in porous materials is outside the primary scope of this book, due to the importance of this topic in modeling outdoor sound propagation, we explore it here from a phenomenological perspective. The goal is to develop equations for propagation in the ground and for the impedance of the ground, in both the frequency- and time-domains. For simplicity, the solid phase of the ground is considered to be perfectly rigid. More detailed expositions of porous media acoustics can be found in a number of other references [1, 13, 16].

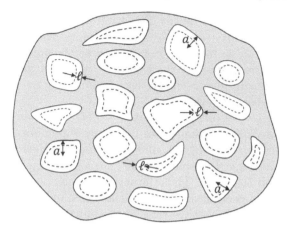

FIGURE 10.1
Boundary layers in a porous medium. The frame material is gray and the pores
are white. The dashed lines indicate the extent of the boundary layers for a
case in which the layer thickness, ℓ, is somewhat smaller than the characteristic
pore size, a.

A key concept in the following discussion is that of *viscous* and *thermal* boundary layers. Such layers are formed when a sound wave inter-
acts with a solid surface. Within these layers, viscous and thermal diffusion have a considerable influence. Well outside these layers, the viscous and
thermal diffusion may be neglected. Pierce [313] shows that the thicknesses
of the boundary layers are given, respectively, by $\ell_{\mathrm{vor}} = (2\mu/\omega\varrho_0)^{1/2}$ and
$\ell_{\mathrm{ent}} = (2\kappa/\omega\varrho_0 c_P)^{1/2} = \ell_{\mathrm{vor}}/(N_{\mathrm{pr}})^{1/2}$, where μ is the bulk viscosity, ϱ_0 the
ambient air density, κ the thermal conductivity, and N_{pr} the Prandtl num-
ber. Here, the subscript vor indicates the so-called vorticity (viscous) mode,
and ent the entropy (thermal) mode. The thickness of the boundary layers is
inversely dependent on the frequency because, at lower frequencies, there is a
longer period for the viscous and thermal diffusion to take place. Figure 10.1
illustrates boundary layers for in a porous medium.

Based on the preceding discussion, it follows that if a is a characteristic
size of the pores, when $a \ll \ell_{\mathrm{vor}}$, the viscous boundary layer will encompass
essentially the entire pore space, and when $a \gg \ell_{\mathrm{vor}}$, the viscous boundary
layer will be small relative to the pore space; that is, the propagation will
be similar to that in open air. Similar statements apply to the limiting cases
$a \ll \ell_{\mathrm{ent}}$ and $a \gg \ell_{\mathrm{ent}}$. Assuming the heat capacity of the frame material is
much larger than the fluid (air) saturating the pores, isothermal and adiabatic
thermodynamics apply, respectively.

As a starting point, let us consider the relatively simple phenomenological model for propagation in porous media originally proposed by Zwikker

and Kosten [445]. The equations of motion and continuity are given by equations (1.23) and (1.22) in that book, respectively. These are:

$$\sigma \mathbf{w} + \frac{\alpha \varrho_0}{\Omega} \frac{\partial \mathbf{w}}{\partial t} = -\nabla p, \qquad (10.23)$$

and

$$\frac{1}{K} \frac{\partial p}{\partial t} = -\nabla \cdot \mathbf{w}, \qquad (10.24)$$

where $\mathbf{w} = \mathbf{w}(\mathbf{R}, t)$ is the acoustic particle velocity (averaged over a slice perpendicular to the macroscopic direction of sound propagation), $p = p(\mathbf{R}, t)$ is the acoustic pressure, t is time, Ω is the porosity (void fraction), and σ, α, and K are coefficients representing acoustical properties of the porous material. Conceptually, equation (10.23) indicates a pressure gradient will accelerate the air in the pores, but there is also a resistance to the flow due to viscous drag. Equation (10.24) indicates that the compressibility of the air in the pores may differ from its adiabatic value in open air, due to heat conduction between the air and solid matrix. This model is functionally the same as one given by Morse and Ingard [264].

Zwikker and Kosten called σ the *resistance constant*, α the *structure constant*, and K the *compression (bulk) modulus*. They actually used the symbol k_s in place of α; we have adopted the latter here for consistency with the modern literature, where α indicates the *dynamic tortuosity* [181, 1], which accounts for the fact that the microscopic flow in the pores does not, in general, align locally with the macroscopic pressure gradient. Rather, the flow must adjust to the orientations of the channels and grains of the solid matrix. For pores at a constant slant angle θ, $\alpha = 1/\cos^2 \theta$. For more complicated pore geometries, α may be defined as the average distance traveled by the flow through the pores divided by the actual thickness of the sample. Many authors alternatively use the symbol $q = \sqrt{\alpha}$, with q still being referred to as the tortuosity.

If the quantities σ, α, and K are considered to be constants, transformation of equations (10.23) and (10.24) from the time to the frequency domain is quite straightforward. Applying a Fourier transform in time, $\partial/\partial t \to -i\omega$, and we find

$$\left(\sigma - \frac{i\omega \alpha \varrho_0}{\Omega} \right) \hat{\mathbf{w}} = -\nabla \hat{p}, \qquad (10.25)$$

and

$$\frac{-i\omega}{K} \hat{p} = -\nabla \cdot \hat{\mathbf{w}}. \qquad (10.26)$$

However, despite the names they assigned to the quantities σ, α, and K, Zwikker and Kosten allowed for the possibility that they are frequency dependent, and considered low- and high-frequency limiting behavior for these quantities. The dynamic tortuosity and bulk modulus may furthermore be complex. When viewed in this context, we will write α as $\hat{\alpha}(\omega)$ and K as $\hat{K}(\omega)$ in equations (10.25) and (10.26), so as to emphasize that they are complex, frequency-dependent quantities. This frequency dependence considerably complicates transformation between the time and frequency domains.

For convenience, a frequency-dependent complex density may be defined as*

$$\hat{\varrho}(\omega) = \frac{i\sigma(\omega)}{\omega} + \frac{\hat{\alpha}(\omega)\varrho_0}{\Omega}. \tag{10.27}$$

Then for equation (10.25) we have simply $-i\omega\hat{\varrho}(\omega)\hat{\mathbf{w}} = -\nabla\hat{p}$, in analogy to propagation in a non-moving, homogeneous fluid. The characteristic impedance Z and complex wavenumber Γ can then be readily derived from equations (10.25) and (10.26). Assuming a plane-wave disturbance propagating in the x-direction, we substitute the trial solution $\hat{p} = Ae^{i\Gamma x}$ and $\hat{w}_x = Be^{i\Gamma x}$. Since Z is, by definition, A/B, we find

$$Z = \sqrt{\hat{\varrho}(\omega)\hat{K}(\omega)}, \ \Gamma = \omega\sqrt{\hat{\varrho}(\omega)/\hat{K}(\omega)}. \tag{10.28}$$

Let us consider in more detail low- and high-frequency limits of σ, $\hat{\alpha}$, and $\hat{K}$. Specifically, based on the preceding discussion, we consider the two extreme cases of $a \ll \ell_{\text{vor}}, \ell_{\text{ent}}$, which is the limit of relatively small pores and low frequency (viscous and thermal boundary layers occupying the entire pore space), and $a \gg \ell_{\text{vor}}, \ell_{\text{ent}}$, which is the limit of relatively large pores and high frequency (thin viscous and thermal boundary layers).

In the case small pores/low frequencies, the thermodynamic process in the pores is essentially isothermal, as the temperature in the pores equalizes to the solid matrix (which is assumed to have much greater heat capacity), so that *within the pores* K takes on its isothermal value, which for an ideal gas is simply the ambient pressure P_0. The compressibility, β, is defined as the inverse of the bulk modulus. (Note that β has a different meaning here than the normalized admittance introduced in section 10.2.1.) Assuming the frame material is incompressible, the overall compressibility of the porous medium at low frequency is thus $\beta_0 = \Omega/P_0$. At low frequency, the term in equation (10.23) involving $\sigma(\omega)$ dominates. The low-frequency limit of $\sigma(\omega)$, which we will designate here as σ_0, is called the *static flow resistivity*; it equals μ/b, where b, the permeability, depends only the geometry of the pores.

Analysis of the structure constant (dynamic tortuosity) for small pores/low frequencies is not quite as straightforward, as it depends on the details of the interaction between the flow and the porous frame geometry. Allard and Champoux's [2] formulation yields the low-frequency limit of $\hat{\alpha}(\omega)$ (using the

---

*Note that the complex density in the convention of some other authors, e.g., reference [2], is Ω times $\hat{\varrho}(\omega)$ as defined here. The reason for this discrepancy is that we are here defining the complex density relative to the velocity for the bulk medium (solid frame and pores), rather than the pore space only.

symbology of this book) as $\alpha_0 = 1/4s^2$, where s is called the shape factor. Generally, s^2 is in the range 0.1 to 10 [181]. Practically speaking, whether one deals with α_0 or s, the complexity of most ground materials compels us to regard these quantities as empirical constants.

In summary, a simple model for porous media for relatively small pores/low frequencies corresponds to equations (10.25) and (10.26) with $\sigma \to \sigma_0$, $\alpha \to \alpha_0$, and $1/K \to \beta_0 = \Omega/P_0$.

For large pores and high frequencies, the boundary layers are thin and follow the surface of the solid frame. Outside of the thin thermal boundary layer, K takes on the adiabatic value for an ideal gas, γP_0, where $\gamma \simeq 1.4$ is the ratio of specific heats. The overall compressibility of the porous medium at low frequency is thus $\beta_\infty = \Omega/\gamma P_0$. Likewise, for a thin viscous boundary layer, the flow resistivity term in equation (10.23) becomes unimportant. The fluid motion is dominated by an effective mass that accelerates in proportion to the macroscopic pressure gradient, much as in open air. As is the case at low frequencies, at a local, microscopic level, the flow does not align with the macroscopic pressure gradient. However, due to the thinner boundary layer (i.e., the relative unimportance of viscous diffusion), the local flow directions are in general different than at low frequency. Hence the effective value of α must differ from α_0. We designate the high-frequency limit as α_∞. Thus, a simple model for porous media with relatively large pores and high frequencies corresponds to equations (10.25) and (10.26) with $\sigma \to 0$, $\alpha \to \alpha_\infty$, and $1/K \to \beta_\infty = \Omega/\gamma P_0$.

Finally, we consider propagation when the boundary layers are somewhat smaller than the pore size, but not so small that they may be neglected, i.e., $a \gtrsim \ell_{\rm vor}, \ell_{\rm ent}$. This situation turns out to have important implications for the acoustical properties of porous media, because it causes significant phase differences to emerge between the fields inside and outside of the boundary layers; these phase differences lead to dissipation of sound energy. Although a detailed treatment of this situation is very complex, we can understand some of the most important physics using the concept of acoustic, vorticity, and entropy modes as described in Chapter 8 of Pierce's [313] text. The total fields consist of the sum of these three modes. The acoustic mode is conceptually present throughout the pore space; however, near solid boundaries, the vorticity and entropy modes make important contributions. In particular, the velocity fields associated with these modes decay with distance from the boundary z as $e^{-(1-i)z/\ell_{\rm vor}}$ and $e^{-(1-i)z/\ell_{\rm ent}}$, respectively, where z is the distance from the boundary. The pressure field associated with the non-acoustic modes is negligible in comparison to the acoustic mode. We thus anticipate that when the velocity fields associated with the non-acoustic modes are integrated over the pore space, they will be proportional to the velocity of the acoustic mode, times factors of $(1+i)\ell_{\rm vor}$ and $(1+i)\ell_{\rm ent}$. The area of the boundary layers, in a cross section perpendicular to the direction of propagation, should go as ℓ times the effective perimeter of the pores.

A rigorous analysis of the structure constant (dynamic tortuosity) for large pores/high frequencies was undertaken by Johnson et al. [181] and by Allard and Champoux [2]. In particular, Allard and Champoux's equation (20), in the limit of large ω, is equivalent to

$$\frac{\hat{\alpha}(\omega)}{\alpha_\infty} \simeq 1 + (1+i)\frac{\ell_{\text{vor}}}{\Lambda_{\text{vor}}} \tag{10.29}$$

where α_∞ is the dynamic tortuosity in the limit $\omega \to \infty$, which is equivalent to q^2 in the works of Attenborough [11] and other authors. The length scale Λ_{vor} is determined by integrating the squared velocity field within the pore volume V and dividing by the integral along the pore wall area A:

$$\Lambda_{\text{vor}} = \frac{2\int_V |w_{\text{ac}}(\mathbf{R})|^2 dV}{\int_A |w_{\text{ac}}(\mathbf{R})|^2 dA}. \tag{10.30}$$

In the preceding, $|w_{\text{ac}}(\mathbf{R})|$ is the magnitude of the particle velocity *for an inviscid fluid* in the pores. The subscript ac indicates the acoustic mode of the fluid, which may be modeled as inviscid [313]. Considering a cross section through the medium, the parameter Λ_{vor} represents twice the effective area of the pore space divided by its perimeter.

Regarding the impact of the thermal boundary layer on the high-frequency compressibility, Allard and Champoux's equation (28) implies

$$\frac{\hat{\beta}(\omega)}{\beta_\infty} \simeq 1 + \frac{\beta_0 - \beta_\infty}{\beta_\infty}(1+i)\frac{\ell_{\text{ent}}}{\Lambda_{\text{ent}}} \tag{10.31}$$

where

$$\Lambda_{\text{ent}} = \frac{2\int_V dV}{\int_A dA} = \frac{2V}{A}, \tag{10.32}$$

which is the same as the definition of Λ_{vor}, although without the weighting by the squared velocity.

For some simple geometries, Λ_{vor} and Λ_{ent} can be readily calculated. For circular cylindrical pores of radius a, aligned with the propagation direction, the area of a pore is πa^2 and the perimeter is $2\pi a$, and thus $\Lambda_{\text{vor}} = \Lambda_{\text{ent}} = a$. But for more complex materials of practical interest, Λ_{vor} and Λ_{ent} must generally be considered empirical constants. Allard and Champoux [2] formally define the shape factor s through the equation $\Lambda_{\text{vor}} = s\sqrt{8\mu\alpha_\infty/\varrho_0\sigma_0}$.

In lieu of the length-scale ratios $\ell_{\text{vor}}/\Lambda_{\text{vor}}$ and $\ell_{\text{end}}/\Lambda_{\text{ent}}$, we could alternatively define frequency-timescale products $\omega\tau_{\text{vor}}$ and $\omega\tau_{\text{ent}}$. Defining these products such that $\omega\tau = (\Lambda/\ell)^2/2$, we find

$$\tau_{\text{vor}} = \frac{\varrho_0}{\mu}\left(\frac{\Lambda_{\text{vor}}}{2}\right)^2, \quad \tau_{\text{ent}} = \frac{\varrho_0 c_p}{\kappa}\left(\frac{\Lambda_{\text{ent}}}{2}\right)^2. \tag{10.33}$$

With these definitions, τ_{vor} is a characteristic time for the viscous relaxation (vorticity-mode diffusion) and τ_{ent} for the thermal relaxation (entropy-mode diffusion).

In general, the acoustical properties of real-world porous ground surfaces, at frequencies intermediate between the low- and high-frequency limits, are very complex and dependent upon detailed geometric properties of the solid matrix and the flow interactions with it. A reasonable approach to dealing with this complexity is to formulate a model that smoothly interpolates between the two limits. For this purpose, we introduce frequency-domain relaxation-type equations for α and β, namely

$$\hat{\alpha}(\omega) = \alpha_\infty + (\alpha_0 - \alpha_\infty)S\left(\omega\tau_{\text{vor}}\right), \tag{10.34}$$

and

$$\hat{\beta}(\omega) = \beta_\infty + (\beta_0 - \beta_\infty)S\left(\omega\tau_{\text{ent}}\right), \tag{10.35}$$

where S is the relaxation function, which should behave such that $S(\xi) \to 1$ in the limit $\xi \to 0$, and $S(\xi) \to 0$ in the limit $\xi \to \infty$. (The relaxation functions for $\hat{\alpha}(\omega)$ and $\hat{\beta}(\omega)$ need not be the same, but practically speaking there is little reason to use different functions.) A desirable choice for the relaxation function will furthermore interpolate between these limits in a physically realistic manner, while also being readily transformable to the time domain. The following simple function has the desired frequency dependence for small and large ξ, as well as the desired $1/\sqrt{-i\xi}$ dependence for moderately large ξ:

$$S\left(\xi\right) = \frac{1}{\sqrt{1 - i\xi}}. \tag{10.36}$$

Alternative forms for the relaxation function may be considered. For example, the following form follows from the exact solution for sound propagation in circular cylindrical pores [410]:

$$S_c(\xi) = \frac{\sqrt{2}J_1\left[(1+i)\sqrt{2\xi}\right]}{(1+i)\sqrt{\xi}J_0\left[(1+i)\sqrt{2\xi}\right]}, \tag{10.37}$$

where J_0 and J_1 are the zeroth and first order cylindrical Bessel functions. The functions (10.36) and (10.37) are compared in figure 10.2. They match for small or large values of the argument ξ, and differ only slightly for moderate values of ξ, where the geometry of the pores has the strongest influence. There does not appear to be a compelling reason to use a more complex relaxation function than equation (10.36).

By applying an inverse Fourier transform to equation (10.36), we can convert this relaxation function to the time domain, with result

$$s(t) = \mathcal{F}^{-1}\left(\frac{1}{\sqrt{1 - i\omega\tau}}\right) = \frac{1}{\sqrt{\pi\tau t}}\exp\left(-\frac{t}{\tau}\right)H\left(t\right), \tag{10.38}$$

in which $H\left(t\right)$ is the Heaviside function (0 for $t < 0$ and 1 for $t \geq 0$). The

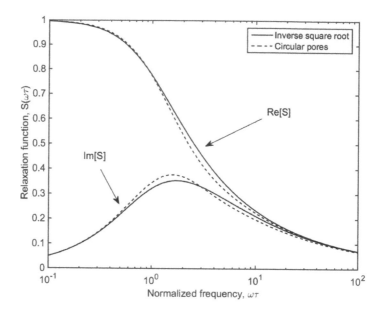

FIGURE 10.2
Comparison of relaxation functions based on a simple inverse square root function $1/\sqrt{1-i\xi}$ (equation (10.36)) and the exact solution for circular cylindrical pores (equation (10.37)). The response is shown as a function of the normalized frequency $\xi = \omega\tau$.

function $s(t)$ can be interpreted as the response of the medium to a step change in the pressure gradient or pore temperature [413].

Elsewhere [410, 413] it has been shown that the relaxation model predictions are nearly indistinguishable from microstructural models (i.e., those derived from solutions for circular cylindrical pores), such as those of Attenborough [11, 12] and Allard et al. [3]. The various microstructural models differ mainly in how adjustments are made to perturb the analytical solution for an idealized pore shape to conform with real-world materials of interest. Equations (10.34)–(10.36) are similar to empirical models suggested by Johnson et al. [181] and Allard and Champoux [2]. Like the relaxation function suggested here, the others use radicals of the form $\sqrt{1-i\omega\tau}$ to interpolate in a simple way between the correct asymptotic behavior at low and high frequencies. However, the functional forms of the complex operators differ somewhat. The equations used here enable straightforward transformation of the model to the time domain, based on equation (10.38). It should also be mentioned that the formulation of the relaxation model in this paper is somewhat different than previous ones [410, 413], in that the relaxation function is applied to

$\hat{a}(\omega)$ rather than to the specific volume $\hat{V}(\omega) = 1/\hat{\varrho}(\omega)$. Although the models behave similarly in the asymptotic limits, the formulation here provides more flexibility and is more consistent with recent literature on sound propagation through porous media, which emphasizes the role of the dynamic tortuosity.

Substituting the full forms of equations (10.34)–(10.36) into the frequency-domain equations (10.25) and (10.26), we have

$$\left\{ \sigma_0 - \frac{i\omega \varrho_0}{\Omega} \left[\alpha_\infty + (\alpha_0 - \alpha_\infty)S(\omega\tau_{\text{vor}}) \right] \right\} \hat{\mathbf{w}} = -\nabla\hat{p} \qquad (10.39)$$

and

$$-i\omega \left[\beta_\infty + (\beta_0 - \beta_\infty)S(\omega\tau_{\text{ent}}) \right] \hat{p} = -\nabla \cdot \hat{\mathbf{w}}. \qquad (10.40)$$

Some additional simplifications to equations (10.39) and (10.40) are reasonable. Since, at low frequencies, the flow resistance term in equation 10.25 will dominate the structure constant term, the value of α_0 has little impact on the model. Thus we could set α_0 to zero or some other convenient value. To mimic equation (10.29), it is actually most convenient to set $\alpha_0 = 2\alpha_\infty$, so that $(\alpha_0 - \alpha_\infty)/\alpha_\infty = 1$. Regarding equation (10.35), because $\beta_0 = \Omega/P_0$ and $\beta_\infty = \Omega/\gamma P_0$ differ only by a factor of $1/\gamma$, we might simply set $\hat{\beta}(\omega)$ to β_0 or β_∞, depending on whether $\omega\tau_{\text{ent}}$ is smaller to or larger than 1. While such an approximation simplifies the model equations, it does not actually remove a free parameter from the model, because γ is known. Further model simplifications will be discussed in section 10.2.3.

Setting $\alpha_0 = 2\alpha_\infty = 2q^2$, the model as described thus far has five distinct parameters dependent upon the geometry of the porous medium (as opposed to parameters for the air). These are σ_0, Ω, q^2, τ_{vor} (or Λ_{vor}), and τ_{ent} (or Λ_{ent}). The complex density and compressibility operators can be written in the following normalized forms:

$$\frac{\hat{\varrho}(\omega)}{\varrho_0 q^2/\Omega} = \frac{i\sigma_0\Omega}{\omega\varrho_0 q^2} + 1 + S(\omega\tau_{\text{vor}}), \qquad (10.41)$$

and

$$\frac{\hat{\beta}(\omega)}{\Omega/\gamma P_0} = 1 + (\gamma - 1)S(\omega\tau_{\text{ent}}). \qquad (10.42)$$

These normalized operators are plotted in figure 10.3. Note that the normalized complex compressibility depends only on $\omega\tau_{\text{ent}}$. The real part equals γ for small values of $\omega\tau_{\text{ent}}$, and 1 for large values. The imaginary part peaks at moderate values of $\omega\tau_{\text{ent}}$. The real part of the normalized complex density depends only on $\omega\tau_{\text{vor}}$, and decreases from 2 at small values of $\omega\tau_{\text{vor}}$, to 1 at large values. The imaginary part, however, also depends on $\omega\varrho_0 q^2/\sigma_0\Omega$. It diminishes as this parameter is increased; it is very large for small values $\omega\tau_{\text{vor}}$, and decreases to zero as $\omega\tau_{\text{vor}}$ increases.

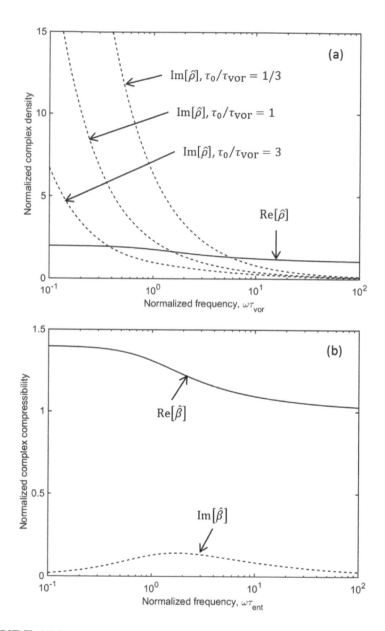

FIGURE 10.3
Complex operators characterizing propagation in a porous medium. (a) Normalized complex density, $\hat{\varrho}(\omega)/(\varrho_0 q^2/\Omega)$, as a function of the normalized frequency for the vorticity mode, $\omega\tau_{\mathrm{vor}}$. The imaginary part of $\hat{\varrho}(\omega)$ is shown for several values of $\tau_0/\tau_{\mathrm{vor}}$, where $\tau_0 = 2\varrho_0 q^2/\sigma_0\Omega$. (b) Normalized complex compressibility, $\hat{\beta}(\omega)/(\Omega/\gamma P_0)$, as a function of the normalized frequency for the entropy mode, $\omega\tau_{\mathrm{ent}}$.

Writing equations (10.39) and (10.40) in terms of these five parameters, we have

$$\left\{ \sigma_0 - \frac{i\omega\varrho_0 q^2}{\Omega}\left[1 + S(\omega\tau_{\text{vor}})\right]\right\}\hat{\mathbf{w}} = -\nabla\hat{p} \tag{10.43}$$

and

$$-\frac{i\omega\Omega}{\gamma P_0}\left[1 + (\gamma - 1)S(\omega\tau_{\text{ent}})\right]\hat{p} = -\nabla\cdot\hat{\mathbf{w}}. \tag{10.44}$$

When the preceding equations are transformed to the time domain, the products between the relaxation function and the acoustic fields become convolutions of inverse transforms, which results in

$$\sigma_0\mathbf{w} + \frac{\varrho_0 q^2}{\Omega}\left\{\frac{\partial\mathbf{w}}{\partial t} + \mathcal{F}^{-1}\left[S(\omega\tau_{\text{vor}})\right] * \frac{\partial\mathbf{w}}{\partial t}\right\} = -\nabla p \tag{10.45}$$

and

$$\frac{\Omega}{\gamma P_0}\left\{\frac{\partial p}{\partial t} + (\gamma - 1)\mathcal{F}^{-1}\left[S(\omega\tau_{\text{ent}})\right] * \frac{\partial p}{\partial t}\right\} = -\nabla\cdot\mathbf{w}, \tag{10.46}$$

where the asterisk indicates convolution. Substituting with equation (10.38) and writing out the convolutions explicitly, we find

$$\sigma_0\mathbf{w} + \frac{\varrho_0 q^2}{\Omega}\left\{\frac{\partial\mathbf{w}}{\partial t} + \frac{1}{\sqrt{\pi\tau_{\text{vor}}}}\int_{-\infty}^t \frac{\partial\mathbf{w}\left(t'\right)/\partial t'}{\sqrt{t-t'}}\exp\left(-\frac{t-t'}{\tau_{\text{vor}}}\right)dt'\right\} = -\nabla p \tag{10.47}$$

and

$$\frac{\Omega}{\gamma P_0}\left\{\frac{\partial p}{\partial t} + \frac{\gamma - 1}{\sqrt{\pi\tau_{\text{ent}}}}\int_{-\infty}^t \frac{\partial p\left(t'\right)/\partial t'}{\sqrt{t-t'}}\exp\left(-\frac{t-t'}{\tau_{\text{ent}}}\right)dt'\right\} = -\nabla\cdot\mathbf{w}. \tag{10.48}$$

The following equations for the characteristic impedance Z and the complex wavenumber Γ result from substituting equations (10.34) and (10.35) into (10.28):

$$\frac{Z}{\varrho_0 c_0} = \frac{q}{\Omega}\left\{\left[\frac{i\sigma_0\Omega}{\omega\varrho_0 q^2} + 1 + S\left(\omega\tau_{\text{vor}}\right)\right]/\left[1 + (\gamma - 1)S\left(\omega\tau_{\text{ent}}\right)\right]\right\}^{1/2}$$

$$= \frac{q}{\Omega}\left[\left(\frac{i\sigma_0\Omega}{\omega\varrho_0 q^2} + 1 + \frac{1}{\sqrt{1 - i\omega\tau_{\text{vor}}}}\right)/\left(1 + \frac{\gamma - 1}{\sqrt{1 - i\omega\tau_{\text{ent}}}}\right)\right]^{1/2} \tag{10.49}$$

and

$$\frac{\Gamma}{(\omega/c_0)} = q\left\{\left[\frac{i\sigma_0\Omega}{\omega\varrho_0 q^2} + 1 + S\left(\omega\tau_{\text{vor}}\right)\right]\left[1 + (\gamma - 1)S\left(\omega\tau_{\text{ent}}\right)\right]\right\}^{1/2}$$

$$= q\left[\left(\frac{i\sigma_0\Omega}{\omega\varrho_0 q^2} + 1 + \frac{1}{\sqrt{1 - i\omega\tau_{\text{vor}}}}\right)\left(1 + \frac{\gamma - 1}{\sqrt{1 - i\omega\tau_{\text{ent}}}}\right)\right]^{1/2}. \tag{10.50}$$

10.2.3   Modeling of outdoor ground surfaces

The previous subsection provided a general theory for sound propagation in porous media. In this section, we consider practical issues involving the application of this theory to outdoor ground surfaces.

Let us begin by considering an alternative formulation of the model, which involves the use of shape factors in lieu of Λ_{vor} and Λ_{ent} (or τ_{vor} and τ_{ent}). Such factors are often introduced to account for the deviations between the actual pore shape and a solution based on an idealized pore shape model. Circular cylindrical pores of constant radius a are typically taken as the reference solution. However, definitions of the shape factors vary and depend upon the frequency range in which the actual acoustical properties are being compared to the reference solution [1, 2, 11, 64]. For present purposes, since $\Lambda_{vor} = \Lambda_{ent} = a$ and $\sigma_0 = 8\mu q^2/\Omega a^2$ for circular cylindrical pores, we have in this case $\Lambda_{vor} = \Lambda_{ent} = \sqrt{8\mu q^2/\Omega\sigma_0}$. Hence it seems reasonable to define shape factors s_{vor} and s_{ent} as

$$s_{vor} = \Lambda_{vor}/\sqrt{8\mu q^2/\Omega\sigma_0}, \quad s_{ent} = \Lambda_{ent}/\sqrt{8\mu q^2/\Omega\sigma_0}. \tag{10.51}$$

With these definitions, s_{vor} and s_{ent} equal 1 for circular cylindrical pores. It can also be shown that

$$\tau_{vor} = s_{vor}^2\tau_0, \quad \tau_{ent} = s_{ent}^2 N_{pr}\tau_0, \tag{10.52}$$

where

$$\tau_0 = \frac{2\varrho_0 q^2}{\sigma_0\Omega}. \tag{10.53}$$

With the shape factor definitions, equations (10.49) and (10.50) become

$$\frac{Z}{\varrho_0 c_0} = \frac{q}{\Omega}\left[\left(\frac{2i}{\omega\tau_0} + 1 + \frac{1}{\sqrt{1 - is_{vor}^2\omega\tau_0}}\right) \Big/ \left(1 + \frac{\gamma - 1}{\sqrt{1 - is_{ent}^2 N_{pr}\omega\tau_0}}\right)\right]^{1/2} \tag{10.54}$$

and

$$\frac{\Gamma}{(\omega/c_0)} = q\left[\left(\frac{2i}{\omega\tau_0} + 1 + \frac{1}{\sqrt{1 - is_{vor}^2\omega\tau_0}}\right)\left(1 + \frac{\gamma - 1}{\sqrt{1 - is_{ent}^2 N_{pr}\omega\tau_0}}\right)\right]^{1/2}. \tag{10.55}$$

Although these equations have a relatively simple appearance, they still involve five parameters characterizing the porous medium (σ_0, Ω, q, s_{vor}, and s_{ent}). Apparently, none of these parameters can be eliminated without compromising the flexibility and accuracy of the model to some extent. Some of the parameters, particularly σ_0, Ω, and q, can be determined by non-acoustical

TABLE 10.1
Representative parameter values for various outdoor ground types: σ_0 is the static flow resistivity, Ω the porosity, q the tortuosity, and τ_0 the characteristic relaxation time, as defined by equation (10.53).

Ground type	σ_0 (Pa s m^{-2})	Ω	q	τ_0 (μs)
asphalt	3×10^7	0.1	1.8	3
compacted soil	1×10^6	0.3	1.4	15
soft soil	2×10^5	0.5	1.3	40
medium gravel	1×10^5	0.4	1.5	160
forest floor	2×10^4	0.7	1.1	200
snow	5×10^3	0.8	1.0	600

methods [1]. References [11, 64, 383, 410] and many others discuss shape factors and their values for various models of porous media. The shape factors are generally of order unity, and empirical data and curve fitting may be required to determine them. Given the difficulty of characterizing the shape factors, we might pragmatically choose $s_{\text{vor}} = s_{\text{ent}} = 1$. We would then be left with a three-parameter model, which retains the correct behavior in the low- and high-frequency limits, but may sacrifice some accuracy around the transition between these limits.

Another possible model simplification is to use empirical relationships to interrelate some of the remaining model parameters. In particular, the tortuosity may be estimated from the formula $q \simeq \Omega^{-0.25}$ [12, 13]; note that tortuosity increases with decreasing porosity. This relationship, combined with setting $s_{\text{vor}} = s_{\text{ent}} = 1$, leaves two parameters, σ_0 and Ω. It does not appear that the number of parameters can be reasonably reduced further. The porosity has a direct effect on the impedance. The static flow resistivity depends on the pore size and porosity, and hence cannot be eliminated from the model without losing the model dependence on pore size, which is important to attenuation and dispersion.

Representative parameter values for σ_0, Ω, q, and τ_0, for several types of outdoor ground, are given in table 10.1. These values are based on information presented in references [11, 14, 107, 241]. (See also Attenborough et al. [13] for a recent discussion and data on the ranges of ground parameters.) Since frequencies below roughly 2 kHz are usually most important for noise control applications, for typical surfaces such as soils $\omega\tau_0 \lesssim 1$, although $\omega\tau_0$ may sometimes exceed 1 for highly porous grounds such as gravel, forest floors, and snow.

Particularly simple equations for the impedance and complex wavenumber result when we attempt only to capture the low- and high-frequency limits correctly. In the limit $\omega\tau_0 \to 0$,

$$\frac{Z}{\varrho_0 c_0} = \frac{q}{\Omega}\sqrt{\frac{2i}{\gamma\omega\tau_0}}, \quad \frac{\Gamma}{(\omega/c_0)} = q\sqrt{\frac{2i\gamma}{\omega\tau_0}}. \tag{10.56}$$

In the limit $\omega\tau_0 \to \infty$,

$$\frac{Z}{\varrho_0 c_0} = \frac{q}{\Omega}, \quad \frac{\Gamma}{(\omega/c_0)} = q. \tag{10.57}$$

The following equations extrapolate smoothly between these limits:

$$\frac{Z}{\varrho_0 c_0} = \frac{q}{\Omega}\sqrt{\frac{2i}{\gamma\omega\tau_0} + 1}, \quad \frac{\Gamma}{(\omega/c_0)} = q\sqrt{\frac{2i\gamma}{\omega\tau_0} + 1}. \tag{10.58}$$

The Delany–Bazley empirical equations [94] are widely used for predicting the acoustical properties of porous media, including the ground. These equations describe $Z/\varrho_0 c_0$ and $\Gamma/(\omega/c)$ as functions of a single parameter, $C = f\varrho_0/\sigma_0$ [11, 94]:

$$\frac{Z}{\varrho_0 c_0} = 1 + 0.0571C^{-0.754} + i0.087C^{-0.732} \tag{10.59}$$

and

$$\frac{\Gamma}{(\omega/c_0)} = 1 + 0.0978C^{-0.693} + i0.189C^{-0.618}. \tag{10.60}$$

The parameter C is proportional to $\omega\tau_0 = 2\omega\varrho_0 q^2/\sigma_0\Omega$. Since Delany and Bazley performed measurements on only fibrous materials for which $\Omega \simeq 1$ and $q \simeq 1$, for this class of materials we find $\omega\tau_0 = 2\omega\varrho_0/\sigma_0 = 4\pi C$. Hence, for high-porosity materials, equations (10.54)–(10.55) and (10.58) can be readily written in a form that gives $Z/\varrho_0 c_0$ and $\Gamma/(\omega/c_0)$ as functions of only C. However, several important facts must be kept in mind with regard to application of the Delany–Bazley equations to porous ground materials. First, the approximations $\Omega \simeq 1$ and $q \simeq 1$ are *not* reasonable for most such materials. The porosity for ground materials is relatively small, typically in the range of 0.2 to 0.5. Second, the range of Delany and Bazley's is confined to $0.01 < C < 1$. Given the high static flow resistivity of typical ground materials, typical values of C are often much less than 0.01. Finally, the Delany–Bazley equations are known to imply non-causal propagation effects [199, 251]. The relaxation model, equations (10.54)–(10.55), as well as many other models for acoustical properties of porous media, do not suffer from these defects. Hence we recommend against use of the Delany–Bazley equations for outdoor sound propagation applications.

Predictions from the relaxation model and Delany–Bazley equations are compared in figures 10.4 and 10.5, for $Z/\varrho_0 c_0$ and $\Gamma/(\omega/c_0)$, respectively. In

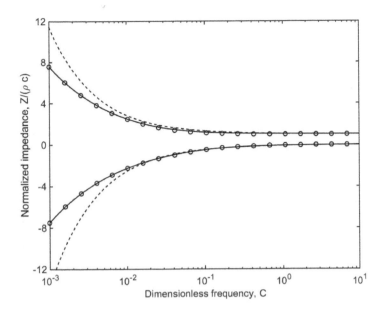

FIGURE 10.4

Comparison of characteristic impedance predictions from the relaxation model (solid line) and Delany–Bazley equations (dashed line). The positive curves show the real part of the normalized impedance (the normalized resistance), whereas the negative curves show the negative of the imaginary part (the negative of the normalized reactance). The circles are a simplified version of the relaxation model, as described in the text.

these predictions, the porosity and tortuosity are both assumed to equal 1. The abscissas are the normalized frequency $C = f\varrho_0/\sigma_0 = \omega\tau_0/4\pi$. Relaxation model predictions from the full model, equations (10.54)–(10.55), as well as the simplified model, equation (10.58), are shown. Within the range of validity of the Delany–Bazley equations, the predictions are quite similar. Outside this range, particularly for smaller values of C, the predictions diverge. Predictions from the simplified relaxation model are nearly indistinguishable from the full model over the entire range of C.

For locally reacting ground surfaces, the ratio of the pressure to the normal velocity at the surface impedance is independent of the angle of incidence, and equal to the characteristic impedance. Hence equations for Z play an important role in modeling boundary conditions for outdoor sound propagation. Based on the preceding discussion, equation (10.58) provides a frequency-domain model for the boundary condition that is both simple and reasonably accurate. Let us now consider the time-domain formulation of this model,

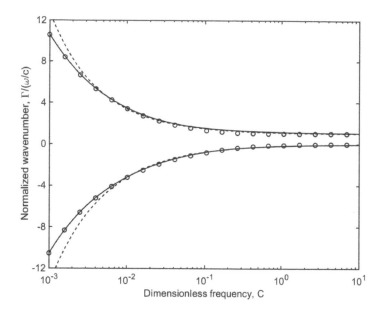

FIGURE 10.5
Comparison of complex wavenumber predictions from the relaxation model and Delany–Bazley equations. The positive curves show the real part of the normalized complex wavenumber (equal to c_0/c_p, where c_p is the phase speed), whereas the negative curves show the negative of the imaginary part (negative of the attenuation, normalized by ω/c_0).

which we will refer to as a time-domain boundary condition, or TDBC. It is convenient to first rewrite equation (10.58) in the form

$$Z = Z_\infty \sqrt{\frac{1 - i\omega\tau}{-i\omega\tau}}, \tag{10.61}$$

where we have introduced the shorthand $\tau = \gamma\tau_0/2$ and $Z_\infty = \varrho_0 c_0 q/\Omega$. In the frequency domain, we have at the surface

$$\hat{p}(\omega) = Z_\infty \sqrt{\frac{1 - i\omega\tau}{-i\omega\tau}} \hat{w}_z(\omega), \tag{10.62}$$

where $\hat{w}_z$ is the velocity component normal to the surface. From standard tables, the inverse Fourier transform of the impedance is found to be

$$\mathcal{F}^{-1}\left(Z_\infty \frac{\sqrt{1 - i\omega\tau}}{\sqrt{-i\omega\tau}}\right) = Z_\infty \left[\delta\left(t\right) + \frac{1}{\tau}g\left(\frac{t}{\tau}\right)\right], \tag{10.63}$$

in which $\delta(t)$ is the Dirac delta function and

$$g(t) = \frac{\exp(-t/2)}{2}\left[I_1\left(\frac{t}{2}\right) + I_0\left(\frac{t}{2}\right)\right]H(t). \tag{10.64}$$

In the preceding, I_n is the modified Bessel function of the first kind of order n, and $H(t)$ is the Heaviside function. As the inverse Fourier transform of a product is the convolution of the factors, we have the TDBC

$$p(t) = Z_\infty\left\{w_z(t) + \frac{1}{\tau}g\left(\frac{t}{\tau}\right) * w_z(t)\right\}. \tag{10.65}$$

Since $H(t) = 0$ for $t < 0$, this TDBC is causal.

The function $g(\bar{t})$ (where $\bar{t} = t/\tau$) describes the reaction of the pressure signal to an impulse in the normal particle velocity. It is plotted in figure 10.6. Of particular importance is the slow asymptotic decay of $g(\bar{t})$: for large values of $\bar{t}$, one has $g(\bar{t}) \simeq \sqrt{1/\pi\bar{t}}$. This slow decay is an inherent property of dissipation in porous materials, which is proportional to $\sqrt{f}$ as described for example by Fellah et al. [114]. It can be problematic numerically, since the velocity needs to be stored over a long duration of time. On the other hand, at high frequencies ($\omega\tau \gg 1$) the contribution of the convolution integral becomes small, and $p(t) \simeq Z_\infty w_z(t)$.

Due to the slow decay of the response function in the TDBC, direct numerical evaluation of the convolution integral may require a very large number (up to hundreds) of time steps, which imposes a substantial computational burden both in terms of calculation time and memory [419]. Recursive algorithms can substantially reduce this computational burden. The main idea behind such algorithms is to approximate the response function $g(t)$ with a summation of decaying exponential functions. This approach is well known in the electromagnetic and seismic literatures. We expand the function $g(\bar{t})$ as an exponential series:

$$g(\bar{t}) \approx \sum_{j=1}^{J} a_j e^{-\gamma_j \bar{t}} H(\bar{t}), \tag{10.66}$$

where J is the number of terms in the series approximation and a_j and γ_j are to be determined by fitting the function $g(\bar{t})$. In reference [294], some example values of a_j and γ_j are determined, based on mean-square error minimization, and are shown to approximate the response function well for $J = 6$.

By transformation of equation (10.66), it can be shown that the approximation for the time-domain response function is equivalent to a pole expansion of the impedance in the frequency domain, namely

$$Z(\omega) \simeq Z_\infty\left[1 + \sum_{j=1}^{J}\frac{a_j}{\gamma_j - i\omega\tau}\right]. \tag{10.67}$$

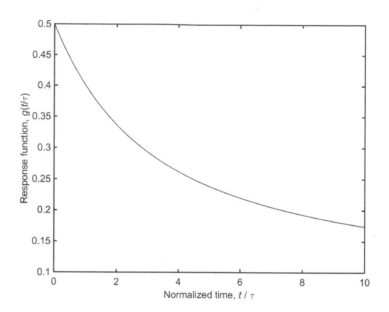

FIGURE 10.6
The function $g(t/\tau)$, which represents the response of the ground in the time domain. Time is normalized by τ as defined in the text.

Substituting the exponential series into equation (10.65), we find

$$p(t) = Z_\infty \left[w(t) + \sum_{j=1}^{J} \int_0^\infty w(t - t') a_j e^{-\gamma_j t'/\tau} \frac{dt'}{\tau} \right]. \tag{10.68}$$

We then discretize this equation at time steps $t = t_n = n\Delta t$, $n = 1, 2, \ldots$. Let us further define the auxiliary variable Ψ_j^n as the value of the integral in equation (10.68) at $t = t_n$. It can then be shown that

$$\Psi_j^n = e^{-\gamma_j \Delta \bar{t}} \Psi_j^{n-1} + \int_0^{\Delta t} w(t_n - t') a_j e^{-\gamma_j t'/\tau} \frac{dt'}{\tau}. \tag{10.69}$$

Here $\Delta \bar{t} = \Delta t/\tau$. If $w(t)$ is approximated as being constant over $t_{n-1} < t \leq t_n$, the integration yields

$$\Psi_j^n = e^{-\gamma_j \Delta \bar{t}} \Psi_j^{n-1} + \frac{a_j w(t_n)}{\gamma_j} \left(1 - e^{-\gamma_j \Delta \bar{t}} \right)$$

$$\simeq e^{-\gamma_j \Delta \bar{t}} \Psi_j^{n-1} + a_j w(t_n) \Delta \bar{t}. \tag{10.70}$$

The final form assumes that the time steps are sufficiently small that $\gamma_j \Delta \bar{t} \ll$

1. The utility of equation (10.70) is that Ψ_j^n can be recalculated from Ψ_j^{n-1}, without knowledge of the field at earlier time steps. Substituting into equation (10.68) and solving for $w(t_n)$, we have

$$w(t_n) = \frac{1}{1 + \sum_{j=1}^{J} a_j \Delta \bar{t}} \left[\frac{p(t_n)}{Z_\infty} - \sum_{j-1}^{J} e^{-\gamma_j \Delta \bar{t}} \Psi_j^{n-1} \right]. \tag{10.71}$$

This result can be used, for example, to implement the TDBC in finite-difference, time-domain calculations, which will be discussed further in Chapter 12.

10.3 Ray tracing in a moving medium

Ray tracing is based on solution of a high-frequency approximation to the full wave equation. Rays can be very helpful for visualizing propagation phenomena such as reflections and refraction. Conceptually, the rays may be regarded as the trajectories of particles moving through the fluid at a velocity determined by the speed of sound and the fluid velocity. The impact of refraction on sound-pressure amplitude can be efficiently estimated from the spacing between adjacent rays. References [155] and [319] exemplify application of ray-tracing methods to the near-ground atmosphere. While ray acoustics can be used to calculate angle-of-arrival fluctuations, approximations inherent to this approach make it inappropriate for small-scale turbulent scattering applications.

Unlike optics, the motion in the fluid (atmospheric wind and turbulence) is quite important in acoustics. The fluid motion makes the ray tracing equations and their solution rather more involved. Ray tracing can readily accommodate many complications pertinent to outdoor sound propagation, such as a spatially and temporally varying atmosphere and irregular terrain features. The main drawback of ray acoustics is its unsuitability for low frequencies and at caustics (crossing of ray paths). Ray tracing cannot be used to directly determine the sound field behind buildings or in shadow zones created by upward refraction. For such situations, ray tracing may be supplemented by the geometric theory of diffraction. In this book, we do not consider extensions to geometric acoustics around caustics and diffraction into shadow regions; readers may refer to Salomons [345] and Pierce [313] for such treatments. Gaussian beam methods have been developed, in part, to address the deficiencies of ray methods [128]. However, beam methods capture diffraction only in a heuristic sense, and must be calibrated on a frequency-by-frequency basis to obtain reasonable results. Generally speaking, the wave-based methods to be described in Chapters 11 and 12 should be used when it is important to calculate diffraction effects.

Ray tracing schemes have been based upon a great variety of equation sets and numerical techniques. Chapter 3 of this book described ray treatments that correctly incorporate the effect of wind on refraction. For a moving medium, equations (3.55) and (3.56) provide a particularly general and convenient starting point, since they represent the temporal evolution of the position and orientation of the ray in a form involving spatial derivatives of only the sound speed (c) and medium velocity ($\mathbf{v}$). These equations are equivalent to (8-1.10a) and (8-1.10b) in Pierce [313]. In Cartesian coordinates, the latter equations can be written as [313]

$$\frac{dR_i}{dt} = \frac{c^2 s_i}{\Omega} + v_i \tag{10.72}$$

and

$$\frac{ds_i}{dt} = -\frac{\Omega}{c}\frac{\partial c}{\partial x_i} - \sum_{j=1}^{3} s_j \frac{\partial v_j}{\partial x_i}. \tag{10.73}$$

In the preceding, $\mathbf{R}$ is the position of the ray, t is time, $\Omega = 1 - \mathbf{s}\cdot\mathbf{v} = c/(\mathbf{n}\cdot\mathbf{u})$ (not to be confused with the porosity in section 10.2), $\mathbf{u} = c\mathbf{n} + \mathbf{v}$ is the ray velocity (group velocity along the ray path), $\mathbf{n}$ is the unit normal to the wavefront, and $\mathbf{s} = \mathbf{k}/\omega$ is wave slowness.*

Equations 10.72 and 10.73 are in a form that can be readily solved with standard numerical methods for ordinary differential equations, such as Runge-Kutta methods. Use of time as the independent variable makes the integration stable around turning points (which is not the case, for example, when z is used as the integration variable). A system of six equations (three components each of $\mathbf{R}$ and $\mathbf{s}$) must be simultaneously integrated. The example ray paths for various regimes of the atmospheric surface layer (ASL), presented in section 3.5.4, were calculated by this approach.

The sound-pressure amplitude for a ray path can be determined using the Blokhintzev invariant, which was given earlier as equation (3.62), and equivalently as equation (8-6.13) in Pierce [313]. Using the preceding definitions, the invariant can be written

$$\frac{p_{\text{ray}}^2(\mathbf{R})u\,d\sigma_{\text{ray}}}{2\Omega\varrho_0 c_0^2} = \text{const}, \tag{10.74}$$

where $p_{\text{ray}}(\mathbf{R})$ is the sound-pressure amplitude for the ray path, $d\sigma_{\text{ray}}$ is the ray tube cross-sectional area, and $u = |\mathbf{u}|$. To calculate transmission loss, we wish to determine the ratio of $p_{\text{ray}}(\mathbf{R})$ to its value at $R = 1$ m in free space (a homogeneous, non-moving medium). In free space, the ray paths are simply straight lines emanating from the source, and the invariant becomes

---

*Note that the symbol $\mathbf{s}$ has a different meaning in Chapter 3 of this book than in Pierce's book. In the latter, $\mathbf{s}$ indicates the wave slowness, which equals $\mathbf{b}/c_0$ in this book. The parameter Ω in Pierce's book equals bc/c_0 here.

$p_{\text{free}}^2(\mathbf{R})d\sigma/2\varrho_0 c_0 = \text{const.}$ The ray-tube area is $d\sigma = R^2\Delta\phi[\sin(\theta + \Delta\theta) - \sin(\theta)]$, where $\Delta\phi$ is the azimuthal extent of the ray tube, and θ and $\theta + \Delta\theta$ bound the ray tube in elevation angle. The elevation angle is taken as zero at horizontal, and positive upwards. Hence we have

$$p_{\text{ray}}^2(\mathbf{R}) = p_{\text{free}}^2(R_0)\frac{\Omega\varrho c^2 R_0^2\Delta\phi[\sin(\theta + \Delta\theta) - \sin(\theta)]}{\varrho_0 c_0 u\, d\sigma_{\text{ray}}}, \tag{10.75}$$

where $R_0 = 1$ m. The primary feature of this equation is the inverse dependence of p_{ray}^2 upon $d\sigma_{\text{ray}}$.

Most ray tracing procedures involve calculating rays at constant increments in the elevation and azimuthal angles, i.e., at $\theta = \theta_n = \theta_0 + n\Delta\theta$, $n = 0, 1, \ldots, N$, and $\phi = \phi_m = \phi_0 + m\Delta\phi$, $m = 0, 1, \ldots, M$. Thus $d\sigma_{\text{ray}}$ is naturally calculated from adjacent rays, such as the four-ray bundle consisting of the rays launched at θ_n, θ_{n+1}, ϕ_m, and ϕ_{m+1}. Three-ray bundles can also be used, e.g., θ_n, θ_{n+1}, ϕ_m. The ray tracing calculation may be regarded as yielding $\mathbf{R}_{mn}(t)$, where t is time. From the ray positions, we can determine $d\sigma_{mn}(t)$. For the three-ray bundle, $d\sigma_{mn}(t)$ can be estimated from the elementary formula $bh/2$, where b is the base and h is the height of the triangle formed from the three vertices. Alternatively, for improved numerical accuracy and robustness (particularly in the presence if multipath), we might calculate the ray tube areas using supplemental rays that are launched at very small increments $\delta\phi$ and $\delta\theta$ from the "main" rays launched at (ϕ_m, θ_n). A disadvantage of this procedure is that it increases the number of rays that must be traced. For example, for a three-ray bundle, we must trace rays at $(\phi_m + \delta\phi, \theta_n)$ and $(\phi_m, \theta_n + \delta\theta)$, in addition to (ϕ_m, θ_n). However, for many ordinary differential equation solvers, the calculation time does *not* increase linearly with the number of rays; that is, a single call to the solver involving several closely spaced rays may involve only a marginally longer calculation time than for a single ray.

The ray paths reaching a particular location $\mathbf{R}$ are termed the *eigenrays*. Many procedures exist for finding eigenrays. For present purposes, we describe a relatively simple procedure that does not involve any particular assumptions about the ray paths. The procedure is described here for a vertical plane (x, z); that is, out-of-plane refraction is neglected. While the ray tracing and eigenray search can be readily generalized to three spatial dimensions, in practice such calculations are computationally intensive.

1. For each azimuthal angle of interest, rays are launched at a discrete set of elevation angles $\theta_n = (n - N/2)\Delta\theta$, $n = 0, 1, \ldots, N$. The positions of the rays are stored at discrete time steps t_i, $i = 1, 2, \ldots$. Hence we have a set of ray positions $[x_n(t_i), z_n(t_i)]$.

2. For each position of interest in the vertical plane, (x, z), find the values of n and i such that the points $[x_n(t_i), z_n(t_i)]$, $[x_{n+1}(t_i), z_{n+1}(t_i)]$, $[x_n(t_{i+1}), z_n(t_{i+1})]$, and $[x_{n+1}(t_{i+1}), z_{n+1}(t_{i+1})]$ form a quadrilateral containing (x, z), as shown in figure 10.7. This

step can be readily handled with many available numerical libraries for determining whether a point lies within an arbitrary polygon. Note that multipath propagation will result in multiple such quadrilaterals.

3. To find the elevation, time-of-arrival, and ray-tube area at (x, z) (the eigenrays), perform an interpolation using *barycentric* coordinates, by partitioning the quadrilateral into two triangles (upper left and lower right, as shown in figure 10.7, or alternatively lower left and upper right).

Regarding the final step above, the interpolation based on the barycentric coordinates is recommended here because it is linear and relatively simply to apply to an unstructured grid, as occurs in the present problem. The barycentric coordinates are given by [407]

$$x = \lambda_1 x_1 + \lambda_2 x_2 + \lambda_3 x_3, \quad z = \lambda_1 z_1 + \lambda_2 z_2 + \lambda_3 z_3, \tag{10.76}$$

where the λ_i are constants, and the (x_i, z_i) are the vertices of the triangle. Solving for the λ_i yields [407]:

$$\lambda_1 = \frac{(z_2 - z_3)(x - x_3) + (x_3 - x_2)(z - z_3)}{(z_2 - z_3)(x_1 - x_3) + (x_3 - x_2)(z_1 - z_3)},$$

$$\lambda_2 = \frac{(z_3 - z_1)(x - x_3) + (x_1 - x_3)(z - z_3)}{(z_2 - z_3)(x_1 - x_3) + (x_3 - x_2)(z_1 - z_3)},$$

$$\lambda_3 = 1 - \lambda_1 - \lambda_2. \tag{10.77}$$

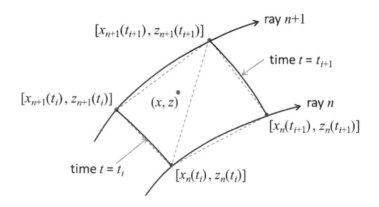

FIGURE 10.7

Geometry for determining eigenrays and interpolating ray properties to a particular point in the vertical plane. The ray paths are indexed by the elevation angle at which they were launched θ_n, and are output at discrete time steps t_i. Properties for two neighboring ray paths, at consecutive time steps, are interpolated.

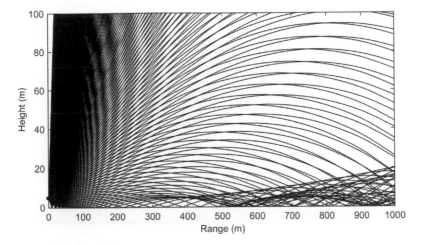

FIGURE 10.8
Ray paths for downward refraction with a linear wind speed profile $v_x(z) = 0.1z$ (m/s). The source height is 5 m.

Looping through the launch angles n and time steps i, and applying the barycentric interpolation within each of the two triangles comprising the quadrilateral, enables calculation of the time-of-arrival $t_j(x, z)$, launch angle $\theta_j(x, z)$, and ray-tube area $d\sigma_j(x, z)$ at each location (x, z) of interest.

A simple approach to handling ground reflections is to continue tracing the rays through the ground, as though it were not present. Then, the absolute value of z, rather than z itself, is plotted, so that the ray paths below the ground correctly appear above the ground. This procedure creates a specular reflection (the angle of incidence equals the angle of reflection). The atmospheric profiles below the ground should be mirror reflections of the profiles above; that is, the profiles depend only on the absolute value of z. In this manner, "downward" refracting conditions correspond to rays always bending toward $z = 0$, and "upward" refracting conditions away from $z = 0$. Each time a ray passes from $z > 0$ to $z < 0$, or vice versa, we multiply the complex pressure by the plane-wave reflection coefficient $\mathcal{R}_p(\theta_j)$, equation (10.21), where θ_j is the angle of incidence of the jth ray path on the ground, which is a constant for the path.

Once the eigenrays have been determined, the total complex pressure field $\hat{p}(x, z)$ can be determined from a coherent summation over the eigenrays. Since the procedure here involves integrating the ray paths in time, we know the phase of the ray along each propagation path, namely $\exp(-i\omega t_{mn})$. Hence

$$\hat{p}(x, z) = \sum_j p_j(x, z) \mathcal{R}_p^{G_j}(\theta_j) e^{-i\omega t_j(x,z)}. \tag{10.78}$$

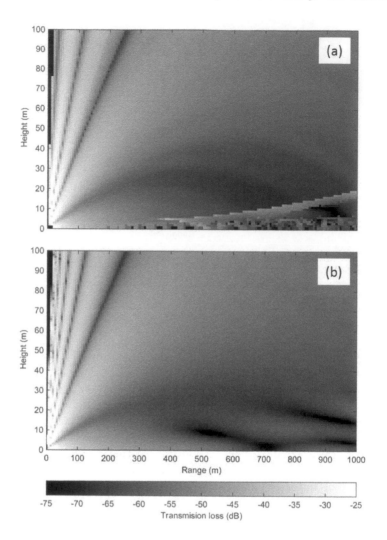

FIGURE 10.9
Transmission loss at 125 Hz for the downward refraction case shown in figure 10.8 as calculated by (a) ray tracing and (b) the wide-angle CNPE.

Here, G_j is the number of ground reflections for the jth path, and $p_j(x, z)$ is calculated from the square root of the right-hand side of equation (10.75) with $\theta = \theta_j(x, z)$ and $d\sigma_{\mathrm{ray}} = d\sigma_j(x, z)$.

Let us consider now ray tracing for downward refraction with a linear wind speed profile, namely $v_x(z) = 0.1z$ (m/s), where 0.1 s^{-1} is the gradient. The sound speed is constant at 340 m/s. Although a linear profile is rather

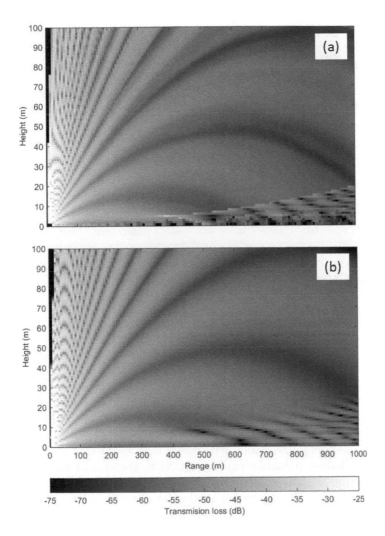

FIGURE 10.10
Same as figure 10.9, except for 500 Hz.

unrealistic for the atmosphere (more realistic profiles are approximately logarithmic, as discussed in section 3.5.4), this case provides a good numerical example because of the pronounced caustics and refraction at higher altitudes. The ray paths for a source height of 5 m are shown in figure 10.8. Rays were launched at elevation angles between $-80°$ and $80°$, in $0.5°$ increments. The MATLAB® ode23 function [357], which implements an explicit Runge-Kutta (2,3) method, was used to step equations (10.72) and (10.73) forward in time

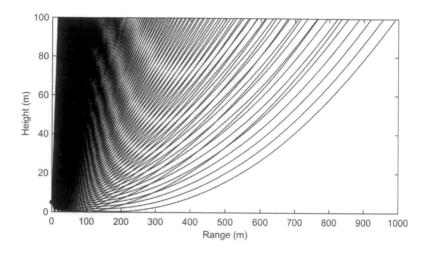

FIGURE 10.11
Ray paths for upward refraction with a linear velocity profile $v_x(z) = -0.1z$
(m/s). The source height is 5 m.

increments of 0.03 s. The relative tolerance was set to 10^{-3} and the absolute
tolerance to 10^{-10}. Note the formation of a caustic starting at a range of about
300 m and height of 5 m, and sloping upward at longer ranges. Caustics at
lower heights are also present. Ray paths with turning points (slopes of zero)
are also clearly evident.

Figures 10.9 and 10.10 show the transmission loss (TL) as calculated from
the rays in figure 10.8, along with the TL as calculated with a wide-angle
Crank–Nicholson parabolic equation (CNPE, sections 11.2.1 and 11.2.2) for
the same profile. The source frequencies for the two figures are 125 and 500
Hz, respectively. The relaxation model described in section 10.2.3 was used
to model the ground properties, with the static flow resistivity $\sigma_0 = 2 \times 10^5$
$\mathrm{Pa\,s\,m^{-2}}$, porosity $\Omega = 0.3$, tortuosity $q = \Omega^{-0.25}$, and shape factors all set
to 1, resulting in $Z/\varrho_0 c_0 = 16.3 + i15.5$ at 125 Hz and $8.8 + i7.2$ at 500 Hz.
The TL predictions from ray tracing are fairly similar to the CNPE at heights
above the caustic formation. Below this height, however, ray tracing does not
describe the interference pattern well, although it is qualitatively better at
500 Hz than at 125 Hz.

Calculations for a linear sound-speed profile $c(z) = 340 + 0.1z$ (m/s) in
a non-moving atmosphere ($v_x = 0.0$), although not shown here, are nearly
indistinguishable from figures 10.8–10.10, thus indicating the suitability of
the effective sound-speed approximation for this example. This topic will be
revisited in section 11.2.2, for a more realistic logarithmic profile.

Figures 10.11 and 10.12 repeat figures 10.8 and 10.10, except that the sign
on the velocity gradient has been changed; i.e., $v_x(z) = -0.1z$ (m/s). This

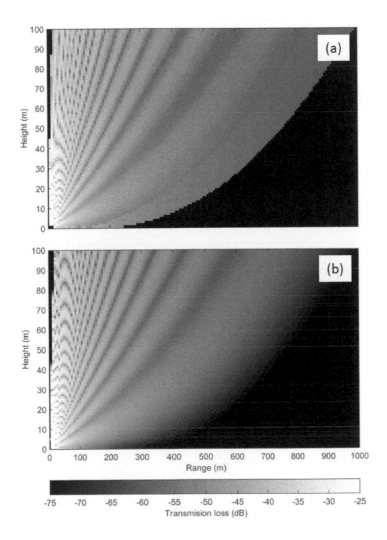

FIGURE 10.12
Transmission loss at 500 Hz for the upward refraction case shown in figure 10.11 as calculated by (a) ray tracing and (b) the wide-angle CNPE.

results in upward refraction and formation of a shadow zone. The TL from the ray tracing agrees fairly well with the CNPE, except in the vicinity of the shadow zone boundary. There, ray tracing predicts an unrealistic, perfectly sharp boundary between the illuminated and shadow regions.

11

Wave-based frequency-domain methods

Many numerical methods for sound propagation are based on solution of the full wave equation, or related equations, after application of a Fourier transform to the time coordinate. We call such methods *frequency-domain* methods. The primary advantage of this approach is that the explicit dimensionality of the problem is reduced: the time-domain problem is replaced by independent solutions calculated at each frequency of interest. When a time-domain or broadband solution is desired, however, the computational benefits of frequency-domain methods are diminished.

Wavenumber integration [100, 220, 224, 321, 332, 345, 404, 411] and the parabolic equation (PE) [35, 89, 137, 135, 345, 403] are two of the most popular frequency-domain methods. The fast-field program (FFP) refers to wavenumber integration in which the integral is calculated with a fast Fourier transform (FFT). These techniques were introduced into atmospheric acoustics in the late 1980s to avoid the high-frequency approximations inherent to ray-tracing methods. Wavenumber integration and PE solvers are both usually (but not always) formulated for one-way wave propagation; that is, waves propagate outward from the source but not back toward the source. Due to this restriction, these methods are not well suited to situations where reflections from buildings or barriers affect the sound level at the receiver locations of interest. Many PE solvers are further restricted to a finite-width beam emanating from the source, and thus are less useful for near-vertical propagation. Wavenumber integration generally assumes a horizontally stratified (layered) propagation medium. Therefore wavenumber integration, in its normal form, cannot be applied to problems involving turbulence or variations in terrain elevation. PEs, on the other hand, can be applied to horizontally inhomogeneous environments and can therefore handle turbulence and uneven terrain; this largely explains the increasing popularity of PEs.

Frequency-domain boundary-element methods (BEM) are widely used in computational acoustics to handle surfaces with complex geometries. While they are best suited to boundaries with a homogeneous medium, Premat and Gabillet [316] devised an approach to BEM capable of including both a noise barrier and refraction. Still, efficient application of BEM to inhomogeneous and moving media remains a difficult problem, and thus this chapter focuses on wavenumber integration and the PE.

Although the frequency-domain methods described in this section are not particularly useful for calculating the sound that is directly scattered and absorbed by vegetation, they can be useful for calculating indirect impacts of vegetation, which are particularly important for lower audible frequencies. Such impacts include modification of the ground impedance, and of the wind and temperature profiles within the canopy [382].

References [17] and [97] provide benchmark calculations which are appropriate for testing FFP and PE calculations in a refractive atmosphere.

11.1   Wavenumber integration

11.1.1   Background

The basis for wavenumber-integration methods is a Fourier transform of the acoustic field with respect to the horizontal spatial variables and time. For the acoustic pressure, for example,

$$P(\kappa_x, \kappa_y, z, \omega) = \frac{1}{8\pi^3} \int \int \int p(x, y, z, t) e^{i\omega t - i\boldsymbol{\kappa} \cdot \mathbf{r}} \, dt \, dx \, dy, \qquad (11.1)$$

where $P(\kappa_x, \kappa_y, z, \omega)$ is the transformed pressure field. The limits on the Fourier transforms are implicitly $-\infty$ to $+\infty$. In this section, we use an uppercase letter (rather than the "hat" used in Chapter 2) to indicate the Fourier transform with respect to the horizontal spatial variables and time. The hat notation, e.g., $\hat{p}$, will be reserved for the Fourier transform with respect to time only. The inverse transform corresponding to equation (11.1) is

$$p(x, y, z, t) = \int \int \int P(\kappa_x, \kappa_y, z, \omega) e^{-i\omega t + i\boldsymbol{\kappa} \cdot \mathbf{r}} \, d\omega \, d\kappa_x \, d\kappa_y. \qquad (11.2)$$

Conventional wavenumber integration assumes that the propagation medium is horizontally stratified and frozen in time, so that the atmospheric fields are constant with respect to the preceding Fourier transform pair. For this situation, it was shown in Chapter 2 that a function $\tilde{P}(\kappa_x, \kappa_y, z, \omega)$ (using the notation $\tilde{P}$ here in place of $\tilde{p}$) can be defined which obeys the following relatively simple ordinary differential equation (2.64):

$$\frac{\partial^2 \tilde{P}}{\partial z^2} + \left[\gamma^2(z) + \frac{f''}{2f} - \frac{3}{4} \left(\frac{f'}{f} \right)^2 \right] \tilde{P} = i\omega (\varrho \varrho_0)^{1/2} Q, \qquad (11.3)$$

where, by comparing equation (2.63) to (11.2), we make the identification

$$P(\kappa_x, \kappa_y, z, \omega) = \frac{\sigma}{\omega} \left(\frac{\varrho}{\varrho_0} \right)^{1/2} \tilde{P}(\kappa_x, \kappa_y, z, \omega). \qquad (11.4)$$

In the preceding equations, primes indicate partial derivatives with respect to z, $\varrho(z)$ is the density profile, ϱ_0 is a reference value for the density, $\mathbf{v}$ is the horizontal wind (the subscript $\perp$ having been dropped for brevity), $\sigma = \omega - \boldsymbol{\kappa} \cdot \mathbf{v}$, $\gamma^2 = \sigma^2/c^2 - \boldsymbol{\kappa} \cdot \boldsymbol{\kappa}$ (equivalent to q^2 in Chapter 2), $Q(\kappa_x, \kappa_y, z, \omega)$ is the Fourier transform of the source strength, and

$$\frac{f''}{2f} - \frac{3}{4}\left(\frac{f'}{f}\right)^2 = -\frac{\boldsymbol{\kappa} \cdot \mathbf{v}''}{\sigma} - 2\left(\frac{\boldsymbol{\kappa} \cdot \mathbf{v}'}{\sigma}\right)^2 + \frac{\varrho'}{\varrho}\frac{\boldsymbol{\kappa} \cdot \mathbf{v}'}{\sigma} + \frac{\varrho''}{2\varrho} - \frac{3}{4}\left(\frac{\varrho'}{\varrho}\right)^2. \quad (11.5)$$

In practice, the temperature profile $T(z)$ is usually specified, rather than the density profile $\varrho(z)$. These profiles are related through equation (2.41). In particular, the density derivatives can be derived as:

$$\frac{\varrho'(z)}{\varrho(z)} = -\left(\frac{T'}{T} + \frac{g}{TR_a}\right) \quad (11.6)$$

and

$$\frac{\varrho''(z)}{\varrho(z)} = -\left(\frac{T'}{T} + \frac{g}{TR_a}\right)^2 - \frac{T''}{T}. \quad (11.7)$$

Thus, given the profiles $T(z)$ and $\mathbf{v}(z)$, we can in principle solve equation (11.3), calculate the transformed pressure from equation (11.4), and then apply an inverse transformation to determine the pressure field.

In the following, we consider a point mass source at $(x, y, z) = (0, 0, z_s)$, where $z = z_s$ is the source height, in which case $q(x, y, z, t) = q(t)\delta(x)\delta(y)\delta(z - z_s)$. Substitution into the definition of the forward transform, equation (11.1), yields

$$Q(\kappa_x, \kappa_y, z, \omega) = Q(\omega)\delta(z - z_s), \quad (11.8)$$

where $Q(\omega) = \hat{q}(\omega)/4\pi^2$ and

$$\hat{q}(\omega) = \frac{1}{2\pi}\int_{-\infty}^{\infty} q(t)\,e^{i\omega t}\,dt. \quad (11.9)$$

The quantity $\hat{q}$ is related to the complex source amplitude $\hat{A}$ by

$$4\pi\hat{A}(\omega) = -i\omega\rho_0\hat{q}(\omega). \quad (11.10)$$

A value of $\hat{A} = 1$ is referred to as a *unit-amplitude* source, and produces a transmission loss of 0 dB at a distance of $R = 1$ m when the source radiates into free space.

11.1.2 Solution for a layered system

Most formulations of wavenumber integration involve partitioning the profiles into a number of layers (elements), as shown in figure 11.1. Each layer has a constant density ϱ, sound speed c, wind velocity $\mathbf{v}$, and (if desired) attenuation

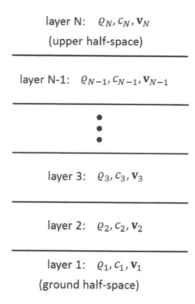

layer N: $\varrho_N, c_N, \mathbf{v}_N$

(upper half-space)

layer N-1: $\varrho_{N-1}, c_{N-1}, \mathbf{v}_{N-1}$

layer 3: $\varrho_3, c_3, \mathbf{v}_3$

layer 2: $\varrho_2, c_2, \mathbf{v}_2$

layer 1: $\varrho_1, c_1, \mathbf{v}_1$

(ground half-space)

FIGURE 11.1

Partitioning of the vertical profiles for density, sound speed, and wind velocity into constant-valued layers.

coefficient. The ground layer has zero velocity. A complex density may be used for the ground layer, so as to represent a rigid, dissipative porous medium (section 10.2.2). Coefficients for the upward and downward propagating waves in each layer are determined by enforcing radiation boundary conditions at the top and bottom of the domain, and continuity conditions at the layer interfaces. After solving a linear system of equations for the coefficients, the result is transformed back to the spatial domain.

Early versions of the FFP were based on a transmission line, or impedance, analogy [220]. Later versions used a somewhat simpler and numerically stable global-matrix method [351]. This section considers the latter approach, as applied to propagation in a horizontally stratified, moving medium. As the starting point, we take Pierce's [314] equations for the velocity quasi-potential, ψ, which appear as equations (2.92) and (2.93) in this book. Within a layer in which the density and sound speed are constant, these equations can be written

$$p = \varrho \frac{d\psi}{dt} \tag{11.11}$$

and

$$\nabla^2 \psi - \frac{1}{c^2} \frac{d^2 \psi}{dt^2} = -q\left(t\right) \delta\left(x\right) \delta\left(y\right) \delta\left(z - z_s\right), \tag{11.12}$$

where $d/dt = \partial/\partial t + \mathbf{v}\cdot\nabla$ and $\mathbf{v} = (v_x, v_y, 0)$ is the ambient fluid velocity, which is assumed here to be steady and directed horizontally. We have also included in equation (11.12) an explicit point-source term at $(x, y, z) = (0, 0, z_s)$, as described previously. The particle displacement $\mathbf{w}$ and potential are related by

$$\frac{d\mathbf{w}}{dt} = \nabla\psi. \tag{11.13}$$

(Note that, in this chapter, $\mathbf{w}$ is used for the acoustic particle displacement rather than velocity.)

The next step involves taking the Fourier transform of equations (11.11)–(11.13) with respect to the horizontal coordinates and time. From equation (11.2), one can show that the partial derivatives transform as $\partial/\partial t \to -i\omega$, $\partial/\partial x \to i\kappa_x$, and $\partial/\partial y \to i\kappa_y$. We thus find [411]

$$P = -i\varrho\sigma\Psi, \tag{11.14}$$

$$\frac{d^2\Psi}{dz^2} + \gamma^2\Psi = -Q(\omega)\,\delta(z - z_s), \tag{11.15}$$

and

$$\mathbf{W} = \frac{1}{\sigma}\left(-\kappa_x, -\kappa_y, i\frac{d}{dz}\right)\Psi. \tag{11.16}$$

Equation (11.15) is recognized as a Helmholtz equation, with homogeneous solution

$$\Psi^h(z) = A^+\exp(i\gamma z) + A^-\exp(-i\gamma z), \tag{11.17}$$

where the terms represent upward and downward traveling waves, respectively, and A^+ and A^- are coefficients to be determined. The particular part of the solution is [264]

$$\Psi^p(z) = \frac{iQ(\omega)}{2\gamma}\exp(i\gamma|z - z_s|). \tag{11.18}$$

The homogeneous solution, equation (11.17), holds in each layer n, with substitution of the parameters c_n, ϱ_n, and $\mathbf{v}_n$ corresponding to that layer. The acoustic pressure in layer n is thus

$$P_n^h(z) = -i\varrho_n\sigma_n\left[A_n^+\exp(i\gamma_n z) + A_n^-\exp(-i\gamma_n z)\right], \tag{11.19}$$

whereas the vertical component of the particle displacement is

$$W_n^h(z) = \frac{\gamma_n}{\sigma_n}\left[A_n^+\exp(i\gamma_n z) - A_n^-\exp(-i\gamma_n z)\right]. \tag{11.20}$$

Since the waves in the lower half-space (layer 1, the ground layer) propagate downward only, and the waves in the upper half-space (layer N) propagate upward only, A_1^+ and A_N^- are zero. With the exclusion of A_1^+ and A_N^-, there are $2N - 2$ coefficients to be determined. These can be arranged in a column vector $\mathbf{A}$ as follows:

$$\mathbf{A} = \begin{bmatrix} A_1^- & A_2^+ & A_2^- & \cdots & A_{N-1}^+ & A_{N-1}^- & A_N^+ \end{bmatrix}^T, \tag{11.21}$$

where the superscript T indicates matrix transposition.

From the homogeneous solutions, we can construct $(N - 1) \times 1$ column vectors $\mathbf{P}_-^h$ and $\mathbf{P}_+^h$ of the pressures immediately below and above the layer interfaces, respectively:

$$\mathbf{P}_-^h = \begin{bmatrix} -i\varrho_1\sigma_1 A_1^- \\ -i\varrho_2\sigma_2 \left(A_2^+ e^{i\gamma h_2} + A_2^- e^{-i\gamma h_2} \right) \\ -i\varrho_3\sigma_3 \left(A_3^+ e^{i\gamma h_3} + A_3^- e^{-i\gamma h_3} \right) \\ \vdots \\ -i\varrho_{N-1}\sigma_{N-1} \left(A_{N-1}^+ e^{i\gamma h_{N-1}} + A_{N-1}^- e^{-i\gamma h_{N-1}} \right) \end{bmatrix} \tag{11.22}$$

and

$$\mathbf{P}_+^h = \begin{bmatrix} -i\varrho_2\sigma_2 \left(A_2^+ + A_2^- \right) \\ -i\varrho_3\sigma_3 \left(A_3^+ + A_3^- \right) \\ \vdots \\ -i\varrho_{N-1}\sigma_{N-1} \left(A_{N-1}^+ + A_{N-1}^- \right) \\ -i\varrho_N\sigma_N A_N^+ \end{bmatrix}. \tag{11.23}$$

Next, we construct analogous column vectors $\mathbf{W}_-^h$ and $\mathbf{W}_+^h$ of the vertical displacements immediately below and above the layer interfaces:

$$\mathbf{W}_-^h = \begin{bmatrix} -\gamma_1 A_1^- /\sigma_1 \\ \gamma_2 \left(A_2^+ e^{i\gamma h_2} - A_2^- e^{-i\gamma h_2} \right) /\sigma_2 \\ \gamma_3 \left(A_3^+ e^{i\gamma h_3} - A_3^- e^{-i\gamma h_3} \right) /\sigma_3 \\ \vdots \\ \gamma_{N-1} \left(A_{N-1}^- e^{i\gamma h_{N-1}} + A_{N-1}^- e^{-i\gamma h_{N-1}} \right) /\sigma_{N-1} \end{bmatrix} \tag{11.24}$$

and

$$\mathbf{W}_+^h = \begin{bmatrix} \gamma_2 \left(A_2^+ - A_2^- \right) /\sigma_2 \\ \gamma_3 \left(A_3^+ - A_3^- \right) /\sigma_3 \\ \vdots \\ \gamma_N \left(A_{N-1}^+ - A_{N-1}^- \right) /\sigma_N \\ \gamma_n A_N^+ /\sigma_N \end{bmatrix}. \tag{11.25}$$

The homogeneous solution can thus be conveniently written in matrix form as

$$\mathbf{P}_-^h = \mathbf{C}_{P-}\mathbf{A}, \tag{11.26}$$

$$\mathbf{P}_+^h = \mathbf{C}_{P+}\mathbf{A}, \tag{11.27}$$

$$\mathbf{W}_-^h = \mathbf{C}_{W-}\mathbf{A}, \tag{11.28}$$

and

$$\mathbf{W}_+^h = \mathbf{C}_{W+}\mathbf{A}, \tag{11.29}$$

where the $\mathbf{C}$'s are $(N-1) \times 2(N-1)$ matrices, the elements of which follow from equations (11.22)–(11.25).

The particular solution for the pressure, within a layer occupied by a source, is determined from equations (11.14) and (11.18):

$$P_n^p(z) = \frac{\varrho_n \sigma_n Q(\omega)}{2\gamma_n} \exp\left(i\gamma_n |z - z_s|\right). \tag{11.30}$$

The particular solution for the vertical displacement is

$$W_n^p(z) = \frac{iQ(\omega)}{2\sigma_n} \text{sign}(z - z_s) \exp\left(i\gamma |z - z_s|\right). \tag{11.31}$$

Similarly to the homogeneous solution, we can construct $(N-1) \times 1$ column vectors $\mathbf{P}_-^p$, $\mathbf{P}_+^p$, $\mathbf{W}_-^p$, and $\mathbf{W}_+^p$ containing the particular solution immediately below and above the interfaces. For a single source, these vectors are zero everywhere, except in the layer of the source. Multiple sources are handled by summing particular solutions for each source.

The sum of the particular and homogeneous solutions for the pressure and vertical particle displacement must be continuous at each interface. The resulting system of $2(N-1)$ equations, for the $2(N-1)$ coefficients, can be conveniently written in matrix form as

$$\left[\begin{array}{c} \mathbf{C}_{P+} - \mathbf{C}_{P-} \\ (\mathbf{C}_{W+} - \mathbf{C}_{W-})\,\omega\varrho_0 c_0 \end{array} \right] [\ \mathbf{A}\ ] = -\left[\begin{array}{c} \mathbf{P}_+^p - \mathbf{P}_-^p \\ (\mathbf{W}_+^p - \mathbf{W}_-^p)\,\omega\varrho_0 c_0 \end{array} \right]. \tag{11.32}$$

The coefficients $\mathbf{A}$ can now be determined using commonly available solvers for linear matrix equations. In the literature, this approach is referred to as the *global-matrix method* [351]. The matrix system is sparse when there are many layers, which enables relatively efficient solution. Once the coefficients have been determined, the sound pressures follow immediately from equation (11.19). In the source layer(s), the particular solution must be added.

An example solution for $P(\kappa_x, \kappa_y, z, \omega)$, for sound propagation in a homogeneous, motionless atmosphere, is shown in figure 11.2. In this case, the solution is radially symmetric; that is, it depends only on $\kappa = (\kappa_x^2 + \kappa_y^2)^{1/2}$. The source has unit amplitude and frequency $f = 125$ Hz. The source height is 1.5 m, and the receiver height 1.2 m. Only three layers are needed for an exact solution of this problem: one for the finite atmospheric layer containing the source (taken somewhat arbitrarily here as 0 m $\leq z \leq$ 100 m), one for the atmospheric above the source extending to infinity ($z \geq$ 100 m), and one for

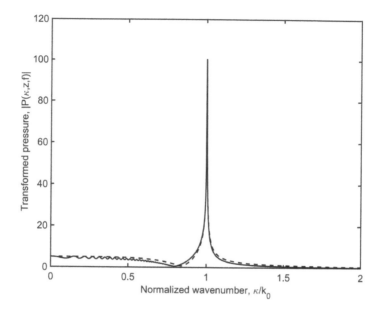

FIGURE 11.2
Magnitude of the transformed pressure in the horizontal wavenumber domain, $|P(\kappa, z, \omega)|$, as a function of the normalized radial wavenumber κ/k_0. The calculation is for propagation through a homogeneous atmosphere above a porous ground at $f = 125$ Hz, with a source height of 1.5 m and receiver height 1.2 m. Solutions are shown for the layered-medium (global-matrix) method (dashed line) and the tridiagonal matrix method (solid line).

the ground half-space ($z \leq 0$ m). Values of $c_0 = 340$ m/s and $\rho_0 = 1.2$ kg/m^3 were used for the atmosphere. Ground properties were modeled with the relaxation model (section 10.2.2), using a static flow resistivity of 200 kPa s/m^2, porosity 0.515, and tortuosity 1.18. These values, which are representative of compact soil, lead to a vorticity relaxation time of $\tau_{\text{vor}} = 0.325$ μs and entropy relaxation time of $\tau_{\text{ent}} = 0.235$ μs. The peak around $\kappa \approx k_0$ in figure 11.2 is the freely propagating part of the solution; lower wavenumbers represent interactions with the ground. Wavenumbers such that $\kappa > k_0$ represent evanescent waves.

Figure 11.4 shows $P(\kappa_x, \kappa_y, z, \omega)$ in the horizontal wavenumber plane (κ_x, κ_y) for a case involving propagation through a windy atmosphere. The scenario is the same as the previous example, except that a logarithmic wind profile with friction velocity $u_* = 0.6$ m/s and roughness length 0.05 m is present, and the frequency has been increased to $f = 250$ Hz. The global-matrix method, with layers 1-m thick, was used. Figure 11.3(a) is for a source height of 1.5 m, whereas figure 11.3(b) is for a source height of 50 m. Plotted in this manner, the calculations appear nearly independent of azimuth. For the lower source height, multiple modes and evanescent waves are evident. For the higher height, there is essentially no energy in the evanescent waves $(\kappa > k_0)$.

Figure 11.4 is for the same case as figure 11.3. Individual curves are shown for three propagation directions, namely upwind $(k_x > 0, \ k_y = 0)$, crosswind $(k_x = 0, \ k_y > 0)$, and downwind $(k_x < 0, \ k_y = 0)$. Some differences in $P(\kappa_x, \kappa_y, z, \omega)$ become clearer when the curves are overlaid in this manner, particularly in the vicinity of $\kappa/k_0 \approx 1$. Later, in section 11.1.4, we will see that these rather subtle differences have important impacts on the propagation.

11.1.3 Tridiagonal matrix solution

In this subsection, we consider an alternative approach to wavenumber integration, based on solving equation (11.3) directly. Unlike the global-matrix method described in the previous section, the one described here leads to a tridiagonal matrix, which is more efficient to solve; it also provides a useful conceptual bridge between the FFP to the Crank–Nicholson PE (CNPE), which will be described in the next section. However, we will see that the finer vertical discretization and handling of the boundary conditions (BCs) characteristic of the tridiagonal solution are significant drawbacks.

We begin by discretizing the vertical coordinate axis into N points with constant spacing Δz; that is, $z_n = n\Delta z$, $n = 1, 2, \ldots, N$, and $\tilde{P}_n = \tilde{P}(z_n)$, etc. The second-order derivative can be approximated as a centered finite difference, namely

$$\frac{\partial^2 \tilde{P}_n}{\partial z^2} \simeq \frac{\tilde{P}_{n-1} - 2\tilde{P}_n + \tilde{P}_{n+1}}{(\Delta z)^2}. \tag{11.33}$$

When the $\tilde{P}_n$ are grouped into an $N \times 1$ column matrix $\tilde{\mathbf{P}}$, the differentiation can be represented in numerical form as $\mathbf{T}\tilde{\mathbf{P}}/(\Delta z)^2$, where $\mathbf{T}$ is a tridiagonal matrix with values of -2 along the main diagonal, and values of one along the sub- and super-diagonals. (Exceptions to this pattern are needed to represent the boundary conditions, as will be discussed shortly.) Let us also define

$$b_n = \gamma^2(z_n) + \frac{f''(z_n)}{2f(z_n)} - \frac{3}{4}\left[\frac{f'(z_n)}{f(z_n)}\right]^2, \tag{11.34}$$

$$\mathbf{B} = \text{diag}\left([b_1 \ b_2 \ \cdots \ b_N]\right), \tag{11.35}$$

where diag($\cdot$) indicates a diagonal matrix formed from the vector argument,

$$q_n = i\omega(\rho\rho_0)^{1/2}\tilde{Q}_n, \tag{11.36}$$

and

$$\mathbf{q} = [q_1 \ q_2 \cdots q_N]^T, \tag{11.37}$$

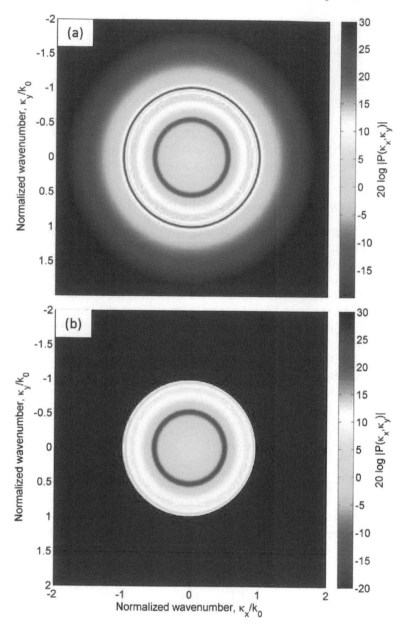

FIGURE 11.3

Magnitude of the transformed pressure in the horizontal wavenumber domain, $20 \log |P(\kappa_x, \kappa_y, z, \omega)|$, for propagation through a windy atmosphere above an impedance ground, based on the global-matrix method. The receiver height is 1.2 m and the frequency is 250 Hz. (a) Source height of 1.5 m. (b) Source height of 50 m.

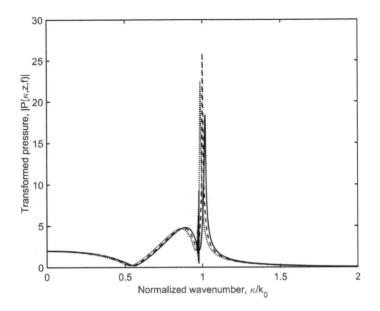

FIGURE 11.4
Same as figure 11.3(a), except that curves are shown for three particular directions: upwind (dashed line), crosswind (solid line), and downwind (dotted line).

where the superscript T indicates the matrix transpose. Then equation (11.3) can be written in the compact matrix form

$$\left[\mathbf{T}(\Delta z)^{-2} + \mathbf{B}\right]\tilde{\mathbf{P}} = \mathbf{q}. \tag{11.38}$$

Application of the finite-difference approximation, equation (11.33), requires knowledge of $\tilde{P}_0 = \tilde{P}(z_0)$ and $\tilde{P}_{N+1} = \tilde{P}(z_{N+1})$ in order to determine the second derivative at $z = z_1$ and $z = z_n$. Hence numerical solution requires BCs in which $\tilde{P}_0$ and $\tilde{P}_{N+1}$ are related to levels at which $\tilde{P}(z)$ is known. At the ground, the numerical BC may be taken to have the following general form, which gives $\tilde{P}_0$ in terms of the $\tilde{P}_1$ and $\tilde{P}_2$:

$$\tilde{P}_0 = \eta_1 \tilde{P}_1 + \eta_2 \tilde{P}_2, \tag{11.39}$$

where η_1 and η_2 are constants. Assuming the wind vanishes at the surface, and the surface is locally reacting in the acoustical sense, the BC is given by equation (4.62) as $\partial \tilde{P}/\partial z = -ik_0\beta(\omega)\tilde{P}$, where $\beta(\omega)$ is the normalized surface admittance. A first-order implementation of the boundary condition may be derived by approximating $\partial \tilde{P}/\partial z$ at $z = 0$ as $(\tilde{P}_1 - \tilde{P}_0)/\Delta z$. Then $\tilde{P}_1 = (1 - ik_0\Delta z\beta)\tilde{P}_0$, from which

$$\eta_1 = (1 - ik_0\Delta z\beta)^{-1}, \quad \eta_2 = 0. \tag{11.40}$$

As suggested by West et al. [403], a second-order accurate BC follows from the approximation $\partial \tilde{P}/\partial z \approx (3/2)(\tilde{P}_1 - \tilde{P}_0)/\Delta z - (1/2)(\tilde{P}_2 - \tilde{P}_1)/\Delta z$. This results in

$$\eta_1 = 4(3 - 2ik_0\Delta z\beta)^{-1}, \quad \eta_2 = -(3 - 2ik_0\Delta z\beta)^{-1}. \tag{11.41}$$

Note that, by assuming a locally reacting boundary, the solution may be less accurate than the one described in the previous section, for which the ground was a homogeneous half-space.

Similarly, at the upper boundary we can consider a condition of the form

$$\tilde{P}_{N+1} = \eta_N \tilde{P}_N + \eta_{N-1}\tilde{P}_{N-1}. \tag{11.42}$$

The top of the domain, $z = z_N$, should be made high enough that it does not significantly impact the solution in the region of interest. It is reasonable to model heights such that $z \geq z_{N-1}$ as having vanishing gradients in $T(z)$ and $\mathbf{v}(z)$. Then, at these heights, equation (11.3) simplifies to $\partial^2 \tilde{P}/\partial z^2 + \gamma_n^2 \tilde{P} = 0$, the solution of which is $\tilde{P}(z) = A_+ e^{i\gamma_N z} + A_- e^{-i\gamma_N z}$, where A_+ and A_- are coefficients to be determined. We set $A_- = 0$, so that only upward traveling waves exist when $z \geq z_N$. Hence it follows $\partial \tilde{P}/\partial z = i\gamma_N \tilde{P}$ at $z = z_{N+1}$. The analysis then proceeds in the same manner as with the lower boundary, except that γ_N replaces $k_0\beta$. We thus have the first-order BC

$$\eta_N = (1 - i\gamma_N\Delta z)^{-1}, \quad \eta_{N-1} = 0, \tag{11.43}$$

and the second-order BC

$$\eta_N = 4(3 - 2i\gamma_N\Delta z)^{-1}, \quad \eta_{N-1} = -(3 - 2i\gamma_N\Delta z)^{-1}. \tag{11.44}$$

With the BC formulation, the matrix $\mathbf{T}$ now takes the following form:

$$\mathbf{T} = \begin{bmatrix} -2+\eta_1 & 1+\eta_2 & 0 & \cdots & & & \\ 1 & -2 & 1 & \cdots & & & \\ 0 & 1 & -2 & 1 & \cdots & & \\ \vdots & \vdots & & \ddots & \ddots & \ddots & \\ & & \cdots & 1 & -2 & 1 \\ & & \cdots & 0 & 1+\eta_{N-1} & -2+\eta_N \end{bmatrix}. \tag{11.45}$$

Because the matrix $\mathbf{T}(\Delta z)^{-2} + \mathbf{B}$ premultiplying $\tilde{P}$ in equation 11.38 is an $N \times N$ tridiagonal matrix, solution of the linear system involves only $O(N)$ floating-point operations, and is thus highly efficient. For accurate solution, Δz should be a small fraction of the wavelength, usually about 0.1λ.

A solution for $P(\kappa_x, \kappa_y, z, \omega)$ based on the tridiagonal method with second-order BCs, for propagation in a homogeneous atmosphere above impedance ground, is compared to the global-matrix solution for the same problem in figure 11.2. The solutions are very similar, although some discrepancies are evident. These are caused primarily by the differences in the BCs. The ripples for small κ/k_0 in the tridiagonal method are artifacts of the upper BC. (When the first-order upper BC is used, the ripples become worse.) There is also a systematic bias for $\kappa/k_0 \lesssim 1$ due to the locally reacting impedance BC. It will be shown in section 11.1.5, however, that these artifacts have little discernible impact on the solution, at least for this example.

Although the solution method described here yields a tridiagonal system, the dimensions of the matrix are rather large. For the computational domain $0 \text{ m} \le z \le 100 \text{ m}$, with $N = (100\,\text{m})/(0.1\lambda) = 368$. Recall that the global-matrix solution for this problem had only 3 layers, and thus involved a much smaller 6×6 matrix (although not a tridiagonal one). The net result was that the computational time for the tridiagonal method was about 4.5 times longer. For more complicated problems involving more atmospheric layers, however, the tridiagonal method may be preferable.

11.1.4 Inverse transform in a horizontal plane

The previous section described efficient solution of the wave equation for a stratified, moving medium in the frequency/horizontal wavenumber domain. In this subsection and the next, transformation of the solution back to the spatial domain is described. Transformation from the frequency to the time domain is not described here (since, in many applications, the frequency-domain calculation is the desired end result), although it can be accomplished by similar methods.

The integral to be solved is

$$\hat{p}(x, y, z, \omega) = \int \int P(\kappa_x, \kappa_y, z, \omega)e^{i\boldsymbol{\kappa}\cdot\mathbf{r}}\, d\kappa_x\, d\kappa_y. \tag{11.46}$$

Generally, the function $P(\kappa_x, \kappa_y, z, \omega)$ will be concentrated at values of κ less than or in the vicinity of ω/c_0. We may then replace the integrations over the full (κ_x, κ_y)-plane by integrations between $-\kappa_{\max} \le (\kappa_x, \kappa_y) < \kappa_{\max}$, where $\kappa_{\max}$ is somewhat greater than ω/c_0. Let us discretize the wavenumbers over this region as follows:

$$\kappa_x = \kappa_{x,m} = -\kappa_{\max} + m\Delta\kappa, \tag{11.47}$$

$$\kappa_y = \kappa_{y,n} = -\kappa_{\max} + n\Delta\kappa, \tag{11.48}$$

where $\Delta\kappa = 2\kappa_{\max}/M$, and the indices m and n range from 0 to $M-1$. The horizontal spatial coordinates are discretized as

$$x = x_j = j\Delta r, \tag{11.49}$$

$$y = y_k = k\Delta r, \tag{11.50}$$

where $\Delta r = \pi/\kappa_{\max}$, and the indices j and k also range from 0 to $M-1$. We can now write equation (11.46) in the form of a discrete 2D inverse Fourier transform:

$$\hat{p}(x_j, y_k, z, \omega) = (-1)^{j+k}(\Delta\kappa)^2 \sum_{m=0}^{M-1}\sum_{n=0}^{M-1} P(\kappa_{x,m}, \kappa_{y,n}, z, \omega)e^{i2\pi(mj+nk)/M}.$$

$$(11.51)$$

The preceding equation can be solved efficiently with a 2D fast Fourier transform (FFT). The calculated $\hat{p}$ will cover the spatial region $0 \leq (x, y) < 2r_{\max}$, where $r_{\max} = 2\pi/\Delta\kappa$. Since the calculation is harmonic with period $r_{\max}$, in effect the source is positioned at each of the four corners of the domain. The source can be moved to the middle of the domain by swapping the left and right sides of the matrix representation of $\hat{p}$, and then the top and bottom.

The maximum usable range in a calculation is one-half $r_{\max}$, or $r_f = \pi M/2\kappa_{\max}$. This distance is analogous to the so-called Nyquist or folding frequency in conventional time-frequency analysis. Since $\kappa_{\max}$ is proportional to the frequency, the maximum range of the computation *decreases* as the frequency *increases*. Thus wavenumber integration, like other wave-based methods, is best suited to low frequencies.

Figure 11.5 shows examples of the TL as calculated in a horizontal plane using a 2D inverse FFT. The calculation is for the same windy atmosphere case with $f = 250$ Hz, as considered in figure 11.3. The 2D inverse FFT was applied to 8192×8192 matrices, which were calculated at a wavenumber resolution of $\Delta\kappa/k_0 = 4.88 \times 10^{-4}$. The top panel is for a source height of 1.5 m; bottom is for 50 m. At the lower source height, there is a minimum in the downwind direction at about 200 m. A strong refractive shadow forms in the upwind direction. At the higher source height, the formation of a caustic is evident at an upwind distance of about 250 m, and the shadow-zone boundary shifts to longer ranges.

11.1.5   Radial approximation

In practice, we are often interested in calculating the sound field for a particular azimuth relative to the source. Or, it may be computationally infeasible to perform a full, 2D calculation in a horizontal plane. In such situations we may approximate the inverse transform, equation (11.46), with a single integral involving the radial distance r and wavenumber κ. As the starting point, we rewrite equation (11.46) in cylindrical coordinates such that $(x, y) = (r\cos\phi, r\sin\phi)$ and $(\kappa_x, \kappa_y) = (\kappa\cos\theta, \kappa\sin\theta)$:

$$\hat{p}(r, \phi, z, \omega) = \int_0^\infty \int_{-\pi}^{\pi} P(\kappa, \theta, z, \omega)e^{i\kappa r\cos(\theta-\phi)}\kappa\, d\theta\, d\kappa. \qquad (11.52)$$

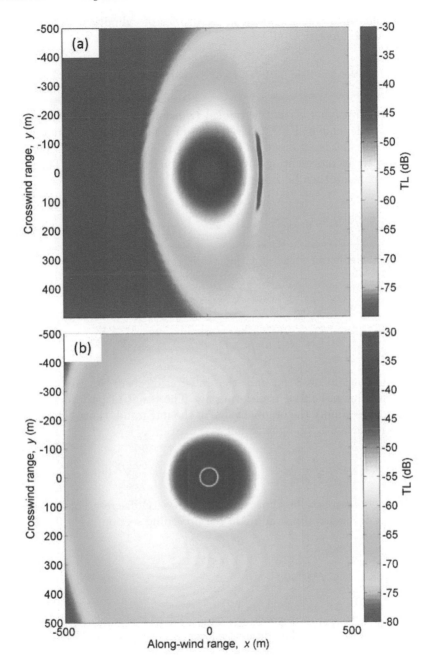

FIGURE 11.5

Transmission loss in a horizontal plane for propagation through a windy atmosphere above an impedance ground, based on the global-matrix method. The receiver height is 1.2 m and the frequency is 250 Hz. The source is in the middle; downwind is to the right, upwind to the left. (a) Source height of 1.5 m. (b) Source height of 50 m.

The exponential can be expanded as a series of cylindrical Bessel functions based on the identity

$$\exp(i\xi\cos\alpha) = J_0(\xi) + 2\sum_{n=1}^{\infty} i^n J_n(\xi)\cos(n\alpha). \tag{11.53}$$

The term involving J_0 is the Hankel transform familiar from the original, azimuth-independent FFP [220]. To obtain a far-field approximation, we replace the Bessel functions with their large-argument approximations, and retain only terms representing outward radiation from the source. The preceding series expansion then becomes

$$\exp(i\xi\cos\alpha) \simeq \sqrt{\frac{1}{2\pi\xi}} e^{i(\xi-\pi/4)} \sum_{n=-\infty}^{\infty} e^{in\alpha}. \tag{11.54}$$

For the inverse transform, equation (11.52), we now have:

$$\hat{p}(r,\phi,z,\omega) \simeq \sqrt{\frac{1}{2\pi r}} e^{-i\pi/4} \sum_{n=-\infty}^{\infty} e^{-in\phi}$$
$$\times \int_0^\infty \int_{-\pi}^\pi P(\kappa,\theta,z,\omega) e^{i(\kappa r+n\theta)} \sqrt{\kappa}\, d\theta\, d\kappa. \tag{11.55}$$

The summation over n, and the integration over θ, comprises a discrete/continuous Fourier transform pair. The result of performing these two operations is simply the original value of the integrand, times the period (2π), with ϕ replacing θ. The final result is

$$\hat{p}(r,\phi,z,\omega) \simeq \sqrt{\frac{1}{2\pi r}} e^{-i\pi/4} \int_0^\infty P(\kappa,\phi,z,\omega) e^{i\kappa r} \sqrt{\kappa}\, d\kappa. \tag{11.56}$$

The integral can now be approximated with a 1D discrete Fourier transform. Following the general procedure in the preceding section, we set

$$\kappa = \kappa_m = m\Delta\kappa, \tag{11.57}$$

where $\Delta\kappa = \kappa_{\max}/M$, and

$$r = r_n = n\Delta r, \ n = 1, 2, \ldots, M, \tag{11.58}$$

where $\Delta r = 2\pi/\kappa_{\max}$. The discretized version of equation (11.56) is then

$$\hat{p}(r_n,\phi,z,\omega) \simeq \sqrt{\frac{1}{2\pi r_n}} e^{-i\pi/4} \Delta\kappa \sum_{m=0}^{M-1} \sqrt{\kappa_m} P(\kappa_m,\phi,z,\omega) e^{i2\pi mn/M}. \tag{11.59}$$

Accurate numerical evaluation of equation (11.59) can present practical difficulties. When the ground is highly reflective and refraction is downward,

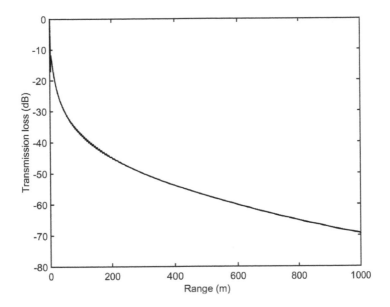

FIGURE 11.6
Transmission loss for propagation through a homogeneous atmosphere above
an impedance ground. The source height is 1.5 m and the receiver height
1.2 m. The frequency is 125 Hz. Shown are solutions by the global-matrix
method (dashed line) and the tridiagonal-matrix method (solid line), and the
theoretical result (dotted line). (The curves are essentially indistinguishable.)

ducted modes may propagate with little energy loss, which can necessitate
incorporation of artificial attenuation to avoid very sharp peaks in P [345].
When there is strong upward refraction, on the other hand, application of an
appropriate window function (such as a Hanning window) prior to evaluation
of the FFT may be needed in order to diminish spurious numerical oscillations,
particularly deep within shadow regions [409].

When the propagation is independent of azimuth (cylindrically symmetric), only the first term in the Bessel series is non-zero, and thus the preceding
approach is exact. Such is the case for a homogeneous atmosphere, as considered for the example shown in figure 11.2. In figure 11.6, we plot the result of
applying the inverse transform to the calculations in figure 11.2. Also plotted
is the theoretical solution [15]. The three curves are essentially indistinguishable.

Shown in figure 11.7 are inverse transforms for the windy atmosphere case
from figures 11.4. Transmission loss curves at $f = 250$ Hz are shown, for
the upwind, crosswind, and downwind directions. The loss is strongest in the
upwind direction, due to upward refraction of the sound and the resulting

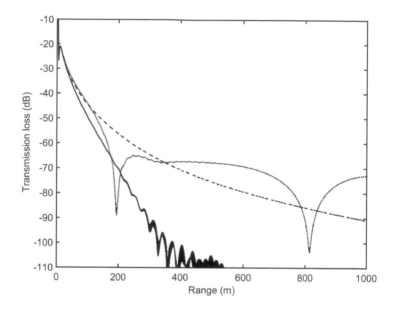

FIGURE 11.7
Transmission loss for propagation through a windy atmosphere above an impedance ground, as calculated by the global-matrix method. The source height is 1.5 m and the receiver height 1.2 m. The frequency is 250 Hz. Shown are calculations for the upwind (dashed line), crosswind (solid line), and downwind (dotted line) directions.

shadow zone. At upwind ranges larger than about 300 m, spurious oscillations appear in the solution; these are numerical artifacts due to windowing effects on the Fourier transform [409], which dominate when the pressure becomes very small. In the downwind direction, the transmission loss is complicated by modal interference.

11.2 Parabolic equation

Initial versions of the PE for atmospheric sound propagation were based on a Crank–Nicholson finite-difference solution [137, 403], known as the CNPE. A faster method based on solving the vertical finite differences with a fast Fourier transform (FFT), known as the Green's function PE (GFPE), was introduced later [135]. The CNPE is described here for its relative simplicity and the ease with which horizontal (range-dependent) atmospheric structure, such as turbulence, can be incorporated.

Numerical PE solutions generally involve a starting (initial) condition for the source field, which is marched forward in the horizontal range coordinate [137, 345, 403]. The wave energy spreads within a confined beam, the angular extent of which depends on the type of parabolic approximation. Narrow-angle approximations (less than $20°$–$30°$) are simplest and most commonly used, but wider angle approximations, which do not depend on an effective sound-speed approximation, have been formulated [89, 228]. PE calculations are normally performed in two-dimensional (2D), vertical planes, at a fixed azimuth from the source, although 3D formulations are available [345]. The primary disadvantage of PEs is that they do not include reflections or scattering in directions outside of the beam. Hence they are unsuitable for problems involving multiple large-angle scattering, in particular.

11.2.1 Narrow-angle Crank–Nicholson solution

As the starting point, we take equation (2.111), although only a 2D (vertical plane) solution is considered here. We thus neglect the horizontal derivatives transverse to the propagation path (the coordinate y) and replace $\Delta_\perp$ with $\partial^2/\partial z^2$. After some rearrangement, the result is:

$$\frac{\partial A}{\partial x} = ik_0 \Psi(x,z) A(x,z), \tag{11.60}$$

in which $A(x,z)$ is the complex amplitude (related to the complex sound pressure through $\hat{p}(x,z) = \exp(ik_0 x) A(x,z)$),

$$\Psi(x,z) = \frac{1}{2k_0^2} \left[\frac{\partial^2}{\partial z^2} + k^2(x,z) - k_0^2 \right], \tag{11.61}$$

and $k^2(x,z) = \omega^2/c_{\text{eff}}^2(x,z)$. The locally reacting boundary condition (BC) at the ground $(z=0)$ is

$$\left. \frac{\partial A(x,z)}{\partial z} \right|_{z=0} = -ik_0 \beta A(x,0), \tag{11.62}$$

in which $\beta = \varrho_0 c_0/Z_s$ is the normalized admittance of the ground, ϱ_0 is the reference density of the air, and Z_s is the ground impedance. Numerical solution of equations (11.60)–(11.62) for a moving medium is essentially the same as for a motionless medium, because of the validity of the effective sound-speed approximation in the narrow-angle PE.

We begin by discretizing the vertical coordinate into N points with constant spacing Δz; that is, $z_n = n\Delta z$, $n = 1, 2, \ldots, N$. We define

$$h(x,z) = \frac{k^2(x,z) - k_0^2}{2k_0^2}, \quad A_n(x) = A(x, z_n). \tag{11.63}$$

Note that ε, as defined in section 2.5, equals $2h(x, z)$. With the centered finite-difference approximation,

$$\left.\frac{\partial^2 A}{\partial z^2}\right|_{z=z_n} \simeq \frac{A_{n-1} - 2A_n + A_{n+1}}{(\Delta z)^2}, \tag{11.64}$$

equation (11.60) becomes

$$\frac{\partial A_n(x)}{\partial x} = \frac{i}{2k_0(\Delta z)^2}\left[A_{n-1}(x) - 2A_n(x) + A_{n+1}(x)\right] + ik_0 h(x, z_n) A_n(x). \tag{11.65}$$

The numerical BC at the ground has the following general form, which gives A_0 in terms of A_1 and A_2:

$$A_0(x) = \eta_1 A_1(x) + \eta_2 A_2(x), \tag{11.66}$$

where η_1 and η_2 are constants. Appropriate first- and second-order expressions for these constants were derived previously for the FFP as equations (11.40) and (11.41), respectively. Those equations apply here, as well, due to the local reaction assumption. For the upper BC, Salomons [345] has suggested applying the second-order BC to the upper boundary, but with $\beta = 1$, as a means of mimicking a radiation condition and thus reducing numerical reflections. Hence

$$A_{N+1}(x) = \eta_N A_N(x) + \eta_{N-1} A_{N-1}(x), \tag{11.67}$$

with

$$\eta_N = 4(3 - 2ik_0\Delta z)^{-1}, \quad \eta_{N-1} = -(3 - 2ik_0\Delta z)^{-1}. \tag{11.68}$$

Equations (11.65) and (11.66) can be combined and recast in the convenient matrix form

$$\frac{\partial \mathbf{A}(x)}{\partial x} = ik_0 \mathbf{\Psi}(x) \mathbf{A}(x), \tag{11.69}$$

in which

$$\mathbf{A}(x) = [A_1(x) \ A_2(x) \ \cdots \ A_N(x)]^T \tag{11.70}$$

and

$$\mathbf{\Psi}(x) = \left[1/2(k_0\Delta z)^2\right] \mathbf{T} + \mathbf{H}(x). \tag{11.71}$$

Here, $\mathbf{H}(x)$ is the diagonal matrix given by

$$\mathbf{H}(x) = \begin{bmatrix} h(x, z_1) & 0 & \cdots & \\ 0 & h(x, z_2) & \cdots & \\ \vdots & \vdots & \ddots & \\ & & & h(x, z_N) \end{bmatrix} \tag{11.72}$$

and $\mathbf{T}$ is the tridiagonal matrix given by equation (11.45).

The final step in numerical solution of the narrow-angle PE is to discretize the range coordinate, $x = x_m = m\Delta x$, $m = 0,\ldots,M$. The Crank–Nicholson method consists of introducing the centered approximations

$$\left.\frac{\partial \mathbf{A}}{\partial x}\right|_{x_{m+1/2}} \simeq \frac{\mathbf{A}\left(x_{m+1}\right) - \mathbf{A}\left(x_m\right)}{\Delta x}, \quad \mathbf{A}\left(x_{m+1/2}\right) \simeq \frac{\mathbf{A}\left(x_{m+1}\right) + \mathbf{A}\left(x_m\right)}{2},$$

$$(11.73)$$

from which we obtain

$$\mathbf{M}^+\mathbf{A}\left(x_{m+1}\right) = \mathbf{M}^-\mathbf{A}\left(x_m\right), \tag{11.74}$$

where

$$\mathbf{M}^+ = \left[\mathbf{I}_N - (ik_0\Delta x/2)\,\mathbf{\Psi}\right], \quad \mathbf{M}^- = \left[\mathbf{I}_N + (ik_0\Delta x/2)\,\mathbf{\Psi}\right], \tag{11.75}$$

and $\mathbf{I}_N$ is the $N \times N$ identity matrix. The usual method for advancing the solution at step x_m to x_{m+1} is to first perform the multiplication $\mathbf{y} = \mathbf{M}^-\mathbf{A}\left(x_m\right)$, thereby obtaining a linear system of the form $\mathbf{M}^+\mathbf{A}\left(x_{m+1}\right) = \mathbf{y}$. Because the matrix $\mathbf{M}^+$ premultiplying $\mathbf{A}\left(x_{m+1}\right)$ is tridiagonal, solution of the linear system involves only $O(N)$ floating-point operations, and is thus highly efficient.

In addition to the upper BC, equation (11.67), an artificial absorbing layer is usually placed at the top of the domain to further reduce numerical reflections. The absorbing layer is typically 40–50 wavelengths thick. West et al. [403] and Salomons [345] suggest using a quadratic height dependence for the absorbing layer, namely adding to $k(x,z)$ a term $ik_0\alpha_t(z-z_t)^2/Z^2$, where α_t is a constant, z_t is the height at which the layer starts, and Z is its thickness. Based on numerical tests, Salomons's [345] suggested values for α_t are 1, 0.5, 0.4, and 0.2 for the frequencies 1000, 500, 125, and 30 Hz, with linear interpolation in between.

Also needed is an initial condition, or *starter*, at $x = 0$. Since this topic is rather involved and outside the scope of this book, interested readers are referred to references [345] and [373] for details. For narrow-angle PEs, the most commonly used function for the starter is Gaussian, given by

$$A(0,z) = \sqrt{ik_0}\exp\left[-k_0^2(z-z_s)^2/2\right], \tag{11.76}$$

where z_s is the source height. When the source is close (within several wavelengths) to the ground, a correction should be added for the effect of the boundary on its strength and directionality. Specifically, we may set

$$A(0,z) = \sqrt{ik_0}\left\{\exp\left[-k_0^2(z-z_s)^2\right] + \mathcal{R}\exp\left[-k_0^2(z+z_s)^2\right]\right\}, \tag{11.77}$$

where $\mathcal{R}$ is the reflection coefficient for the ground. The simplest choice is to use the plane-wave reflection coefficient, namely $\mathcal{R} = \mathcal{R}_p = (1-\beta)/(1+\beta)$.

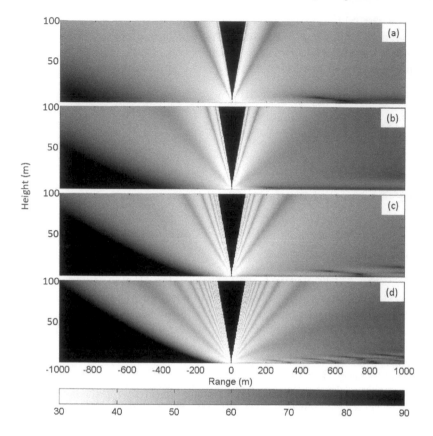

FIGURE 11.8
Narrow-angle CNPE calculations for varying frequencies. All calculations are
for a mostly cloudy condition with moderate wind ($u_* = 0.3$ m/s, $Q_H = Q_E =$
0 W/m^2). Source height is 1.5 m. Upwind is to the left, and downwind to the
right. The gray scale represents the transmission loss (TL) in dB. (a) 125 Hz.
(b) 250 Hz. (c) 500 Hz. (d) 1000 Hz.

Let us now consider some example calculations with the narrow-angle
CNPE, which illustrate the impact of atmospheric conditions on near-ground
sound propagation. For all of the calculations, the source height is 1.5 m.
Ground properties are characteristic of loose soil with tall grass; from table 10.1, the static flow resistivity is $\sigma_0 = 20$ kPa s m^{-2} and the porosity is
$\Omega = 0.5$. A relaxation model, as described in section 10.2, was used to calculate the ground admittance from σ_0 and Ω. The atmospheric profiles are
modeled using the Monin–Obukhov similarity theory (MOST), as described
in section 2.2.3, with a roughness length $z_0 = 0.05$ m.
    The first sequence of calculations, shown in figure 11.8, is for mostly cloudy
(neutral) conditions as a function of frequency. In all of these calculations,

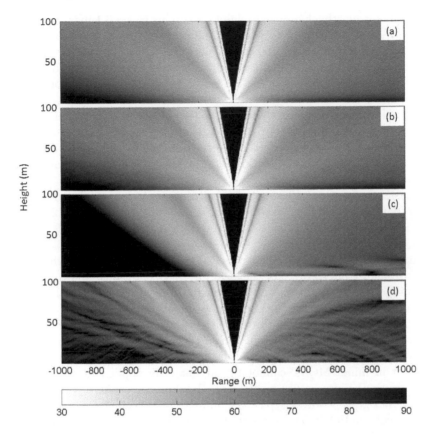

FIGURE 11.9
Narrow-angle CNPE calculations for varying wind speeds and turbulence conditions. All calculations are for mostly cloudy conditions ($Q_H = Q_E = 0$ W/m^2). Source frequency is 250 Hz and the height is 1.5 m. (a) Low wind ($u_* = 0.1$ m/s), no turbulence. (b) Low wind, with turbulence. (c) High wind ($u_* = 0.6$ m/s), no turbulence. (d) High wind, with turbulence.

$u_* = 0.3$ m/s, $Q_H = Q_E = 0$ W/m^2, $z_i = 1000$ m, and no turbulence is included. The frequencies, from top to bottom, are 125, 250, 500, and 1000 Hz. The figures show propagation in the upwind and downwind directions (to the left and right, respectively). As the frequency is increased, the upwind shadow zone boundary becomes more pronounced. In the downwind direction, there is a transition from a single quiet height [399] at low frequencies, to a more complicated modal interference pattern at higher frequencies. (The vertically oriented black isosceles triangle, with vertex near the origin, is a computational artifact of the narrow-angle parabolic approximation.)

The next sequence of calculations, figure 11.9, shows the effect of varying wind speed and turbulence. The frequency is 250 Hz in each case. The top two cases are for mostly cloudy, light wind conditions ($u_* = 0.1$ m/s, $Q_H = Q_E = 0$ W/m^2, and $z_i = 1000$ m), whereas the bottom two are for mostly cloudy, high wind conditions ($u_* = 0.6$ m/s, $Q_H = Q_E = 0$ W/m^2, and $z_i = 1000$ m). Calculations with and without turbulence, as simulated by the inhomogeneous spectral method described in section 9.1.2, are shown. As would be expected, the upwind/downwind calculations differ much more dramatically as the wind speed is increased. Because there is little turbulence in the low wind case, the calculations with and without turbulence are very similar. In the high wind case, however, the turbulence has a very dramatic impact, both upwind and downwind. In the upwind direction, the turbulence scatters substantial sound energy into the shadow zone.

Lastly, in figure 11.10 are shown calculations in low-wind conditions, for clear daytime (unstable stratification, $u_* = 0.1$ m/s, $Q_H = 200$ W/m^2, $Q_E = 50$ W/m^2, and $z_i = 1000$ m) and clear nighttime (stable stratification, $u_* = 0.1$ m/s, $Q_H = -4$ W/m^2, $Q_E = -1$ W/m^2, and $z_i = 1000$ m) stratification. Again, the frequency is 250 Hz, and calculations with and without turbulence are shown. For the clear daytime conditions, upward refraction prevails in both the upwind and downwind directions. Even with low wind, turbulent scattering is significant, due to the buoyantly produced turbulence. For the clear nighttime conditions, there is strong ducting downwind, but a shadow zone still forms upwind. Turbulent scattering is minimal. However, the limitations of MOST for this situation, as discussed in section 3.5.4, must be kept in mind. MOST cannot be satisfactorily applied to strongly stable stratification (temperature inversions). Random atmospheric variations such as internal gravity waves typically dominate over turbulence, and may have a very substantial impact on sound scattering that is not captured in figure 11.10(d).

11.2.2 Wide-angle Crank–Nicholson solution

Equations (2.126)–(2.128) serve as the starting point for the wide-angle PE solution. These equations may be viewed as a generalization of (11.60) and (11.61) in this chapter. When they are adapted to two dimensions, by omitting all derivatives in the transverse horizontal coordinate y, they can be written in the following form:

$$\Psi_1(x, z)\frac{\partial A}{\partial x} = ik_0\Psi_2(x, z)A(x, z), \tag{11.78}$$

where

$$\Psi_1(x, z) = h_{1,0} + \frac{h_{1,1}}{k_0}\frac{\partial}{\partial z} + \frac{h_{1,2}}{k_0^2}\frac{\partial^2}{\partial z^2}, \tag{11.79}$$

$$\Psi_2(x, z) = h_{2,0} + \frac{h_{2,1}}{k_0}\frac{\partial}{\partial z} + \frac{h_{2,2}}{k_0^2}\frac{\partial^2}{\partial z^2} + \frac{h_{2,3}}{k_0^3}\frac{\partial^3}{\partial z^3}. \tag{11.80}$$

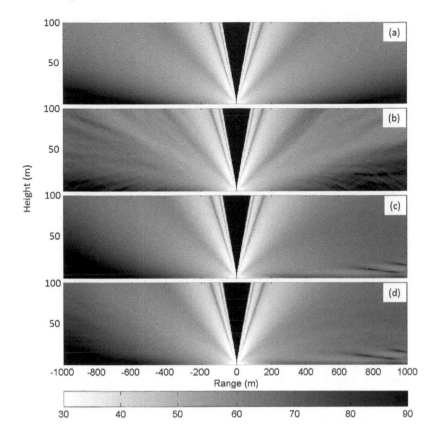

FIGURE 11.10
Narrow-angle CNPE calculations for varying stratification and turbulence conditions. All calculations are for a light wind condition ($u_* = 0.1$ m/s). Source frequency is 250 Hz and the height is 1.5 m. (a) Clear daytime ($Q_H = 200$ W/m^2 and $Q_E = 50$ W/m^2), no turbulence. (b) Clear daytime, with turbulence. (c) Clear nighttime ($Q_H = -4$ W/m^2 and $Q_E = -1$ W/m^2), no turbulence. (d) Clear nighttime, with turbulence.

In the preceding,

$$h_{1,0}(x, z) = a_3 + a_4\varepsilon + \frac{2ia_4}{\omega}\frac{\partial v_x}{\partial x} + \frac{ia_2}{k_0}\frac{\partial \ln(\varrho/\varrho_0)}{\partial x} + \frac{2a_2 v_x}{c_0}, \qquad (11.81)$$

$$h_{1,1}(x, z) = \frac{2ia_4 v_z}{c_0} - \frac{2a_2}{\omega}\left(\frac{\partial v_x}{\partial z} + \frac{\partial v_z}{\partial x}\right) - \frac{a_4}{k_0}\frac{\partial \ln(\varrho/\varrho_0)}{\partial z}, \qquad (11.82)$$

$$h_{1,2}(x, z) = a_4\left[1 + \frac{2i}{\omega}\left(\frac{\partial v_x}{\partial x} - \frac{\partial v_z}{\partial z}\right)\right], \qquad (11.83)$$

$$h_{2,0}(x, z) = a_1 - a_3 + (a_2 - a_4)\left[\varepsilon + \frac{2i}{\omega}\frac{\partial v_x}{\partial x} - \frac{i}{k_0}\frac{\partial \ln(\varrho/\varrho_0)}{\partial x} - \frac{2v_x}{c_0}\right], \quad (11.84)$$

$$h_{2,1}(x,z) = (a_2 - a_4)\left[\frac{2iv_z}{c_0} + \frac{2}{\omega}\left(\frac{\partial v_x}{\partial z} + \frac{\partial v_z}{\partial x}\right) - \frac{1}{k_0}\frac{\partial \ln(\varrho/\varrho_0)}{\partial z}\right], \quad (11.85)$$

$$h_{2,2}(x,z) = (a_2 - a_4)\left[1 + \frac{2i}{\omega}\left(\frac{\partial v_x}{\partial x} - \frac{\partial v_z}{\partial z}\right)\right] + a_4\left[\frac{i}{k_0}\frac{\partial \ln(\varrho/\varrho_0)}{\partial x} + \frac{2v_x}{c_0}\right],$$
$$(11.86)$$

and

$$h_{2,3}(x,z) = -\frac{2a_4}{\omega}\left(\frac{\partial v_x}{\partial z} + \frac{\partial v_z}{\partial x}\right). \quad (11.87)$$

If we consider a high-frequency approximation as in section 2.5 ($k_0\ell \gg 1$, where ℓ is a characteristic scale for the variations in the fields), these equations simplify considerably:

$$h_{1,0}(x,z) = a_3 + a_4\varepsilon + 2a_2\frac{v_x}{c_0}, \quad (11.88)$$

$$h_{1,1}(x,z) = 2ia_4\frac{v_z}{c_0}, \quad (11.89)$$

$$h_{1,2}(x,z) = a_4, \quad (11.90)$$

$$h_{2,0}(x,z) = a_1 - a_3 + (a_2 - a_4)\left(\varepsilon - \frac{2v_x}{c_0}\right), \quad (11.91)$$

$$h_{2,1}(x,z) = 2i(a_2 - a_4)\frac{v_z}{c_0}, \quad (11.92)$$

$$h_{2,2}(x,z) = a_2 - a_4 + 2a_4\frac{v_x}{c_0}, \quad (11.93)$$

and

$$h_{2,3}(x,z) = 0. \quad (11.94)$$

For a motionless medium without density gradients, the preceding equations reduce to $\Psi_1 = a_3 + a_4 s$ and $\Psi_2 = a_1 - a_3 + (a_2 - a_4)s$, where $s = \varepsilon + k_0^{-2}(\partial^2/\partial z^2) = k_0^{-2}(k^2 - k_0^2 + \partial^2/\partial z^2)$. As discussed in section 2.5.4, the conventional Padé (1, 1) approximation involves setting $a_1 = 1$, $a_2 = 3/4$, $a_3 = 1$, and $a_4 = 1/4$, which yields, for equation (11.78),

$$(1 + s/4)\frac{\partial A}{\partial x} = ik_0(s/2)A(x,z). \quad (11.95)$$

This is equivalent to the wide-angle PEs for a motionless medium derived in references [345] and [403]. Comparison to equation (11.61) indicates that the narrow-angle approximation for a motionless medium can be recovered by setting $a_1 = 1$, $a_2 = 1/2$, $a_3 = 1$, and $a_4 = 0$.

From a numerical perspective, the wide-angle PE is complicated by the introduction of terms involving the odd-order derivatives $\partial/\partial z$ and $\partial^3/\partial z^3$. For the narrow-angle PE, we already dealt with the second-order derivative, $\partial^2/\partial z^2$, which was represented in a discrete form as the tridiagonal matrix $\mathbf{T}$, equation (11.45). For present purposes, we will designate the matrices discretizing the derivatives as $\mathbf{T}_n$, where n is the order of the derivative. Hence $\mathbf{T}$ from the previous section will be designated $\mathbf{T}_2$, and we must now derive $\mathbf{T}_1$ and $\mathbf{T}_3$. Let us thus introduce the centered finite-difference approximations

$$\left.\frac{\partial A}{\partial z}\right|_{z=z_n} \simeq \frac{A_{n+1/2} - A_{n-1/2}}{\Delta z} \tag{11.96}$$

and

$$\left.\frac{\partial^3 A}{\partial z^3}\right|_{z=z_n} \simeq \frac{A_{n+3/2} - 3A_{n+1/2} + 3A_{n-1/2} - A_{n-3/2}}{(\Delta z)^3}. \tag{11.97}$$

These approximations cannot be applied directly, however, because the right sides involve half intervals in the heights. A standard approach to addressing this problem is to average over an interval, e.g., $A_{n+1/2} = (A_{n+1} + A_n)/2$. We then have

$$\left.\frac{\partial A}{\partial z}\right|_{z=z_n} \simeq \frac{A_{n+1} - A_{n-1}}{2\Delta z} \tag{11.98}$$

and

$$\left.\frac{\partial^3 A}{\partial z^3}\right|_{z=z_n} \simeq \frac{A_{n+2} - 2A_{n+1} + 2A_{n-1} - A_{n-2}}{2(\Delta z)^3}. \tag{11.99}$$

Using equations (11.66), (11.67), and (11.98), we find

$$\mathbf{T}_1 = \frac{1}{2}\begin{bmatrix} -\eta_1 & 1-\eta_2 & 0 & \cdots & & & \\ -1 & 0 & 1 & \cdots & & & \\ 0 & -1 & 0 & 1 & \cdots & & \\ \vdots & \vdots & \ddots & \ddots & & \ddots & \\ & & \cdots & -1 & 0 & 1 \\ & & \cdots & 0 & \eta_{N-1}-1 & \eta_N \end{bmatrix}. \tag{11.100}$$

Derivation of $\mathbf{T}_3$ is complicated somewhat by the fact that A_n is needed at two levels above and below $z = z_n$ to evaluate the derivative. In particular, at the bottom ($z = z_1$) and top ($z = z_N$) of the numerical domain, we would need to know A_{n-1} and A_{N+2}, respectively, which are unavailable. To address this problem, we are compelled to use uncentered approximations at these heights; specifically, we approximate $\partial^3 A/\partial z^3$ at $z = z_1$ by its value at $z = z_2$, and similarly at the top of the domain. This yields

$$
\mathbf{T}_3 = \frac{1}{2}
\begin{bmatrix}
2-\eta_1 & -\eta_2 & -2 & 1 & \cdots & & & \\
2-\eta_1 & -\eta_2 & -2 & 1 & \cdots & & & \\
-1 & 2 & 0 & -2 & 1 & \cdots & & \\
& \ddots & \ddots & \ddots & \ddots & \ddots & & \\
& \cdots & -1 & 2 & 0 & -2 & 1 & \\
& & \cdots & -1 & 2 & \eta_{N-1} & \eta_N - 2 & \\
& & \cdots & -1 & 2 & \eta_{N-1} & \eta_N - 2 &
\end{bmatrix}.
\tag{11.101}
$$

With these matrix approximations for the finite derivatives, we can now write equation (11.78) in the matrix form

$$
\mathbf{\Psi}_1(x)\frac{\partial \mathbf{A}}{\partial x} = ik_0 \mathbf{\Psi}_2(x)\mathbf{A}(x),
\tag{11.102}
$$

where

$$
\mathbf{\Psi}_1(x) = \mathbf{H}_{1,0} + (k_0\Delta z)^{-1}\mathbf{H}_{1,1}\mathbf{T}_1 + (k_0\Delta z)^{-2}\mathbf{H}_{1,2}\mathbf{T}_2
\tag{11.103}
$$

and

$$
\mathbf{\Psi}_2(x) = \mathbf{H}_{2,0} + (k_0\Delta z)^{-1}\mathbf{H}_{2,1}\mathbf{T}_1 + (k_0\Delta z)^{-2}\mathbf{H}_{2,2}\mathbf{T}_2 + (k_0\Delta z)^{-3}\mathbf{H}_{2,3}\mathbf{T}_3.
\tag{11.104}
$$

Here the $\mathbf{H}_{i,j}$ are diagonal matrices such that

$$
\mathbf{H}_{i,j}(x) =
\begin{bmatrix}
h_{i,j}(x,z_1) & 0 & \cdots & \\
0 & h_{i,j}(x,z_2) & \cdots & \\
\vdots & \vdots & \ddots & \\
& & & h_{i,j}(x,z_N)
\end{bmatrix}.
\tag{11.105}
$$

Solution by the Crank–Nicholson method proceeds in the same manner as the previous section; instead of equation (11.75), we find the following:

$$
\mathbf{M}^+ = \left[\mathbf{\Psi}_1 - (ik_0\Delta x/2)\,\mathbf{\Psi}_2\right], \quad \mathbf{M}^- = \left[\mathbf{\Psi}_1 + (ik_0\Delta x/2)\,\mathbf{\Psi}_2\right].
\tag{11.106}
$$

Because $\mathbf{T}_3$ is a pentadiagonal matrix, so is $\mathbf{\Psi}_2$. Hence a tridiagonal solver cannot be applied to equation (11.106), and numerical solution is considerably less efficient than the narrow-angle PE.

As the starter for the wide-angle PE, Salomons [345] recommends the following:

$$
A(0,z) = \sqrt{ik_0}\left[\alpha_0 + \alpha_2 k_0^2(z - z_s)^2\right]\exp\left[-k_0^2(z - z_s)^2/\zeta\right],
\tag{11.107}
$$

where $\alpha_0 = 1.3717$, $\alpha_2 = -0.3701$, and $\zeta = 3$. A second term (of the same functional form) can also be added for the ground reflection, as discussed in section 11.2.1.

Example calculations with a wide-angle CNPE, based on equations (11.102)–(11.107), are shown in figure 11.11. In the upper panel, figure 11.11(a), equations (11.88)–(11.94) were used to calculate the factors $h_{i,j}$. The numerical methodology is in other aspects the same as described in section 11.2.1. In the next panel, figure 11.11(b), an effective sound-speed approximation has been made by replacing ε with $\varepsilon - 2v_x/c_0$ in equations (11.88)–(11.94), and setting $v_x = 0$ otherwise. The next panel, figure 11.11(c), shows the difference between these wide-angle calculations with and without the effective sound-speed approximation. The most significant discrepancies, which may exceed ± 10 dB, occur in the downwind direction, at heights below roughly 20 m. Apparently, the effective sound-speed approximation impacts the relative phases of the ducted modes in this direction. Strong differences are also observed in the upwind shadow region; however, the sound levels there are very low and the discrepancies are due to relatively unimportant numerical effects such as weak reflections from the upper absorbing layer. Lastly, the lower panel, figure 11.11(d), is the difference between the narrow-angle calculation for the same case (figure 11.9(c), which inherently uses an effective sound-speed approximation) and the wide-angle PE without the effective sound-speed approximation (figure 11.11(a)). Near the ground, the errors are essentially the same as figure 11.11(c), thus demonstrating that these errors are due to the effective sound-speed approximation rather than the angular approximation. Errors attributable to the angular approximation are evident only at steeper angles above the source.

It should be kept in mind that the details of the interference pattern in the downwind direction are very sensitive to the shape of the mean wind profile. In reality, the profiles are usually not known accurately enough to predict the locations of the various interference minima and maxima exactly [420], and, even if they were, turbulence causes the interference pattern to shift randomly. Thus, practically speaking, the effective sound-speed approximation seems quite reasonable for situations involving nearly horizontal sound propagation. Issues related to uncertainty in propagation predictions will be further considered in Chapter 13.

11.2.3 Moment equations and moment screens

The most common approach to numerical calculations of wave propagation through a medium with random spatial variations is to solve the parabolic equation (or other solution method) with realizations of the entire 2D or 3D spatial structure of the incorporated directly into the calculation. The formulations described in sections 11.2.1 and 11.2.2 can be applied in this manner. Typically, an ensemble consisting of many 2D or 3D random realizations, or snapshots, of the medium is generated. The PE is then solved independently for each realization, and the desired wavefield statistics are estimated from this

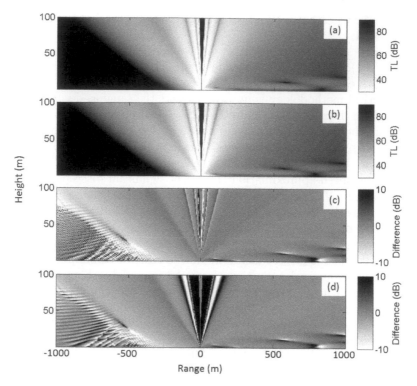

FIGURE 11.11
Wide-angle CNPE calculations for a mostly cloudy, strong wind condition ($u_* = 0.6$ m/s, $Q_H = Q_E = 0$ W/m^2), without turbulence. The source frequency is 250 Hz and the height is 1.5 m. (a) Calculation without the effective sound-speed approximation. (b) Calculation with the effective sound-speed approximation. (c) Calculation in panel (b) minus that shown in (a). (d) Narrow-angle calculation (figure 11.9(c)) minus that shown in (a).

ensemble.* This basic approach has been widely adopted for wave propagation in various environments with refraction and random scattering, including sound propagation through atmospheric turbulence [75, 136, 344] and ocean internal waves [70, 118, 379]. In this section, we describe an alternative approach based on directly solving equations for the statistical moments of the sound field, which were derived in Chapter 7.

    The first moment of the sound field, $\langle A(x) \rangle$, represents the unscattered part of the field. In Chapter 8, equation (8.40), the following equation for the

---

*This is an example of a Monte Carlo method, which will be discussed in more detail in Chapter 13.

first moment was derived using the narrow-angle and Markov approximations:

$$\frac{\partial \langle A \rangle}{\partial x} - \frac{i}{2k_0} \left[\frac{\partial^2}{\partial z^2} + k^2(z) - k_0^2 \right] \langle A \rangle + \frac{k_0^2}{8} b_{\text{eff}}(z, z) \langle A \rangle = 0. \tag{11.108}$$

This equation differs from equations (11.60) and (11.61) only in the presence of the final term representing the turbulence effect. The wavenumber $k(z) = k_0 c_0 / [c(z) + V_x(z)]$ is calculated from the mean vertical profiles for the sound speed, $c(z)$, and wind speed in the nominal propagation direction x, namely $V_x(z)$. Defining

$$b_{nn'} = \frac{k_0^2}{8} b_{\text{eff}}(z_n; z_{n'}) \tag{11.109}$$

and

$$\mathbf{F} = \text{diag}\left([b_{11} \ b_{22} \ \dots b_{NN}]\right), \tag{11.110}$$

we can now rewrite equation (11.108) in the matrix form

$$\frac{\partial \langle \mathbf{A} \rangle}{\partial x} = (ik_0 \boldsymbol{\Psi} - \mathbf{F}) \langle \mathbf{A} \rangle, \tag{11.111}$$

in which

$$\langle \mathbf{A}(x) \rangle = [\langle A_1(x) \rangle \ \langle A_2(x) \rangle \ \dots \ \langle A_N(x) \rangle]^T \tag{11.112}$$

and $\boldsymbol{\Psi}$ is defined by equation (11.71). Proceeding in the same manner as in section 11.2.1, we again arrive at equation (11.74), but with

$$\mathbf{M}^+ = [\mathbf{I}_N - (\Delta x/2)(ik_0 \boldsymbol{\Psi} - \mathbf{F})], \quad \mathbf{M}^- = [\mathbf{I}_N + (\Delta x/2)(ik_0 \boldsymbol{\Psi} - \mathbf{F})]. \tag{11.113}$$

Because $\mathbf{F}$ (like $\boldsymbol{\Psi}$) is a diagonal matrix, the same efficient method for the narrow-angle PE, based on a tridiagonal solver, can be employed for the first-moment solution as well.

The following PE for the second moment was derived in Chapter 8, equation (8.44), again using the narrow-angle and Markov approximations:

$$\frac{\partial P(x, z, z')}{\partial x} - \frac{i}{2k_0} \left[\frac{\partial^2}{\partial z^2} - \frac{\partial^2}{\partial z'^2} + k^2(z) - k^2(z') \right] P(x, z, z')$$

$$+ \frac{k_0^2}{8} d_{\text{eff}}(z, z') P(x, z, z') = 0, \tag{11.114}$$

in which

$$d_{\text{eff}}(z, z') = b_{\text{eff}}(z, z) + b_{\text{eff}}(z', z') - 2b_{\text{eff}}(z, z'). \tag{11.115}$$

The impedance BC implies

$$\left.\frac{\partial P\left(x, z, z'\right)}{\partial z}\right|_{z=0} = -ik_0\beta P\left(x, 0, z'\right), \qquad \left.\frac{\partial P\left(x, z, z'\right)}{\partial z'}\right|_{z'=0} = ik_0\beta^* P\left(x, z, 0\right).$$

$$(11.116)$$

The initial condition is $P\left(x, z, z'\right) = p_0\left(z\right)p_0^*\left(z'\right)$, where $p_0\left(z\right)$ is the (non-random) starter field.

In practice, we are often most interested in the mean-square pressure, namely $P\left(x, z, z'\right)$ with $z = z'$. Unfortunately, since equation (11.114) involves derivatives with respect to both z and z', $P\left(x, z, z\right)$ cannot be propagated in isolation; the entire function $P\left(x, z, z'\right)$ must be advanced in x.

Applying the discretizations $z = z_n = n\Delta z$, $n = 1, \dots, N$, and $z' = z_{n'} = n'\Delta z$, $n' = 1, \dots, N$, the second moment $P\left(x, z, z'\right)$ becomes an $N \times N$ matrix $\mathbf{P}\left(x\right)$, for which (n, n')th element is $\langle p\left(x, z_n\right)p^*\left(x, z_{n'}\right)\rangle$. Equation (11.114) can then be recast in the following matrix form [422]:

$$\frac{\partial \mathbf{P}\left(x\right)}{\partial x} = ik_0\left[\boldsymbol{\Psi}\left(x\right)\mathbf{P}\left(x\right) + \mathbf{P}\left(x\right)\widetilde{\boldsymbol{\Psi}}\left(x\right)\right] - \mathbf{G}\odot\mathbf{P}\left(x\right), \qquad (11.117)$$

where the tilde indicates the matrix conjugate and transpose, and $\odot$ indicates Hadamard (element-by-element) matrix multiplication. The elements of the $N \times N$ matrix $\mathbf{G}$ are

$$\left[\mathbf{G}\right]_{nn'} = \frac{k_0^2}{8}d_{\text{eff}}\left(z_n, z_{n'}\right). \qquad (11.118)$$

Equation (11.117) differs from the conventional narrow-angle PE, equation (11.69), in that it involves pre- and postmultiplication of $\mathbf{P}\left(x\right)$ by $ik_0\boldsymbol{\Psi}$. These terms represent refraction. Equation (11.117) also has the term involving Hadamard multiplication by $\mathbf{G}$, which represents the turbulence effect. If only the turbulent effect were present, the solution would be an exponential decay for each element of $\mathbf{P}\left(x\right)$. When $\mathbf{P}\left(x\right)$ is rearranged, column by column, into an $N^2 \times 1$ vector $\overline{\mathbf{P}}\left(x\right)$, equation (11.117) can actually be recast in the same functional form as equation (11.69). One has [422]

$$\left[\mathbf{I}_{N^2} - \frac{\Delta x}{2}\left(ik_0\overline{\boldsymbol{\Psi}} - \overline{\mathbf{G}}\right)\right]\overline{\mathbf{P}}\left(x + \Delta x\right) = \left[\mathbf{I}_{N^2} + \frac{\Delta x}{2}\left(ik_0\overline{\boldsymbol{\Psi}} - \overline{\mathbf{G}}\right)\right]\overline{\mathbf{P}}\left(x\right),$$

$$(11.119)$$

where

$$\overline{\boldsymbol{\Psi}} = \mathbf{I}_N \otimes \boldsymbol{\Psi} + \boldsymbol{\Psi}^* \otimes \mathbf{I}_N, \qquad (11.120)$$

$$\overline{\mathbf{G}} = \text{diag}\left(\mathbf{G}\right) \qquad (11.121)$$

and $\otimes$ is the Kronecker matrix product. Here the $N \times N$ tridiagonal transition matrix $\boldsymbol{\Psi}$ in equation (11.69) is replaced by $\overline{\boldsymbol{\Psi}} - \overline{\mathbf{G}}$, which is an $N^2 \times N^2$

tridiagonal matrix with fringes: the 0, ± 1, and $\pm N$ diagonals are non-zero. The increased size of the transition matrix as well as the fringes make a direct solution, based for example on the Crank–Nicholson method, highly computationally intensive [427]. To avoid the large matrices and computational times implied by equation (11.119), references [65] and [422] examined approximate, iterative solutions. The general accuracy and stability of these methods remain to be thoroughly assessed.

An alternative approach [427] to solving the second-moment PE involves factoring $\mathbf{P}(x)$ into two matrices $\mathbf{A}(x)$ and $\mathbf{B}(x)$. For the time being, we do not specify the type of factorization, only that it exists:

$$\mathbf{P}(x) = \mathbf{A}(x)\,\mathbf{B}(x).\tag{11.122}$$

Consider the following auxiliary equations:

$$\frac{\partial \mathbf{A}}{\partial x} = ik_0 \mathbf{\Psi}\mathbf{A}\tag{11.123}$$

and

$$\frac{\partial \widetilde{\mathbf{B}}}{\partial x} = ik_0 \mathbf{\Psi}\widetilde{\mathbf{B}}.\tag{11.124}$$

Postmultiplying equation (11.123) by $\mathbf{B}$, taking the conjugate transpose of equation (11.124) and premultiplying by $\mathbf{A}$, and then summing the two equations, we find

$$\frac{\partial \mathbf{A}}{\partial x}\mathbf{B} + \mathbf{A}\frac{\partial \mathbf{B}}{\partial x} = ik_0\left(\mathbf{\Psi}\mathbf{A}\mathbf{B} + \mathbf{A}\mathbf{B}\widetilde{\mathbf{\Psi}}\right)\tag{11.125}$$

or

$$\frac{\partial \mathbf{P}}{\partial x} = ik_0\left(\mathbf{\Psi}\mathbf{P} + \mathbf{P}\widetilde{\mathbf{\Psi}}\right).\tag{11.126}$$

Hence, by separately solving equations (11.123) and (11.124), we can determine $\mathbf{P}(x)$ in the absence of random scattering ($\mathbf{G} = \mathbf{0}$). The significance of this result is that equations (11.123) and (11.124) can be solved on a column-by-column basis, using the familiar Crank–Nicholson approach.

The random scattering can be treated in an approximate manner using a split-step approach, by which the scattering and refraction are calculated alternately. With the random scattering only, equation (11.117) becomes

$$\frac{\partial \mathbf{P}(x)}{\partial x} = -\mathbf{G}\odot\mathbf{P}(x).\tag{11.127}$$

The solution to this equation over the interval δ_x is simply

$$\mathbf{P}(x + \delta_x) = \mathbf{P}(x)\odot\exp\left(-\mathbf{G}\delta_x\right).\tag{11.128}$$

To combine the scattering and refraction effects, suppose we use equations (11.123) and (11.124) to advance the solution from x to $x + \delta_x - \epsilon$, where ϵ is a very small number. Next, concentrating the random scattering effect for the interval from x to $x + \delta_x$ in the thin screen of thickness ϵ, we have

$$\mathbf{P}\left(x + \delta_x\right) \simeq \mathbf{P}\left(x + \delta_x - \epsilon\right) \odot \exp\left(-\mathbf{G}\delta_x\right). \tag{11.129}$$

Except for trivial cases, the factors $\mathbf{A}\left(x + \delta_x - \epsilon\right)$ and $\mathbf{B}\left(x + \delta_x - \epsilon\right)$ are *not* simply related to $\mathbf{A}\left(x + \delta_x\right)$ and $\mathbf{B}\left(x + \delta_x\right)$. The matrix $\mathbf{P}\left(x + \delta_x\right)$ must generally be refactored before again advancing the solution with equations (11.123) and (11.124). This process, of factoring $\mathbf{P}\left(x\right)$, calculating the refraction by solving equations (11.123) and (11.124), multiplying the factors to reform $\mathbf{P}\left(x\right)$, and then applying the turbulence screen with equation (11.129), is thus repeated until the desired final range. This procedure is analogous to phase-screen approaches, which will be described in the next section. The main distinctions are that the moments are being propagated directly and the random phase screen is replaced by the statistical attenuation factor $\exp\left(-\mathbf{G}\delta_x\right)$. We thus call the operation in equation (11.129) a *moment screen*.

Many types of factorizations may be considered. Since $\mathbf{P}\left(x\right)$ is a correlation matrix, it is positive-definite and Hermitian. Thus it possesses a Cholesky factorization, such that $\mathbf{P}\left(x\right) = \mathbf{C}\left(x\right)\widetilde{\mathbf{C}}\left(x\right)$, where $\mathbf{C}\left(x\right)$ is a lower triangular matrix. The first column of the Cholesky factorization corresponds to the part of the pressure field that is coherent with z_1, the second column to the part that is coherent with z_2 but not z_1, etc. Another possibility is an eigenfactorization, namely $\mathbf{P}\left(x\right) = \mathbf{V}\left(x\right)\widetilde{\mathbf{V}}\left(x\right)$, where $\mathbf{V}\left(x\right)$ contains the scaled eigenvectors (eigenvectors multiplied by the square roots of the eigenvalues) as its columns. For either the Cholesky or eigenfactorization, equations (11.123) and (11.124) match, so that only one of them needs to be solved. This would not be true for other possible factorizations such as a QR (orthogonal-triagular) factorization. It should also be recognized that the initial factorization is not normally preserved with range. For example, more non-zero elements of $\mathbf{C}\left(x\right)$ will become populated at each range step, so that it is no longer a triangular matrix. Nonetheless, the matrix remains a valid factorization of $\mathbf{P}\left(x\right)$.

The Cholesky factorization is particularly efficient if we intend to propagate the entire second-moment matrix $\mathbf{P}\left(x\right)$. In this approach, solution of equation (11.123) at each range step involves N tridiagonal systems with N elements, and hence requires $O\left(N^2\right)$ operations. The Cholesky factorization itself can be shown to require $O\left(N^3\right)$ operations. Hence lengthening δ_x helps to decrease computation time, by minimizing the number of times $\mathbf{P}\left(x\right)$ must be refactored. However, increasing δ_x to a very large value would lead to an inaccurate solution. Alternatively, we could attempt to find an approximate factorization of $\mathbf{P}\left(x\right)$ that efficiently compresses the amount of information that is propagated (i.e., the number of columns of $\mathbf{A}$), without significantly impacting the accuracy. For example, we might apply an eigenfactorization, and propagate only the eigenvectors associated with the largest eigenvalues.

Let us apply these ideas for solving the second-moment PE to a numerical example. As with several earlier examples in this chapter, the atmosphere is modeled as mostly cloudy (neutral) with strong wind, namely with friction velocity $u_* = 0.6$ m/s, heat fluxes $Q_H = Q_E = 0$ W/m^2, and roughness length $z_0 = 0.05$ m. The source height is 1.5 m and the frequency is 125 Hz. The ground parameters correspond to the compact-soil case from table 10.1, which, based on the relaxation model from section 10.2, results in a normalized impedance $Z/\varrho_0 c_0 = 35.6 + i35.3$. Although the second-moment PE solution includes information on coherence between points at different heights, here we examine only the square roots of the diagonal elements, i.e., the root-mean-square pressure at each height. Shown in figure 11.12(a) is the TL calculation without turbulence based on these elements; i.e., the prediction for the logarithmic wind profile and no turbulence. As would be expected, in the upwind direction a strong shadow zone forms; in the downwind direction ducting and interference are evident near the ground. Figure 11.12(b)–(d) shows a sequence in which progressively more accurate factorizations are propagated: an eigenfactorization with the 32 strongest eigenvectors, an eigenfactorization with the 128 strongest eigenvectors, and the full Cholesky factorization, respectively. Refactorizations were performed at increments δ_x equal to five wavelengths. Recall that the full Cholesky factorization is exact in the sense that no information is discarded upon each refactorization, and thus may be regarded as the correct numerical solution of the second-moment PE. When 32 eigenvectors are propagated, a substantial amount of the energy scattered into the upwind shadow zone is missed. On the other hand, a careful comparison of the figures shows that the 32 eigenvectors suffice to correctly predict scattering into the interference minima in the downwind direction. Inclusion of 128 eigenvectors enables correct prediction of the upwind scattering, except at locations very deep in the shadow zone.

Although calculation times vary greatly among computer systems and software libraries, it is instructive to examine the relative efficiency of the various methods considered here. (More detailed discussion and comparisons can be found in reference [427].) For the preceding example, the narrow-angle PE calculation without turbulence (which is equivalent to propagation of a single eigenvector of the second-moment matrix), when performed on a laptop computer with a program written in Matlab®, required about 0.63 s for each propagation direction. The calculation based on the full Cholesky factorization, which in this case involved a 1030×1030 matrix, required 405 s in each direction. About 47 s of this calculation time involved refactorization of the second-moment matrix; the remaining time was spent propagating the factors. The eigenfactorization was significantly faster, namely 18 s in each direction for 32 eigenvectors, and 82 s for 128 eigenvectors.

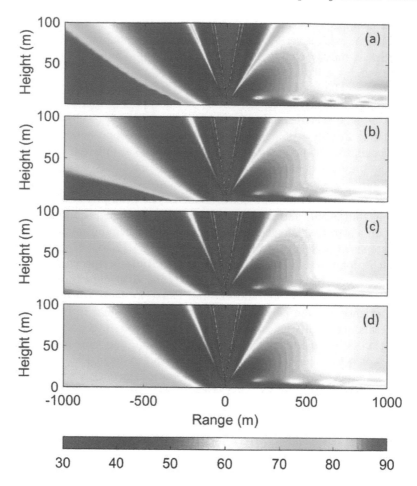

FIGURE 11.12

Solutions of the second-moment PE by various methods. All calculations show the TL for a mostly cloudy condition with strong wind ($u_* = 0.6$ m/s, $Q_H = Q_E = 0$ W/m², $z_0 = 0.05$ m). Source height is 1.5 m. Upwind is to the left, and downwind to the right. (a) Calculation without turbulence. (b) Calculation with turbulence, based on propagating the first 32 eigenvectors of the second-moment matrix factorization. (c) Calculation with turbulence, based on propagating the first 128 eigenvectors of the second-moment matrix factorization. (d) Calculation with turbulence, based on propagating the entire Cholesky factorization of the second-moment matrix.

11.2.4   Phase screens

Phase screens have been widely used to simulate wave propagation in random media [55, 242, 345]. In the customary split-step approach, the wavefield is multiplied by random phase factors at a series of fixed range intervals. Between the phase screens, the wavefield is propagated as though there were no turbulence. The standard deviation of the random phases is proportional to the index-of-refraction fluctuations in the medium. In this section, we derive an appropriate specification of the phase factors based on matching the desired second-order statistics of the propagating wavefield.

The basic premise of the phase screen is that the effect of the random medium over the interval between x and $x + \delta_x$ (where δ_x is the interval between screen applications) can be represented by multiplying the field by a set of random phase factors; that is,

$$p(x + \delta_x, z) = \exp[i\mu(z)] p(x, z), \tag{11.130}$$

where $\mu(z)$ is a real-valued, random variable. Note that we have written the random phase as a function of the vertical coordinate z only. As is consistent with the Markov approximation used in the derivation of equation (11.114), the assumption is made that the phase screens are uncorrelated with respect to the range coordinate x. Multiplying each side of equation (11.130) by the conjugate of the equation evaluated at $z = z'$, we have

$$p(x + \Delta_s x, z) p^*(x + \delta_x, z') = \exp[i\mu(z) - i\mu(z')] p(x, z) p^*(x, z'). \tag{11.131}$$

Next we take the ensemble average of each side. By the Markov approximation, the phase factors are independent of $p(x, z)$, thus yielding

$$P(x + \delta_x, z, z') = \langle \exp[i\mu(z) - i\mu(z')] \rangle P(x, z, z'). \tag{11.132}$$

We assume the $\mu(z)$ are normally distributed and zero mean. For such a random variable y, $\langle e^y \rangle = \exp(\langle y^2 \rangle / 2)$. Hence

$$\langle \exp[i\mu(z) - i\mu(z')] \rangle \simeq \exp\left[-\langle \mu^2(z) + \mu^2(z') - 2\mu(z)\mu(z') \rangle / 2\right]. \tag{11.133}$$

On an element-by-element basis, equation (11.129) is equivalent to

$$P(x + \delta_x, z, z') = \exp\left[-\frac{k_0^2 \delta_x}{8} d_{\text{eff}}(z, z')\right] P(x, z, z'). \tag{11.134}$$

We can now identify

$$\frac{k_0^2 \delta_x}{8} d_{\text{eff}}(z, z') \simeq \langle \mu^2(z) + \mu^2(z') - 2\mu(z)\mu(z') \rangle / 2 \tag{11.135}$$

or, equivalently,

$$R_\mu(z, z') = \langle \mu(z)\mu(z') \rangle \simeq \frac{k_0^2 \delta_x}{4} b_{\text{eff}}(z, z'). \tag{11.136}$$

In summary, if the random variations of $\mu(z)$ are normally distributed with cross-correlation function specified by equation (11.136), the phase screen calculations will be statistically equivalent to applying equation (11.129).

When statistics of the propagation medium depend on height, we can use an eigenfactorization method to generate random values for $\mu(z)$ possessing the desired correlation function $R_\mu(z, z')$, as discussed in chapter 9. Specifically, an $N \times N$ matrix $\mathbf{R}_b$ is created whose elements are the $b_{\text{eff}}(z, z')$ evaluated at the discrete heights $z = z_n$ and $z' = z_{n'}$. Let $\mathbf{V}_b$ be the scaled eigenvectors for this matrix. (These eigenvectors generally bear no simple relationship to the eigenvectors of $\mathbf{P}$ mentioned earlier.) An $N \times 1$ column vector $\boldsymbol{\mu}$, containing a realization of the $\mu(z_n)$ possessing the desired cross correlations, is given by

$$\boldsymbol{\mu} = \frac{k_0 \delta_x^{1/2}}{2} \mathbf{V}_b \mathbf{N}, \tag{11.137}$$

where $\mathbf{N}$ is a diagonal matrix whose elements are independent, random Gaussian variables with zero mean and unit variance. The equivalence of equations (11.136) and (11.137) can be verified by postmultiplying equation (11.136) by $\widetilde{\mu}$, taking the expected value, and recognizing that $\langle \mathbf{N} \mathbf{N} \rangle = \mathbf{I}$. The factorization of the $\mathbf{R}_b$ matrix needs to be performed only once for a particular random medium model. Phase-screen realizations based on equation (11.137) are straightforward, general, and consistent with the Markov approximation used to derive equation (11.114). When the correlations of $\mathbf{R}_b$ depend only on the separation between the vertical observation points, $|z - z'|$, the eigenvector method reduces to random-field synthesis with harmonic functions, which has been widely used for sound propagation in turbulence [75, 136, 345].

In practice, the phase-screen method becomes more efficient than the moment-screen methods described in the previous section when the sound waves are strongly scattered, such as in a refractive shadow zone [427]. For the example shown in figure 11.12, phase-screen based calculations, in which the screens were applied every 5 wavelengths, became essentially indistinguishable from the Cholesky factorization method after averaging 128 random propagation trials. The total calculation time for the 128 trials was 115 s, as compared to 405 s for the Cholesky factorization.

11.2.5   Terrain elevation variations

PE methods can be very useful for simulating propagation over hills, particularly if the elevation variations are gradual. Although topographic effects are not considered in detail here, we review some of the pertinent literature.

Sack and West [340] provide a general solution based on a non-orthogonal, terrain-following coordinate transformation. Lihoreau et al. [226] and van Renterghem et al. [388] adopt a distinctly different approach, based on solving the PE over constant-slope segments, and rotating the coordinates at each break in slope. The main effort involves performing an interpolation at each break, to initialize the starting field for the next segment. Parakkal et al. [307] introduced a relatively simple approach based on the Beilis–Tappert transformation, which is valid for local radii of curvature much larger than an acoustic wavelength, and has angular accuracy consistent with narrow-angle PE approximation.

PEs are particularly convenient for simultaneously incorporating atmospheric effects and propagation over a barrier. An example is reference [343], which demonstrates an interesting effect by which the effectiveness of a barrier is reduced by enhanced downward refraction as the wind accelerates over the top of the barrier.

As mentioned in the introduction to this chapter, one of the primary challenges of realistic modeling of sound propagation over variable topography lies in adequate characterization of the atmospheric flow. Wind and temperature vary with the topography, as well, and the current state of atmospheric computational fluid dynamics generally does not provide resolution sufficient for accurate acoustical modeling. In particular, flows in inhomogeneous and mountainous terrain, and strong radiational cooling (which leads to drainage flow down hillsides and pooling of cool air in valleys), are difficult to simulate accurately. References [169, 226] describe the utilization of atmospheric mesoscale simulations in complex topography for input to PE calculations.

12

Wave-based time-domain methods

Finite-difference, time-domain (FDTD) methods were adopted for outdoor sound propagation more than 10 years after the frequency-domain methods such as the parabolic equation (PE) and fast-field program (FFP) (see Chapter 11) came into widespread usage. Like the PE, FDTD can incorporate refraction by sound-speed and wind gradients [38, 303], as well as scattering by turbulence [67, 105, 372]. FDTD methods can also incorporate many real-world complications that the frequency-domain methods often cannot (or only with much difficulty), including transient source mechanisms, dynamic atmospheric variations, sound-wave interactions with heterogeneous ground surfaces, and backscatttering and multiple scattering in complex terrain such as forested and urban environments. The main disadvantages of FDTD methods are computational intensiveness and the difficulty of implementing an impedance (reactive) boundary condition in the time domain. On desktop computers, only 2D spatial calculations, at frequencies low in the audible range, are currently practical. However, since FDTD calculations are readily parallelizable by decomposing the overall spatial domain into subdomains, 3D calculations can be implemented on clusters and supercomputers.

Besides FDTD, other wave-based, time-domain methods have been developed. Hornikx et al. [171] demonstrated a pseudospectral time-domain (PSTD) simulation involving a noise barrier and ground. However, PSTD is most efficient when the medium can be partitioned into a small number of homogeneous subdomains. In the current state of development, FDTD is the approach most amenable for application to inhomogeneous moving media, and thus our focus in this chapter. Collier et al. [83] and Cheinet et al. [65] incorporated turbulence into FDTD calculations and demonstrated close agreement with analytical solutions given in Part II of this book.

One of the main practical attractions of FDTD methods is their suitability for simulating complex interactions of sound waves with hills, vegetation, and buildings. Ketcham et al. [196] demonstrated 3D calculations with a domain encompassing dozens of city blocks, and showed that the resulting field statistics are similar to those for scattering by turbulence. Liu and Albert [229] performed FDTD calculations with various simple combinations of walls and corners, and compared the results to experimental data. Heimann and Karle [158] formulated an FDTD solution for hilly terrain using a coordinate transformation. Symons et al. [372] incorporated hilly terrain into FDTD calculations more directly, by varying the material properties on the computational grid.

Vegetation has many important practical effects, both direct and indirect, on sound propagation [317]. The vegetation itself can scatter and absorb sound. At frequencies above several hundred Hz, tree trunks may scatter sound back toward the source. Generally, this phenomenon is described with analytical models [369], because it is difficult or impossible to incorporate directly into less intensive numerical methods, such as the PE or FFP. However, with growing computing capabilities, the scattering from individual trees can be incorporated into FDTD calculations [350].

Coupling of FDTD with frequency-domain methods also holds much promise. Van Renterghem et al. [389] coupled an FDTD calculation in the vicinity of an urban street canyon to a PE over an open region. Ovenden et al. [305] similarly coupled an analytical calculation around a highway noise barrier to a PE.

12.1   Finite-difference equations for a moving medium

FDTD methods generally involve discretization and solution of first-order differential equations, rather than the wave equation itself. As the starting point, we take equations (2.80) and (2.81) from Chapter 2 of this book. These equations were derived from the linearized fluid dynamic equations with approximations appropriate to sound waves. Isolating the partial derivatives with respect to time on the left side of these equations, we have

$$\frac{\partial p}{\partial t} = -(\mathbf{v} \cdot \nabla)p - K\nabla \cdot \mathbf{w} + KQ, \tag{12.1}$$

and

$$\frac{\partial \mathbf{w}}{\partial t} = -(\mathbf{v} \cdot \nabla)\mathbf{w} - (\mathbf{w} \cdot \nabla)\mathbf{v} - b\nabla p + b\mathbf{D}, \tag{12.2}$$

where $b = 1/\varrho_0$ is the mass buoyancy, $K = \varrho_0 c_0^2 = \gamma P_0$ is the adiabatic bulk modulus, and Q and $\mathbf{D}$ are mass and force sources, respectively. Writing out the derivatives for the two-dimensional case, we have

$$\frac{\partial p}{\partial t} = -v_x \frac{\partial p}{\partial x} - v_y \frac{\partial p}{\partial y} - K\frac{\partial w_x}{\partial x} - K\frac{\partial w_y}{\partial y} + KQ, \tag{12.3}$$

$$\frac{\partial w_x}{\partial t} = -v_x \frac{\partial w_x}{\partial x} - v_y \frac{\partial w_x}{\partial y} - w_x \frac{\partial v_x}{\partial x} - w_y \frac{\partial v_x}{\partial y} - b\frac{\partial p}{\partial x} + bD_x, \tag{12.4}$$

and

$$\frac{\partial w_y}{\partial t} = -v_x \frac{\partial w_y}{\partial x} - v_y \frac{\partial w_y}{\partial y} - w_x \frac{\partial v_y}{\partial x} - w_y \frac{\partial v_y}{\partial y} - b\frac{\partial p}{\partial y} + bD_y. \tag{12.5}$$

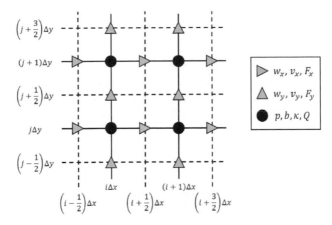

FIGURE 12.1
Two-dimensional spatial finite-difference grid for acoustics.

Although, for generality, we have written these equations explicitly in terms of the coordinates x and y, in many of the following examples we will apply them in a vertical plane, in which case y should be replaced by the vertical coordinate z for consistency with earlier chapters.

12.2 Spatial discretization

Finite-difference solution involves discretizing the spatial and temporal derivatives in the preceding equations. The spatial discretization is the focus of this section; temporal discretization will be considered in section 12.3. Suitable approaches to the spatial discretization, *in the absence of the nonlinear terms involving the medium velocity* **v**, are well established. It would seem reasonable to take such approaches as a starting point for moving media, in the hope that they remain stable and accurate. The most common finite-difference grid for motionless media (e.g., references [41, 439]) involves staggering (offsetting) the acoustic particle velocity components from the pressure grid, as shown in figure 12.1. (Although a 2D spatial grid is depicted, generalization to a 3D grid is straightforward.) Here, the acoustic pressure is stored at integer node positions, namely $(x, y) = (i\Delta x, j\Delta y)$, where i and j are integers, and Δx and Δy are the grid intervals in the x and y directions, respectively. The x-component of the acoustic particle velocity, w_x, is offset by $\Delta x/2$ in the x-direction, and w_y is similarly offset by $\Delta y/2$.

One of the main motivations for basing finite-difference schemes on figure 12.1 is that it provides a convenient basis for constructing stable, centered

spatial differences for many of the derivatives in equations (12.1) and (12.2).
Centered approximations of equations (12.1) and (12.2) require an evaluation
of each of the terms of the right sides of these equations at the grid nodes
where the field variable on the left side is stored. For example, since $\partial p/\partial t$ in
equation (12.1) is evaluated at the integer nodes, we need approximations for
the terms on the right side of this equation at the same nodes. The following second-order accurate, centered approximation for the velocity derivative
$\partial w_x/\partial x$ has this property:

$$\frac{\partial w_x}{\partial x}\left(i\Delta x, j\Delta y, k\Delta z, t\right) \simeq$$
$$\frac{w_x\left[(i+1/2)\,\Delta x, j\Delta y, k\Delta z, t\right] - w_x\left[(i-1/2)\,\Delta x, j\Delta y, k\Delta z, t\right]}{\Delta x}. \quad (12.6)$$

The derivatives $\partial w_y/\partial y$ and $\partial w_z/\partial z$ follow similarly. For the pressure derivatives appearing in equation (12.2), we can form centered differences across
a single grid interval as needed at the half-integer locations of the velocity
components. For the x-component of equation (12.2),

$$\frac{\partial p}{\partial x}\left[(i+1/2)\,\Delta x, j\Delta y, k\Delta z, t\right] \simeq$$
$$\frac{p\left[(i+1)\,\Delta x, j\Delta y, k\Delta z, t\right] - p\left[i\Delta x, j\Delta y, k\Delta z, t\right]}{\Delta x} \quad (12.7)$$

and likewise for the y- and z-components of equation (12.2). Still focusing on
the case of a motionless medium, the remaining terms to be addressed are
KQ in equation (12.1) and $b\mathbf{D}$ in equation (12.2). If we store K and Q at the
pressure nodes, then KQ is directly evaluated where needed. Furthermore, if
we store the components of $\mathbf{D}$ at the corresponding velocity nodes, they are
also directly evaluated where needed. When the scalar quantity b is stored at
the pressure nodes, it can still be readily determined at the locations needed for
the components of equation (12.2) by averaging two neighboring grid points.
For the x-component of equation (12.2), for example,

$$b\left[(i+1/2)\,\Delta x, j\Delta y, k\Delta z, t\right] \simeq$$
$$\left\{b\left[(i+1)\,\Delta x, j\Delta y, k\Delta z, t\right] + b\left[i\Delta x, j\Delta y, k\Delta z, t\right]\right\}/2. \quad (12.8)$$

Evaluation of the remaining terms, which are particular to propagation in
a moving medium, is somewhat more complicated. Let us store the medium
velocity components v_x, v_y, and v_z at the same grid locations as the corresponding components of the acoustic particle velocity, as shown in figure 12.1.
With this approach, however, the derivatives of the pressure field in equation (12.1) cannot be centered at the integer nodes using approximations
across a single grid interval. Centered approximations occur, rather, at the

velocity nodes. For example,

$$\left(v_x\frac{\partial p}{\partial x}\right)[(i+1/2)\,\Delta x, j\Delta y, k\Delta z, t] \simeq v_x\,[(i+1/2)\,\Delta x, j\Delta y, k\Delta z, t]$$

$$\times \frac{p\,[(i+1)\,\Delta x, j\Delta y, k\Delta z, t] - p\,[i\Delta x, j\Delta y, k\Delta z, t]}{\Delta x}. \quad (12.9)$$

We can then average this result with one determined in the same manner at $[(i-1/2)\,\Delta x, j\Delta y, k\Delta z, t]$, in order to find $v_x(\partial p/\partial x)$ at the desired integer grid location, with result:

$$\left(v_x\frac{\partial p}{\partial x}\right)[i\Delta x, j\Delta y, k\Delta z, t] \simeq$$

$$\frac{1}{2\Delta x}\Big\{v_x\,[(i+1/2)\,\Delta x, j\Delta y, k\Delta z, t]$$

$$\times\,(p\,[(i+1)\,\Delta x, j\Delta y, k\Delta z, t] - p\,[i\Delta x, j\Delta y, k\Delta z, t])$$

$$+\,v_x\,[(i-1/2)\,\Delta x, j\Delta y, k\Delta z, t]$$

$$\times\,(p\,[i\Delta x, j\Delta y, k\Delta z, t] - p\,[(i-1)\,\Delta x, j\Delta y, k\Delta z, t])\Big\}.$$

$$(12.10)$$

Evaluation of $v_y(\partial p/\partial y)$ and $v_z(\partial p/\partial z)$ is handled similarly.

Turning now to equation (12.2), for the x-component of this equation we need $(v_x\partial/\partial x + v_y\partial/\partial y + v_z\partial/\partial z)w_x$ evaluated at $[(i+1/2)\,\Delta x, j\Delta y, k\Delta z, t]$. The same term, but with the roles of $\mathbf{v}$ and $\mathbf{w}$ reversed, is also needed. Referring to figure 12.1, a centered difference for $v_x\partial w_x/\partial x$ can be calculated in two steps. First, we evaluate this quantity at $(x, y, z) = [(i+1)\Delta x, j\Delta y, k\Delta z, t]$, by averaging v_x across an interval of Δx centered on this point, and differencing w_x across the same interval. The same procedure is then applied at $(x, y, z) = (i\Delta x, j\Delta y, k\Delta z, t)$, and the results averaged to obtain $v_x\partial w_x/\partial x$ at the desired locations, thus yielding:

$$\left(v_x\frac{\partial w_x}{\partial x}\right)[(i+1/2)\,\Delta x, j\Delta y, k\Delta z, t] \simeq \frac{1}{4\Delta x}$$

$$\Big\{\,(v_x\,[(i+3/2)\,\Delta x, j\Delta y, k\Delta z, t] + v_x\,[(i+1/2)\,\Delta x, j\Delta y, k\Delta z, t])$$

$$\times\,(w_x\,[(i+3/2)\,\Delta x, j\Delta y, k\Delta z, t] - w_x\,[(i+1/2)\,\Delta x, j\Delta y, k\Delta z, t])$$

$$+\,(v_x\,[(i+1/2)\,\Delta x, j\Delta y, k\Delta z, t] + v_x\,[(i-1/2)\,\Delta x, j\Delta y, k\Delta z, t])$$

$$\times\,(w_x\,[(i+1/2)\,\Delta x, j\Delta y, k\Delta z, t] - w_x\,[(i-1/2)\,\Delta x, j\Delta y, k\Delta z, t])\Big\}.$$

$$(12.11)$$

A centered difference for $v_y\partial w_x/\partial y$ can be calculated by the same approach. First, we evaluate this quantity at $(x, y, z) = [(i+1/2)\Delta x, (j+1/2)\Delta y, k\Delta z, t]$, by averaging v_y across the separation in Δx, and then differencing w_x across the separation in Δy. The same operation is done at $(x, y, z) =$

$[(i + 1/2)\Delta x, (j - 1/2)\Delta y, k\Delta z, t]$, and we then average the results to obtain $v_y \partial w_x / \partial y$ at the desired locations:

$$\left(v_y \frac{\partial w_x}{\partial y} \right) [(i + 1/2)\,\Delta x, j\Delta y, k\Delta z, t] \simeq$$

$$\frac{1}{4\Delta y} \Big\{ (v_y \, [(i + 1)\,\Delta x, (j + 1/2)\,\Delta y, k\Delta z, t] + v_y \, [i\Delta x, (j + 1/2)\,\Delta y, k\Delta z, t])$$

$$\times (w_x \, [(i + 1/2)\,\Delta x, (j + 1)\,\Delta y, k\Delta z, t] - w_x \, [(i + 1/2)\,\Delta x, j\Delta y, k\Delta z, t])$$

$$+ (v_y \, [(i + 1)\,\Delta x, (j - 1/2)\,\Delta y, k\Delta z, t] + v_y \, [i\Delta x, (j - 1/2)\,\Delta y, k\Delta z, t])$$

$$\times (w_x \, [(i + 1/2)\,\Delta x, j\Delta y, k\Delta z, t] - w_x \, [(i + 1/2)\,\Delta x, (j - 1)\,\Delta y, k\Delta z, t]) \Big\}.$$

$$(12.12)$$

This same procedure can be used to derive $v_z \partial w_x / \partial z$. The moving-media terms appearing in the y- and z-components of equation (12.2) are obtained similarly.

12.3   Temporal discretization

Let us define the function f_p as the right side of equation (12.1), and f_x, f_y, and f_z as the three vector components on the right side of equation (12.2). Thus

$$\frac{\partial p \, (i\Delta x, j\Delta y, k\Delta z, t)}{\partial t} = f_p \, (i\Delta x, j\Delta y, k\Delta z, t), \tag{12.13}$$

where

$$f_p \, (x, y, z, t) = -(\mathbf{v} \cdot \nabla)p - K\nabla \cdot \mathbf{w} + KQ, \tag{12.14}$$

and

$$\frac{\partial w_x \, [(i + 1/2)\,\Delta x, j\Delta y, k\Delta z, t]}{\partial t} = f_x \, [(i + 1/2)\,\Delta x, j\Delta y, k\Delta z, t], \tag{12.15}$$

where

$$f_x \, (x, y, z, t) = -(\mathbf{v} \cdot \nabla)w_x - (\mathbf{w} \cdot \nabla)v_x - b\nabla p + bD_x. \tag{12.16}$$

Finite-difference approximation of equation (12.15), centered at $t = (l + 1/2)\,\Delta t$, yields:

$$w_x \, [(i + 1/2)\,\Delta x, j\Delta y, k\Delta z, (l + 1/2)\,\Delta t] \tag{12.17}$$

$$\simeq w_x \, [(i + 1/2)\,\Delta x, j\Delta y, k\Delta z, (l - 1/2)\,\Delta t] \tag{12.18}$$

$$+ \Delta t f_x \, [(i + 1/2)\,\Delta x, j\Delta y, k\Delta z, l\Delta t] \tag{12.19}$$

and likewise for the y and z components. For a motionless medium, f_x, f_y, and f_z depend on $\mathbf{w}$, b, and $\mathbf{D}$. For brevity, let us define $\mathbf{P}_l$ as the vector comprising the acoustic pressures at all grid locations at time $t = l\Delta t$, and $\mathbf{W}_l$ as the vector comprising acoustic particle velocities (all components) at all grid locations at time $t = l\Delta t$. $\dot{\mathbf{P}}_l$ and $\dot{\mathbf{W}}_l$ indicate the time derivatives (f-functions) for these quantities. Equation (12.19) then becomes

$$\mathbf{W}_{l+1/2} = \mathbf{W}_{l-1/2} + \Delta t \dot{\mathbf{W}}_l\left(\mathbf{P}_l\right). \tag{12.20}$$

Likewise, finite-difference approximation of equation (12.13), centered at $t = l\Delta t$, yields:

$$\begin{aligned}
p\left[i\Delta x, j\Delta y, k\Delta z, (l+1)\Delta t\right] &\simeq p\left[i\Delta x, j\Delta y, k\Delta z, l\Delta t\right] \\
&+ \Delta t f_p\left[i\Delta x, j\Delta y, k\Delta z, (l+1/2)\Delta t\right].
\end{aligned} \tag{12.21}$$

For a motionless medium, f_p depends only on $\mathbf{w}$, K, and Q. The vectorized equation is thus

$$\mathbf{P}_{l+1} = \mathbf{P}_l + \Delta t \dot{\mathbf{P}}_{l+1/2}(\mathbf{W}_{l+1/2}). \tag{12.22}$$

Taken together, equations (12.20) and (12.22) indicate that the pressure is needed only at the integer time steps $t = l\Delta t$, but not at the half-integer time steps $t = (l+1/2)\Delta t$. On the other hand, $\mathbf{w}$ is needed at the half-integer, but not at the integer, time steps. This observation motivates the widely used method for advancing the solution forward in time using a staggered temporal grid, in which the pressures are stored at the integer time steps and the particle velocities at the half-integer time steps. The fields are then updated in an alternating "leapfrog" fashion [41, 439]. This temporal grid is shown in figure 12.2. The fields from the previous time step can be overwritten in place, thus requiring storage of the fields at only a single time level.

Considering now the full equations for a moving medium, the time-centered finite-difference approximation of the term $-(\mathbf{v} \cdot \nabla)\mathbf{w}$ in equation (12.19) requires the velocity at integer time steps. Likewise, the term $-(\mathbf{v} \cdot \nabla)p$ in equation (12.21) requires the pressure field at the half-integer time steps. Instead of equations (12.20) and (12.22), we therefore have

$$\mathbf{W}_{l+1/2} = \mathbf{W}_{l-1/2} + \Delta t \dot{\mathbf{W}}_l\left(\mathbf{P}_l, \mathbf{W}_l\right) \tag{12.23}$$

and

$$\mathbf{P}_{l+1} = \mathbf{P}_l + \Delta t \dot{\mathbf{P}}_{l+1/2}(\mathbf{P}_{l+1/2}, \mathbf{W}_{l+1/2}). \tag{12.24}$$

Hence, for a moving medium, the customary staggered leapfrog approach does not lead to a scheme providing explicit updates for the acoustic fields at all time steps where they are needed. However, alternative solution schemes are available for which all terms are evaluated to the same order of accuracy, regardless of whether they are specific to a moving medium. Several of these schemes will now be described.

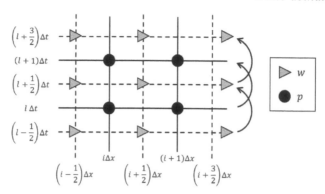

FIGURE 12.2
Temporally staggered grid for alternately advancing the acoustic pressure and particle velocity fields. (Only one spatial dimension is shown.)

Unstaggered leapfrog. The most straightforward approach is to abandon the staggered temporal grid and form centered finite differences over an interval of *two* integer time steps, as shown in figure 12.3. The resulting updating equations are

$$\mathbf{W}_{l+1} = \mathbf{W}_{l-1} + 2\Delta t \dot{\mathbf{W}}_l(\mathbf{P}_l, \mathbf{W}_l) \tag{12.25}$$

and

$$\mathbf{P}_{l+1} = \mathbf{P}_{l-1} + 2\Delta t \dot{\mathbf{P}}_l(\mathbf{P}_l, \mathbf{W}_l). \tag{12.26}$$

When the solution is advanced to $t = (l+1)\Delta t$, the pressure and particle velocities from $t = (l-1)\Delta t$ can be overwritten with the new values. However, the values at $t = l\Delta t$ must be retained for the next time step. Hence this scheme requires twice the memory of the staggered leapfrog. Additionally, the numerical dispersion and instability characteristics are inferior to those of the conventional staggered scheme due to the advancement of the wavefield variables over two time steps instead of one. On the other hand, the unstaggered leapfrog does provide a simple second-order scheme that is not specialized to low Mach number flows.

Aldridge method. D. F. Aldridge (personal communication; see also Symons et al. [372]) proposed a scheme that maintains temporal staggering of the pressure and particle velocity, yet still provides explicit updates using second-order, centered temporal finite differences. The velocity and pressure update equations are, respectively,

$$\mathbf{W}_{l+1/2} = \mathbf{W}_{l-3/2} + 2\Delta t \dot{\mathbf{W}}_{l-1/2}[(\mathbf{P}_{l-1} + \mathbf{P}_l)/2, \mathbf{W}_{l-1/2}] \tag{12.27}$$

and

$$\mathbf{P}_{l+1} = \mathbf{P}_{l-1} + 2\Delta t \dot{\mathbf{P}}_l[\mathbf{P}_l, (\mathbf{W}_{l-1/2} + \mathbf{W}_{l+1/2})/2]. \tag{12.28}$$

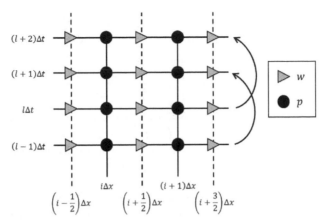

FIGURE 12.3
Temporally unstaggered grid for advancing the acoustic pressure and particle velocity fields. (Only one spatial dimension is shown.)

Like the unstaggered leapfrog, this scheme requires storage of the fields at two time levels.

Heun's method. Heun's method, also known as second-order Runge-Kutta, is a two-stage adaptive method. It has the following form:

$$\mathbf{K}_{w,1} = \dot{\mathbf{W}}_l(\mathbf{P}_l, \mathbf{W}_l), \tag{12.29}$$

$$\mathbf{K}_{p,1} = \dot{\mathbf{P}}_l(\mathbf{P}_l, \mathbf{W}_l), \tag{12.30}$$

$$\mathbf{K}_{w,2} = \dot{\mathbf{W}}_{l+1}(\mathbf{P}_l + \Delta t \mathbf{K}_{p,1}, \mathbf{W}_l + \Delta t \mathbf{K}_{w,1}), \tag{12.31}$$

and

$$\mathbf{K}_{p,2} = \dot{\mathbf{P}}_{l+1}(\mathbf{P}_l + \Delta t \mathbf{K}_{p,1}, \mathbf{W}_l + \Delta t \mathbf{K}_{w,1}). \tag{12.32}$$

The updates are given by

$$\mathbf{W}[(l+1)\,\Delta t] = \mathbf{W}_x[l\Delta t] + (\Delta t/2)(\mathbf{K}_{w,1} + \mathbf{K}_{w,2}) \tag{12.33}$$

and

$$\mathbf{P}[(l+1)\,\Delta t] = \mathbf{P}[l\Delta t] + (\Delta t/2)(\mathbf{K}_{p,1} + \mathbf{K}_{p,2}). \tag{12.34}$$

Many other methods from the literature can also be applied to this problem. In particular, higher-order methods, such as the fourth-order Runge-Kutta method, can be employed, but place more demands on computer memory. Another alternative is provided in reference [386], which uses a perturbative solution based on the assumption that the flow velocity is small.

For numerical stability of the FDTD calculation, the time step Δt and grid spacing Δr must satisfy the Courant condition, $C < 1$, where the Courant number is defined as

$$C = \frac{u \Delta t}{\Delta r}. \tag{12.35}$$

Here, u is the speed at which the sound energy propagates. For a nonuniform grid, $\Delta r = [(\Delta x)^{-2} + (\Delta y)^{-2} + (\Delta z)^{-2}]^{-1/2}$. Since the grid spacing must generally be a small fraction of a wavelength for good numerical accuracy, the Courant condition in practice imposes a limitation on the maximum time step possible for stable calculations. An even smaller time step may be necessary for satisfactory accuracy (low numerical dispersion).

For propagation in a uniform flow, u is determined by a combination of the sound speed and wind velocity. In the downwind direction, we have $u = u_+ = c + v$. In the upwind direction, $u = u_- = c - v$. The wavelengths in these two directions are $\lambda_+ = (c + v)/f$ and $\lambda_- = (c - v)/f$, respectively, where f is the frequency. Since the wavelength is shortest in the *upwind* direction, the value of λ_- dictates the grid spacing. We set

$$\Delta r = \frac{\lambda_-}{N} = \frac{\lambda}{N}(1 - M), \tag{12.36}$$

where N is the number of grid points per wavelength in the upwind direction, $M = v/c$ is the Mach number, and $\lambda = c/f$ is the wavelength for the medium at rest. If N is to be fixed at a constant value, a finer grid is required as M increases. Regarding the time step, the Courant condition implies

$$\Delta t < \frac{\lambda_-}{Nu}. \tag{12.37}$$

This condition is most difficult to meet when u is largest, which is the case in the *downwind* direction. Therefore we should use u_+ in the preceding inequality if we are to have accurate results throughout the domain; specifically, we must set

$$\Delta t < \frac{\lambda_-}{Nu_+} = \frac{1}{Nf}\frac{1 - M}{1 + M}. \tag{12.38}$$

Thus the time step must also be shortened as M increases. For example, the time step at $M = 1/3$ must be $1/2$ the value necessary at $M = 0$. At $M = 2/3$, the time step must be $1/5$ the value at $M = 0$. The reduction of the required time step and grid spacing combine to make calculations at large Mach numbers computationally expensive.

The ability of the FDTD methodology to simulate scattering by turbulence, without any approximations involving small scattering angles or effective sound speed, can be useful for validating theories of wave propagation through turbulence. An example of such a simulation, to assess the impact of turbulent velocity fluctuations on the spatial coherence of acoustic signals, was

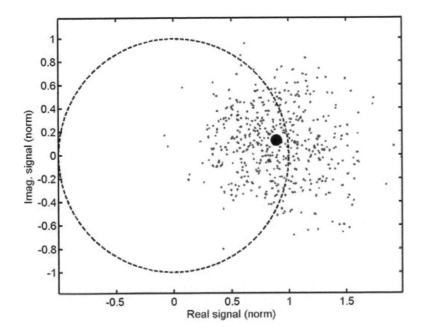

FIGURE 12.4
Samples of the complex pressure signal from the numerical FDTD experiment
of Collier et al. [83]. The horizontal axis is the real part of the normalized
signal; vertical axis is the imaginary part. The unit circle is shown as a dashed
line. The large dark circle is the mean of the complex samples. The lighter
dots are 500 individual samples. The source had a frequency of 100 Hz and
the signals propagated through turbulent velocity fluctuations to a ring-like
array of sensors at a distance of 400 m.

performed by Collier et al. [83]. The simulation domain was $1051 \times 1051 \times 401$
nodes (a total of 442 million) with a uniform grid spacing $\Delta r = 0.5$ m. A
point source was placed in the middle of the domain, and absorbing boundary
conditions along the edges mimicked a free-space radiation condition (section 12.4.3). Sensors were configured in two concentric, ring-like surfaces at
fixed distances of 200 m and 400 m from the source. Each ring had 1050 virtual
sensor positions. The source was a 100-Hz tapered harmonic signal and was
located at the origin (centers of the wheels). Quasi-wavelets (QWs) were used
to synthesize the turbulence fields, as described in section 9.2. The outer scale
was 4 m and the inner scale 0.5 m, thus producing realizations and a spectrum
as shown in figures 9.10 and 9.11, respectively. The turbulent kinetic energy
dissipation rate (section 9.2.1) was 10 m^2 s^{-3}. A total of 4 million QWs were

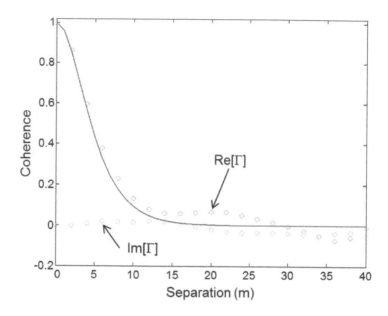

FIGURE 12.5
Simulated mutual coherence from the numerical FDTD experiment of Collier et al. [83]. The circles are the real and imaginary parts of the normalized second moment from the sensor ring at 400 m. The solid line is the theoretical prediction.

generated to synthesize the turbulence. The test required 32,000 total time steps, which were completed in 32 hrs on a 121-processor supercomputer.

Samples of the complex pressure signal, at 500 sensor locations along the 400-m ring, are shown in figure 12.4. The samples are normalized by the theoretical mean, as calculated with equations (7.150) and (7.151). In principle, the mean of the samples should be 1; while the actual value is close to 1, it deviates somewhat due to random sampling effects. The mutual coherence for a spherical wave, as a function of the turbulence spectrum, was derived earlier as equation (7.183). This theory is compared to results from the simulation in figure 12.5, again for the sensor ring at 400 m. Agreement is close, particularly when the coherence is in the range of about 0.4–1; random sampling errors tend to bias the results at smaller values of the coherence.

12.4   Boundary conditions and ground interactions

As mentioned in the introduction to this chapter, one of the strengths of the FDTD method is its ability to simulate propagation in complex environments including reflections and multiple scattering. In this section, we overview the merits and drawbacks of various approaches to incorporating reflective surfaces and variable material properties into FDTD calculations. We also consider the related topic of constructing layers that reduce numerical reflections at the simulation boundaries.

12.4.1   Rigid and impedance boundaries

Since the density of most solid objects is much higher than air, they appear as rigid surfaces from the standpoint of waves propagating in air. Implementation of a rigid boundary condition (BC) is very straightforward when the boundaries are aligned with the Cartesian coordinate system on which the FDTD calculation is performed. A rigid boundary implies that the velocity component normal to the boundary is zero, for the acoustic field as well as for the wind. The boundary is thus naturally placed coincident with nodes where the normal velocity components vanish. For a rigid boundary on a line or surface of constant y, for example, w_y and v_y are zero; assuming the boundary is aligned with the dashed line in figure 12.1 at $y = (j - 1/2)\Delta y$, we would need to explicitly calculate fields at $y = j\Delta y$ and above. In the second-order spatial scheme described in section 12.1, the fields below the boundary are not needed to evaluate any of the terms, when we apply the appropriate conditions on w_y and v_y; hence closure of the equations is obtained at the boundary.

Conversely, a pressure-release surface should be aligned with the pressure nodes; for example, if the surface is aligned with $x = i\Delta x$ in figure 12.1. In the second-order spatial scheme, the velocities tangential to the boundary must be taken as continuous in order to obtain closure at the boundary.

As mentioned in the introduction to this chapter, one of the main practical difficulties of time-domain methods is the formulation of a counterpart to the familiar frequency-domain impedance BC. The imaginary part of the impedance implies a reactive response; that is, when an impulsive signal is incident upon the boundary, the reflection will have a finite decay time. Time-domain boundary conditions (TDBCs) for porous ground surfaces and their efficient implementation in time-domain calculations were discussed in section 10.2.3. References [294, 419] derived and demonstrated implementation of such TDBCs in FDTD calculations. Like the rigid BC discussed above, it is assumed that the surface is straight and aligns with the nodes on the computational grid corresponding to velocities normal to the surface. In this case, we must infer the pressure at the surface. A simple procedure, similar to that used for the CNPE as described in section 11.2.1, is to extrapolate the pressure from the grid levels above the surface. Constructing a second-degree Lagrange polynomial around the values of the three pressure nodes closest to the surface $y = y_s$, we derive the extrapolation formula [419]

$$p\left(x, y_s, t\right) \simeq \frac{15}{8} p\left(x, y_s + \frac{\Delta y}{2}, t\right) - \frac{5}{4} p\left(x, y_{s,t} + \frac{3\Delta y}{2}\right)$$
$$+ \frac{3}{8} p\left(x, y_s + \frac{5\Delta y}{2}, t\right), \tag{12.39}$$

where Δy is the spatial grid increment. With this estimate for the pressure at the surface and the particle velocities stored from the previous time steps, $w_y\left(x, y_s, t\right)$ can be inferred from the TDBC, thus closing the equations at the boundary.

Let us now consider a 2D numerical example involving refraction by a wind profile and scattering by turbulent velocity fluctuations. The atmosphere is neutral with friction velocity $u_* = 0.6$ m/s and roughness length $z_0 = 0.01$ m. The wind profile is logarithmic, as described in section 2.2.3. Turbulence is simulated by the inhomogeneous spectral method described in section 9.1.2. The computational domain is 1200 m horizontally and 600 m vertically. The spatial discretization has 10 grid points per wavelength in air. The source is harmonic, with a frequency of $f = 125$ Hz. It is positioned at a height of 5 m, at the horizontal midpoint of the domain. The ground, at the bottom edge of the domain, is perfectly rigid, and perfectly matched layers (section 12.4.3) are positioned at the left, right, and top edges of the domain.

Shown in figure 12.6(a) is the pressure field from the FDTD calculation at time $t = t_f = 1.8$ s. Figure 12.6(b) shows the transmission loss (TL, see section 10.1) as determined from the FDTD, namely $-20 \log_{10}[p_{max}(\mathbf{R})/p_{free}(R_0)]$, where p_{free} is the sound pressure in free space, $R_0 = 1$ m is the reference distance, and p_{max} is the maximum value of $|p|$ as recorded over the last ten wave cycles of the simulation (the time window $[t_f - 10/f, t_f]$) at each location in the computational domain. Finally, the bottom panel shows the TL from a wide-angle CNPE calculation (section 11.2.2) for the same case. (For purposes of comparison, we have not multiplied the pressure field from the CNPE by $1/\sqrt{r}$, as is normally done to estimate the TL in 3D, so that the TL represents a true 2D calculation.) The FDTD and CNPE calculations agree reasonably well within an elevation angle of 45°, which is the approximate region of validity of the wide-angle PE. Some discrepancies are also evident close to the ground.

12.4.2 Propagation in the ground and other materials

Objects with density and bulk modulus differing from air can be readily incorporated into FDTD calculations, simply by adjusting the values of b and K in equations (12.1) and (12.2). However, some caution is necessary when mixing media with varying properties, particularly when the sound speed is

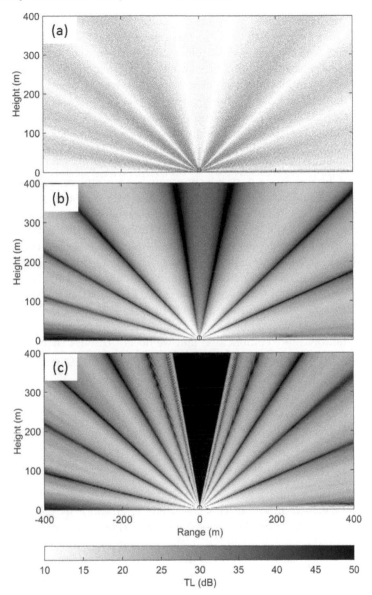

FIGURE 12.6
FDTD simulation of sound propagation in a turbulent atmosphere above a
rigid ground. The source is harmonic with a frequency of 125 Hz, and is at
a height of 5 m. The friction velocity is $u_* = 0.6$ m/s. (a) Snapshot of a
2D FDTD simulation after 1.33 s. Dark gray represents the positive phase of
the waves. (b) Transmission loss (TL) for the FDTD as calculated from the
maximum absolute pressure during the last 10 wave cycles. (c) TL from a
wide-angle CNPE calculation for the same case. The gray scale is for the TL
in panels (b) and (c).

varied. If the sound speed is increased relative to air, the Courant condition requires a diminishment of the time step, in order to maintain stability. If the sound speed is decreased, the wavelength decreases, and a higher grid resolution becomes necessary. Stability of the FDTD calculation can sometimes be improved by smoothing the variation of the density and bulk modulus at material boundaries.

Explicit FDTD calculation of propagation in the ground and other porous media is complicated by the dissipative properties of these materials. It is, however, usually reasonable to assume that there is no ambient flow within the porous medium. The Zwikker–Kosten phenomenological model, as discussed in Chapter 10, is particularly amenable to implementation in FDTD calculations, if one assumes that the flow resistivity σ, tortuosity α, and bulk modulus K in equations (10.23) and (10.24) can be approximated as real-valued constants in the frequency range of the acoustic disturbance. As such, these equations provide an equivalent fluid model for a rigid-framed porous material, which enables explicit time-domain calculations of propagation in the ground, and thus coupled calculations involving both the atmosphere and ground. Application of these equations to FDTD calculations has been previously demonstrated by several authors [346, 387, 423]. FDTD calculations can also be performed with a poroelastic ground [101], although such calculations are considerably more complicated and we do not discuss them further here.

Isolating the time derivatives in the Zwikker–Kosten equations (10.23) and (10.24) on the left side, we have

$$\frac{\partial p}{\partial t} = -K\nabla \cdot \mathbf{w} \tag{12.40}$$

and

$$\frac{\partial \mathbf{w}}{\partial t} = -\frac{\Omega}{\alpha\rho_0}\nabla p - \frac{\Omega\sigma}{\alpha\rho_0}\mathbf{w}. \tag{12.41}$$

Comparing these equations to (12.1) and (12.2), we see that they are the same, but with the following changes: there is no ambient velocity ($\mathbf{v} = 0$), source terms are not included, $b \to \Omega/\alpha\rho_0$, and there is an additional term in equation (12.41) involving the flow resistivity. Hence, from the standpoint of numerical implementation, the only novelty of these equations is the new term involving σ. This term is fairly trivial to implement. Referring to figure 12.1, we can store σ on the pressure nodes, and the velocity components can be determined by averaging the adjacent half-integer nodes.

Since FDTD calculations are normally performed at low frequencies, it is usually reasonable to set σ, α, and K to their low-frequency limiting values as discussed in section 10.2.3, namely $\sigma \to \sigma_0$, $\alpha \to \alpha_0 = 2q^2$, and $K \to K_0 = P_0/\Omega$, where σ_0 is the static flow resistivity, q^2 is the tortuosity, Ω is the porosity, and P_0 is the ambient pressure. The preceding equations then become

$$\frac{\partial p}{\partial t} = -\frac{P_0}{\Omega}\nabla \cdot \mathbf{w} \tag{12.42}$$

and

$$\frac{\partial \mathbf{w}}{\partial t} = -\frac{1}{\sigma_0 \tau_0} \nabla p - \frac{1}{\tau_0} \mathbf{w}, \qquad (12.43)$$

where $\tau_0 = 2\rho_0 q^2 / \sigma_0 \Omega$ is a time constant. Values of τ_0 for typical ground surfaces were given in table 10.1.

The immersed-boundary method [254] provides a systematic, stable approach to handling boundaries between materials, particularly when they do not conform to the underlying coordinate grid. Xu et al. [438] performed such simulations in which a Zwikker–Kosten medium (representing a wind screen) was embedded in an ambient flow.

An important practical issue of coupled FDTD with the air and ground is that the wavelength of the sound field in a porous medium becomes very short in the ground, particularly at low frequencies. This necessitates a very large numerical grid, with much finer spacing than would be necessary for simulation in air only. An appropriate TDBC, as described in section 12.4.1, circumvents this problem.

The time-domain relaxation model given by equations (10.47) and (10.48) potentially provides more exact calculations of sound propagation in the ground than do the Zwikker–Kosten equations. However, the full relaxation model involves convolutions between the acoustic fields and the relaxation functions, which are expensive to evaluate numerically. Implementation of this theory is described in more detail in reference [423]. In principle, the computational burden can be mitigated by approximating the response with an exponential series, as described in section 10.2.3 with regard to the TDBC.

The Zwikker–Kosten equations have also been used for simulating infrasonic propagation in a forest [370]. In this frequency range, the wavelengths are much larger than the size of the tree trunks and the spacing between them. Hence the forest may be modeled as an effective porous medium.

12.4.3 Perfectly matched layers

Often we wish to mimic a free-space radiation condition at the edges of the computational domain. However, simply truncating the computational domain at a particular location leads to spurious numerical reflections. For this reason, various types of absorbing boundary conditions, or ABCs, have been developed to gradually damp the outwardly propagating waves without reflection. The Zwikker–Kosten equations, discussed in the previous subsection, can be used for this purpose. The porosity and tortuosity should be set to 1, and the static flow resistivity increased gradually with distance from the boundary, in order to avoid numerical reflections.

Unfortunately, ABCs normally require a region many wavelengths wide to accomplish the desired effect and hence can add significantly to the size of the overall domain. An efficient and practical approach alternative to implementing the free-space condition is provided by the *perfectly matched layer*. The original PML formulation is due to Bérenger [31]. More recent PML formulations [132, 341] are based on a general and relatively straightforward complex coordinate transform. Johnson [184] provides an accessible discussion of the various PML formulations.

Let us first consider the coordinate transformation for a PML in one dimension. In the frequency domain, the transformation involves simply replacing the operator $\partial/\partial x$ where it occurs with [184]

$$\frac{\partial}{\partial x} \rightarrow \left[\frac{1}{1 + i\phi(x)/\omega}\right]\frac{\partial}{\partial x},$$ (12.44)

where the function $\phi(x)$ is zero outside of the PML. In one dimension, equations (12.1)–(12.2) (without the fluid velocity and source terms) are, when transformed to the frequency domain,

$$-i\omega\hat{p} = -K\frac{\partial\hat{w}_x}{\partial x}, \quad -i\omega\hat{w}_x = -b\frac{\partial\hat{p}}{\partial x}.$$ (12.45)

As before, the "hat" indicates a Fourier transform with respect to time. Applying the coordinate transformation (12.44), we have

$$-i\omega\left(1 + \frac{i\phi}{\omega}\right)\hat{p} = -K\frac{\partial\hat{w}_x}{\partial x}, \quad -i\omega\left(1 + \frac{i\phi}{\omega}\right)\hat{w}_x = -b\frac{\partial\hat{p}}{\partial x}.$$ (12.46)

Transforming back to the time domain results in

$$\frac{\partial p}{\partial t} = -K\frac{\partial w_x}{\partial x} - \phi p$$ (12.47)

and

$$\frac{\partial w_x}{\partial t} = -b\frac{\partial p}{\partial x} - \phi w_x.$$ (12.48)

These equations have an interesting functional similarity to the Zwikker–Kosten equations (12.40) and (12.41). However, the term $-\phi p$ on the right side of equation (12.47) is new. This modification turns out to be a very important one, as can be seen by determining the characteristic impedance and complex wavenumber implied by these equations. Substituting a trial solution for harmonic plane waves, $p = Ae^{i\Gamma x - i\omega t}$ and $w_x = Be^{i\Gamma x - i\omega t}$, we find that the characteristic impedance, $Z = A/B$, equals ρc. The complex wavenumber is $\Gamma = (\omega/c)(1 + i\phi/\omega)$. Hence the medium defined by equations (12.47) and (12.48) has the same impedance and phase speed as free space, while also incorporating an attenuation factor ϕ/c. The impedance match means that there will be no reflection, and the match in phase speed means the spatial sampling need not be increased in the layer. In principle, ϕ can be made as large as desired, and thus provide rapid attenuation of any wave incident upon the layer.

PML implementation in higher spatial dimensions is somewhat more complicated, in that it involves solution of auxiliary equations. Let us consider the 2D case, as represented by equations (12.3)–(12.5). The transformations for the x and y coordinates are

$$\frac{\partial}{\partial x} \rightarrow \left[\frac{1}{1 + i\phi_x(x)/\omega}\right]\frac{\partial}{\partial x}, \quad \frac{\partial}{\partial y} \rightarrow \left[\frac{1}{1 + i\phi_y(y)/\omega}\right]\frac{\partial}{\partial y}. \qquad (12.49)$$

Proceeding as before leads to

$$\left(1 + \frac{i\phi_x}{\omega}\right)\left(1 + \frac{i\phi_y}{\omega}\right)(-i\omega\hat{p}) =$$
$$-K\left(1 + \frac{i\phi_y}{\omega}\right)\frac{\partial\hat{w}_x}{\partial x} - K\left(1 + \frac{i\phi_x}{\omega}\right)\frac{\partial\hat{w}_y}{\partial y}, \qquad (12.50)$$

$$\left(1 + \frac{i\phi_x}{\omega}\right)(-i\omega\hat{w}_x) = -b\frac{\partial\hat{p}}{\partial x}, \qquad (12.51)$$

$$\left(1 + \frac{i\phi_y}{\omega}\right)(-i\omega\hat{w}_y) = -b\frac{\partial\hat{p}}{\partial y}. \qquad (12.52)$$

Equations (12.51) and (12.52) can be handled just as in the 1D calculation. However, equation (12.50) has terms in which $-i\omega$ occurs in the *denominator*; in the time domain, these terms correspond to integrations in time. To handle these terms, we define the auxiliary variables ψ_x, ψ_y, and ψ_{xy} through the following equations:

$$\hat{\psi}_x = -\left(\frac{i\phi_y K}{\omega}\right)\frac{\partial\hat{w}_x}{\partial x}, \quad \hat{\psi}_y = -\left(\frac{i\phi_x K}{\omega}\right)\frac{\partial\hat{w}_y}{\partial y}, \quad \hat{\psi}_{xy} = \left(\frac{i\phi_x\phi_y}{\omega}\right)\hat{p}.$$
$$(12.53)$$

We then have a system of six equations to solve in the time domain:

$$\frac{\partial p}{\partial t} = -K\frac{\partial w_x}{\partial x} - K\frac{\partial w_y}{\partial y} + \psi_x + \psi_y - \psi_{xy} - \phi_x p - \phi_y p, \qquad (12.54)$$

$$\frac{\partial w_x}{\partial t} = -b\frac{\partial p}{\partial x} - \phi_x w_x, \qquad (12.55)$$

$$\frac{\partial w_y}{\partial t} = -b\frac{\partial p}{\partial y} - \phi_y w_y, \qquad (12.56)$$

$$\frac{\partial\psi_x}{\partial t} = -\phi_y K\frac{\partial w_x}{\partial x}, \qquad (12.57)$$

$$\frac{\partial\psi_y}{\partial t} = -\phi_x K\frac{\partial w_y}{\partial y}, \qquad (12.58)$$

and

$$\frac{\partial\psi_{xy}}{\partial t} = \phi_x\phi_y p. \qquad (12.59)$$

Although solution of the six-equation system involves more computation time and memory than the basic three-equation system for sound propagation in a motionless medium, this system need only be solved within the PML, where the auxiliary variables are non-zero. In practice, since the PML can be very thin, there should be a substantial reduction in the computational burden. The auxiliary variables are naturally stored at the pressure nodes.

While the PML equations were explicitly derived here for a homogeneous, motionless medium, the added terms can be readily (but non-rigorously) incorporated into the equations for propagation in an inhomogeneous, moving medium, namely equations (12.3)–(12.5). One simply adds $\psi_x + \psi_y - \psi_{xy} - \phi_x p - \phi_y p$ to the right side of equation (12.3), $-\phi_x w_x$ to the right side of equation (12.4), and $-\phi_y w_y$ to the right side of equation (12.5). The auxiliary equations (12.57)–(12.59) are implemented within the PML as usual. In our experience, performance of the PML is not substantially degraded in comparison to a homogeneous, motionless medium.

Let us consider a 2D numerical example involving propagation above a porous ground with a rigid barrier, as shown in figure 12.7. The atmosphere is assumed to be homogeneous and motionless. The size of the computational domain is 50 m horizontally and 30 m vertically. The source is placed at a height of 5 m, at the midpoint of the domain in the horizontal direction ($x = 25$ m). The barrier is placed 12.5 m to the right of the source and is 10 m tall. The source signal is a once-differentiated Gaussian with positive initial phase:*

$$Q\left(t\right) = \left(t - t_0\right)\omega_0^2 \exp\left[-\left(t - t_0\right)^2 \omega_0^2/2\right], \tag{12.60}$$

where $f_0 = \omega_0/2\pi$ is the center frequency and t_0 the time of zero-crossing. In this example, $t_0 = 2/f_0$, where $f_0 = 100$ Hz and $t_0 = 0.0125$ s.

The bottom 5 m of the domain consists of a Zwikker–Kosten medium with parameters characteristic of snow, as given in table 10.1. A rigid boundary is positioned below the snow layer. The spatial discretization has 160 grid points per wavelength in air at 100 Hz, which yields a resolution of 0.0214 m. Such a dense grid is necessary because the source signal is broadband; furthermore, as discussed earlier, the wavelength in the ground is much smaller than in air. For most other porous ground surfaces besides snow, the wavelength becomes too short for a simulation with a regular Cartesian grid to remain practical. The top, right, and left edges of the domain are configured with perfectly matched layers having a thickness of 480 grid nodes (4 wavelengths).

Shown in figure 12.7 are six snapshots of the pressure field at various times up to 0.1 s. The first snapshot, at $t = 0.02$ s, shows the source signal before it has interacted with the ground or barrier. At $t = 0.04$, the direct wave, ground reflection, and transmission into the ground can be clearly distinguished. At-

---

*In two dimensions, this input signal produces two positive peaks surrounding a single negative peak. This behavior is a consequence of the multidimensional nature of the propagation, rather than numerical error.

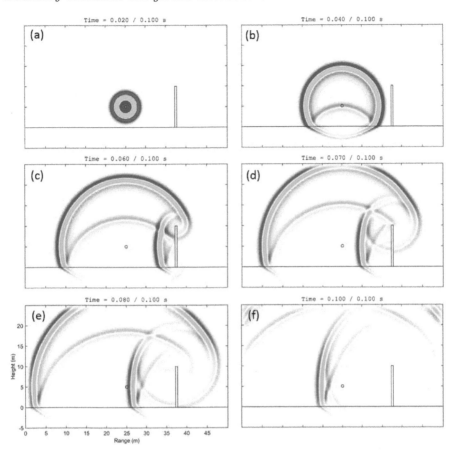

FIGURE 12.7
FDTD simulation of sound propagation in a homogeneous atmosphere above a porous ground, with properties characteristic of snow. The source is at $(25\,\text{m}, 5\,\text{m})$, as indicated by the open circle. The horizontal line at $y = 0\,\text{m}$ is the ground boundary. The rectangle at $x = 37.5\,\text{m}$ is a 10-m tall rigid barrier. Red indicates a positive pressure variation and green is negative. Shown are snapshots at (a) $t = 0.02$, (b) $t = 0.04$, (c) $t = 0.06$, (d) $t = 0.07$, (e) $t = 0.08$, and (f) $t = 0.1$ s.

tenuation and shortening of the wavelength in the ground are evident. At $t = 0.06$ s, the direct wave is just clearing the top of the barrier, and a reflection from the barrier is seen. At $t = 0.07$ s, the diffraction over the barrier is well developed. The final two snapshots, at $t = 0.08$ s and $t = 0.1$ s, illustrate the effectiveness of the PMLs in preventing reflections from the top, right, and left boundaries.

13

Uncertainty in sound propagation and its quantification

Parts I and II of this book described the complex manner in which an inhomogeneous moving medium influences sound propagation. The previous chapters in Part III subsequently described a number of computational approaches capable of incorporating these complexities, with varying degrees of generality and fidelity. In this chapter, we explore the uncertainties inherent to application of such computational modeling.

Although the underlying physics of sound propagation is, in a fundamental sense, well understood, modeling capabilities are typically inadequate to capture all of the intricacies of sound propagation in real environments such as the atmosphere or ocean, which are impacted by phenomena occurring on a variety of spatial and temporal scales. Even if it were possible to develop a numerical model capable of perfectly simulating the propagation physics, the resolution of environmental data used in the models often greatly limits their accuracy. The lurking presence of uncertainties also presents a challenge to the design and execution of appropriately controlled experiments for testing new theories and calculation methods, as well as the proper attribution of the causes of discrepancies when they are observed. Predictive methods can be employed more effectively when their capabilities, relative to the inputs provided to them and the uncertainties inherent to sound propagation, are well understood and quantified.

The important role of uncertainty and randomness in outdoor sound propagation has, in a sense, been recognized since systematic scientific experiments of outdoor propagation were first undertaken. Knudsen [202] recorded a 4000-Hz tone propagating over a 100-ft path, and found that "the tone ... fluctuated violently over a range of more than 10 db, with short periods (a tenth second or less) and long periods (several seconds) all jumbled together." Ingard [174] similarly observed random variations in sound levels of 10–20 dB and attributed this to the gustiness of the wind. Modern simulations of atmospheric boundary-layer turbulence, when coupled with numerical sound propagation calculations [417], enable visualization of the gustiness and fluctuations identified by Knudsen and Ingard. Many long-term experimental studies of sound propagation outdoors have subsequently reported strong variability in sound levels and revealed the challenge of directly relating the signal variations to atmospheric observations [208, 275, 353, 421, 444]. Predictive uncertainties and

449

the highly nonlinear dependence on model parameters have also been studied in the context of ocean acoustics [102, 115].

The first section in this chapter provides basic statistical background pertinent to the description of uncertainty; types of uncertainty are examined in the context of outdoor sound propagation, with a simple illustrative example incorporating turbulence and uncertainty in the wind velocity. The second section describes random sampling methods useful for incorporating and quantifying uncertainty in predictions. The third section provides a comprehensive application of these methods to near-ground sound propagation. The chapter closes with a brief discussion of practical issues related to atmospheric modeling in outdoor sound propagation calculations.

13.1   Parametric uncertainties

Uncertainties are inherently introduced when a real, physical system is represented with a model. Many taxonomies have been devised to categorize these uncertainties [18]. For present purposes, we distinguish between those associated with the *model* itself and with the *parameters* (inputs) to the model. The emphasis in the literature on outdoor sound propagation has typically been on reducing the model uncertainties by incorporating more realistic physics and by developing more accurate numerical methods. Some important sources of model uncertainty, many of which relate to topics explored in earlier chapters of this book, are listed in the top part of figure 13.1. In practice, these uncertainties may be quite challenging to quantify and mitigate. For example, modeling of the turbulence spectrum in the energy-containing subrange remains an essentially unsolved problem in physics. On the other hand, many numerical approximations, such as discretization of continuous differential equations in space and time, have been studied thoroughly and much is known about their impacts.

As models improve in capability and fidelity, their accuracy is often increasingly limited by parametric uncertainties. The parameters may be difficult to measure, or they may be difficult to characterize at the spatial and temporal resolution required for an accurate prediction. In order to assess the accuracy of sound propagation predictions, it is important to systematically identify and quantify the impacts of such parametric uncertainties.

Parametric uncertainties may be categorized *epistemic* (alternatively, *structural* or *reducible*) or as *aleatory* (alternatively, *statistical* or *irreducible*). Significant sources of epistemic and aleatory parametric uncertainty are shown in the bottom part of figure 13.1. The epistemic uncertainties represent limitations in our knowledge of the problem, and may include the mean vertical wind and temperature profiles, ground properties such as porosity and static flow resistivity, and large-scale variations in terrain elevation (such as hills and

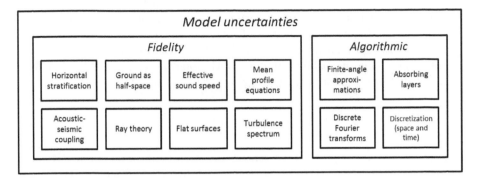

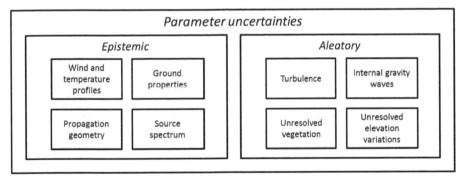

FIGURE 13.1

Top: Important sources of model uncertainty in outdoor sound propagation, as categorized into model fidelity and algorithmic (computational or numerical) types. Bottom: Important sources of parametric uncertainty in outdoor sound propagation, as categorized into epistemic and aleatory types.

buildings). Atmospheric turbulence and internal gravity waves are normally regarded as random and unpredictable motions, unresolvable with available atmospheric observations and weather models, and thus aleatory uncertainties. Depending on the problem, other sources of aleatory uncertainty besides turbulence may be present, such as small-scale variations in the terrain elevation, ground properties, vegetation, and construction details on buildings. The distinction between epistemic and aleatory uncertainty is often a matter of perspective, dependent upon the predictive methodology and available measurements. For example, if meteorological towers were employed at many different locations around an area of interest, turbulence and other variations at space and time scales resolved by the towers would be considered epistemic, whereas the smaller, unresolved variations would be aleatory.

To model the impacts of uncertainties on a prediction, a probability distribution [218, 408] for the uncertain variables must be deduced in some manner, either empirically or from theoretical considerations. For example, if the wind direction is uncertain, we could specify a distribution for it, randomly sample from the distribution, and then perform a sound propagation calculation for each sample direction. Quantities of interest, such as the mean and variance of the squared sound pressure, can then be estimated from the ensemble of predictions. The probability that a continuous random variable X will take on a value in the interval $[x, x + \Delta x]$ is

$$\Pr[x \leq X \leq x + \Delta x] = \int_x^{x+\Delta x} g(\xi) \, d\xi, \tag{13.1}$$

where $g_X(x)$ is the probability density function (pdf). The cumulative distribution function (cdf), $G_X(x)$, is the probability that X will take on a value less than or equal to x, namely

$$G_X(x) = \Pr[X \leq x] = \int_{-\infty}^x g_X(\xi) \, d\xi. \tag{13.2}$$

Differentiation of the preceding integral yields the relationship $g_X(x) = dG_X(x)/dx$.

These definitions can be conveniently extended to multiple random variables by grouping them as a vector. For M variables, we write $\mathbf{X} = [X_1, X_2, \ldots, X_M]$. The joint pdf of these variables, $g_{\mathbf{X}}$, is defined such that

$$\Pr[\mathbf{X} \in \Omega'] = \int_{\Omega'} g_{\mathbf{X}}(x_1, \ldots, x_M) \, dx_1 \ldots dx_M, \tag{13.3}$$

where Ω' is a subdomain within the overall domain Ω upon which $\mathbf{X}$ is defined ($\Omega' \in \Omega$). The cdf is defined as

$$G(x_1, \ldots, x_M) = \Pr[X_1 \leq x_1 \cdots X_M \leq x_M]. \tag{13.4}$$

If the variables in $\mathbf{X}$ are independent, one has $g_{\mathbf{X}}(x_1, \ldots, x_M) = g_{X_1}(x_1) \cdots g_{X_M}(x_m)$.

One might fairly argue that, in practice, the distribution is rarely known exactly, and thus by positing one we are introducing an additional source of uncertainty. Indeed, the accuracy of an analysis is always limited by the validity of such assumptions. It should be kept in mind, however, that the alternative, of assuming that the variables are perfectly known, is often a much more drastic simplification than positing a reasonable pdf, and provides no information on the impacts of uncertainties upon the predictions. The best practice is generally to employ the simplest distribution consistent with the known constraints and observations, and then to estimate its parameters as well as possible. The robustness of the predictions to the selection of the distribution and its parameters should be explored and quantified when feasible.

The type of distribution and parameter values also depend greatly on the application. For example, if one is attempting to model the impact of random wind gusts on sound propagation at a particular time and location, wind speed variations of 20% to 40% and directional variations of 30° to 60° might be typical. On the other hand, if the purpose is to estimate a distribution of sound levels over the course of an entire year, a wide range of speeds, and essentially all directions, must be represented based on a climatology.

The uniform, beta, normal (Gaussian), log-normal, exponential, and Rayleigh distributions are particularly useful for continuous random variables, and often appear in problems involving wave propagation and scattering. The uniform distribution represents a variable X that takes on all values within an interval $[a, b]$ with equal probability, and with zero probability outside this interval. Such a distribution would apply, for example, to a signal phase that varies randomly between 0 and 2π. The uniform pdf is given by

$$g_X(x) = \begin{cases} 1/(b-a) & \text{for } a \leq x \leq b, \\ 0 & \text{elsewhere.} \end{cases} \tag{13.5}$$

The mean of the uniform distribution is $\mu = (a+b)/2$, whereas the variance is $\sigma^2 = (b-a)^2/12$. As described in Chapter 9, scattering by turbulence has often been handled in outdoor sound propagation by a Monte Carlo approach, in which a number of random realizations of a turbulence field are generated by randomizing the phases of the spectral components. In effect, a large number of independent, uniformly distributed random variables (one for the phase of each Fourier mode) are sampled for each realization.

By the central limit theorem, the normal pdf applies to a variable determined by summing many other random variables, regardless of the distributions of the original variables (with some caveats). It is given by

$$g_X(x) = \frac{1}{\sigma\sqrt{2\pi}} e^{-\frac{(x-\mu)^2}{2\sigma^2}}. \tag{13.6}$$

Figure 13.2(a) illustrates application of the normal pdf to modeling uncertainty in the wind direction.

For positive definite variables (such as the friction velocity, roughness length, and static flow resistivity of the ground), a log-normal pdf may be appropriate. It is the same as the normal pdf, except that $\ln x$, rather than x itself, is distributed normally. Hence

$$g_X(x) = \frac{1}{xs\sqrt{2\pi}} e^{-\frac{(\ln x - m)^2}{2s^2}}, \quad x > 0. \tag{13.7}$$

The parameters m and s are termed here the log-mean and log-deviation, respectively. The mean of x is $e^{m+s^2/2}$, and the variance $(e^{s^2} - 1)e^{2m+s^2}$. Figure 13.2(b) illustrates application of the log-normal pdf to modeling uncertainty in the friction velocity.

The exponential distribution pertains to the *sum of the squares* of two independent, normally distributed variables with zero mean and equal variance. It applies, for example, to the intensity of a strongly scattered signal, for which the real and imaginary parts of the complex pressure satisfy the

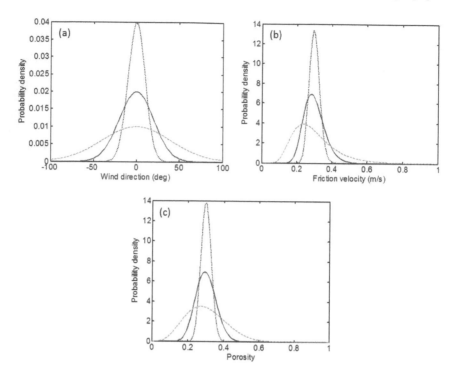

FIGURE 13.2

Illustrative probability density functions (pdfs) for modeling parametric uncertainties in outdoor sound propagation. (a) Uncertainty in the wind direction, as modeled with a normal pdf with a mean of $0°$ and standard deviation equal to $10°$ (dotted), $20°$ (solid), or $40°$ (dashed). (b) Uncertainty in the friction velocity, as modeled with a log-normal pdf having a mean of 0.3 m/s and log-deviation equal to 0.1 (dotted), 0.2 (solid), or 0.4 (dashed). (c) Uncertainty in the ground porosity, as modeled with a beta pdf having a mean of 0.3 and parameter ν equal to 256 (dotted), 64 (solid), or 16 (dashed).

conditions just mentioned. The exponential pdf is

$$g_X(x) = \lambda e^{-\lambda x}. \tag{13.8}$$

Here, λ is called the rate parameter, which equals the inverse of the mean of X. The variance is λ^{-2}.

The Rayleigh distribution is closely related to the exponential, except that it pertains to the *square root* of the sum of the squares; for strong scattering, it thus applies to the amplitude rather than the intensity. It is given by

$$g_X(x) = \frac{x}{\sigma^2} e^{-\frac{x}{2\sigma^2}}, \tag{13.9}$$

where σ^2 is the variance of each of the original normally distributed variables.

Like the uniform distribution, the beta distribution applies to a continuous variable confined to a particular interval. However, the probability is not constant within this interval. Such a distribution might be appropriate for the porosity of the ground, which is bounded between 0 and 1, but would typically peak at a mode between these limits. For convenience, the beta pdf is usually specified for random variables X confined to the interval $[0, 1]$. A variable X' on the interval $[a, b]$ can be readily normalized to this range using the linear transformation $X = (X' - a)/(b - a)$. The beta pdf is given by

$$g_X(x) = \frac{1}{B(\alpha, \beta)} x^{\alpha-1} (1 - x)^{\beta-1}, \ 0 \le x \le 1, \tag{13.10}$$

where α and β are termed shape parameters, and $B(\alpha, \beta)$ is the beta function. The mean is $\mu = \alpha/(\alpha+\beta)$, and the variance is $\sigma^2 = \alpha\beta/[(\alpha+\beta)^2(\alpha+\beta+1)]$. When $\alpha = \beta = 1$, the beta distribution reduces to the uniform distribution. For given values of μ and σ^2, the shape parameters follow from $\alpha = \mu\nu$ and $\beta = (1 - \mu)\nu$, where $\nu = \mu(1 - \mu)/\sigma^2 - 1$. Since the shape parameters must be positive, ν must also be positive, and hence $\sigma^2 < \mu(1 - \mu)$. Figure 13.2(c) illustrates application of the beta pdf to modeling uncertainty in the porosity.

The expected value of a function $I(x)$ can be calculated from the integral [29, 408]

$$\langle I \rangle = \int_{-\infty}^{\infty} I(x) g_X(x) \, dx. \tag{13.11}$$

Defining the new random variable $Y = G_X(X)$ (where $y = G_X(x)$ is the cdf, and thus $dy = g_X(x)dx$), equation (13.11) can be written as

$$\langle I \rangle = \int_0^1 I[x(y)] \, dy. \tag{13.12}$$

Here $x(y) = G_X^{-1}(y)$, where G_X^{-1} is the inverse cdf. The variable Y has a uniform pdf on the interval $[0, 1]$. In effect, the transformation recasts the integration into equal-probability intervals, which can greatly simplify simulations.

To illustrate the incorporation of uncertainties, let us consider a relatively simple example with epistemic uncertainty in the wind speed and direction, and aleatory uncertainty in the form of scattering by turbulent velocity fluctuations. The wind speed is proportional to the friction velocity u_*, for which we choose a log-normal pdf, since it is positive definite. The pdf, as shown in figure 13.2(b), has a mean $\mu = 0.3$ m/s and log-deviation $s = 0.4$, thus implying a log-mean of $m = \ln\mu - s^2/2 = -1.284$. Uncertainty in the wind direction θ is described by a normal pdf, with a mean of $0°$ and standard deviation of $40°$, as shown in figure 13.2(a). The variations in u_* and θ are assumed to be uncorrelated.

Based on these distributions and parameters, the following eight random samples for $\Theta = (u_*, \theta)$ were simulated with a random-number generator (u_*

in m/s, θ in deg): $(0.221, 88.5)$, $(0.148, 42.5)$, $(0.229, -19.6)$, $(0.414, -41.0)$, $(0.140, 12.3)$, $(0.370, 40.5)$, $(0.308, -10.6)$, and $(0.402, 5.70)$. CNPE calculations (section 11.2.1) of the transmission loss (TL, as defined by equation (10.9)) based on these values are overlaid in figure 13.3. The source frequency used in these calculations is 200 Hz, the source height is 5 m, and the receiver height is 1.5 m. The first series of calculations (top) has only epistemic uncertainties; that is, one curve is calculated for each of the preceding samples of (u_*, θ), without turbulent scattering. The second series of calculations (middle) has only turbulent scattering (aleatory uncertainty); (u_*, θ) is held constant at the mean value, $(0.3, 0)$. The final series (bottom) uses the same random samples of (u_*, θ) as the first, and also includes turbulent scattering. In this calculation, the epistemic and aleatory uncertainties both introduce variations in the TL of roughly 10 dB. However, the variations are qualitatively different: uncertainty in the wind causes relatively smooth variations, with a shifting of the interference minimum in the downwind direction and of the shadow zone boundary in the upwind direction. With turbulence only, the refraction characteristics are relatively stable, but there are rapid spatial variations in the sound level. The calculations with both types of uncertainty combine these characteristics.

Figure 13.4 shows the mean TL for the same calculation as in figure 13.3. The number of trials was $N = 4096$. The mean TL is remarkably similar regardless of whether parametric (epistemic) uncertainty or turbulent scattering (or both) is present. It is particularly interesting that variability in the wind speed and direction impacts the mean TL in a manner very similar to turbulent scattering. Overlaid on figure 13.4 are also pdfs of normalized signal amplitude (specifically, $|\hat{g}| = 10^{TL/20}$; see equation (10.9)) at receiver locations 1000 m upwind (deep in the shadow), 500 m upwind (near the shadow boundary), 500 m downwind (near an interference null), and 1000 m downwind (deep within the duct). When only turbulent scattering is present, the amplitude pdfs resemble a Rayleigh distribution. When only parametric (epistemic) uncertainty is present, the pdfs are rather more complicated, particularly in the downwind direction, where they exhibit multiple peaks.

In practice, the mean TL, or other statistic of interest, must be estimated from some finite number of samples N. In the previous example, we set $N = 4096$, but such a large number of samples may be impractical when the computational models are time consuming. So, it is important to assess the dependence of the accuracy of our estimates on N, for example by determining a confidence interval. Bootstrap methods (e.g., reference [408]) are very useful for this purpose, as they enable an estimation of the intervals with a relatively small number of initial samples. The process begins by generating an ensemble of N_e realizations of the TL. The bootstrap involves randomly selecting, with replacement,* N_b sets of N samples from the original N_e sam-

---

*Sampling *with replacement* means that the probability of choosing a particular member of the ensemble does not depend on whether it has been chosen previously. In sampling *without replacement*, an ensemble member, once selected, cannot be selected a second time.

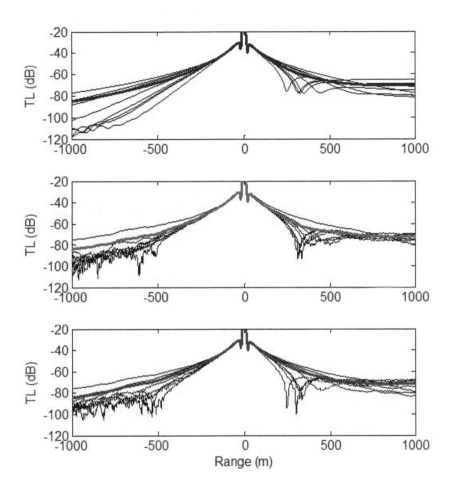

FIGURE 13.3
Randomized calculations of transmission loss (TL) vs. range at 200 Hz in a neutral, turbulent atmosphere. Shown are 8 independent realizations; the thick lines show the mean-square pressure from the 8 realizations. Downwind is to the right, upwind to the left. (Top) Epistemic uncertainties only (no turbulent scattering). (Middle) Turbulent scattering only (no epistemic uncertainty). (Bottom) Both epistemic uncertainties and turbulent scattering.

ples. The sample mean is then calculated for each of these N_b bootstrap trials. From the quantiles of the distribution of the N_b estimates of the mean TL, we can then estimate the confidence interval [408] corresponding to that value of N. Figure 13.5 shows the 90% confidence intervals determined by such a

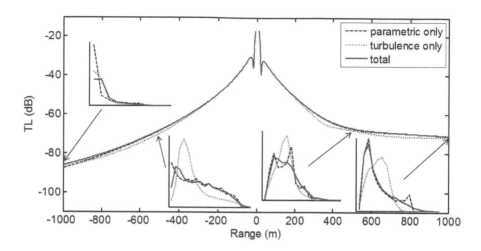

FIGURE 13.4
Mean TL for sound propagation at 200 Hz in a neutral, turbulent atmosphere, as calculated with parametric (epistemic) uncertainties only, with turbulent scattering only, and with both parametric uncertainties and turbulent scattering. Also shown are the corresponding pdfs of the normalized signal amplitude at four different receiver locations.

procedure. Results are presented for estimation of the mean TL using $N = 8$ or $N = 64$ samples, for the same case considered in figure 13.3. The initial number of realizations was $N_e = 4096$, and $N_b = 16,384$ bootstrap trials were used to estimate the intervals.

13.2   Stochastic integration and sampling

When sound propagation models have a highly complex and nonlinear dependence upon their input parameters, as is typically the case, conventional methods for relating uncertainties in the model input parameters to uncertainties in the output predictions, based on small perturbations and linear expansions, do not apply. Methods for randomly sampling the input parameter space, known as stochastic or Monte Carlo integration, are a valuable and widely employed tool in such circumstances. The simple example in the previous section, in which a mean sound level was estimated from a number of random samples, provided an introduction to these techniques. We will now explore stochastic integration in more detail, beginning with a integral formulation of sound-field prediction, and then followed by a description of

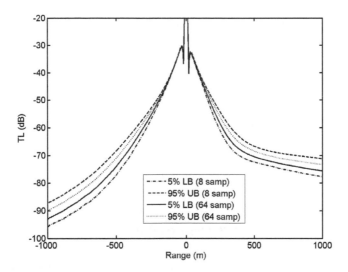

FIGURE 13.5
90% confidence intervals (5% and 95% lower (LB) and upper (UB) bounds)
for mean TL as estimated from $N = 8$ or $N = 64$ samples. The intervals were
estimated from $N_e = 4096$ randomized PE calculations at 200 Hz, which were
then resampled into $N_b = 16{,}384$ bootstrap ensembles of size N.

stochastic approaches to approximating such integrals. Many approaches to
stochastic integration have been devised [104, 109]; the suitability of a particular approach depends on the application. In this section, a few basic techniques
particularly pertinent to sound propagation are considered.

13.2.1 Formulation

The general problem of incorporating uncertainties into a sound-propagation
calculation can be formulated as an extension of equation (13.11) to M multiple, independent random variables:

$$\langle I(\mathbf{R}) \rangle = \int_{\Omega} I(\mathbf{R}, \boldsymbol{\theta}) g_{\boldsymbol{\Theta}}(\boldsymbol{\theta}) \, d\boldsymbol{\theta}. \tag{13.13}$$

Here, we have made explicit the dependence on position $\mathbf{R}$. The integrand
$I(\mathbf{R}, \boldsymbol{\theta})$ represents the squared magnitude of the sound pressure (power),
transmission loss, or other quantity of interest. The parameter set $\boldsymbol{\theta}$ incorporates the uncertain (stochastic) parameters, and may additionally incorporate
non-stochastic parameters impacting the sound-field calculation, if desired.
The M parameters comprising $\boldsymbol{\theta}$ are defined in the parameter space Ω. The
joint pdf of the parameters is $g_{\boldsymbol{\Theta}}(\boldsymbol{\theta})$. If desired, independent variables can be

transformed to uniform distributions in the range $[0, 1]$, as described in connection with equation (13.12). Setting $\psi_m(\theta_m) = G_{\Theta_m}(\theta_m)$, where $G_{\Theta_m}(\theta_m)$ is the cdf for the mth parameter $(m = 1, \ldots, M)$, we have

$$\langle I(\mathbf{R}) \rangle = \int_U I(\mathbf{R}, \boldsymbol{\theta}(\boldsymbol{\psi})) \, d\boldsymbol{\psi}. \tag{13.14}$$

Here U indicates the M-dimensional unit volume. With this transformed version of the integration, the actual values of the variables are determined from the inverse cdfs, i.e., $\theta_m = G_{\Theta_m}^{-1}(\psi_m)$.

When the integrand represents a spectral density and the goal is to determine the total sound power, an additional integration over frequency is specified as follows:

$$\langle I(\mathbf{R}) \rangle = \int_{f_{\min}}^{f_{\max}} \int_\Omega I(\mathbf{R}, \boldsymbol{\theta}, f) g_\Theta(\boldsymbol{\theta}) \, d\boldsymbol{\theta} \, df. \tag{13.15}$$

The frequency integration can be readily mapped to a unit axis by setting $f' = (f - f_{\min})/(f_{\max} - f_{\min})$, which then enables use of equation (13.14). Alternatively, acoustical calculations are often performed on a logarithmic frequency axis (e.g., in octave or one-third octave bands). In this case, we can apply the following transformation to the frequency:

$$\nu = (\ln f - \ln f_{\min}) / \Delta_f, \tag{13.16}$$

where $\Delta_f = \ln f_{\max} - \ln f_{\min}$. Then equation (13.14) becomes

$$\langle I(\mathbf{R}) \rangle = \int_U I(\mathbf{R}, \boldsymbol{\theta}(\boldsymbol{\psi}), f(\nu)) f(\nu) \Delta_f \, d\boldsymbol{\psi} \, d\nu, \tag{13.17}$$

where $f(\nu) = \exp(\Delta_f \nu + \ln f_{\min})$ and U is now the $(M + 1)$-dimensional unit volume. The normalized frequency or log-frequency may be incorporated into Θ as an additional parameter. This simplifies the integrals and, as will be demonstrated later in this chapter, can lead to very efficient broadband calculations.

Many science and engineering problems involve a large number of parameters, thus making equation (13.13) and its variants highly multidimensional [109, 280]. A conventional numerical approach to integrating equation (13.13), such as the midpoint or trapezoidal method, would involve discretizing the integral into a finite number of intervals along each of the M variable axes. However, the factorial increase in the number of evaluations of the integrand becomes computationally expensive when dealing with many parameters. Suppose there are 8 parameters (for example, frequency, source and receiver height, three atmospheric parameters, and two ground parameters), and the integration for each of these is partitioned into 64 discrete intervals. That would amount to $64^8 = 2.8 \times 10^{14}$ calculations (e.g., solutions of a parabolic equation). If each of these calculations takes an average of 1 s, almost 9 million years would be needed to finish! A further disadvantage of this approach is that the integrand is repeatedly evaluated for the same value of each variable. As a somewhat extreme example, if just one of the 8 variables has a substantial impact on the end result, in effect only 64 evaluations of the integrand are important; the remaining $64^8 - 64$ provide substantially redundant information.

Stochastic techniques are particularly suitable to such situations involving multidimensional integrals [104, 109]. All sources of uncertainty can be sampled simultaneously. The basic process begins by generating N samples of the parameter set Θ (Θ_n, $n = 1, 2, \ldots, N$). The function $I(\mathbf{R}, \Theta)$ is evaluated for each of these samples. The average of these N evaluations, $\hat{I}(\mathbf{R})$, is then an estimate for the mean:

$$\langle I(\mathbf{R}) \rangle \simeq \hat{I}(\mathbf{R}) = \frac{1}{N} \sum_{n=1}^{N} I(\mathbf{R}, \Theta_n). \tag{13.18}$$

When the random samples Θ_n are drawn independently from the joint pdf $g(\Theta)$, each evaluation of the integrand is an equally likely value, and $\hat{I}(\mathbf{R})$ is thus an unbiased approximation to $\langle I(\mathbf{R}) \rangle$. By the statistical law of large numbers, $\hat{I}(\mathbf{R})$ will converge to $\langle I(\mathbf{R}) \rangle$ as N increases. The rate of convergence can be examined by calculating variance of $\tilde{I}$, which is defined as:

$$\sigma_{\hat{I}}^2(\mathbf{R}) = \left\langle \left[\hat{I}(\mathbf{R}) - \langle I(\mathbf{R}) \rangle \right]^2 \right\rangle. \tag{13.19}$$

Substituting equation (13.18) into equation (13.19), we find

$$\sigma_{\hat{I}}^2(\mathbf{R}) = \frac{1}{N^2} \left\langle \left\{ \sum_{n=1}^{N} [I(\mathbf{R}, \Theta_n) - \langle I(\mathbf{R}) \rangle] \right\}^2 \right\rangle. \tag{13.20}$$

For generality, we will first derive a result for $\sigma_{\hat{I}}^2$ that does *not* assume $\hat{I}(\mathbf{R})$ is an unbiased estimator for $\langle I(\mathbf{R}) \rangle$. (This result will be useful later, when we discuss stratified sampling.) The derivation involves replacing $[I(\mathbf{R}, \Theta_n) - \langle I(\mathbf{R}) \rangle]$ in the preceding equation with $[I(\mathbf{R}, \Theta_n) - \langle I(\mathbf{R}, \Theta_n) \rangle] + [\langle I(\mathbf{R}, \Theta_n) \rangle - \langle I(\mathbf{R}) \rangle]$. When the square is written explicitly as a double summation, cross terms of $[I(\mathbf{R}, \Theta_n) - \langle I(\mathbf{R}, \Theta_n) \rangle]$ for differing values of n evaluate to zero under the assumption of independence. After some algebra, we find

$$\sigma_{\hat{I}}^2(\mathbf{R}) = \frac{1}{N^2} \sum_{n=1}^{N} \sigma^2(\mathbf{R}, \Theta_n) + \frac{1}{N^2} \left[\sum_{n=1}^{N} b(\mathbf{R}, \Theta_n) \right]^2, \tag{13.21}$$

where

$$\sigma^2(\mathbf{R}, \Theta) = \left\langle \left[I(\mathbf{R}, \Theta) - \langle \hat{I}(\mathbf{R}, \Theta) \rangle \right]^2 \right\rangle \tag{13.22}$$

and

$$b(\mathbf{R}, \mathbf{\Theta}) = \langle I(\mathbf{R}, \mathbf{\Theta}) \rangle - \langle I(\mathbf{R}) \rangle \tag{13.23}$$

are the variance and bias of the integrand as a function of $\mathbf{\Theta}$, respectively. Note that the overall bias of the estimate is

$$b_{\hat{I}}(\mathbf{R}) = \frac{1}{N} \sum_{n=1}^{N} b(\mathbf{R}, \mathbf{\Theta}_n) = \langle \hat{I}(\mathbf{R}) \rangle - \langle I(\mathbf{R}) \rangle. \tag{13.24}$$

Now, turning our attention again to ordinary Monte Carlo integration, the assumption that the $\mathbf{\Theta}_n$ are drawn independently from $g(\mathbf{\Theta})$ (from anywhere within the domain Ω) removes the dependence of any expectations of $I(\mathbf{R}, \mathbf{\Theta}_n)$ on $\mathbf{\Theta}_n$; that is, $\langle I(\mathbf{R}, \mathbf{\Theta}_n) \rangle = \langle I(\mathbf{R}) \rangle$ for all $\mathbf{\Theta}_n$, and hence the estimator is unbiased. Furthermore, $\sigma^2(\mathbf{R}, \mathbf{\Theta}_n)$ also becomes independent of $\mathbf{\Theta}_n$, and equation (13.21) reduces to

$$\sigma_{\hat{I}}^2(\mathbf{R}) = \frac{\sigma^2(\mathbf{R})}{N}, \tag{13.25}$$

in which $\sigma^2(\mathbf{R}) = \left\langle [I(\mathbf{R}) - \langle I(\mathbf{R}) \rangle]^2 \right\rangle$. The root-mean-square (rms) error for an estimate based on N realizations is thus proportional to $1/\sqrt{N}$. Examining figure 13.3, we can estimate $\sigma(\mathbf{R})$ to be in the range of 4–10 dB at distances greater than a few hundred meters from the source. (A value of 6 dB is characteristic of strong scattering [118].) With 64 evaluations of the integrand, an error in the range of 0.5–1.2 dB is obtained. It should be kept in mind, though, that the actual error is random. Deterministic methods for solving integrals by discretization also have errors, but for a given integration algorithm and integrand, they are repeatable bias errors, rather than random ones.

The $1/\sqrt{N}$ dependence indicated by equation (13.25) is tested in figure 13.6. Here, σ_N is plotted as a function of N for the 200-Hz propagation calculation and four receiver locations shown in figure 13.4. Parametric uncertainties and turbulent scattering are both included. The means were calculated from $N_b = 16{,}384$ bootstrap trials of N samples each, as described in connection with figure 13.5. The dashed lines are proportional to $1/\sqrt{N}$. The calculated variances follow this trend nearly perfectly.

As discussed in Chapter 9, the main procedure for incorporating turbulent scattering into sound propagation calculations is to generate random realizations of turbulence fields [75, 136]. The sound power is then calculated for each realization, and the calculations averaged. This procedure exemplifies the application of equation (13.18). Conceptually, we might view the initial parameter set $\mathbf{\Theta}$ in equation (13.13) as comprising the phases of the Fourier modes of the turbulence spectrum, which are randomized uniformly over the interval $[0, 2\pi)$ to create the turbulence realizations. Typically, thousands of such modes are included in the randomization, so that equation (13.13) is indeed a

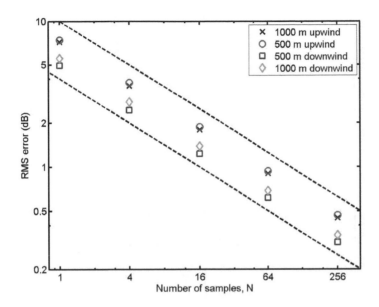

FIGURE 13.6
Root-mean-square error $\sigma_{\hat{I}}$ for mean TL estimates as a function of the number of samples N used to estimate the mean. The frequency is 200 Hz and four different receiver locations are shown, as in figure 13.4. The lower and upper dashed lines correspond to $(4 \text{ dB})/\sqrt{N}$ and $(10 \text{ dB})/\sqrt{N}$, respectively.

many-dimensional integral. Direct solution of equation (13.13) is thus infeasible, and the value of the Monte Carlo approximation becomes quite apparent. But it has not been as widely appreciated that, when such random realizations of the turbulence are generated, there is negligible additional computational cost to simultaneously sampling epistemic uncertainties, such as those due to uncertainty in the wind speed and direction, temperature stratification, and ground properties. This powerful idea enables efficient quantification of the impacts of all pertinent sources of uncertainty.

13.2.2 Stratified and Latin hypercube sampling

The examples considered thus far have used ordinary Monte Carlo sampling (MCS), in which each sample Θ_n is drawn from $g_\Theta(\theta)$ independently of the other samples. MCS is illustrated in figure 13.7(a), for a case involving 16 samples of two independent random variables, each distributed uniformly over the range $[0, 1]$. A drawback of MCS is that it tends to randomly undersample certain parts of the input parameter space, while oversampling others.

Stratified sampling is commonly employed to ensure that important re-

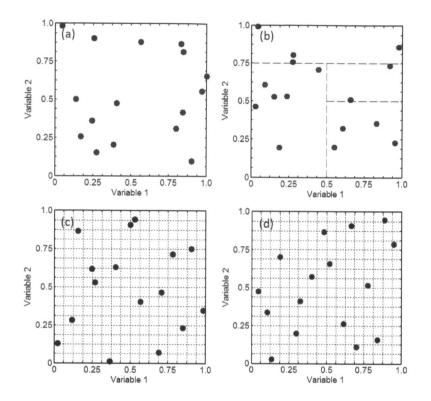

FIGURE 13.7

Comparison of several sample strategies. The depiction is for 16 samples of two independent, uniform random variables. (a) Ordinary Monte Carlo sampling (MCS). (b) Stratified sampling with four unequal strata. (c) Ordinary Latin hypercube sampling (LHS). (d) LHS with an iterative maximin criterion.

gions of the parameter space are adequately sampled. This approach involves partitioning the overall parameter space into strata (subvolumes) and then sampling each stratum independently. The strata should be mutually exclusive and exhaustive, meaning that each possible parameter combination belongs to one and only one stratum. The overall statistics are then determined by appropriately weighting statistics as calculated from the individual strata. When the individual strata have relatively homogeneous properties, the variance of the overall prediction can be reduced.

Stratified sampling is also a natural approach when a comparison between distinctive categories of propagation conditions (e.g., upward vs. downward refraction, times of day, or seasons) is desired. Previously, sound propagation conditions have often been categorized into classes, based on the wind speed, propagation direction relative to the wind, and atmospheric stability. For example, Marsh [240] and Zouboff et al. [444] have formulated extensions of

Pasquill [308] stability classes (which are widely used in micrometeorology) for acoustical applications. Raspet and Wolf [322] partitioned calculations by wind speed and direction intervals. Heimann and Salomons [159] developed a scheme based on 25 classes of a logarithmic/linear approximation to the effective sound-speed profile. In essence, these are all stratified sampling schemes.

Suppose there are K strata, designated Ω_k (where $\Omega_k \in \Omega$ and $k = 1, \ldots, K$), with p_k being the probability of the parameters occurring within stratum k. Within each stratum, N_k random samples of Θ are drawn, which we designate $\Theta_{k,n}$ ($n = 1, \ldots, N_k$, where $\sum_k N_k = N$ is the total number of samples). The estimate for the mean within stratum k is

$$\hat{I}_k(\mathbf{R}) = \frac{1}{N_k} \sum_{n=1}^{N_k} I(\mathbf{R}, \Theta_{k,n}). \tag{13.26}$$

The overall estimate for $\langle I(\mathbf{R}) \rangle$ is then

$$\hat{I}(\mathbf{R}) = \sum_{k=1}^{K} p_k \hat{I}_k(\mathbf{R}). \tag{13.27}$$

In most applications of stratified sampling, *proportionate* sampling is used, meaning that N_k is set to $p_k N$. Figure 13.7(b) illustrates this approach. Here, the parameter space for two variables has been partitioned into four strata ($K = 4$). As drawn in the figure, the strata have probability $p_1 = 1/4$, $p_2 = 1/8$, $p_3 = 1/4$, and $p_4 = 3/8$ (starting at top, and moving clockwise). Hence, with 16 total samples, proportionate sampling requires $N_1 = 4$, $N_2 = 2$, $N_3 = 4$, and $N_4 = 6$.

The variance for stratified sampling is derived by first substituting equation (13.27) into (13.19), with result

$$\sigma_{\hat{I}}^2(\mathbf{R}) = \left\langle \left\{ \sum_{k=1}^{K} p_k \left[\hat{I}_k(\mathbf{R}) - \langle I(\mathbf{R}) \rangle \right] \right\}^2 \right\rangle. \tag{13.28}$$

The next step involves replacing $[\hat{I}_k(\mathbf{R}) - \langle I(\mathbf{R}) \rangle]$ with $[\hat{I}_k(\mathbf{R}) - \langle \hat{I}_k(\mathbf{R}) \rangle] + [\langle \hat{I}_k(\mathbf{R}) \rangle - \langle I(\mathbf{R}) \rangle]$. As before, cross terms of $[\hat{I}_k(\mathbf{R}) - \langle \hat{I}_k(\mathbf{R}) \rangle]$ for differing values of k evaluate to zero under the assumption that the strata are sampled independently, and we find

$$\sigma_{\hat{I}}^2(\mathbf{R}) = \sum_{k=1}^{K} p_k^2 \sigma_{\hat{I}_k}^2(\mathbf{R}) + \left[\sum_{k=1}^{K} p_k b_{\hat{I}_k}(\mathbf{R}) \right]^2, \tag{13.29}$$

where

$$\sigma_{\hat{I}_k}^2(\mathbf{R}) = \left\langle \left[\hat{I}_k(\mathbf{R}) - \langle \hat{I}_k(\mathbf{R}) \rangle \right]^2 \right\rangle \tag{13.30}$$

and

$$b_{\hat{I}_k}(\mathbf{R}) = \langle \hat{I}_k(\mathbf{R}) \rangle - \langle I(\mathbf{R}) \rangle \tag{13.31}$$

are the variance and bias, respectively, for sampling of the kth stratum. The overall bias of the estimator is

$$b_{\hat{I}}(\mathbf{R}) = \sum_{k=1}^{K} p_k b_{\hat{I}_k}(\mathbf{R}) = \sum_{k=1}^{K} p_k \langle \hat{I}_k(\mathbf{R}) \rangle - \langle I(\mathbf{R}) \rangle. \tag{13.32}$$

Within each stratum, equations (13.21)–(13.23) apply, so that

$$\sigma_{\hat{I}_k}^2(\mathbf{R}) = \frac{1}{N_k^2} \sum_{n=1}^{N_k} \sigma^2(\mathbf{R}, \boldsymbol{\Theta}_{k,n}) + \frac{1}{N_k^2} \left[\sum_{n=1}^{N_k} b(\mathbf{R}, \boldsymbol{\Theta}_{k,n}) \right]^2. \tag{13.33}$$

Two approaches to stratified sampling are of particular interest. First, when ordinary MCS is applied *within* each stratum, the estimates within the stratum will be unbiased, and the $\sigma^2(\mathbf{R}, \boldsymbol{\Theta}_{k,n})$ will be a constant, $\sigma_k^2(\mathbf{R})$, within each stratum. Then, $\sigma_{\hat{I}_k}^2(\mathbf{R}) = \sigma_k^2(\mathbf{R})/N_k$, and from equation (13.28) we have, for proportionate sampling,

$$\sigma_{\hat{I}}^2(\mathbf{R}) = \frac{1}{N^2} \sum_{k=1}^{K} N_k \sigma_k^2(\mathbf{R}). \tag{13.34}$$

Should the σ_k^2 happen to all equal the same value, σ^2, this result reduces to equation (13.25) for ordinary MCS applied throughout the domain; in this situation, stratified sampling provides no benefit. The best possible case would be strata for which $\sigma_k^2 = 0$; then, so long as each stratum is sampled at least once, $\hat{I}$ is exactly $\langle I \rangle$. However, in most practical problems, stratified sampling would lead to variances between these two extremes, and thus have the benefit of reducing $\sigma_{\hat{I}}^2$ relative to ordinary MCS.

As an example of this first approach to stratified sampling, let us consider the following simple function:

$$I(\theta_1, \theta_2) = \begin{cases} 0, & 0 \le \theta_2 < 0.25, \\ 1, & 0.25 \le \theta_2 < 0.5, \\ 2, & 0.5 \le \theta_2 < 0.75, \\ 3, & 0.75 \le \theta_2 \le 1. \end{cases} \tag{13.35}$$

This function is sampled using the strata as drawn in figure 13.7(b). One can then calculate the means for the strata as (again, starting at the top stratum and proceeding clockwise) $\hat{I}_1 = \langle I \rangle_1 = 3$, $\hat{I}_2 = \langle I \rangle_2 = 2$, $\hat{I}_3 = \langle I \rangle_3 = 1/2$, and $\hat{I}_4 = \langle I \rangle_4 = 1$. Thus, from equation (13.27), $\langle I \rangle = (1/4)(3) + (1/8)(2) + (1/4)(1/2) + (3/8) = 3/2$. The variances for samples from the four strata $\sigma_1^2 = 0$, $\sigma_2^2 = 0$, $\sigma_3^2 = ((-1/2)^2 + (1/2)^2)/2 = 1/4$, and $\sigma_4^2 = ((-1)^2 + 0 +$

$1^2)/3 = 2/3$. Applying equation (13.34) with proportionate sampling, we have $\sigma_{\hat{I}}^2 = (0+0+4(1/4)+6(2/3))/256 = 5/256$. In comparison, the variance for a sample with ordinary MCS is $[(-3/2)^2 + (-1/2)^2 + (1/2)^2 + (3/2)^2]/4 = 5/4$, and thus from equation (13.25), $\sigma_{\hat{I}}^2 = 5/64$. Hence, the stratified sampling reduces the variance of the estimate by a factor of $1/4$.

The second approach to stratified sampling is that in which $\hat{I}_k(\mathbf{R})$ is estimated on the basis of a single sample, $\mathbf{\Theta}_{k,1}$, within that stratum ($N_k = 1$ for all k). This is the approach by which stratified sampling has often been applied to outdoor sound propagation, as in several of the references mentioned earlier in this section. For example, the calculation might be performed at the center of the parameter ranges for that stratum. In this case, $\hat{I}_k(\mathbf{R}) = \langle \hat{I}_k(\mathbf{R})\rangle$, so that the sampling variances for the strata as given by equation 13.30 are zero. The error for estimating $\langle I(\mathbf{R})\rangle$ consists entirely of bias, and is given by equation (13.32).

Latin hypercube sampling (LHS) is a particular type of stratified sampling. It provides a general and relatively simple method for improving sample dispersion [161]. LHS, unlike most stratified sampling procedures, can be readily applied to situations involving many variables, and programs utilizing MCS can usually be converted to LHS with minimal effort. The range for each of the M variables is first partitioned into N equal-probability intervals (where N is the number of samples to be drawn, as before). The resulting grid has $K = N^M$ strata. One and only sample is then randomly drawn from each of the N intervals for each variable. In two dimensions, for example, each row and each column in the grid is sampled just once. Note that all of the strata are *not* sampled; in fact, the probability of any given stratum being sampled is N^{1-M}, which becomes very small when M is large. LHS is illustrated in figures 13.7(c) and (d). The first of these depicts ordinary LHS; it was generated using the MATLAB® lhsdesign function [244] with the criterion argument set to "none". The second was generated with the criterion argument set to "maximin", which applies an iterative procedure to maximize the minimum distance between the sample points. The latter approach provides a particularly well dispersed set of sample points.

Like ordinary MCS, the LHS estimate $\hat{I}(\mathbf{R})$ for $\langle I(\mathbf{R})\rangle$ can be calculated from equation (13.18). The estimate is unbiased, since the strata are equal-probability subspaces of Ω and sampled with equal likelihood. However, unlike MCS, the samples of $\mathbf{\Theta}_n$ are not drawn independently. Hence, equation (13.25) no longer strictly gives the variance of $\hat{I}(\mathbf{R})$; the purpose of applying LHS is to reduce the variance below that of ordinary MCS. A comparison between MCS and LHS for outdoor sound propagation will be provided in section 13.3.

13.2.3 Importance sampling

Importance sampling (e.g., references [28, 104]) is motivated by the simple idea that the stochastic integration will converge faster if regions where the integrand is largest (and thus contribute most strongly to the final result) are sampled preferentially. Formally, importance sampling (IS) involves rewriting equation (13.13) in the equivalent form

$$\langle I(\mathbf{R}) \rangle = \int_\Omega I(\mathbf{R}, \boldsymbol{\Theta}) w_{\boldsymbol{\Theta}}(\boldsymbol{\theta}) h_{\boldsymbol{\Theta}}(\boldsymbol{\theta}) \, d\boldsymbol{\theta}, \tag{13.36}$$

where $h_{\boldsymbol{\Theta}}(\boldsymbol{\theta})$ is the *importance sampling function* and

$$w_{\boldsymbol{\Theta}}(\boldsymbol{\theta}) = \frac{g_{\boldsymbol{\Theta}}(\boldsymbol{\theta})}{h_{\boldsymbol{\Theta}}(\boldsymbol{\theta})} \tag{13.37}$$

is the *weighting function*. While this recasting of equation (13.13) is mathematically trivial, the interpretation of the integrand and resulting stochastic integration are quite different. Namely, the integral is now approximated by drawing random samples of $\boldsymbol{\Theta}$ from $h_{\boldsymbol{\Theta}}(\boldsymbol{\theta})$ (rather than from $g_{\boldsymbol{\Theta}}(\boldsymbol{\theta})$), and then weighting by $w_{\boldsymbol{\Theta}}(\boldsymbol{\theta})$ to unbias the estimate. Instead of equation (13.18), we thus have:

$$\hat{I}(\mathbf{R}) = \frac{1}{N} \sum_{n=1}^{N} I(\mathbf{R}, \boldsymbol{\Theta}_n) w_{\boldsymbol{\Theta}}(\boldsymbol{\Theta}_n). \tag{13.38}$$

The random samples may be drawn from $h_{\boldsymbol{\Theta}}(\boldsymbol{\theta})$ using MCS, LHS, or other valid approach. A diagram of IS would differ from figure 13.7 only with regard to how the axes are *interpreted*; namely, they would correspond to partitioning into equally *weighted* intervals based on $h_{\boldsymbol{\Theta}}(\boldsymbol{\theta})$ rather than *equal probability* intervals based on $g_{\boldsymbol{\Theta}}(\boldsymbol{\theta})$.

A good importance sampling function generally mimics $I(\mathbf{R}, \boldsymbol{\theta}) g_{\boldsymbol{\Theta}}(\boldsymbol{\theta})$, i.e., it is large where the function being integrated is largest and for values of $\boldsymbol{\theta}$ that are most probable. A less obvious, but nonetheless important, condition is that $h_{\boldsymbol{\Theta}}(\boldsymbol{\theta})$ should have asymptotes that decay more slowly than $g_{\boldsymbol{\Theta}}(\boldsymbol{\theta})$. Numerical instabilities occur in the evaluation of $w_{\boldsymbol{\Theta}}(\boldsymbol{\theta})$ when this condition is not met. To avoid this problem, the prescribed form of $h_{\boldsymbol{\Theta}}(\boldsymbol{\theta})$ can be mixed with a uniform pdf. That is, for the original $h_{\boldsymbol{\Theta}}(\boldsymbol{\theta})$ we substitute $(1 - \delta) h_{\boldsymbol{\Theta}}(\boldsymbol{\theta}) + \delta$, where $\delta < 1$. (A value $\delta = 1$ would correspond to ordinary MCS.) In effect, this procedure ensures that, should the importance of certain parameter ranges be severely underestimated by the selection of $h_{\boldsymbol{\Theta}}(\boldsymbol{\theta})$, some samples will still be drawn for these parameter ranges.

Typically, IS will be applied to only one or a few variables for which $I(\mathbf{R}, \boldsymbol{\theta})$ has important systematic dependencies. For example, if IS is applied only to θ_1, we have, assuming statistical independence of the parameters,

$$h_{\boldsymbol{\Theta}}(\boldsymbol{\theta}) = h_{\Theta_1}(\theta_1) g_{\Theta_2}(\theta_2) \cdots g_{\Theta_M}(\theta_M) \tag{13.39}$$

and hence $w_{\boldsymbol{\Theta}}(\boldsymbol{\theta}) = w_{\Theta_1}(\theta_1) = g_{\Theta_1}(\theta_1)/h_{\Theta_1}(\theta_1)$.

For outdoor sound propagation, IS can focus computational effort on frequently occurring propagation conditions and/or those leading to relatively high sound levels. In particular, conditions of downward refraction, as caused by downwind propagation or a temperature inversion, often dominate the sound level when it is averaged over a long time interval involving many weather conditions. Upward refraction conditions have comparatively little impact on the average level. Hence, an importance sampling function might be chosen to focus computational effort on downward refraction. However, one must remember to apply equation (13.38) with the weighting factor, to correct for the actual prevalence of each propagation condition.

But, how do we select an appropriate importance function $h_\Theta(\boldsymbol{\theta})$ *prior* to evaluating the model and performing the stochastic integration? Typically, a suitable importance function would be selected based on some prior physical or mathematical insights into the problem. One possibility is to use a relatively simple and computationally efficient model to estimate the integrand $I(\mathbf{R}, \boldsymbol{\Theta})$. This approximating model should realistically capture the important sensitivities of the propagation to $\boldsymbol{\theta}$, although it may not be necessary for the approximating model to capture dependencies on less important parameters. In essence, a simpler, more efficient model drives calculations made with a more accurate, computationally intensive model. In the context of outdoor sound propagation, for example, we might use an impedance-plane model to estimate the frequency dependence of the sound pressure, and incorporate this information into the importance function and subsequent selection of the parameter samples $\boldsymbol{\Theta}_n$. Then, the $I(\mathbf{R}, \boldsymbol{\Theta}_n)$ can be calculated with a model capable of including atmospheric refraction effects, such as the PE. This hierarchical approach requires setting a receiver location $\mathbf{R} = \mathbf{R}_h$ at which to evaluate the importance function. The results may become unsatisfactory at distances far from this location.

The international standard method for predicting outdoor sound propagation [177] may be regarded as a simple implementation of IS. The method endeavors to predict sound levels characteristic of moderate downward refraction conditions, i.e., downwind propagation with a moderate shear, or a moderate temperature inversion. The underlying assumption is that sound levels are highest in conditions of downward refraction, whereas conditions of very strong downward refraction are rare. Hence, conditions of moderate downward refraction are expected to have the most important impact on the mean sound level. In effect, the standard thus predicts the overall sound level on the basis of a single importance sample, although no probabilistic weighting function is formally used.

The importance sampling function can also be refined in response to information obtained through repeated sampling of the integrand. This class of techniques, known as *adaptive* importance sampling, was applied to outdoor sound propagation in reference [430].

TABLE 13.1

Low-, medium-, and high-uncertainty cases considered in the calculations. Uncertainties in six parameters (as described in the text) are modeled with the indicated probability density functions (pdfs) and statistical parameters.

Parameter	pdf	mean	low	medium	high
Ω	beta	0.27	$\nu = 256$	$\nu = 64$	$\nu = 16$
σ_0 $(\mathrm{Pa\,s\,m^{-2}})$	log-normal	2×10^6	$s = 0.1$	$s = 0.2$	$s = 0.4$
z_0 (m)	log-normal	0.01	$s = 0.1$	$s = 0.2$	$s = 0.4$
u_* $(\mathrm{m\,s^{-1}})$	log-normal	0.6	$s = 0.1$	$s = 0.2$	$s = 0.4$
θ (deg)	normal	0	$\sigma = 10$	$\sigma = 20$	$\sigma = 40$
z_s (m)	log-normal	5	$s = 0.1$	$s = 0.2$	$s = 0.4$

13.3 Application to near-ground propagation

In this section, we explore application of stochastic sampling methods to sound propagation in the near-ground atmosphere with refraction by the mean wind profile and scattering by turbulence. The atmospheric properties are characteristic of windy, neutrally stratified conditions, whereas the ground properties are characteristic of grass-covered soil. Monin–Obukhov and mixed-layer similarity theories are used to relate the atmospheric parameters to the mean vertical profiles and turbulence spectrum, as described in sections 2.2.3 and 6.2.4, respectively. The calculations were all performed with a narrow-angle Crank–Nicholson parabolic equation (CNPE, as described in section 11.2.1), with turbulence incorporated by the method of phase screens (section 11.2.4). The source height is $z_s = 5$ m, and the receiver height is $z_r = 1.5$ m.

Epistemic uncertainty is considered in association with six parameters: the friction velocity u_*, the wind direction θ, the ground porosity Ω, the static flow resistivity σ_0, the surface roughness z_0, and the source height z_s. Aleatory uncertainty is present in the form of random, turbulent fluctuations in the wind velocity. Low-, medium-, and high-uncertainty cases are considered with the pdfs and parameters as listed in table 13.1.

13.3.1 Narrowband calculations

In this section, CNPE calculations are considered at several discrete frequencies, namely 100 Hz, 200 Hz, and 400 Hz. The sound power at the receiver, for a unit-amplitude source, was estimated by averaging calculations made from $N = 8$, 16, 32, or 64 random samples of the parameters shown in table 13.1. Each of these calculations also included a random realization of the turbulence field (phase screens), based on the sample atmospheric parameters. The generation of the N samples was then repeated over a series $N_t = 256$ random trials, and the rms error (standard deviations of the predicted means about the actual means) for these 256 trials was calculated. The total number of CNPE calculations performed for each numerical experiment is the number of random samples for each trial (N), times the number of trials (N_t).

Figure 13.8(a)–(d) compares rms errors for various situations. The first of these shows the impact of varying the degree of epistemic uncertainty, for propagation at 100 Hz. As would be expected, the errors increase when greater uncertainty is present. Deep within the upwind shadow zone (large negative ranges in the figure), however, the low- and medium-uncertainty cases result in similar errors, likely because turbulent scattering (an aleatory uncertainty) dominates there. The medium-uncertainty case also exhibits a peak in the error around the shadow zone boundary.

The impact of varying the frequency, for the high uncertainty case with $N = 16$ MCS samples, is shown in figure 13.8(b). At 100 Hz, the errors increase steadily, in both the upwind and downwind directions. The errors become relatively large at long ranges, indicating that it is challenging to accurately predict sound levels at long ranges for low frequencies. The errors at 200 Hz and 400 Hz appear to saturate in the upwind shadow zone as well as at long ranges downwind, likely indicating the dominance of strong turbulent scattering at these locations. The lack of such a saturation effect at 100 Hz perhaps results from the dominance of diffraction over turbulent scattering.

Results for varying the number of samples with MCS, for the high uncertainty case at 100 Hz, are shown in figure 13.8(c). As would be expected, increasing the number of samples decreases the errors. The plot shows, for example, that 32 samples are needed to predict mean sound levels with errors less than about 3 dB. Increasing the number of samples by a factor of 4 roughly halves the rms error, as would be expected if the error goes inversely as $\sqrt{N}$.

The various sampling methods are compared in figure 13.8(d), for calculations at 200 Hz with 16 samples. (Note that the sampling method is varied only with regard to the epistemically uncertain parameters; in effect, MCS is always used for the turbulent fluctuations.) LHS outperforms MCS by up to 1 dB, out to distances of about 500 m. Ordinary LHS and LHS with the maximin criterion yield essentially identical results upwind, but the latter performs somewhat better downwind. The improvement from LHS is obtained with no additional computational effort in exercising the CNPE. At longer ranges, turbulent scattering dominates so that the sampling strategy becomes unimportant.

13.3.2 Broadband calculations

For noise control and other applications, we generally wish to predict sound levels from sources emitting a broad range of frequencies. Typically, this is done by running frequency-domain models at multiple frequencies, such as the center frequencies in a set of octave or one-third octave bands, and then

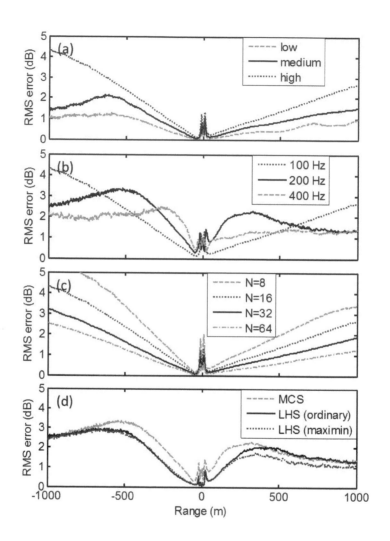

FIGURE 13.8
Root-mean-square (rms) errors for prediction of the sound power at single
frequencies. The source is at zero range, with upwind propagation negative
(to the left) and downwind positive (to the right). (a) Comparison of the low-,
medium-, and high-uncertainty cases at 100 Hz, for MCS with 16 samples.
(b) Comparison of calculations at 100, 200, and 400 Hz, high uncertainty
case, for MCS with 16 samples. (c) Comparison of MCS for various numbers
of samples N, high uncertainty case at 100 Hz. (d) Comparison of several
sampling methods (MCS, ordinary LHS, and LHS with the maximin criterion),
for $N = 16$ samples, 200 Hz, and high uncertainty.

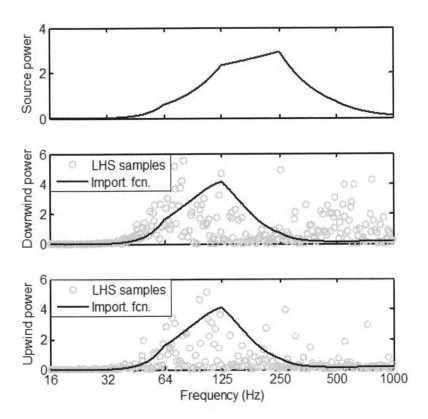

FIGURE 13.9
(a) Normalized source power spectral density used for the broadband calculation. (b) Importance function based on the impedance-ground model, for a receiver range of $x_h = 500$ m (solid line). Circles are 256 LHS samples of CNPE calculations, including parametric uncertainties and turbulence. (c) Same as (b), except the LHS samples are for upwind propagation. All curves and the LHS samples have been normalized such that they integrate, on a normalized logarithmic frequency axis, to 1.

integrating the signal power over frequency. Let us now consider an example calculation similar to the previous section, except that the goal is to predict the received sound level from a broadband source. The spectrum, which is shown in figure 13.9(a), peaks between 125 and 250 Hz and is representative of traffic noise. For convenience in generating the random samples, a logarithmic frequency axis from 16 Hz to 1000 Hz, normalized to an interval from 0 to 1, has been used (equation (13.16)).

In the following, predictions are made based on MCS, LHS with the maximin criterion, and IS based on an impedance-plane ground model. For the latter model, the complex sound-pressure field $\hat{p}(\mathbf{R}, f)$ has two terms, representing sound originating from the actual source position and from an image source, as given by equations (8.1)–(8.3). From this estimate for the sound pressure follows the received power spectral density, $P(\mathbf{R}, f) = |\hat{p}(\mathbf{R}, f)|^2/2$. We then set the importance sampling function $h(\nu) = P(\mathbf{R}_h, f(\nu))f(\nu)\Delta_f$, as evaluated at $\mathbf{R}_h = (x_h, 0, z_h)$. We set $x_h = 500$ m (half the maximum propagation range) and $z_h = z_r = 1.5$ m (the receiver height). Finally, a uniform pdf is mixed with the specified $h(\nu)$ at a level $\delta = 0.2$, as described in section 13.2.3.

Figures 13.9(b) and (c) compare CNPE calculations of the power spectrum received 500 m downwind and upwind, respectively, as a function of normalized log-frequency ν, to the importance sampling function, prior to mixing with the uniform pdf. The CNPE calculations involved 256 Latin hypercube samples, for the high uncertainty case shown in table 13.1, as well as random scattering by turbulence. To facilitate the comparison, the power spectrum was normalized by the mean with respect to ν, so that the integral over ν is 1. The importance function predicts the behavior of the CNPE calculations reasonably well, although the randomness in the latter is quite evident. Agreement is better in the upwind direction; in the downwind direction, ducting appears to enhance sound levels at high frequencies, above that predicted by the impedance-plane model.

Errors for prediction of the broadband received level (RL, as defined in section 10.1), as a function of range, are shown in figure 13.10. Compared are the low- and high-uncertainty cases, various values of N (8, 16, 32, and 64), and the MCS, LHS, and IS approaches to sampling. Regardless of the sampling approach, sound levels can be predicted somewhat more accurately in the downwind direction than upwind. Generally speaking, MCS results in the largest errors. An exception is the upwind direction with low uncertainty, for which IS performs most poorly, likely because the importance function does not correctly select the dominant frequency range of the turbulent scattering. LHS generally performs better than both MCS and IS, with exception that it is outperformed by IS in the upwind direction for the high uncertainty case.

The rms errors for the various sampling procedures are plotted against the calculation times in figure 13.11. Shown are results for MCS, LHS, and IS, each with $N = 8, 16, 32$, and 64 samples. The presented errors are averages between the ranges 500 and 1000 m. A total of $N_t = 256$ trials were conducted (each with N samples) to determine the rms error. Calculation times are the total for completing all N CNPE calculations, as averaged over all trials. Since the methods select different calculation frequencies, the total calculation times are not simply proportional to N. The dashed line in each figure corresponds to an error which is inversely proportional to the square root of calculation time. Methods providing the best trade-off between calculation time and error are farthest to the left and below this line. IS based on the impedance ground is relatively fast as it avoids calculations above several hundred Hz (figure 13.9(b)

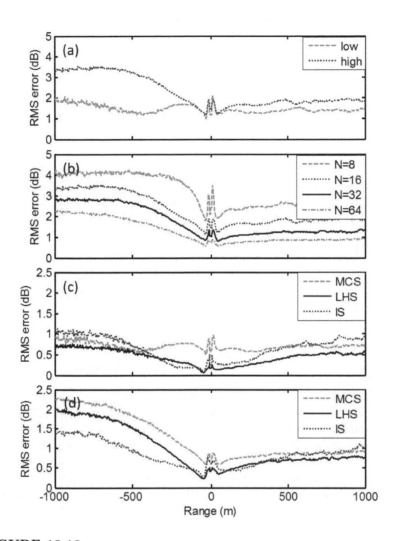

FIGURE 13.10

Root-mean-square errors for prediction of the broadband received level (RL). The source is at zero range, with upwind propagation negative (to the left) and downwind positive (to the right). (a) Comparison of the low- and high-uncertainty cases, for MCS with 16 samples. (b) Comparison of MCS for various numbers of samples N, high-uncertainty case. (c) Comparison of several sampling methods (MCS, LHS with the maximin criterion, and IS based on the impedance ground-plane model), for $N = 64$ samples and low uncertainty. (d) Same as (c), except for high uncertainty.

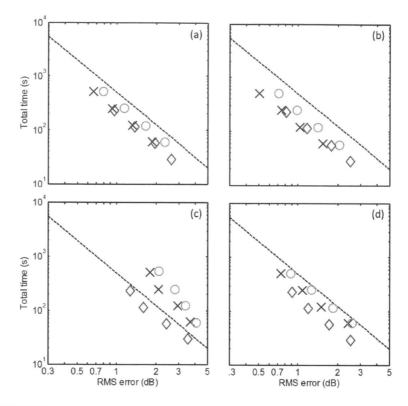

FIGURE 13.11
Root-mean-square error vs. calculation time for MCS (circles), LHS (x's), and IS (diamonds). Calculation times are the total for $N = 8, 16, 32$, and 64 CNPE runs at frequencies between 16 and 1000 Hz, out to a range of 1 km. The errors are averages over ranges between 500 and 1000 m. The slope of the dashed line corresponds to an error which is inversely proportional to the square root of calculation time. (a) Upwind, low uncertainty. (b) Downwind, low uncertainty. (c) Upwind, high uncertainty. (d) Downwind, high uncertainty.

and (c)). In all cases, MCS provides the least satisfactory performance. For the low uncertainty case, LHS and IS perform similarly. For the high uncertainty case, IS performs best. In fact, for upwind propagation with high uncertainty, the performance of both LHS and MCS is *worse* than the dashed line, thus indicating that as N is increased, the samples do not provide independent calculations.

13.4 Representation of the atmosphere: practical issues

Previous sections in this chapter have described how uncertainties in the representation of the atmosphere and terrain lead to uncertainties in sound propagation predictions. While such uncertainties can be mitigated, it is generally impossible to measure atmospheric and terrain conditions with spatial and temporal resolution sufficient to accurately predict the sound field in a deterministic sense. Many options are available for characterizing the atmospheric structure, for example: wind and temperature sensors on a tower, weather balloons (tethersondes and radiosondes), remote sensing systems (radar, sodar, and lidar), and numerical weather predictions (NWP). The availability of so many alternatives motivates important practical questions regarding what types of environmental data will lead to the best propagation predictions, how to best process the data, and what is the accuracy of predictions for particular sources of data.

To address these issues, it is helpful to distinguish between *event* and *mean* prediction. An event is defined here a single transmission (or perhaps multiple, rapid transmissions) spanning a time interval substantially shorter than the decorrelation time for sound scattering by random atmospheric fluctuations, such as turbulence and gravity waves. The event could correspond to an explosion, overflight of a rapidly moving aircraft, or sound produced by a steady source but observed over a short interval. Sound fields have been found, based on both experimental data [276] and atmospheric simulations [417], to decorrelate substantially over intervals as short as a few seconds. *Mean* prediction refers to an average over a time interval long enough to effectively remove the variability associated with random atmospheric fluctuations. A mean prediction approximates the ensemble or expected value. Strong variations in acoustic signals may be produced by the largest eddies in the turbulent boundary layer, which have time scales of several minutes or longer [186]. Time intervals of roughly 30 min or longer are needed for good estimates of first-order statistics in signals responding to atmospheric boundary layer turbulence [186, 221]. The longest time interval that may be used for averaging depends on statistical stationarity of the atmosphere. The atmosphere may be reasonably stationary for intervals as long as several hours during relatively steady daytime or nighttime conditions; however, stationarity may break down entirely at times such as the transitions around sunrise and sunset.

Suppose, for example, we wish to predict the sound exposure level associated with an event, such as an explosion. On the basis of figure 13.6, the level can be predicted with an accuracy in the range 4–10 dB, depending on the propagation direction and distance. This is based on the assumption that there is no knowledge of the random turbulent fluctuations at the time of the event. It might be anticipated that predictions would best be based upon a single, "instantaneous" set of vertical profiles, recorded at the same time as

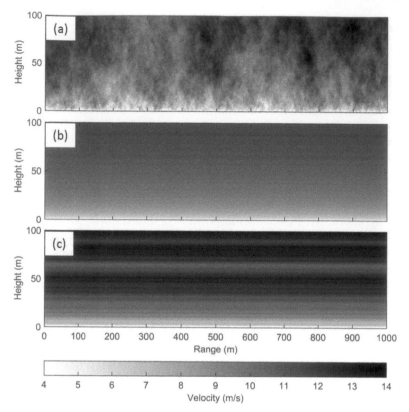

FIGURE 13.12
Example representations of the atmospheric velocity field for incorporation
into sound-propagation calculations (vertical plane). (a) Full realization of
turbulent velocity fluctuations plus the mean vertical profile. (b) Mean vertical
profile for the velocity only. (c) "Plywood" model based on extending the
fluctuations and vertical profile at zero range to the entire domain.

the event, since such profiles would incorporate information on the random
turbulence present at the time of the propagation event. However, by making
predictions from such profiles, in effect one is assuming that the turbulence
structure varies only in the vertical direction; that is, the turbulent eddies are
of infinite horizontal extent. J. Wyngaard (personal communication; see also
Ref. [437]) aptly termed this a "plywood" atmosphere.

Figure 13.12 illustrates this concept. The top panel shows a random turbulent velocity field in neutral atmospheric conditions. The field was synthesized using the methods described in section 9.1.2, with $u_* = 0.6$ m/s and
$z_0 = 0.05$ m. The turbulent fluctuations were added to the logarithmic mean
profile. The middle panel shows the velocity field when there are no turbulent
fluctuations, i.e., there is only a mean profile. The bottom panel shows the

result of extending the fluctuations and mean profile at a particular location (in this case, zero range) throughout the horizontal domain, i.e., a plywood atmosphere. The artificially large horizontal extent of the turbulence structure in the plywood atmosphere can result in overprediction of the scattering strength. In reference [417], it was found that predictions of event propagation have rms errors of 8–10 dB when they are based on instantaneous local vertical profile data synchronized to the propagation event, which corresponds to the plywood atmosphere. This error depends only weakly on the location, relative to the propagation path, at which the vertical profiles were sampled. However, the errors decrease by several dB when predictions are instead made from the mean vertical profiles. Thus some smoothing of the random fluctuations in the vertical profiles can actually improve the accuracy of predictions of event sound levels.

At the other extreme, predictions of mean sound levels from mean vertical profiles have relatively low rms errors (typically less than 2 dB), except in shadow zones and interference minima, where sound levels are substantially underpredicted when scattering by turbulence is neglected [420]. A von Kármán turbulence spectrum (section 6.2) is adequate for determining the scattered signal energy [420]. Although it might initially seem reasonable that predictions would be most accurate when the sound level and atmospheric data are averaged over the same time interval, this is not necessarily so, since the relationship between the atmospheric fields and the sound propagation is nonlinear. Generally, the more averaging applied to the atmospheric data, the better, so long as the atmospheric statistics remain approximately stationary during the averaging interval.*

Despite the potential pitfalls for misusing meteorological data in sound propagation predictions, references [417] and [420] indicate that predictive skill does indeed generally improve, in a statistical sense, through the use of available meteorological data, even when such data are limited. Mean vertical profiles produce the most accurate predictions of both mean and event sound levels, because they are most representative of typical conditions along the propagation path. In downward refraction, however, predictive skill for event-type predictions does not depend substantially on knowledge of the meteorological profile. Interference patterns in downward refraction do not appear to be predictable in detail due to random propagation effects.

Monin–Obukhov similarity theory (MOST) can be beneficial for inferring the mean vertical profiles from meteorological data collected at the surface or at two heights on a short tower. Numerical weather predictions (NWP) may also be useful for inferring mean profiles. When data from weather balloons (rawinsonde or tethersonde) are used, care should be taken to avoid the

---

*Recall that the average output of a nonlinear system does *not* equal the output calculated from the average of the inputs. Even when the propagation is linear in the sense that the amplitude is small enough to satisfy the linear wave equation, the sound propagation is nonetheless a highly nonlinear function of the sound speed and other environmental parameters.

above-described plywood atmosphere, by smoothing fine-scale variations in the profiles prior to their use in propagation modeling. Future research will hopefully provide more insights into the most suitable procedures for incorporating atmospheric data into sound propagation predictions and a better quantitative understanding of the resulting accuracy of the predictions.

References

[1] J. F. Allard and N. Atalla. *Propagation of Sound in Porous Media: Modelling Sound Absorbing Materials*. Wiley, Chichester, United Kingdom, second edition, 2009.

[2] J.-F. Allard and Y. Champoux. New empirical equations for sound propagation in rigid frame fibrous materials. *J. Acoust. Soc. Am.*, 91(6):3346–3353, 1992.

[3] J. F. Allard, C. Depollier, J. Nicolas, W. Lauriks, and A. Cops. Propriétés acoustiques des matériaux poreux saturés d'air et théorie de biot. *Journal d'Acoustique*, 3:29–38, 1990.

[4] L. S. Al'perovich, B. O. Vugmester, and M. B. Gokhberg. On the experiment for modelling magneto-ionosphere effects caused by seismic phenomena. *Trans. Acad. Sci. USSR, Earth Science*, 269(3):573, 1983.

[5] N. N. Andreev and I. G. Rusakov. *Acoustics in a Moving Medium*. GTTI, Leningrad, 1934 (in Russian).

[6] L. C. Andrews and R. L. Phillips. *Laser Beam Propagation through Random Media*. SPIE, Bellingham, WA, 2005.

[7] V. I. Arabadgi. Zones of abnormal audibility. *Meteorolgy and Hydrology*, 5:21, 1946 (in Russian).

[8] D. F. J. Arago. Resultats des experiences faites en 1822, par ordre du bureau des longitudes, pour la determination de la vitesse du son dans l'atmosphere. *Ann. Chim. Phys.*, 20:210, 1822.

[9] K. Arnold, A. Ziemann, and A. Raabe. Tomographic monitoring of wind and temperature at different heights above the ground. *Acust. Acta Acust.*, 87(6):703–708, 2001.

[10] K. Arnold, A. Ziemann, A. Raabe, and G. Spindler. Acoustic tomography and conventional meteorological measurements over heterogeneous surfaces. *Meteor. Atmos. Phys.*, 85(1):175–186, 2004.

[11] K. Attenborough. Acoustical characteristics of rigid fibrous absorbents and granular materials. *J. Acoust. Soc. Am.*, 73(3):785–799, 1983.

[12] K. Attenborough. Acoustical impedance models for outdoor ground surfaces. *J. Sound Vib.*, 99(4):521–544, 1985.

[13] K. Attenborough, I. Bashir, and S. Taherzadeh. Outdoor ground impedance models. *J. Acoust. Soc. Am.*, 129(5):2806–2819, 2011.

[14] K. Attenborough and O. Buser. On the application of rigid-porous models to impedance data for snow. *J. Sound Vib.*, 124(2):315–327, 1988.

[15] K. Attenborough, S. I. Hayek, and J. M. Lawther. Propagation of sound above a porous half-space. *J. Acoust. Soc. Am.*, 68(5):1493–1501, 1980.

[16] K. Attenborough, K. M. Li, and K. Horoshenkov. *Predicting Outdoor Sound.* Taylor & Francis, New York, 2007.

[17] K. Attenborough, S. Taherzadeh, H. E. Bass, X. Di, R. Raspet, G. R. Becker, A. Güdesen, A. Chrestman, G. A. Daigle, A. L'Espérance, Y. Gabillet, K. E. Gilbert, Y. L. Li, M. J. White, P. Naz, J. M. Noble, and H. A. J. M. van Hoof. Benchmark cases for outdoor sound propagation models. *J. Acoust. Soc. Am.*, 97(1):173–191, 1995.

[18] B. M. Ayyub and R. J. Chao. Uncertainty modeling in civil engineering with structural and reliability applications. In B. M. Ayyub, editor, *Uncertainty Modeling and Analysis in Civil Engineering*, pages 3–32. CRC Press, Boca Raton, FL, 1998.

[19] H. D. Baehr. *Thermodynamik.* Springer, Berlin, 1962.

[20] C. Bailly, P. Lafon, and S. Candel. A stochastic approach to compute noise generation and radiation of free turbulent flows. In *First Joint CEAS/AIAA Aeroacoustics Conference*, pages 669–674 (paper 95–092), 1995.

[21] N. K. Balachandron, W. L. Donn, and D. H. Rind. Concord sonic booms as an atmospheric probe. *Science*, 17(4298):47–49, 1977.

[22] M. Barth, A. Raabe, K. Arnold, C. Resack, and R. du Puits. Flow field detection using acoustic travel time tomography. *Meteorol. Z.*, 16(4):443–450, 2007.

[23] E. H. Barton. On the refraction of sound by wind. *Philos. Mag.*, 1(1):159–165, 1901.

[24] H. E. Bass, L. N. Bolen, R. Raspet, W. McBride, and J. Noble. Acoustic propagation through a turbulent atmosphere: Experimental characterization. *J. Acoust. Soc. Am.*, 90(6):3307–3313, 1991.

[25] G. K. Batchelor. *The Theory of Homogeneous Turbulence.* Cambridge University Press, Cambridge, 1953.

[26] H. Bateman. The influence of meteorological conditions on the propagation of sound. *Mon. Weather Rev.*, 42(5):258–265, 1914.

[27] N. R. Beers. Meteorological thermodynamics and atmospheric statistics. In F. A. Berry, E. Bollay, and N. R. Beers, editors, *Handbook of Meteorology*, pages 314–409. McGraw-Hill, New York, 1973.

[28] I. Beichl and F. Sullivan. The importance of importance sampling. *Comput. Sci. Eng*, 1:71–73, 1999.

[29] J. S. Bendat and A. G. Piersol. *Random Data: Analysis and Measurement Procedures*. Wiley, New York, 2011.

[30] M. J. Beran and A. M. Whitman. Effect of random velocity fluctuations on underwater scattering. *J. Acoust. Soc. Am.*, 81(3):647–649, 1987.

[31] J.-P. Bérenger. A perfectly matched layer for the absorption of electromagnetic waves. *J. Comp. Phys.*, 114(1):185–200, 1994.

[32] C. Besset and E. Blanc. Propagation of vertical shock waves in the atmosphere. *J. Acoust. Soc. Am.*, 95(4):1830–1839, 1994.

[33] R. Betchov and W. O. Kriminale. *Stability of Parallel Flows*. Academic Press, New York, 1967.

[34] P. Blanc-Benon. Moment d'ordre deux en deux points d'une onde acoustique sphérique après traversée d'une turbulence cinématique. *Revue du CETHEDEC-Ondes et Signal*, 79(2):21–29, 1984.

[35] P. Blanc-Benon, L. Dallois, and D. Juvé. Long range sound propagation in a turbulent atmosphere within the parabolic approximation. *Acust. Acta Acust.*, 87(6):659–669, 2001.

[36] P. Blanc-Benon, J. Wasier, D. Juvé, and V. E. Ostashev. Experimental studies of sound propagation through thermal turbulence near a boundary. In *29th International Congress on Noise Control Engineering*, 2000.

[37] D. I. Blokhintzev. *Acoustics of an Inhomogeneous Moving Medium*. Physics Dept. Brown Univ., Providence, 1956.

[38] R. Blumrich and D. Heimann. A linearized Eulerian sound propagation model for studies of complex meteorological effects. *J. Acoust. Soc. Am.*, 112(2):446–455, 2002.

[39] A. Y. Bogushevich and N. P. Krasnenko. Doppler-effect in the acoustics of an inhomogeneous moving medium. *Sov. Phys. Acoust.*, 34(4):345–347, 1988.

[40] G. Borne. Über die Verbreitung die Dynamitexplosion zu Forde in Westfallen verursachten Schallphenomene. *Erdebenwarte*, 4:1, 1904.

[41] D. Botteldooren. Acoustical finite-difference time-domain simulation in a quasi-Cartesian grid. *J. Acoust. Soc. Am.*, 95(5):2313–2319, 1994.

[42] P. Boulanger, R. Raspet, and H. E. Bass. Sonic boom propagation through a realistic turbulent atmosphere. *J. Acoust. Soc. Am.*, 98(6):3412–3417, 1995.

[43] V. Bovsheverov and G. Karyukin. Influence of wind on the accuracy of determination of the temperature structure parameter by acoustic sounding. *Izv. Acad. Sci. USSR, Atmos. Oceanic Phys.*, 17(2):151–153, 1981.

[44] S. Bradley. *Atmospheric Acoustic Remote Sensing: Principles and Applications.* CRC Press, Boca Raton, FL, 2010.

[45] H. Braun and A. Hauck. Tomographic reconstruction of vector fields. *IEEE Trans. Signal Proces.*, 39:464–471, 1991.

[46] L. M. Brekhovskikh, editor. *Oceanic Acoustics.* Nauka, Moscow, 1974 (in Russian).

[47] L. M. Brekhovskikh and O. A. Godin. *Acoustics in Layered Media.* Nauka, Moscow, 1989 (in Russian).

[48] L. M. Brekhovskikh and O. A. Godin. *Acoustics of Layered Media II: Point Sources and Bounded Beams.* Springer, Berlin, 1999.

[49] L. M. Brekhovskikh and Y. P. Lysanov. *Fundamentals of Ocean Acoustics.* Springer, Berlin, 1982.

[50] R. I. Brent, M. J. Jacobson, and W. L. Siegmann. Sound propagation through random currents using parabolic approximations. *J. Acoust. Soc. Am.*, 84(5):1765–1776, 1988.

[51] F. P. Bretherton and C. J. R. Garrett. Wavetrains in inhomogeneous moving media. *Proc. R. Soc. London*, A302:529–554, 1969.

[52] E. H. Brown and F. F. Hall. Advances in atmospheric acoustics. *Rev. Geophys. Space Phys.*, 16(1):47–110, 1978.

[53] E. H. Brown, C. G. Little, and W. M. Wright. Echosonde interferometer for atmospheric research. *J. Acoust. Soc. Am.*, 63(3):694–699, 1978.

[54] E. Buckingham. On physically similar systems; illustrations of the use of dimensional equations. *Phys. Rev.*, 4(4):345–376, 1914.

[55] R. Buckley. Diffraction by a random phase-changing screen: A numerical experiment. *J. Atmos. Terr. Phys.*, 37(11):1431–1446, 1975.

[56] M. Buret, K. M. Li, and K. Attenborough. Diffraction of sound due to moving sources by barriers and ground discontinuities. *J. Acoust. Soc. Am.*, 120(3):1274–1283, 2006.

[57] S. D. Burk. Refractive index structure parameters: time dependent calculations using a numerical boundery-layer model. *J. Appl. Meteorol.*, 19:562–575, 1980.

[58] S. D. Burk. Temperature and humidity effects on refractive index fluctuations in upper regions of the convective boundary layer. *J. Appl. Meteorol.*, 20(6):717–721, 1981.

[59] G. A. Bush, A. I. Grachev, E. A. Ivanov, S. N. Kulichkov, and M. I. Mordukhovich. On anomalous sound propagation in the atmosphere. *Izv. Acad. Sci. USSR, Atmos. Oceanic Phys.*, 22(1):68–70, 1986.

[60] G. A. Bush, E. A. Ivanov, S. N. Kulichkov, A. V. Kuchayev, and M. V. Pedanov. Acoustic sounding of the fine structure of the upper atmosphere. *Izv. Acad. Sci. USSR, Atmos. Oceanic Phys.*, 25(4):251–256, 1989.

[61] S. M. Candel. Acoustic conservation principles and an application to plane and modal propagation in nozzles and diffusers. *J. Sound Vib.*, 41(2):207–232, 1975.

[62] R. H. Cantrell and R. W. Hart. Interaction between sound and flow in acoustic cavities: mass, momentum and energy considerations. *J. Acoust. Soc. Am.*, 36(4):697–706, 1964.

[63] C. F. Cassini. *Sur la propagation du son.* Mem. de l'Acad. Paris, 1738.

[64] Y. Champoux and M. R. Stinson. On acoustical models for sound propagation in rigid frame porous materials and the influence of shape factors. *J. Acoust. Soc. Am.*, 92(2):1120–1131, 1992.

[65] S. Cheinet. A numerical approach to sound levels in near-surface refractive shadows. *J. Acoust. Soc. Am.*, 131(3):1946–1958, 2012.

[66] S. Cheinet. Long-term, global-scale statistics of sound propagation. *J. Acoust. Soc. Am.*, 135(5):2581–2590, 2014.

[67] S. Cheinet, L. Ehrhardt, D. Juvé, and P. Blanc-Benon. Unified modeling of turbulence effects on sound propagation. *J. Acoust. Soc. Am.*, 132(4):2198–2209, 2012.

[68] S. Cheinet and A. P. Siebesma. Variability of local structure parameters in the convective boundary layer. *J. Atmos. Sci.*, 66(4):1002–1017, 2009.

[69] H. Y. Chen and I. T. Lu. Matched-mode processing schemes of a moving point source. *J. Acoust. Soc. Am.*, 92(4):2039–2050, 1992.

[70] T. Chen, P. Ratilal, and N. C. Makris. Mean and variance of the forward field propagated through three-dimensional random internal waves in a continental-shelf waveguide. *J. Acoust. Soc. Am.*, 118(6):3560–3574, 2005.

[71] L. A. Chernov. The acoustics of a moving medium. *Sov. Phys. Acoust.*, 4(4):311–318, 1958.

[72] L. A. Chernov. *Waves in Randomly Inhomogeneous Media.* Nauka, Moscow, 1975 (in Russian).

[73] C. I. Chessel. Three-dimensional acoustic-ray tracing in an inhomogeneous anisotropic atmosphere using Hamilton's equations. *J. Acoust. Soc. Am.*, 53(1):83–87, 1973.

[74] C. I. Chessel. Propagation of noise along a finite impedance boundary. *J. Acoust. Soc. Am.*, 62(4):825–834, 1977.

[75] P. Chevret, P. Blanc-Benon, and D. Juvé. A numerical model for sound propagation through a turbulent atmosphere near the ground. *J. Acoust. Soc. Am.*, 100(6):3587–3599, 1996.

[76] S. V. Chibisov. Time of sound ray propagation in the atmosphere. *Izv. Acad. Scienc. USSR, Geogr. Geophiz.*, 1:33–118, 1940 (in Russian).

[77] I. P. Chunchuzov. Field of a low-frequency point source of sound in an atmosphere with a nonuniform wind-height distribution. *Sov. Phys. Acoust.*, 30(4):323–326, 1984.

[78] I. P. Chunchuzov. Field of a point sound source in the ground layer of the atmosphere. *Sov. Phys. Acoust.*, 31(1):78–79, 1985.

[79] I. P. Chunchuzov, S. N. Kulichkov, V. Perepelkin, A. Ziemann, K. Arnold, and A. Kniffka. Mesoscale variations in acoustic signals induced by atmospheric gravity waves. *J. Acoust. Soc. Am.*, 125(2):651–664, 2009.

[80] I. P. Chunchuzov, S. N. Kulichkov, O. E. Popov, R. Waxler, and J. Assink. Infrasound scattering from atmospheric anisotropic inhomogeneities. *Izv. Atmos. Oceanic Phys.*, 47(5):540–557, 2011.

[81] J. F. Claerbout. *Fundamentals of Geophysical Data Processing.* McGraw-Hill, New York, 1976.

[82] S. F. Clifford and R. J. Lataitis. Turbulence effects on acoustic wave propagation over a smooth surface. *J. Acoust. Soc. Am.*, 73(5):1545–1550, 1983.

[83] S. L. Collier, D. K. Wilson, V. E. Ostashev, N. P. Symons, D. F. Aldridge, and D. H. Marlin. FDTD acoustic wave propagation in complex atmospheric environments: Final year results. In *Military Sensing Symposia (MSS) Specialty Group on Battlefield Acoustic & Seismic Sensing, Magnetic & Electric Field Sensors*, 2005.

[84] E. F. Cox. Far transmission of air blast waves. *Phys. Fluids*, 1:95–101, 1958.

[85] A. P. Crary. Stratosphere winds and temperatures from acoustical propagation studies. *J. Meteorol.*, 7(3):233–242, 1950.

[86] G. A. Daigle. Effects of atmospheric turbulence on the interference of sound waves above a finite impedance boundary. *J. Acoust. Soc. Am.*, 65:45–49, 1979.

[87] G. A. Daigle, J. E. Piercy, and T. F. W. Embleton. Effects of atmospheric turbulence on the interference of sound waves near a hard boundary. *J. Acoust. Soc. Am.*, 64(2):622–630, 1978.

[88] G. A. Daigle, J. E. Piercy, and T. F. W. Embleton. Line-of-sight propagation through atmospheric turbulence near the ground. *J. Acoust. Soc. Am.*, 74(5):1505–1513, 1983.

[89] L. Dallois, P. Blanc-Benon, and D. Juvé. A wide-angle parabolic equation for acoustic waves in inhomogeneous moving media: Applications to atmospheric sound propagation. *J. Comp. Acoustics*, 9(2):477–494, 2001.

[90] C. D. de Groot-Hedlin, M. A. H. Hedlin, and D. P. Drob. Atmospheric variability and infrasound monitoring. In A. Le Pichon, E. Blanc, and A. Hauchecorne, editors, *Infrasound Monitoring for Atmospheric Studies*, pages 475–507. Springer, Dordrecht, 2010.

[91] J. A. De Santo, editor. *Oceanic Acoustics*. Springer, Berlin, 1979.

[92] D. A. de Wolf. A random-motion model of fluctuations in a nearly transparent medium. *Radio Sci.*, 18(2):138–142, 1983.

[93] M. E. Delany. Sound propagation in the atmosphere: a historical review. *Acust. Acta Acust.*, 38(4):201–233, 1977.

[94] M. E. Delany and E. N. Bazley. Acoustical properties of fibrous absorbent materials. *Appl. Acoust.*, 3(2):105–116, 1970.

[95] F. Delaroche. Sur l'influence que le vent exerce dans la propagation du son, sous le rapport de son intensite. *Ann. Chim.*, 1:176, 1816.

[96] W. Derham. Experimenta et observationes de soni motu. *Philos. Trans. R. Soc. London*, 26:2–35, 1708.

[97] X. Di and K. E. Gilbert. An exact Laplace transform formulation for a point source above a ground surface. *J. Acoust. Soc. Am.*, 93(2):714–720, 1993.

[98] D. Di Iorio and D. M. Farmer. Two-dimensional angle of arrival fluctuations. *J. Acoust. Soc. Am.*, 100(2):814–824, 1996.

[99] D. Di Iorio and D. M. Farmer. Separation of current and sound speed in the effective refractive index for a turbulent environment using reciprocal acoustic transmission. *J. Acoust. Soc. Am.*, 103(1):321–329, 1998.

[100] F. R. DiNapoli. *A Fast Field Program for Multilayered Media.* NUSC Tech Report 4103, Naval Underwater Systems Center, Washington, 1971.

[101] H. Dong, A. M. Kaynia, C. Madshus, and J. M. Hovem. Sound propagation over layered poro-elastic ground using a finite-difference model. *J. Acoust. Soc. Am.*, 108(2):494–502, 2000.

[102] S. E. Dosso, P. M. Giles, G. H. Brooke, D. F. McCammon, S. Pecknold, and P. C. Hines. Linear and nonlinear measures of ocean acoustic environmental sensitivity. *J. Acoust. Soc. Am.*, 121(1):42–45, 2007.

[103] P. Duckert. Über die Ausbreitung von Explosionswellen in der Erdatmosphäre. *Ergeb. Kosm. Physik*, 1:236–290, 1931.

[104] W. L. Dunn and J. K. Shultis. *Exploring Monte Carlo Methods.* Elsevier, Amsterdam, 2011.

[105] L. Ehrhardt, S. Cheinet, and D. Juvé. Finite-difference time-domain simulation of sound propagation through turbulent atmosphere. *J. Acoust. Soc. Am.*, 131(4):3332–3332, 2012.

[106] T. F. W. Embleton. Tutorial on sound propagation outdoors. *J. Acoust. Soc. Am.*, 100(1):31–48, 1996.

[107] T. F. W. Embleton, J. E. Piercy, and G. A. Daigle. Effective flow resistivity of ground surfaces determined by acoustical measurements. *J. Acoust. Soc. Am.*, 74(4):1239–1244, 1983.

[108] R. Emden. Beiträge zur thermodynamik der atmosphäre. *Meteorol. Z.*, 35:13–29, 74–81, 112–123, 1918.

[109] M. Evans and T. Swartz. *Approximating Integrals Via Monte Carlo and Deterministic Methods.* Oxford University Press, Oxford, 2000.

[110] R. Ewert and W. Schröder. Acoustic perturbation equations based on flow decomposition via source filtering. *J. Comp. Phys.*, 188(2):365–398, 2003.

[111] C. W. Fairall. The humidity and temperature sensitivity of clear-air radars in the convective boundary layer. *J. Appl. Meteorol.*, 30(8):1064–1074, 1991.

[112] R. F. Fante. Mutual coherence function and frequency spectrum of a laser beam propagating through atmospheric turbulence. *J. Opt. Soc. Am.*, 64(5):592–598, 1974.

[113] D. M. Farmer, S. F. Clifford, and J. A. Verrall. Scintillation structure of a turbulent tidal flow. *J. Geophys. Res.*, 92(C5):5369–5382, 1985.

[114] Z. E. A. Fellah, S. Berger, W. Lauriks, C. Depollier, C. Aristegui, and J.-Y. Chapelon. Measuring the porosity and the tortuosity of porous materials via reflected waves at oblique incidence. *J. Acoust. Soc. Am.*, 113(5):2424–2433, 2003.

[115] S. Finette. Embedding uncertainty into ocean acoustic propagation models. *J. Acoust. Soc. Am.*, 117(3):997–1000, 2005.

[116] S. M. Flatte. Wave propagation through random media: Contributions from ocean acoustics. *Proc. IEEE*, 11:1267–1294, 1983.

[117] S. M. Flatté, R. Dashen, W. H. Munk, K. Watson, and F. Zachariasen. *Sound Transmission through a Fluctuating Ocean.* Cambridge University Press, New York, 1979.

[118] S. M. Flatté and M. D. Vera. Comparison between ocean-acoustic fluctuations in parabolic-equation simulations and estimates from integral approximations. *J. Acoust. Soc. Am.*, 114(2):697–706, 2003.

[119] E. R. Franchi and M. J. Jacobson. Effect of hydrodynamic variations on sound transmission across a geostrophic flow. *J. Acoust. Soc. Am.*, 54(5):1302–1311, 1963.

[120] E. R. Franchi and M. J. Jacobson. Ray propagation in a channel with depth-variable sound speed and current. *J. Acoust. Soc. Am.*, 52(1):316–331, 1972.

[121] E. R. Franchi and M. J. Jacobson. An environmental-acoustics model for sound propagation in a geostrophic flow. *J. Acoust. Soc. Am.*, 53(3):835–847, 1973.

[122] R. Frehlich. Effects of global intermittency on laser propagation in the atmosphere. *Appl. Opt.*, 33(24):5764–5769, 1994.

[123] U. Frisch, P.-L. Sulem, and M. Nelkin. A simple dynamical model of intermittent fully developed turbulence. *J. Fluid Mech.*, 87(4):719–736, 1978.

[124] S. Fujiwhara. On the abnormal propagation of sound waves in the atmosphere. *Bull. Centr. Meteor. Observ. Japan*, 2(1):1–82, 1912.

[125] S. Fujiwhara. On the abnormal propagation of sound waves in the atmosphere. *Bull. Centr. Meteor. Observ. Japan*, 2(4):1–13, 1916.

[126] I. Fuks, M. Charnotskii, and K. Naugolnykh. A multifrequency scintillation method for ocean flow measurements. *J. Acoust. Soc. Am.*, 109(6):2730–2738, 2001.

[127] J. Fung, J. Hunt, N. Malik, and R. Perkins. Kinematic simulation of homogeneous turbulence by unsteady random Fourier modes. *J. Fluid Mech.*, 236(1):281–318, 1992.

[128] Y. Gabillet, H. Schroeder, G. A. Daigle, and A. L'Espérance. Application of the Gaussian beam approach to sound propagation in the atmosphere: theory and experiments. *J. Acoust. Soc. Am.*, 93(6):3105–3116, 1993.

[129] B. Galperin and S. A. Orszag. *Large Eddy Simulation of Complex Engineering and Geophysical Flows*. Cambridge University Press, Cambridge, 1993.

[130] J. R. Garratt. *The Atmospheric Boundary Layer*. Cambridge University Press, Cambridge, 1994.

[131] V. G. Gavrilenko and L. A. Zelekson. Sound amplification in nonuniform flow. *Sov. Phys. Acoust.*, 23(6):497–500, 1977.

[132] S. D. Gedney. An anisotropic perfectly matched layer-absorbing medium for the truncation of FDTD lattices. *IEEE Trans. Antenn. Propag.*, 44(12):1630–1639, 1996.

[133] A. V. Generalov and I. S. Zaguzov. Investigation of the characteristics of the sound field of a moving source, with application to the analysis of aircraft noise. In *Acoustics of Turbulent Flows*, pages 77–86. Nauka, Moscow, 1983 (in Russian).

[134] T. M. Georges and S. F. Clifford. Acoustic sounding in a refracting atmosphere. *J. Acoust. Soc. Am.*, 52(5):1397–1405, 1972.

[135] K. E. Gilbert and X. Di. A fast Green's function method for one-way sound propagation in the atmosphere. *J. Acoust. Soc. Am.*, 94:2343–2352, 1993.

[136] K. E. Gilbert, R. Raspet, and X. Di. Calculation of turbulence effects in an upward-refracting atmosphere. *J. Acoust. Soc. Am.*, 87:2428–2437, 1990.

[137] K. E. Gilbert and M. J. White. Application of the parabolic equation to sound propagation in a refracting atmosphere. *J. Acoust. Soc. Am.*, 85:630–637, 1989.

[138] O. A. Godin. Wave equation for sound in a medium with slow current. *Trans. Acad. Sci. USSR, Earth Science*, 293(1):63–67, 1987.

[139] O. A. Godin. An effective quiescent medium for sound propagating through an inhomogeneous, moving medium. *J. Acoust. Soc. Am.*, 112(4):1269–1275, 2002.

[140] O. A. Godin. Wide-angle parabolic equations for sound in a 3D inhomogeneous moving medium. *Doklady Physics*, 47(9):643–646, 2002.

[141] G. H. Goedecke and H. J. Auvermann. Acoustic scattering by atmospheric turbules. *J. Acoust. Soc. Am.*, 102:759–771, 1997.

[142] G. H. Goedecke, V. E. Ostashev, D. K. Wilson, and H. J. Auvermann. Quasi-wavelet model of von Kármán spectrum of turbulent velocity fluctuations. *Bound. Layer Meteorol.*, 112(1):33–56, 2004.

[143] G. H. Goedecke, D. K. Wilson, and V. E. Ostashev. Quasi-wavelet models of turbulent temperature fluctuations. *Bound. Layer Meteorol.*, 120(1):1–23, 2006.

[144] M. E. Goldstein. *Aeroacoustics*. McGraw-Hill, New York, 1976.

[145] V. P. Goncharov. Emission of low-frequency sound by a point source in a shear flow. *Izv. Acad. Sci. USSR, Atmos. Oceanic Phys.*, 20(4):325–327, 1984.

[146] E. E. Gossard and W. H. Hoke. *Waves in the Atmosphere*. Elsevier, Amsterdam, 1975.

[147] J. Gozani. Clarifying the concepts of wave propagation through intermittent media. *Opt. Lett.*, 24:108–110, 1999.

[148] M. J. Griffiths, J. A. H. Oates, and D. Lord. The propagation of sound from quarry blasting. *J. Sound Vib.*, 69(3):359–370, 1978.

[149] N. S. Grigor'eva and M. I. Yavor. Influence on the sound field in the ocean of a large-scale ocean current that qualitatively alters the nature of guided-wave sound propagation. *Sov. Phys. Acoust.*, 32(6):482–485, 1986.

[150] M. Grigoriu. A spectral representation based model for Monte Carlo simulation. *Prob. Eng. Mech.*, 15(4):365–370, 2000.

[151] G. V. Groves. Geometrical theory of sound propagation in the atmosphere. *J. Atmos. Terr. Phys.*, 7:113–127, 1955.

[152] A. S. Gurvich and A. I. Kon. The backscattering from anisotropic turbulent irregularities. *J. Electromagnet. Wave*, 6(1):107–118, 1992.

[153] A. S. Gurvich, A. I. Kon, V. L. Mironov, and S. S. Khmelevtsov. *Laser Radiation in the Turbulent Atmosphere*. Nauka, Moscow, 1976 (in Russian).

[154] B. Gutenberg. Sound propagation in the atmosphere. In *Compendium of Meteorology*, pages 366–375. Amer. Meteorol. Soc., Boston, 1951.

[155] B. Hallberg, C. Larsson, and S. Israelsson. Numerical ray tracing in the atmospheric surface layer. *J. Acoust. Soc. Am.*, 83:2059–2068, 1988.

[156] D. I. Havelock, X. Di, G. A. Daigle, and M. R. Stinson. Spatial coherence of a sound field in a refractive shadow: Comparison of simulation and experiment. *J. Acoust. Soc. Am.*, 98(4):2289–2302, 1995.

[157] W. D. Hayes. Energy invariant for geometric acoustics in a moving medium. *Phys. Fluids*, 11(8):1654–1656, 1968.

[158] D. Heimann and R. Karle. A linearized Euler finite-difference time-domain sound propagation model with terrain-following coordinates. *J. Acoust. Soc. Am.*, 119:3813–3821, 2006.

[159] D. Heimann and E. Salomons. Testing meteorological classifications for the prediction of long-term average sound levels. *Appl. Acoust.*, 65(10):925–950, 2004.

[160] G. S. Heller. Propagation of acoustic discontinuities in an inhomogeneous moving liquid medium. *J. Acoust. Soc. Am.*, 25(5):950–951, 1953.

[161] J. C. Helton, J. D. Johnson, C. J. Sallaberry, and C. B. Storlie. Survey of sampling-based methods for uncertainty and sensitivity analysis. *Reliab. Eng. Syst. Safe.*, 91:1175–1209, 2006.

[162] R. F. Henrick, M. J. Jacobson, and W. L. Siegmann. General effects of currents and sound-speed variations on short-range acoustic transmission in cyclonic eddies. *J. Acoust. Soc. Am.*, 67(1):121–134, 1980.

[163] R. F. Henrick, W. Siegmann, and M. J. Jacobson. General analysis of ocean eddy effects for sound transmission applications. *J. Acoust. Soc. Am.*, 62(4):860–870, 1977.

[164] J. Henry. *Report of the Lighthouse Board of the United States*. U.S. Government Printing Office, Washington, DC, 1874.

[165] H. G. E. Hentschel and I. Procaccia. Passive scalar fluctuatations in intermittent turbulence with applications to wave propagation. *Phys. Rev. A*, 28:417–426, 1983.

[166] A. S. Hersh and I. Catton. Effect of shear-flow on sound propagation in rectangular ducts. *J. Acoust. Soc. Am.*, 50(3):992–1003, 1971.

[167] J. Hinze. *Turbulence.* McGraw-Hill, New York, 1975.

[168] J. Højstrup. Velocity spectra in the unstable planetary boundary layer. *J. Atmos. Sci.*, 39(10):2239–2248, 1982.

[169] L. R. Hole and H. M. Mohr. Modeling of sound propagation in the atmospheric boundary layer: Application of the MIUU mesoscale model. *J. Geophys. Res.-Atmos.*, 104(D10):11891–11901, 1999.

[170] P. Holstein, A. Raabe, R. Müller, M. Barth, D. Mackenzie, and E. Starke. Acoustic tomography on the basis of travel-time measurements. *Meas. Sci. Technol.*, 15:1420–1428, 2004.

[171] M. Hornikx, R. Waxler, and J. Forssén. The extended Fourier pseudospectral time-domain method for atmospheric sound propagation. *J. Acoust. Soc. Am.*, 128(4):1632–1646, 2010.

[172] K. J. Howell, M. J. Jacobson, and S. W. L. Parabolic approximation predictions of underwater acoustic effects due to source and receiver motions. *J. Acoust. Soc. Am.*, 93(1):293–301, 1994.

[173] V. I. Il'ichev and Y. V. Khokha. Sound field of a radiating domain of finite dimensions, moving with a variable subsonic velocity. *Sov. Phys. Dokl.*, 31(1):45–47, 1986.

[174] U. Ingard. A review of the influence of meteorological conditions on sound propagation. *J. Acoust. Soc. Am.*, 25(3):405–411, 1953.

[175] U. Ingard and G. C. J. Malling. On the effect of atmospheric turbulence on sound propagation over ground. *J. Acoust. Soc. Am.*, 35(7):1056–1058, 1963.

[176] U. Ingard and V. K. Singhal. Upstream and downstream sound radiation into a moving fluid. *J. Acoust. Soc. Am.*, 54(5):1343–1346, 1973.

[177] International Standards Organization. Acoustics – attenuation of sound during propagation outdoors – part 2: General method of calculation. Technical Report ISO 9613-2:1996, International Standards Organization, 1996.

[178] A. Ishimaru. *Wave Propagation and Scattering in Random Media.* IEEE Press, New York, 1997.

[179] S. Itzikowitz, M. J. Jacobson, and W. L. Siegmann. Short-range acoustic transmission through cyclonic eddies between a submerged source and receiver. *J. Acoust. Soc. Am.*, 71(5):1131–1144, 1982.

[180] S. Itzikowitz, M. J. Jacobson, and W. L. Siegmann. Modeling of long-range acoustic transmission through cyclonic and anticyclonic eddies. *J. Acoust. Soc. Am.*, 73(5):1556–1566, 1983.

[181] D. L. Johnson, J. Koplik, and R. Dashen. Theory of dynamic permeability and tortuosity in fluid-saturated porous media. *J. Fluid Mech.*, 176:379–402, 1987.

[182] M. A. Johnson, R. Raspet, and M. T. Bobak. A turbulence model for sound propagation from an elevated source above level ground. *J. Acoust. Soc. Am.*, 81(3):638–646, 1987.

[183] S. A. Johnson, J. F. Greenleaf, C. R. Hansen, W. F. Samayoa, M. Tanaka, A. Lent, D. A. Christensen, and R. L. Woolley. Reconstructing three-dimensional fluid velocity vector fields from acoustic transmission measurements. In L. W. Kessler, editor, *Acoustical Holography*, pages 307–326. Plenum, Berlin, 1977.

[184] S. G. Johnson. Notes on Perfectly Matched Layers (PMLs). Lecture Notes, 2010. http://math.mit.edu/~stevenj/18.369/pml.pdf.

[185] I. Jovanovic, L. Sbaiz, and M. Vetterli. Acoustic tomography for scalar and vector fields: theory and application to temperature and wind estimation. *J. Atmos. Ocean. Tech.*, 26:1475–1492, 2009.

[186] J. C. Kaimal and J. J. Finnigan. *Atmospheric Boundary Layer Flows: Their Structure and Measurement.* Oxford University Press, New York, 1994.

[187] M. A. Kallistratova. Experimental investigation of sound wave scattering in the atmosphere. *Tr. Inst. Fiz. Atmos., Atmos. Turbulentnost,* 4:203–256, 1961 (in Russian).

[188] M. A. Kallistratova and A. I. Kon. *Radioacoustic Sounding of the Atmosphere.* Nauka, Moscow, 1994 (in Russian).

[189] A. J. Kantor and A. E. Cole. Zonal and meridional winds to 120 kilometers. *J. Geophys. Res.*, 69(24):5131–5140, 1964.

[190] V. N. Karavainikov. Fluctuations of amplitude and phase in a spherical wave. *Sov. Phys. Acoust.*, 3:175–186, 1957.

[191] M. Karweit, P. Blanc-Benon, D. Juvé, and G. Comte-Bellot. Simulation of the propagation of an acoustic wave through a turbulent velocity field: a study of phase variance. *J. Acoust. Soc. Am.*, 89(1):52–62, 1991.

[192] G. A. Karyukin. Influence of wind on operation of radar-acoustic atmospheric sounding systems. *Izv. Acad. Sci. USSR, Atmos. Oceanic Phys.*, 18(1):26–30, 1982.

[193] L. Kazandjian and L. Leviandier. A normal mode theory of air-to-water sound transmission by a moving source. *J. Acoust. Soc. Am.*, 96(3):1732–1740, 1994.

[194] J. B. Keller and J. S. Papadakis, editors. *Wave Propagation and Underwater Acoustics*. Springer, New York, 1977.

[195] G. Kerry, D. J. Saunders, and A. G. Sills. The use of meteorological profiles to predict the peak sound-pressure level at distance from small explosions. *J. Acoust. Soc. Am.*, 81(4):888–896, 1987.

[196] S. A. Ketcham, D. K. Wilson, M. W. Parker, and H. H. Cudney. Signal fading curves from computed urban acoustic wave fields. In *SPIE Defense and Security Symposium*, page 6963OK. International Society for Optics and Photonics, 2008.

[197] N. G. Kikina and D. S. Sannikov. Reflection of sound waves from a moving plane-parallel layer. *Sov. Phys. Acoust.*, 15(4):472–474, 1969.

[198] L. E. Kinsler, A. R. Frey, A. B. Coppens, and J. V. Sanders. *Fundamentals of Acoustics*. Wiley, New York, 1999.

[199] R. Kirby. On the modification of Delany and Bazley fomulae. *Appl. Acoust.*, 86:47–49, 2014.

[200] T. Kitagawa and T. Nomura. A wavelet-based method to generate artificial wind fluctuation data. *J. Wind Eng. Ind. Aerod.*, 91(7):943–964, 2003.

[201] G. H. Knightly, D. Lee, and D. F. S. Mary. A higher-order parabolic wave equation. *J. Acoust. Soc. Am.*, 82(2):580–587, 1987.

[202] V. O. Knudsen. The propagation of sound in the atmosphere — attenuation and fluctuations. *J. Acoust. Soc. Am.*, 18(1):90–96, 1946.

[203] A. N. Kolmogorov. The local structure of turbulence in incompressible viscous fluid for very large Reynolds numbers. *Dokl. Akad. Nauk SSSR*, 30(4):299–303, 1941.

[204] A. N. Kolmogorov. A refinement of previous hypotheses concerning the local structure of turbulence in a viscous incompressible fluid at high Reynolds number. *J. Fluid Mech.*, 13(1):82–85, 1962.

[205] S. Kolouri, M. R. Azimi-Sadjadi, and A. Ziemann. A statistical-based approach for acoustic tomography of the atmosphere. *J. Acoust. Soc. Am.*, 135(1):104–114, 2014.

[206] P. Kolykhalov. Amplification of acoustic perturbations in reflection from a critical layer in supersonic flows. *Sov. Phys. Dokl.*, 30(1):52–53, 1985.

[207] A. I. Kon. Qualitative theory of amplitude and phase fluctuations in medium with anisotropic turbulent irregularities. *Wave Random Complex*, 4:297–306, 1994.

[208] K. Konishi and Z. Maekawa. Interpretation of long term data measured continuously on long range sound propagation over sea surfaces. *Appl. Acoust.*, 62(10):1183–1210, 2001.

[209] E. T. Kornhauser. Ray theory for moving fluids. *J. Acoust. Soc. Am.*, 25(5):945–949, 1953.

[210] R. H. Kraichnan. The scattering of sound in a turbulent medium. *J. Acoust. Soc. Am.*, 25(6):1096–1104, 1953.

[211] L. Kristensen, D. H. Lenschow, P. Kirkegaard, and M. Courtney. The spectral velocity tensor for homogeneous boundary-layer turbulence. *Bound. Layer Meteorol.*, 47:149–193, 1989.

[212] S. N. Kulichkov. The coefficient of reflection of acoustic waves from the upper stratosphere. *Izv. Acad. Sci. USSR, Atmos. Oceanic Phys.*, 25(7):506–511, 1989.

[213] S. N. Kulichkov and G. A. Bush. Rapid variations in infrasonic signals at long distances from one-type explosions. *Izv. Atmos. Oceanic Phys.*, 37:306–313, 2001.

[214] V. P. Kuznetsov. On sound scattering by temperature inhomogeneities in the ocean. *Trans. Acad. Sci. USSR, Earth Science*, 290(5):1081–1084, 1986.

[215] J. S. Lamancusa and P. A. Daroux. Ray tracing in a moving medium with two-dimensional sound-speed variation and application to sound propagation over terrain discontinuities. *J. Acoust. Soc. Am.*, 93(4):1716–1726, 1993.

[216] L. D. Landau and E. M. Lifshitz. *The Classical Theory of Fields*. Pergamon Press, New York, 1975.

[217] L. D. Landau and E. M. Lifshitz. *Fluid Mechanics*. Pergamon Press, New York, 1987.

[218] L. L. Lapin. *Statistics: Meaning and Method*. Harcourt Brace Jovanovic, New York, 1980.

[219] A. Le Pichon, E. Blanc, and A. Hauchecorne, editors. *Infrasound Monitoring for Atmospheric Studies*. Springer, Dordrecht, 2010.

[220] S. Lee, N. Bong, W. Richards, and R. Raspet. Impedance formulation of the fast field program for acoustic wave propagation in the atmosphere. *J. Acoust. Soc. Am.*, 79:628–634, 1986.

[221] D. Lenschow, J. Mann, and L. Kristensen. How long is long enough when measuring fluxes and other turbulence statistics? *J. Atmos. Sci.*, 11(3):661–673, 1994.

[222] A. L'Espérance, J. Nicolas, D. Wilson, D. Thomson, Y. Gabillet, and G. Daigle. Sound propagation in the atmospheric surface layer: Comparison of experiment with FFP predictions. *Appl. Acoust.*, 40(4):325–346, 1993.

[223] K. M. Li. A high-frequency approximation of sound propagation in a stratified moving atmosphere above a porous ground surface. *J. Acoust. Soc. Am.*, 95(4):1840–1852, 1994.

[224] Y. L. Li, M. J. White, and S. J. Franke. New fast field programs for anisotropic sound propagation through an atmosphere with a wind velocity profile. *J. Acoust. Soc. Am.*, 95:718–726, 1994.

[225] M. J. Lighthill. Sound generated aerodynamically. *Proc. R. Soc. London*, A267(2):147–182, 1962.

[226] B. Lihoreau, B. Gauvreau, M. Bérengier, P. Blanc-Benon, and I. Calmet. Outdoor sound propagation modeling in realistic environments: application of coupled parabolic and atmospheric models. *J. Acoust. Soc. Am.*, 120:110–119, 2006.

[227] P. H. Lim and J. M. Ozard. On the underwater acoustic field of a moving point source. ii. Range-dependent environment. *J. Acoust. Soc. Am.*, 95(1):138–151, 1994.

[228] J. F. Lingevitch, M. D. Collins, D. K. Dacol, D. P. Drob, J. C. W. Rogers, and W. L. Siegmann. A wide angle and high Mach number parabolic equation. *J. Acoust. Soc. Am.*, 111:729–734, 2002.

[229] L. Liu and D. G. Albert. Acoustic pulse propagation near a right-angle wall. *J. Acoust. Soc. Am.*, 119:2073–2083, 2006.

[230] M. Lowson. The sound field for singularities in motion. *Proc. R. Soc. London*, A286(1407):559–572, 1965.

[231] J. L. Lumley. *Stochastic Tools in Turbulence*. Academic Press, New York, 2007.

[232] L. M. Lyamshev. Theory of sound propagation in a moving layered inhomogeneous medium. *Sov. Phys. Acoust.*, 28(3):217–221, 1982.

[233] L. Mahrt. Intermittency of atmospheric turbulence. *J. Atmos. Sci.*, 46(1):79–95, 1989.

[234] R. Makarewicz. Air absorption affected by Doppler shift. *J. Acoust. Soc. Am.*, 79(5):1339–1344, 1986.

[235] B. B. Mandelbrot. Intermittent turbulence in self-similar cascades: divergence of high moments and dimension of the carrier. *J. Fluid Mech.*, 62(2):331–358, 1974.

[236] B. B. Mandelbrot. *Fractals: Form, Chance, and Dimension.* W.H. Freeman, San Francisco, 1977.

[237] J. Mann. The spatial structure of neutral atmospheric surface layer turbulence. *J. Fluid Mech.*, 273:141–168, 1994.

[238] J. Mann. Wind field simulation. *Prob. Eng. Mech.*, 13:269–282, 1998.

[239] G. I. Marchuk. *Methods of Numerical Mathematics.* Springer, New York, 1975.

[240] K. Marsh. The CONCAWE model for calculating the propagation of noise from open-air industrial plants. *Appl. Acoust.*, 15(6):411–428, 1982.

[241] M. Martens, L. Van der Heijden, H. Walthaus, and W. Van Rens. Classification of soils based on acoustic impedance, air flow resistivity, and other physical soil parameters. *J. Acoust. Soc. Am.*, 78(3):970–980, 1985.

[242] J. M. Martin and S. M. Flatté. Intensity images and statistics from numerical simulation of wave propagation in 3-d random media. *Appl. Opt.*, 27(11):2111–2126, 1988.

[243] P. J. Mason. Large-eddy simulation of the convective atmospheric boundary layer. *J. Atmos. Sci.*, 46(11):1492–1516, 1989.

[244] MATLAB. *Statistics Toolbox User's Guide R2013b.* MathWorks, Natick, MA, 2013.

[245] L. T. Matveev. *Course of General Meteorology. Atmospheric Physics.* Gidrometeoisdat, Leningrad, 1976 (in Russian).

[246] L. G. McAllister. Acoustic sounding of the lower troposphere. *J. Atmos. Terr. Phys.*, 30(7):1439–1440, 1968.

[247] W. E. McBride, H. E. Bass, R. Raspet, and K. E. Gilbert. Scattering of sound by atmospheric turbulence: a numerical simulation above a complex impedance boundary. *J. Acoust. Soc. Am.*, 90(6):3314–3325, 1991.

[248] W. E. McBride, H. E. Bass, R. Raspet, and K. E. Gilbert. Scattering of sound by atmospheric turbulence: predictions in the refractive shadow zone. *J. Acoust. Soc. Am.*, 91(3):1336–1340, 1992.

[249] V. Mellert and B. Schwarz-Rohr. Correlation and coherence measurements of a spherical sound wave traveling in the atmospheric boundary layer. In *Seventh International Symposium on Long Range Sound Propagation*, pages 391–405, 1996.

[250] D. Menemenlis. Line-averaged measurement of velocity fine structure in the ocean using acoustical reciprocal transmission. *Int. J. Remote Sens.*, 15(2):267–281, 1994.

[251] Y. Miki. Acoustical properties of porous materials — Modifications of Delany-Bazley models. *J. Acoust. Soc. Jpn. (E)*, 11(1):19–24, 1990.

[252] J. W. Miles. On the reflection of sound at an interface of relative motion. *J. Acoust. Soc. Am.*, 29(2):226–228, 1957.

[253] E. A. Milne. Sound waves in the atmosphere. *Philos. Mag.*, 42:96–114, 1921.

[254] R. Mittal and G. Iaccarino. Immersed boundary methods. *Ann. Rev. Fluid Mech.*, 37:239–261, 2005.

[255] C.-H. Moeng. A large-eddy-simulation model for the study of planetary boundary-layer turbulence. *J. Atmos. Sci.*, 41(13):2052–2062, 1984.

[256] W. Möhring. Energy flux in duct flow. *J. Sound Vib.*, 18(1):101–109, 1971.

[257] W. Möhring. On energy, group velocity and small damping of sound waves in ducts with shear flow. *J. Sound Vib.*, 29(1):93–101, 1973.

[258] P. Moin and R. D. Moser. Characteristic-eddy decomposition of turbulence in a channel. *J. Fluid Mech.*, 200(41):471–509, 1989.

[259] A. S. Monin. Characteristics of the scattering of sound. *Sov. Phys. Acoust.*, 7(4):370–373, 1961.

[260] A. S. Monin, editor. *Oceanology, Volume I: Ocean Hydrodynamics.* Nauka, Moscow, 1978 (in Russian).

[261] A. S. Monin and R. V. Ozmidov. *Oceanic Turbulence.* Gidrometeoisdat, Leningrad, 1981 (in Russian).

[262] A. S. Monin and A. M. Yaglom. *Statistical Fluid Mechanics, Part II.* MIT Press, Cambridge, MA, 1981.

[263] C. L. Morfey. Acoustic energy in non-uniform flows. *J. Sound Vib.*, 14(2):159–170, 1971.

[264] P. M. Morse and K. U. Ingard. *Theoretical Acoustics.* McGraw-Hill, New York, 1968.

[265] A. G. Munin. *Sources of Aerodynamic Noise*. Mashinostroenie, Moscow, 1981 (in Russian).

[266] W. Munk, P. Worcester, and C. Wunsch. *Ocean Acoustic Tomograpy*. Cambridge University Press, New York, 1995.

[267] W. Munk and C. Wunsch. Ocean acoustic tomography: a scheme for large scale monitoring. *Deep-Sea Res.*, 26(2):123–161, 1979.

[268] K. Murawski. Influence of coherent and random flows on the solar *f*-mode. *Astrophys. J.*, 537:495–502, 2000.

[269] K. Murawski and E. N. Pelinovsky. The effect of random flow on solar acoustic waves. *Astron. Astrophys.*, 359:759–765, 2000.

[270] A. H. Nayfeh, J. E. Kaiser, and D. P. Telionis. Acoustics of aircraft engine duct systems. *AIAA J.*, 13(2):130–153, 1975.

[271] L. Nghiem-Phu and F. Tappert. Parabolic equation modeling of the effects of ocean currents on sound transmission and reciprocity in the time domain. *J. Acoust. Soc. Am.*, 78(2):642–648, 1985.

[272] L. Nijs and C. P. A. Wapenaar. The influence of wind and temperature gradients on sound propagation, calculated with the two way wave equations. *J. Acoust. Soc. Am.*, 87:1987–1998, 1990.

[273] M. A. Nobile and S. I. Hayek. Acoustic propagation over an impedance plane. *J. Acoust. Soc. Am.*, 78(4):1325–1336, 1985.

[274] J. M. Noble, H. E. Bass, and R. Raspet. The effect of large-scale atmospheric inhomogeneities on acoustic propagation. *J. Acoust. Soc. Am.*, 92(2):1040–1046, 1992.

[275] D. E. Norris, D. K. Wilson, and D. W. Thomson. Atmospheric scattering for varying degrees of saturation and intermittency. *J. Acoust. Soc. Am.*, 109:1871–1880, 2001.

[276] D. E. Norris, D. K. Wilson, and D. W. Thomson. Correlations between acoustic travel-time fluctuations and turbulence in the atmospheric surface layer. *Acust. Acta Acust.*, 87(6):677–684, 2001.

[277] V. N. Obolenskii. *Meteorology. Part II.* Gidrometeorologiya, Moscow, 1939 (in Russian).

[278] A. M. Obukhov. On propagation of a sound wave through a flow with vorticity. *Dokl. Akad. Nauk SSSR*, 39(2):46–48, 1943 (in Russian).

[279] A. M. Obukhov. Some specific features of atmospheric turbulence. *J. Geophys. Res.*, 67(8):3011–3014, 1962.

[280] D. P. O'Leary. Multidimensional integration: partition and conquer. *Comput. Sci. Eng*, Nov-Dec:58–66, 2004.

[281] V. E. Ostashev. Wave description of sound propagation in a stratified moving atmosphere. *Sov. Phys. Acoust.*, 30(4):311–314, 1984.

[282] V. E. Ostashev. On the sound field of a point source in a statified moving two-component medium. *Izv. Acad. Sci. USSR, Atmos. Oceanic Phys.*, 21(9):731–735, 1985.

[283] V. E. Ostashev. On the equations for acoustic and internal waves in a stratified medium. In *Diffraction and Wave Propagation*, pages 75–81. MPTI Press, Moscow, 1985 (in Russian).

[284] V. E. Ostashev. The azimuthal dependence of the sound pressure in a moving stratified medium. *Izv. Acad. Sci. USSR, Atmos. Oceanic Phys.*, 22(6):488–493, 1986.

[285] V. E. Ostashev. Wavequide propagation of a high-frequency acoustic field in a stratified moving medium near an impedance boundary. *Izv. Acad. Sci. USSR, Atmos. Oceanic Phys.*, 22(11):936–941, 1986.

[286] V. E. Ostashev. Equation for acoustic and gravity waves in a stratified moving medium. *Sov. Phys. Acoust.*, 33(1):95–96, 1987.

[287] V. E. Ostashev. High-frequency acoustic field of a point source lying above an impedance surface in a stratified moving medium. *Izv. Acad. Sci. USSR, Atmos. Oceanic Phys.*, 23(5):370–377, 1987.

[288] V. E. Ostashev. On sound wave propagation in a three-dimensional inhomogeneous moving medium. In *Diffraction and Wave Propagation in Inhomogeneous Media*, pages 42–49. MPTI Press, Moscow, 1987 (in Russian).

[289] V. E. Ostashev. Doppler effect in a moving medium and variation of the direction of propagation of sound radiated by a moving source. *Sov. Phys. Acoust.*, 34(4):402–405, 1988.

[290] V. E. Ostashev. *Acoustics in Moving Inhomogeneous Media.* E&FN SPON, London, 1997.

[291] V. E. Ostashev, P. Blanc-Benon, and D. Juvé. Coherence function of a spherical acoustic wave after passing through a turbulent jet. *Comptes Rendus de l'Académie des Sciences*, 326(Serie II b):39–45, 1998.

[292] V. E. Ostashev, I. P. Chunchuzov, and D. K. Wilson. Sound propagation through and scattering by internal gravity waves in a stably stratified atmosphere. *J. Acoust. Soc. Am.*, 118(6):3420–3429, 2005.

[293] V. E. Ostashev, S. L. Collier, and D. K. Wilson. Transverse-longitudinal coherence function of a sound field for line-of-sight propagation in a turbulent atmosphere. *Wave Random Complex*, 19(4):670–691, 2009.

[294] V. E. Ostashev, S. L. Collier, D. K. Wilson, D. F. Aldridge, N. P. Symons, and D. Marlin. Padé approximation in time-domain boundary conditions of porous surfaces. *J. Acoust. Soc. Am.*, 122(1):107–112, 2007.

[295] V. E. Ostashev, T. M. Georges, S. F. Clifford, and G. H. Goedecke. Acoustic sounding of wind velocity profiles in a stratified moving atmosphere. *J. Acoust. Soc. Am.*, 109(6):2682–2692, 2001.

[296] V. E. Ostashev, D. Juvé, and P. Blanc-Benon. Derivation of a wide-angle parabolic equation for sound waves in inhomogeneous moving media. *Acust. Acta Acust.*, 83(3):455–460, 1997.

[297] V. E. Ostashev, E. M. Salomons, S. F. Clifford, R. J. Lataitis, D. K. Wilson, P. Blanc-Benon, and D. Juvé. Sound propagation in a turbulent atmosphere near the ground: a parabolic equation approach. *J. Acoust. Soc. Am.*, 109:1894–1908, 2001.

[298] V. E. Ostashev, M. V. Scanlon, D. K. Wilson, and S. N. Vecherin. Source localization from an elevated acoustic sensor array in a refractive atmosphere. *J. Acoust. Soc. Am.*, 124(6):3413–3420, 2008.

[299] V. E. Ostashev and D. K. Wilson. Relative contributions from temperature and wind velocity fluctuations to the statistical moments of a sound field in a turbulent atmosphere. *Acust. Acta Acust.*, 86(2):260–268, 2000.

[300] V. E. Ostashev and D. K. Wilson. Log-amplitude and phase fluctuations of a plane wave propagating through anisotropic, inhomogeneous turbulence. *Acust. Acta Acust.*, 87(6):685–694, 2001.

[301] V. E. Ostashev and D. K. Wilson. Coherence function and mean field of plane and spherical sound waves propagating through inhomogeneous anisotropic turbulence. *J. Acoust. Soc. Am.*, 115(2):497–506, 2004.

[302] V. E. Ostashev, D. K. Wilson, and G. H. Goedecke. Spherical wave propagation through inhomogeneous, anisotropic turbulence: studies of log-amplitude and phase fluctuations. *J. Acoust. Soc. Am.*, 115(1):120–130, 2004.

[303] V. E. Ostashev, D. K. Wilson, L. Liu, D. F. Aldridge, N. P. Symons, and D. Marlin. Equations for finite-difference, time-domain simulation of sound propagation in moving inhomogeneous media and numerical implementation. *J. Acoust. Soc. Am.*, 117(2):503–517, 2005.

[304] V. E. Ostashev, D. K. Wilson, S. N. Vecherin, and S. L. Collier. Spatial-temporal coherence of acoustic signals propagating in a refractive, turbulent atmosphere. *J. Acoust. Soc. Am.*, 136(5):2414–2431, 2014.

[305] N. C. Ovenden, S. R. Shaffer, and H. J. S. Fernando. Impact of meteorological conditions on noise propagation from freeway corridors. *J. Acoust. Soc. Am.*, 126(1):25–35, 2009.

[306] H. A. Panofsky and J. A. Dutton. *Atmospheric Turbulence: Models and Methods for Engineering Applications*. Wiley, New York, 1984.

[307] S. Parakkal, K. E. Gilbert, X. Di, and H. E. Bass. A generalized polar coordinate method for sound propagation over large-scale irregular terrain. *J. Acoust. Soc. Am.*, 128(5):2573–2580, 2010.

[308] F. Pasquill. The estimation of the dispersion of windborne material. *Meteorol. Mag.*, 90(1063):33–49, 1961.

[309] G. Peters, C. Wamser, and H. Hinzpeter. Acoustic Doppler and angle of arrival wind detection and comparisons with direct measurements at a 300 m mast. *J. Appl. Meteorol.*, 17(8):1171–1178, 1978.

[310] P. D. Phillips, H. Richner, and W. Nater. Layer model for assessing acoustic refraction effects in echosounding. *J. Acoust. Soc. Am.*, 62(2):277–285, 1977.

[311] A. D. Pierce. Propagation of acoustic-gravity waves in a temperature- and wind-stratified atmosphere. *J. Acoust. Soc. Am.*, 37(2):218–227, 1965.

[312] A. D. Pierce. Statistical theory of atmospheric turbulence effects on sonic-boom rise times. *J. Acoust. Soc. Am.*, 49(3):906–924, 1971.

[313] A. D. Pierce. *Acoustics: An Introduction to its Physical Principles and Applications*. American Institute of Physics, New York, 1989.

[314] A. D. Pierce. Wave equation for sound in fluids with unsteady inhomogeneous flow. *J. Acoust. Soc. Am.*, 87(6):2292–2299, 1990.

[315] A. D. Pierce, J. W. Posey, and E. F. Iliff. Variation of nuclear explosion generated acoustic-gravity wave forms with burst height and with energy yield. *J. Geophys. Res.*, 76(21):5025–5041, 1971.

[316] E. Premat and Y. Gabillet. A new boundary-element method for predicting outdoor sound propagation and application to the case of a sound barrier in the presence of downward refraction. *J. Acoust. Soc. Am.*, 108(6):2775–2783, 2000.

[317] M. A. Price, K. Attenborough, and N. W. Heap. Sound attenuation through trees: measurements and models. *J. Acoust. Soc. Am.*, 84(5):1836–1844, 1988.

[318] D. C. Pridmore-Brown. Sound propagation in a temperature and wind stratified medium. *J. Acoust. Soc. Am.*, 34(4):438–443, 1962.

[319] J. M. Prospathopoulos and S. G. Voutsinas. Determination of equivalent sound speed profiles for ray tracing in near-ground sound propagation. *J. Acoust. Soc. Am.*, 122(3):1391–1403, 2007.

[320] A. Raabe, K. Arnold, A. Ziemann, F. Beyrich, J. P. Leps, J. Bange, P. Zittel, T. Spiess, T. Foken, M. Göckede, M. Schröter, and S. Raasch. STINHO — Structure of turbulent transport under inhomogeneous conditions — part 1: the micro-α scale field experiment. *Meteorol. Z.*, 14(3):315–327, 2005.

[321] R. Raspet, S. W. Lee, E. Kuester, D. C. Chang, W. F. Richards, R. Gilbert, and N. Bong. A fast-field program for sound propagation in a layered atmosphere above an impedance ground. *J. Acoust. Soc. Am.*, 77(2):345–352, 1985.

[322] R. Raspet and R. K. Wolf. Application of the fast field program to the prediction of average noise levels around sources. *Appl. Acoust.*, 27(3):217–226, 1989.

[323] R. Raspet, L. Yao, S. J. Franke, and M. J. White. Comments on "The influence of wind and temperature gradients on sound propagation, calculated with the two way wave equations" [J. Acoust. Soc. Am. 87, 1987–1998 (1990)]. *J. Acoust. Soc. Am.*, 91(1):498–500, 1992.

[324] L. Rayleigh. *The Theory of Sound.* Dover, New York, 1945.

[325] A. V. Razin. Calculation of acoustic fields in the atmospheric refraction wave guide. *Izv. Acad. Sci. USSR, Atmos. Oceanic Phys.*, 21(7):544–548, 1985.

[326] A. V. Razin. Effect of nonuniformity of air temperature and wind on the field of an acoustic point source in the atmospheric surface layer. In *Preprint No. 223*. NIRFI Press, Gor'kiy, 1987.

[327] J. W. Reed and K. G. Adams. Sonic boom waves — calculation of atmospheric refraction. *Aerospace Eng.*, 21(3):101–105, 1962.

[328] G. Reinhold. Über die Wirksamkeit von Schallschirmen an Straßen unter Berücksichtigung meteorologischer Einflüsse. *Kampf Lärm*, 22(6):162–166, 1975.

[329] O. Reynolds. On the refraction of sound by the atmosphere. *Proc. R. Soc. London*, 22:531–548, 1874.

[330] O. Reynolds. On the refraction of sound by the atmosphere. *Philos. Trans. R. Soc. London*, 166(1):315–324, 1876.

[331] S. Ribner. Reflection, transmission, and amplification of sound by a moving medium. *J. Acoust. Soc. Am.*, 29(4):435–441, 1957.

[332] T. L. Richards and K. Attenborough. Accurate FFT-based Hankel transforms for predictions of outdoor sound propagation. *J. Sound Vib.*, 109(1):157–167, 1986.

[333] W. S. Richardson, W. J. Schmitz, and P. P. Niiler. The velocity structure of the Florida current from the Straits of Florida to Cape Fear. *Deep-Sea Res.*, 16:225–231, 1969.

[334] C. L. Rino. *The Theory of Scintillation with Applications in Remote Sensing.* IEEE Press, Piscataway, NJ, 2011.

[335] J. S. Robertson, W. L. Siegmann, and M. J. Jacobson. Current and current shear effects in the parabolic approximation for underwater sound channels. *J. Acoust. Soc. Am.*, 77(5):1768–1780, 1985.

[336] J. S. Robertson, W. L. Siegmann, and M. J. Jacobson. Acoustical effects of ocean current shear structures in the parabolic approximation. *J. Acoust. Soc. Am.*, 82(2):559–573, 1987.

[337] M. N. Rychagov and H. Ermert. Reconstruction of fluid motion in acoustic diffraction tomography. *J. Acoust. Soc. Am.*, 99(5):3029–3035, 1996.

[338] S. M. Rytov. Diffraction of light by ultrasonic waves. *Izv. Akad. Nauk SSSR, Ser. Fiz.*, 2:223–259, 1937 (in Russian).

[339] S. M. Rytov, Y. A. Kravtsov, and V. I. Tatarskii. *Principles of Statistical Radio Physics. Part 4, Wave Propagation through Random Media.* Springer, Berlin, 1989.

[340] R. A. Sack and M. West. A parabolic equation for sound propagation in two dimensions over any smooth terrain profile: the generalised terrain parabolic equation (GT-PE). *Appl. Acoust.*, 45(2):113–129, 1995.

[341] Z. S. Sacks, D. M. Kingsland, R. Lee, and J. F. Lee. A perfectly matched anisotropic absorber for use as an absorbing boundary condition. *IEEE Trans. Antenn. Propag.*, 43(12):1460–1463, 1995.

[342] E. Salomons, D. Van Maercke, J. Defrance, and F. de Roo. The Harmonoise sound propagation model. *Acust. Acta Acust.*, 97(1):62–74, 2011.

[343] E. M. Salomons. Reduction of the performance of a noise screen due to screen-induced wind-speed gradients. Numerical computations and wind-tunnel experiments. *J. Acoust. Soc. Am.*, 105(4):2287–2293, 1999.

[344] E. M. Salomons. Fluctuations of spherical waves in a turbulent atmosphere: effect of the axisymmetric approximation in computational methods. *J. Acoust. Soc. Am.*, 108(4):1528–1534, 2000.

[345] E. M. Salomons. *Computational Atmospheric Acoustics*. Kluwer, Dordrecht, 2001.

[346] E. M. Salomons, R. Blumrich, and D. Heimann. Eulerian time-domain model for sound propagation over a finite-impedance ground surface. Comparison with frequency-domain models. *Acust. Acta Acust.*, 88(4):483–492, 2002.

[347] E. M. Salomons, V. E. Ostashev, S. F. Clifford, and R. J. Lataitis. Sound propagation in a turbulent atmosphere near the ground: an approach based on the spectral representation of refractive-index fluctuations. *J. Acoust. Soc. Am.*, 109(5):1881–1893, 2001.

[348] T. B. Sanford. Observations of strong current shears in the deep ocean and some applications on sound rays. *J. Acoust. Soc. Am.*, 56(4):1118–1121, 1974.

[349] H. Sato, M. C. Fehler, and T. Maeda. *Wave Propagation and Scattering in the Heterogeneous Earth*. Springer, Berlin, 2012.

[350] A. Schady, D. Heimann, and J. Feng. Acoustic effects of trees simulated by a finite-difference time-domain model. *Acust. Acta Acust.*, 100(6):1112–1119, 2014.

[351] H. Schmidt and G. Tango. Efficient global matrix approach to the computation of synthetic seismograms. *Geophys. J. R. Astr. Soc.*, 84(2):331–359, 1986.

[352] P. D. Schomer. Noise monitoring in the vicinity of general aviation airports. *J. Acoust. Soc. Am.*, 74(6):1764–1772, 1983.

[353] P. D. Schomer. A statistical description of ground-to-ground propagation. *Noise Control Eng. J.*, 51(2):69–89, 2003.

[354] L. K. Schubert. Numerical study of sound refraction by a jet flow. I. Ray acoustics. *J. Acoust. Soc. Am.*, 51(2):439–446, 1972.

[355] L. K. Schubert. Numerical study of sound refraction by a jet flow. II. Wave acoustics. *J. Acoust. Soc. Am.*, 51(2):447–463, 1972.

[356] B. D. Seckler and J. B. Keller. Asymptotic theory of diffraction in inhomogeneous media. *J. Acoust. Soc. Am.*, 31(2):206–216, 1959.

[357] L. F. Shampine and M. W. Reichelt. The Matlab ODE suite. *SIAM J. Sci. Comput.*, 18(1):1–22, 1997.

[358] M. Shinozuka and G. Deodatis. Simulation of multi-dimensional Gaussian stochastic fields by spectral representation. *Appl. Mech. Rev.*, 49(1):29–53, 1996.

[359] S. P. Singal. Acoustic sounding stability studies. In *Encyclopedia of Environment Control Technology*, volume 2, Air Pollution Control, pages 1003–1061. Gulf Publishing, 1989.

[360] J. L. Spiesberger. Locating animals from their sounds and tomography of the atmosphere: experimental demonstration. *J. Acoust. Soc. Am.*, 106:837–846, 1999.

[361] J. L. Spiesberger and K. M. Fristrup. Passive localization of calling animals and sensing of their acoustic environment using acoustic tomography. *Amer. Nat.*, 135:107–153, 1990.

[362] K. R. Sreenivasan. Fractals and multifractals in fluid turbulence. *Ann. Rev. Fluid Mech.*, 23(1):539–604, 1991.

[363] L. A. Stallworth and M. J. Jacobson. Acoustic propagation in an isospeed channel with uniform tidal current and depth change. *J. Acoust. Soc. Am.*, 48(1):382–391, 1970.

[364] L. A. Stallworth and M. J. Jacobson. Sound transmission in an isospeed ocean channel with depth-dependent current. *J. Acoust. Soc. Am.*, 51(5):1738–1750, 1972.

[365] Standard Publishing House. *Standard Atmosphere: Parameters*. Standard Publishing House, Moscow, 1981 (in Russian).

[366] G. G. Stokes. On the effect of wind on the intensity of sound. *Brit. Assoc. Report*, page 22, 1857. (Reprinted in *Mathematical and Physical Papers of G.G. Stokes*, (1904) **4**, Cambridge University Press, New York, pages 110-111.).

[367] R. B. Stull. *An Introduction to Boundary Layer Meteorology*. Kluwer, Dordrecht, 1988.

[368] L. C. Sutherland and G. A. Daigle. Atmospheric sound propagation. In M. Crocker, editor, *Handbook in Acoustics*, pages 305–329. Wiley, New York, 1998.

[369] M. E. Swearingen and M. J. White. Influence of scattering, atmospheric refraction, and ground effect on sound propagation through a pine forest. *J. Acoust. Soc. Am.*, 122(1):113–119, 2007.

[370] M. E. Swearingen, M. J. White, S. A. Ketcham, and M. H. McKenna. Use of a porous material description of forests in infrasonic propagation algorithms. *J. Acoust. Soc. Am.*, 134(4):2647–2659, 2013.

[371] M. A. Swinbanks. The sound field generated by a source distribution in a long duct carrying sheared flow. *J. Sound Vib.*, 40(1):51–76, 1975.

[372] N. P. Symons, D. F. Aldridge, D. K. Wilson, D. H. Marlin, S. L. Collier, and V. E. Ostashev. Finite-difference simulation of atmospheric acoustic sound through a complex meteorological background over a topographically complex surface. In *EuroNoise 2006*, 2006.

[373] F. D. Tappert. The parabolic approximation method. In J. B. Keller and J. S. Papadakis, editors, *Wave Propagation and Underwater Acoustics*, pages 224–287. Springer, Berlin, 1977.

[374] V. I. Tatarskii. *The Effects of the Turbulent Atmosphere on Wave Propagation*. Israel Program for Scientific Translation, Jerusalem, 1971.

[375] V. I. Tatarskii. On the theory of sound propagation in a stratified atmosphere. *Izv. Acad. Sci. USSR, Atmos. Oceanic Phys.*, 15(11):795–801, 1979.

[376] V. I. Tatarskii and V. U. Zavorotnyi. Wave propagation in random media with fluctuating turbulent parameters. *J. Opt. Soc. Am. A*, 2:2069–2076, 1985.

[377] B. J. Tester. The propagation and attenuation of sound in lined ducts containing uniform or "plug" flow. *J. Sound Vib.*, 28(2):151–203, 1973.

[378] R. J. Thompson. Ray-acoustic intensity in a moving medium, II. A stratified medium. *J. Acoust. Soc. Am.*, 55(4):733–737, 1974.

[379] D. Tielbürger, S. Finette, and S. Wolf. Acoustic propagation through an internal wave field in a shallow water waveguide. *J. Acoust. Soc. Am.*, 101(2):789–808, 1997.

[380] C. Torrence and G. P. Compo. A practical guide to wavelet analysis. *B. Am. Meteorol. Soc.*, 79(1):61–78, 1998.

[381] A. A. Townsend. *The Structure of Turbulent Shear Flow*. Cambridge University Press, Cambridge, 1980.

[382] A. Tunick. Calculating the micrometeorological influences on the speed of sound through the atmosphere in forests. *J. Acoust. Soc. Am.*, 114:1796–1806, 2003.

[383] D. Turo and O. Umnova. Time domain modelling of sound propagation in porous media and the role of shape factors. *Acust. Acta Acust.*, 96(2):225–238, 2010.

[384] P. Ugincius. Ray acoustics and Fermat's principle in a moving inhomogeneous medium. *J. Acoust. Soc. Am.*, 51(5):1759–1763, 1972.

[385] D. Van Maercke and J. Defrance. Development of an analytical model for outdoor sound propagation within the Harmonoise project. *Acust. Acta Acust.*, 93(2), 2007-03-01T00:00:00.

[386] T. Van Renterghem. *De eindige-differenties-in-het-tijdsdomeinmethode voor de simulatie van geluidspropagatie in een bewegend medium (The finite-difference time-domain method for simulation of sound propagation in a moving medium)*. Doctoral dissertation, Universiteit Gent, 2003.

[387] T. Van Renterghem and D. Botteldooren. Numerical simulation of the effect of trees on downwind noise barrier performance. *Acust. Acta Acust.*, 89(5):764–778, 2003.

[388] T. Van Renterghem, D. Botteldooren, and P. Lercher. Comparison of measurements and predictions of sound propagation in a valley-slope configuration in an inhomogeneous atmosphere. *J. Acoust. Soc. Am.*, 121(5):2522–2533, 2007.

[389] T. Van Renterghem, E. M. Salomons, and D. Botteldooren. Efficient FDTD-PE model for sound propagation in situations with complex obstacles and wind profiles. *Acust. Acta Acust.*, 91(4):671–679, 2005.

[390] C. J. M. Van Ruiten. Dutch railway noise prediction schemes. *J. Sound Vib.*, 120(2):371–379, 1988.

[391] S. N. Vecherin, V. E. Ostashev, G. H. Goedecke, D. K. Wilson, and A. G. Voronovich. Time-dependent stochastic inversion in acoustic travel-time tomography of the atmosphere. *J. Acoust. Soc. Am.*, 119(5):2579–2588, 2006.

[392] S. N. Vecherin, V. E. Ostashev, and D. K. Wilson. Three-dimensional acoustic travel-time tomography of the atmosphere. *Acust. Acta Acust.*, 94(3):349–358, 2008.

[393] S. N. Vecherin, V. E. Ostashev, and D. K. Wilson. Assessment of systematic errors in acoustic tomography of the atmosphere. *J. Acoust. Soc. Am.*, 134:1802–1813, 2013.

[394] S. N. Vecherin, V. E. Ostashev, A. Ziemann, D. K. Wilson, K. Arnold, and M. Barth. Tomographic reconstruction of atmospheric turbulence with the use of time-dependent stochastic inversion. *J. Acoust. Soc. Am.*, 122(3):1416–1425, 2007.

[395] S. S. Voit. Propagation of waves induced by a sounding disk in a moving medium. *Applied Mathematics Mechanics*, 16(6):699–705, 1952 (in Russian).

510 *Acoustics in Moving Inhomogeneous Media*

[396] S. S. Voit. Reflection and refraction of spherical sound waves by an interface between nonmoving and moving media. *Applied Mathematics Mechanics*, 17(2):157, 1953 (in Russian).

[397] F. Walkden and M. West. Prediction of enhancement factor for small explosive sources in a stratified moving atmosphere. *J. Acoust. Soc. Am.*, 84(1):321–326, 1988.

[398] R. Waxler, K. E. Gilbert, and C. Talmadge. A theoretical treatment of the long range propagation of impulsive signals under strongly ducted nocturnal conditions. *J. Acoust. Soc. Am.*, 124(5):2742–2754, 2008.

[399] R. Waxler, C. L. Talmadge, S. Dravida, and K. E. Gilbert. The near-ground structure of the nocturnal sound field. *J. Acoust. Soc. Am.*, 119(1):86–95, 2006.

[400] S. Weinbrect, S. Raasch, A. Ziemann, K. Arnold, and A. Raabe. Comparison of large-eddy simulation data with spatially averaged measurements obtained by acoustic tomography — presuppositions and first results. *Bound. Layer Meteorol.*, 111:441–465, 2004.

[401] A. R. Wenzel and J. B. Keller. Propagation of acoustic waves in a turbulent medium. *J. Acoust. Soc. Am.*, 50(3):911–920, 1971.

[402] M. L. Wesely. The combined effect of temperature and humidity fluctuations on refractive index. *J. Appl. Meteorol.*, 15(1):43–49, 1976.

[403] M. West, K. Gilbert, and R. A. Sack. A tutorial on the parabolic equation (PE) model used for long range sound propagation in the atmosphere. *Appl. Acoust.*, 37(1):31–49, 1992.

[404] M. West, R. A. Sack, and F. Walkden. The Fast Field Program (FFP). A second tutorial: application to long range sound propagation in the atmosphere. *Appl. Acoust.*, 33(3):199–228, 1991.

[405] A. D. Wheelon. *Electromagnetic Scintillation. Vol. 1: Geometrical Optics.* Cambridge University Press, New York, 2001.

[406] A. D. Wheelon. *Electromagnetic Scintillation, Vol. 2: Weak Scattering.* Cambridge University Press, New York, 2003.

[407] Wikipedia. Barycentric coordinate system — Wikipedia, The Free Encyclopedia, 2014. [Online; accessed 30-November-2014].

[408] D. S. Wilks. *Statistical Methods in the Atmospheric Sciences.* Elsevier, Oxford, third edition, 2006.

[409] D. K. Wilson. Use of wave-number-domain windows in fast field programs. *J. Acoust. Soc. Am.*, 89:448–450, 1991.

[410] D. K. Wilson. Relaxation-matched modeling of propagation through porous media, including fractal pore structure. *J. Acoust. Soc. Am.*, 94(2):1136–1145, 1993.

[411] D. K. Wilson. Sound field computations in a stratified, moving medium. *J. Acoust. Soc. Am.*, 94:400–407, 1993.

[412] D. K. Wilson. Empirical orthogonal function analysis of the weakly convective atmospheric boundary layer. Part I: Eddy structures. *J. Atmos. Sci.*, 53(6):801–823, 1996.

[413] D. K. Wilson. Simple, relaxational models for the acoustical properties of porous media. *Appl. Acoust.*, 50(3):171–188, 1997.

[414] D. K. Wilson. Calculated coherence and extinction of sound waves propagating through anisotropic, shear-induced turbulent velocity fluctuations. *J. Acoust. Soc. Am.*, 105(2):658–671, 1999.

[415] D. K. Wilson. A turbulence spectral model for sound propagation in the atmosphere that incorporates shear and buoyancy forcings. *J. Acoust. Soc. Am.*, 108:2021–2038, 2000.

[416] D. K. Wilson. An alternative function for the wind and temperature gradients in unstable surface layers. *Bound. Layer Meteorol.*, 99:151–158, 2001.

[417] D. K. Wilson, E. L. Andreas, J. W. Weatherly, C. L. Pettit, E. G. Patton, and P. P. Sullivan. Characterization of uncertainty in outdoor sound propagation predictions. *J. Acoust. Soc. Am.*, 121(5):EL177–EL183, 2007.

[418] D. K. Wilson, J. G. Brasseur, and K. E. Gilbert. Acoustic scattering and the spectrum of atmospheric turbulence. *J. Acoust. Soc. Am.*, 105(1):30–34, 1999.

[419] D. K. Wilson, S. L. Collier, V. E. Ostashev, D. F. Aldridge, N. P. Symons, and D. H. Marlin. Time-domain modeling of the acoustic impedance of porous surfaces. *Acust. Acta Acust.*, 92(6):965–975, 2006.

[420] D. K. Wilson, M. S. Lewis, J. W. Weatherly, and E. L. Andreas. Dependence of predictive skill for outdoor narrowband and broadband sound levels on the atmospheric representation. *Noise Control Eng. J.*, 56(6):465–477, 2008.

[421] D. K. Wilson, J. M. Noble, and M. A. Coleman. Sound propagation in the nocturnal boundary layer. *J. Atmos. Sci.*, 60:2473–2486, 2003.

[422] D. K. Wilson and V. E. Ostashev. Statistical moments of the sound field propagating in a random, refractive medium near an impedance boundary. *J. Acoust. Soc. Am.*, 109:1909–1922, 2001.

[423] D. K. Wilson, V. E. Ostashev, S. L. Collier, N. P. Symons, D. F. Aldridge, and D. H. Marlin. Time-domain calculations of sound interactions with outdoor ground surfaces. *Appl. Acoust.*, 68(2):173–200, 2007.

[424] D. K. Wilson, V. E. Ostashev, and G. H. Goedecke. Sound-wave coherence in atmospheric turbulence with intrinsic and global intermittency. *J. Acoust. Soc. Am.*, 124:743–757, 2008.

[425] D. K. Wilson, V. E. Ostashev, and G. H. Goedecke. Quasi-wavelet formulations of turbulence and other random fields with correlated properties. *Prob. Eng. Mech.*, 24(3):343–357, 2009.

[426] D. K. Wilson, V. E. Ostashev, G. H. Goedecke, and H. J. Auvermann. Quasi-wavelet calculations of sound scattering behind barriers. *Appl. Acoust.*, 65:605–627, 2004.

[427] D. K. Wilson, V. E. Ostashev, and M. S. Lewis. Moment-screen method for wave propagation in a refractive medium with random scattering. *Wave Random Complex*, 19(3):369–391, 2009.

[428] D. K. Wilson, V. E. Ostashev, and M. Mungiole. Categorization schemes for near-ground sound propagation. In *18th International Congress on Acoustics*, 2004.

[429] D. K. Wilson, S. Ott, G. H. Goedecke, and V. E. Ostashev. Quasi-wavelet formulations of turbulence and wave scattering. *Meteorol. Z.*, 18(3):237–252, 2009.

[430] D. K. Wilson, C. L. Pettit, V. E. Ostashev, and S. N. Vecherin. Description and quantification of uncertainty in outdoor sound propagation calculations). *J. Acoust. Soc. Am.*, 136(3):1013–1028, 2014.

[431] D. K. Wilson, C. L. Pettit, and S. N. Vecherin. Wavelet-based cascade model for intermittent structure in terrestrial environments. *ArXiv e-prints*, Dec. 2013.

[432] D. K. Wilson and D. W. Thomson. Acoustic propagation through anisotropic, surface-layer turbulence. *J. Acoust. Soc. Am.*, 96(2):1080–1095, 1994.

[433] D. K. Wilson and D. W. Thomson. Acoustic tomographic monitoring of the atmospheric surface layer. *J. Atmos. Ocean. Tech.*, 11:751–769, 1994.

[434] D. K. Wilson, J. C. Wyngard, and D. I. Havelock. The effect of turbulent intermittency on scattering into an acoustic shadow zone. *J. Acoust. Soc. Am.*, 99:3393–3400, 1996.

[435] D. K. Wilson, A. Ziemann, V. E. Ostashev, and A. G. Voronovich. An overview of acoustic travel-time tomography in the atmosphere and its potential applications. *Acust. Acta Acust.*, 87:721–730, 2001.

[436] P. F. Worcester. Reciprocal acoustic transmission in a mid-ocean environment. *J. Acoust. Soc. Am.*, 62(4):895–905, 1977.

[437] J. C. Wyngaard, N. Seaman, S. J. Kimmel, M. Otte, X. Di, and K. E. Gilbert. Concepts, observations, and simulation of refractive index turbulence in the lower atmosphere. *Radio Sci.*, 36(4):643–669, 2001.

[438] Y. Xu, Z. C. Zheng, and D. K. Wilson. A computational study of the effect of windscreen shape and flow resistivity on turbulent wind noise reduction. *J. Acoust. Soc. Am.*, 129(4):1740–1747, 2011.

[439] K. S. Yee. Numerical solution of initial boundary value problems involving Maxwell's equations in isotropic media. *IEEE Trans. Antenn. Propag.*, 14(3):302–307, 1966.

[440] C. Yeh. A further note on the reflection and transmission of sound waves by a moving fluid layer. *J. Acoust. Soc. Am.*, 43(6):1454–1455, 1968.

[441] A. Ziemann, K. Arnold, and A. Raabe. Acoustic tomography as a method to identify small-scale land surface characteristics. *Acust. Acta Acust.*, 87(6):731–737, 2001.

[442] A. Ziemann, K. Arnold, and A. Raabe. Acoustic tomography as a remote sensing method to investigate the near surface atmospheric boundary layer in comparison with in situ measurements. *J. Atmos. Ocean. Tech.*, 19(8):1208–1215, 2002.

[443] W. E. Zorumski and W. L. Willshire. The acoustic field of a point source in a uniform boundary layer over an impedance plane. In *AIAA 10th Aeroacoustics Conference*, 1986 (paper AIAA-86-1923).

[444] V. Zouboff, Y. Brunet, M. Berengier, and E. Sechet. A qualitative approach of atmospherical effects on long range sound propagation. In *Sixth International Symposium on Long-Range Sound Propagation*, pages 251–269, 1994.

[445] C. Zwikker and C. W. Kosten. *Sound Absorbing Materials*. Elsevier, New York, 1949.

Index